Progress in Mathematics
Volume 199

Series Editors

H. Bass
J. Oesterlé
A. Weinstein

Rational Points on Algebraic Varieties

Emmanuel Peyre
Yuri Tschinkel
Editors

Springer Basel AG

Editors:

Emmanuel Peyre
Institut Fourier
UFR de Mathématiques, UMR 5582
Université de Grenoble I et CNRS
B.P. 74
38402 Saint-Martin d'Hères
France

e-mail: Emmanuel.Peyre@ujf.grenoble.fr

Yuri Tschinkel
Department of Mathematics
Princeton University
Washington Road
Princeton, NJ 08544-1000
USA

e-mail: ytschink@math.princeton.edu

2000 Mathematics Subject Classification 14G05, 11G35

A CIP catalogue record for this book is available from the Library of Congress, Washington D.C., USA

Deutsche Bibliothek Cataloging-in-Publication Data

Rational Points on Algebraic Varieties / Emmanuel Peyre ... ed.. – Basel ; Boston ; Berlin : Birkhäuser, 2001
  (Progress in mathematics ; Vol. 199)
  ISBN 978-3-7643-6612-4   ISBN 978-3-0348-8368-9 (eBook)
  DOI 10.1007/978-3-0348-8368-9

ISBN 978-3-7643-6612-4

This work is subject to copyright. All rights are reserved, whether the whole or part of the material is concerned, specifically the rights of translation, reprinting, re-use of illustrations, broadcasting, reproduction on microfilms or in other ways, and storage in data banks. For any kind of use whatsoever, permission from the copyright owner must be obtained.

© 2001 Springer Basel AG
Originally published by Birkhäuser Verlag in 2001

Member of the BertelsmannSpringer Publishing Group
Printed on acid-free paper produced of chlorine-free pulp. TCF ∞

ISBN 978-3-7643-6612-4

9 8 7 6 5 4 3 2 1   www.birkhäuser-science.com

INTRODUCTION

This book is devoted to the study of rational and integral points on higher-dimensional algebraic varieties. It contains research papers addressing the arithmetic geometry of varieties which are not of general type, with an emphasis on how rational points are distributed with respect to the classical, Zariski and adelic topologies.

The book gives a glimpse of the state of the art of this rapidly expanding domain in arithmetic geometry. The techniques involve explicit geometric constructions, ideas from the minimal model program in algebraic geometry as well as analytic number theory and harmonic analysis on adelic groups.

In recent years there has been substantial progress in our understanding of the arithmetic of algebraic surfaces. Five papers are devoted to cubic surfaces: Basile and Fisher study the existence of rational points on certain diagonal cubics, Swinnerton-Dyer considers weak approximation and Broberg proves upper bounds on the number of rational points on the complement to lines on cubic surfaces. Peyre and Tschinkel compare numerical data with conjectures concerning asymptotics of rational points of bounded height on diagonal cubics of rank $\geqslant 2$. Kanevsky and Manin investigate the composition of points on cubic surfaces. Satgé constructs rational curves on certain Kummer surfaces. Colliot-Thélène studies the Hasse principle for pencils of curves of genus 1. In an appendix to this paper Skorobogatov produces explicit examples of Enriques surfaces with a Zariski dense set of rational points.

Poonen constructs a (conditional) example of a 3-dimensional complete intersection which violates the Hasse principle but for which all known obstructions to the Hasse principle vanish. The paper of Hassett-Tschinkel is influenced by ideas from the log-minimal model program. It is devoted to the "potential" density of integral points on quasi-projective algebraic varieties.

The remaining papers are more analytic. Chambert-Loir and Tschinkel investigate the asymptotics of rational points on compactifications of torsors under linear algebraic groups. Similar fibrations appear in the theory of "partial" Eisenstein series initiated by Strauch. Wooley's analysis of certain exponential sums arising in the circle method allow him to significantly improve asymptotic results concerning the number of integral solutions of sums of binary forms of fixed degree. The paper of Peyre extends the first steps of the classical circle method to hypersurfaces in Fano varieties by lifting the counting of rational points on a variety to the counting of integral points on its universal torsor.

We hope the book conveys some of the excitement shared by participants of the conference at Luminy in September 1999, which was the starting point of this project. Finally, we are very grateful to CIRM and the European network "Arithmetic Geometry" for their support.

CONTENTS

Introduction v

Abstracts xi

CARMEN LAURA BASILE & THOMAS ANTHONY FISHER — *Diagonal cubic equations in four variables with prime coefficients* 1
   References 11

NIKLAS BROBERG — *Rational points on cubic surfaces* 13
   Introduction 13
   1. Notations and preliminaries 15
   2. Ternary quadratic forms 20
   3. Proof of the main theorem 28
   References 34

ANTOINE CHAMBERT-LOIR & YURI TSCHINKEL — *Torseurs arithmétiques et espaces fibrés* 37
   Introduction 37
   Notations et conventions 40
   1. Torseurs arithmétiques 40
   2. Espaces fibrés 51
   Références 69

ANTOINE CHAMBERT-LOIR & YURI TSCHINKEL — *Fonctions zêta des hauteurs des espaces fibrés* 71
   Introduction 71
   Notations et conventions 74
   3. Fonctions holomorphes dans un tube 75
   4. Variétés toriques 87

5. Application aux fibrations en variétés toriques 101
   Appendice A. Un théorème taubérien 107
   Appendice B. Démonstration de quelques inégalités 109
   Références 114

JEAN-LOUIS COLLIOT-THÉLÈNE — *Hasse principle for pencils of curves of genus one whose Jacobians have a rational 2-division point, close variation on a paper of Bender and Swinnerton-Dyer* 117
   Statement of the Theorems 119
   1. Selmer groups associated to a degree 2 isogeny 125
   2. Proof of Theorem A 142
   3. Proof of Theorem B 156
   References 160

ALEXEI SKOROBOGATOV — *Enriques surfaces with a dense set of rational points, Appendix to the paper by J.-L. Colliot-Thélène* 163
   References 168

BRENDAN HASSETT & YURI TSCHINKEL — *Density of integral points on algebraic varieties* 169
   Introduction 169
   1. Generalities 171
   2. Geometry 172
   3. The fibration method and nondegenerate multisections 177
   4. Approximation techniques 181
   5. Conic bundles and integral points 183
   6. Potential density for log K3 surfaces 193
   References 195

DIMITRI KANEVSKY & YURI MANIN — *Composition of points and the Mordell–Weil problem for cubic surfaces* 199
   1. Introduction 199
   2. Cardinality of generators of subgroups in a reflection group 202
   3. Structure of universal equivalence 206
   4. A group–theoretic description of universal equivalence 208
   5. Birationally trivial cubic surfaces: a finiteness theorem 213
   References 218

EMMANUEL PEYRE — *Torseurs universels et méthode du cercle* 221
   Introduction 221

1. Une version raffinée d'une conjecture de Manin 223
2. Passage au torseur universel 233
3. Intersections complètes 254
4. Conclusion 271
Références 272

EMMANUEL PEYRE & YURI TSCHINKEL — *Tamagawa numbers of diagonal cubic surfaces of higher rank* 275
Introduction 275
1. Description of the conjectural constant 277
2. The Galois module $\text{Pic}(\overline{V})$ 280
3. Euler product for the good places 286
4. Density at the bad places 288
5. The constant $\alpha(V)$ 291
6. Some statistical formulae 297
7. Presentation of the results 298
References 304

BJORN POONEN — *The Hasse principle for complete intersections in projective space* 307
References 310

PHILIPPE SATGÉ — *Une construction de courbes k-rationnelles sur les surfaces de Kummer d'un produit de courbes de genre 1.* 313
Introduction 313
1. Relèvement des courbes de $\mathbf{P}_{1,k} \times \mathbf{P}_{1,k}$ sur la surface de Kummer .. 316
2. Exemples 320
Références 333

MATTHIAS STRAUCH — *Arithmetic Stratifications and Partial Eisenstein Series* 335
Introduction 335
1. The fibre bundles: geometric-arithmetic preliminaries 338
2. Height zeta functions 342
3. Arithmetic stratification 351
References 355

SIR PETER SWINNERTON-DYER — *Weak Approximation and R-equivalence on Cubic Surfaces* 357
1. Introduction 358

2. Geometric background 361
3. Approximation at an infinite prime 370
4. Approximation at a finite prime 371
5. The lifting process 380
6. The dense lifting process 386
7. Adelic results 394
8. Surfaces $X_1^3 + X_2^3 + X_3^3 - dX_0^3 = 0$ 395
References 403

TREVOR D. WOOLEY — *Hua's lemma and exponential sums over binary forms* 405
  1. Introduction 405
  2. Preliminary reductions 411
  3. Integral points on affine plane curves 421
  4. The inductive step 430
  5. The completion of the proof of Theorem 1.1 441
  References 445

ABSTRACTS

Diagonal cubic equations in four variables with prime coefficients
CARMEN LAURA BASILE & THOMAS ANTHONY FISHER 1

    The aim of this paper is to give an alternative proof of a theorem of R. Heath-Brown regarding the existence of non-zero integral solutions of the equation
$$p_1 X_1^3 + p_2 X_2^3 + p_3 X_3^3 + p_4 X_4^3 = 0,$$
where the p_j are prime integers congruent to 2 modulo 3.

Rational points on cubic surfaces
NIKLAS BROBERG 13

    Let k be an algebraic number field and $F(x_0, x_1, x_2, x_3)$ a non-singular cubic form with coefficients in k. Suppose that the projective cubic k–surface $X \subset \mathbb{P}_k^3$ given by $F = 0$ contains three coplanar lines defined over k, and let $U(k)$ be the set of those k–rational points on X which do not lie on any line on X. We show that the number of points in $U(k)$, with height at most B, is $O_{F,\varepsilon}(B^{4/3+\varepsilon})$ for any $\varepsilon > 0$.

Torseurs arithmétiques et espaces fibrés
ANTOINE CHAMBERT-LOIR & YURI TSCHINKEL 37

    We study the compatibility of Manin's conjecture with natural geometric constructions, like fibrations induced from torsors under linear algebraic groups. The main problem it to understand the variation of metrics from fiber to fiber. For this we introduce the notions of "arithmetic torsors", "adelic torsion" and "Arakelov L-functions". We discuss concrete examples, like horospherical varieties and equivariant compactifications of semi-abelian varieties. These techniques are applied to prove "going up" and "descent" theorems for height zeta functions on such fibrations.

Fonctions zêta des hauteurs des espaces fibrés
ANTOINE CHAMBERT-LOIR & YURI TSCHINKEL 71

In this paper we study the compatibility of Manin's conjectures concerning asymptotics of rational points on algebraic varieties with certain natural geometric constructions. More precisely, we consider locally trivial fibrations constructed from torsors under linear algebraic groups. The main problem is to understand the behaviour of the height function as one passes from fiber to fiber - a difficult problem, even though all fibers are isomorphic. We will be mostly interested in fibrations induced from torsors under split tori. Asymptotic properties follow from analytic properties of height zeta functions. Under reasonable assumptions on the analytic behaviour of the height zeta function for the base we establish analytic properties of the height zeta function of the total space.

Hasse principle for pencils of curves of genus one whose Jacobians have a rational 2-division point, close variation on a paper of Bender and Swinnerton-Dyer
JEAN-LOUIS COLLIOT-THÉLÈNE 117

Une série d'articles exploite une nouvelle technique qui mène à des conditions suffisantes d'existence et de densité des points rationnels sur certaines surfaces fibrées en courbes de genre un au-dessus de la droite projective. Dans les premiers articles de cette série (plusieurs articles de Swinnerton-Dyer, un article en collaboration de Skorobogatov, Swinnerton-Dyer et l'auteur), la jacobienne de la fibre générique des surfaces considérées a tous ses points d'ordre 2 rationnels. Un article récent de Bender et Swinnerton-Dyer traite de cas où cette jacobienne possède seulement un point d'ordre 2 non trivial (pour que la méthode fonctionne, il semble nécessaire que la jacobienne possède un point de torsion rationnel non trivial). Le présent article est une réécriture de celui de Bender et Swinnerton-Dyer. La principale contribution est une reformulation plus abstraite des hypothèses principales des théorèmes.

La première hypothèse est formulée de façon entièrement algébrique (certains groupes de Selmer algébrico-géométriques sont supposés petits) et la seconde hypothèse est simplement : "Il n'y a pas d'obstruction de Brauer-Manin verticale". Comme dans la plupart des articles de cette série, les résultats dépendent de deux conjectures difficiles : l'hypothèse de Schinzel et la finitude des groupes de Tate-Shafarevich.

Enriques surfaces with a dense set of rational points, Appendix to the paper by J.-L. Colliot-Thélène
ALEXEI SKOROBOGATOV 163

Density of integral points on algebraic varieties
BRENDAN HASSETT & YURI TSCHINKEL 169
    We study the Zariski density of integral points on quasi-projective algebraic varieties.

Composition of points and the Mordell–Weil problem for cubic surfaces
DIMITRI KANEVSKY & YURI MANIN 199
    Let V be a plane smooth cubic curve over a finitely generated field k. The Mordell–Weil theorem for V states that there is a finite subset $P \subset V(k)$ such that the whole $V(k)$ can be obtained from P by drawing secants and tangents through pairs of previously constructed points and consecutively adding their new intersection points with V. Equivalently, the group of birational transformations of V generated by reflections with respect to k–points is finitely generated. In this paper, we establish a Mordell–Weil type finite generation result for some birationally trivial cubic surfaces W. To the contrary, we prove that the birational automorphism group generated by reflections cannot be finitely generated if $W(k)$ is infinite.

Torseurs universels et méthode du cercle
EMMANUEL PEYRE 221
    Ce texte décrit les premières étapes d'une généralisation de la méthode du cercle au cas d'une hypersurface lisse dans une variété presque de Fano.
    En effet, sous certaines conditions, il est possible d'exprimer dans ce cas les deux membres d'une version raffinée de la conjecture de Manin sur le comportement asymptotique du nombre de points de hauteur bornée

de l'hypersurface en termes des torseurs universels de la variété ambiante qui jouent, dans ce cadre, le rôle de l'espace affine.

Tamagawa numbers of diagonal cubic surfaces of higher rank
EMMANUEL PEYRE & YURI TSCHINKEL 275

We consider diagonal cubic surfaces defined by an equation of the form
$$ax^3 + by^3 + cz^3 + dt^3 = 0.$$
Numerically, one can find all rational points of height $\leqslant B$ for B in the range of up to 10^5, thanks to a program due to D. J. Bernstein. On the other hand, there are precise conjectures concerning the constants in the asymptotics of rational points of bounded height due to Manin, Batyrev and the authors. Changing the coefficients one can obtain cubic surfaces with rank of the Picard group varying between 1 and 4. We check that numerical data are compatible with the above conjectures. In a previous paper we considered cubic surfaces with Picard groups of rank one with or without Brauer-Manin obstruction to weak approximation. In this paper, we test the conjectures for diagonal cubic surfaces with Picard groups of higher rank.

The Hasse principle for complete intersections in projective space
BJORN POONEN 307

Assuming the existence of a smooth geometrically integral complete intersection X of dimension ≥ 3 in $\mathbf{P}^n$ over a number field K, such that the Zariski closure of $X(K)$ is nonempty but of codimension ≥ 2 in X, we construct a 3-dimensional smooth geometrically integral complete intersection X' in $\mathbf{P}^n_K$ that violates the Hasse principle. Such a violation could not be explained by the Brauer-Manin obstruction or Skorobogatov's generalization thereof.

Une construction de courbes k-rationnelles sur les surfaces de Kummer d'un produit de courbes de genre 1.
PHILIPPE SATGÉ 313

k étant un corps de caractéristique différente de 2, nous décrivons une méthode permettant de construire des courbes k-rationnelles (i.e. k-birationnellement équivalentes à la droite projective) sur les surfaces de Kummer associées à un produit de courbes de genre 1 munies d'involutions k-hyperelliptiques. Nous ramenons ce problème à un problème de géométrie énumérative sur le produit $\mathbf{P}_{1,k} \times \mathbf{P}_{1,k}$ de la droite projective par elle même. Bien que la résolution générale du problème de géométrie énumérative auquel nous arrivons soit hors de portée des méthodes que nous connaissons, la recherche de solutions particulières dans des systèmes linéaires convenablement choisis permet d'obtenir des exemples interessants. On constate par exemple que l'on retrouve ainsi, de manière assez systématique, plusieurs résultats qui apparaissent de manière isolée dans la littérature.

Arithmetic Stratifications and Partial Eisenstein Series
MATTHIAS STRAUCH 335

Let $P\backslash G$ and $Q\backslash H$ be generalized flag varieties over a number field F. In this paper we study certain locally trivial fibre bundles Y_η over $P\backslash G$ having $Q\backslash H$ as general fibre, and determine the arithmetic stratification of Y_η with respect to a line bundle. The arithmetic stratification is defined in terms of height zeta functions and the height zeta function of a stratum is of the form

$$\sum_{\gamma \in P(F)\backslash G(F)} e^{\langle s\lambda, H_P(\gamma)\rangle} E_Q^{Qw^{-1}Q_0}(s\mu, \eta(p_\gamma)),$$

where $E_Q^{Qw^{-1}Q_0}$ is a "partial Eisenstein series" associated to the Schubert cell $Q\backslash Qw^{-1}Q_0$. The computation of the constant term of these gives estimates that allow one to determine the abcissa of convergence of the height zeta function of the stratum.

Weak Approximation and R-equivalence on Cubic Surfaces
SIR PETER SWINNERTON-DYER 357

Let V be a nonsingular cubic surface defined over an algebraic number field K, and assume that V has points in every completion K_v. There is a long-standing problem of finding the obstructions to the Hasse principle and to weak approximation on V, the conjecture in each case (due to Colliot-Thélène and Sansuc) being that the obstruction is just the Brauer-Manin obstruction. The latter is known to be computable, though the algorithm is somewhat ugly and a heuristic process is usually preferable. Another way of phrasing the same problem is to ask what is the adelic closure of the set $V(K)$.

A partial answer to this question is given by the following theorem: to each place v of bad reduction one can associate a finite disjoint union

$$V(K_v) = \cup W_j^{(v)}$$

which is easily computable in any particular case. The v-adic closure of any R-equivalence class in $V(K)$ is a set U_{iv} which is the union of some of the $W_j^{(v)}$; and the adelic closure of any R-equivalence class is of the form $\prod' U_{iv} \times \prod'' V(K_v)$, where i depends on v, the first product is over all places of bad reduction and the second product is over all places of good reduction for V. Thus the adelic closure of $V(K)$ is a union of sets $\prod' W_j^{(v)} \times \prod'' V(K_v)$. For specific V a search program will give those products which contain a point of $V(K)$ and in which points of $V(K)$ are therefore everywhere dense. For a product which appears not to contain a point of $V(K)$, it is reasonable to hope that there is a Brauer-Manin obstruction. For all V for which this process has been used, it turns out that one can indeed find the exact adelic closure of $V(K)$ in this way. This is illustrated in the final section.

Hua's lemma and exponential sums over binary forms
TREVOR D. WOOLEY 405

We establish mean value estimates for exponential sums over binary forms of strength comparable with the bounds attainable via classical, single variable estimates for diagonal forms. These new mean value estimates strengthen earlier bounds of the author when the degree d of the form satisfies $5 \leq d \leq 10$, the improvements stemming from a basic lemma which provides uniform estimates for the number of integral points on affine plane curves in mean square. Exploited by means of the Hardy-Littlewood method, these estimates permit one to establish asymptotic formulae for the number of integral zeros of equations defined as sums of binary forms of the same degree d, provided that the number of variables exceeds $\frac{17}{16}2^d$, improving significantly on what is attainable either by classical additive methods, or indeed the general methods of Birch and Schmidt.

Rational points on algebraic varieties
(E. PEYRE, Y. TSCHINKEL, ed.), p. 1–12

DIAGONAL CUBIC EQUATIONS IN FOUR VARIABLES WITH PRIME COEFFICIENTS

Carmen Laura Basile[*]

Department of Mathematics, Imperial College of Science, Technology and Medicine, London • *E-mail* : laura.basile@ic.ac.uk

Thomas Anthony Fisher

Department of Pure Mathematics and Mathematical Statistics, Sidney Sussex College, Cambridge • *E-mail* : T.A.Fisher@dpmms.cam.ac.uk

Abstract. — The aim of this paper is to give an alternative proof of a theorem of R. Heath-Brown regarding the existence of non-zero integral solutions of the equation

$$p_1 X_1^3 + p_2 X_2^3 + p_3 X_3^3 + p_4 X_4^3 = 0,$$

where the p_j are prime integers congruent to 2 modulo 3.

We start by presenting the main result of this paper. This result has been proved by Roger Heath-Brown [3] under the conjecture that the difference $s(A) - r(A)$ between the Selmer rank and the arithmetic rank of the elliptic curve $X^3 + Y^3 = AZ^3$ is even. In this note we will show that we do not need this assumption and give a detailed proof of this result.

Theorem 1. — *Let p_1, p_2, p_3, p_4 be prime integers such that $p_i \equiv 2 \pmod{3}$ for $1 \leq i \leq 4$. Then the equation*

$$p_1 X_1^3 + p_2 X_2^3 + p_3 X_3^3 + p_4 X_4^3 = 0$$

has non-zero integral solutions, assuming the conjecture that the Tate-Shafarevich group of the elliptic curve $X^3 + Y^3 = AZ^3$ over $\mathbb{Q}$ is finite.

---

[*]Supported by INdAM (Istituto Nazionale di Alta Matematica "F.Severi"), Italy.

A much stronger result has been recently proved by Sir Peter Swinnerton-Dyer [7].

Note that in order to prove the above theorem, it is sufficient to prove that the equations
$$p_1 X_1^3 + p_2 X_2^3 = p$$
$$p_3 X_3^3 + p_4 X_4^3 = -p$$
have non-zero rational solutions for some prime integer p. So it suffices to prove that the equation $p_1 X^3 + p_2 Y^3 = p$ has non-zero solutions in $\mathbb{Q}$ or, equivalently, in a quadratic extension of $\mathbb{Q}$.

Notation. — Let ω be a primitive cube root of unity and let $k = \mathbb{Q}(\omega)$. Let $A \in \mathbb{Z} \setminus \{-1, 0, 1\}$ be a cube-free integer. We denote by E_A the elliptic curve
$$E_A \colon X^3 + Y^3 = AZ^3.$$
For any $\alpha \in k^*$, let $C_{A,\alpha}$ be the smooth projective curve given by the equation
$$C_{A,\alpha} \colon \alpha X^3 + \alpha^{-1} Y^3 = AZ^3.$$

For $\alpha \in k^*$, the curves $C_{A,\alpha}$ are principal homogeneous spaces over E_A. Moreover, it is clear that if α and β belong to the same class modulo $(k^*)^3$ then the curves $C_{A,\alpha}$ and $C_{A,\beta}$ are isomorphic.

Let us consider the *multiplication-by-$\sqrt{-3}$* endomorphism on E_A (see [1]), given by
$$\sqrt{-3} \colon \quad E_A \longrightarrow E_A$$
$$(X, Y, Z) \longmapsto (\omega^2 X^3 + \omega Y^3 - AZ^3, \ \omega X^3 + \omega^2 Y^3 - AZ^3, \ -3XYZ),$$
and the following diagram with exact row
$$0 \longrightarrow E_A(k)/\sqrt{-3}E_A(k) \longrightarrow H^1(k, E_A[\sqrt{-3}]) \longrightarrow WC(E_A/k)[\sqrt{-3}] \longrightarrow 0$$
$$\searrow \qquad \downarrow$$
$$\prod_v WC(E_A/k_v)[\sqrt{-3}]$$
where $WC(E_A/k)$ is the Weil-Châtelet group of E_A/k and v runs over all the places of k. It is readily verified that the group $E_A(\bar{k})[\sqrt{-3}]$ is isomorphic to the group $\mu_3(\bar{k})$ of cube roots of unity, as a $\mathrm{Gal}(\bar{k}/k)$-module. It follows from Kummer theory that $H^1(k, E_A[\sqrt{-3}])$ is isomorphic to $k^*/(k^*)^3$ and so we get the exact sequence
$$0 \longrightarrow E_A(k)/\sqrt{-3}E_A(k) \longrightarrow k^*/(k^*)^3 \xrightarrow{f} WC(E_A/k)[\sqrt{-3}] \longrightarrow 0$$
$$g \searrow \qquad \downarrow$$
$$\prod_v WC(E_A/k_v)[\sqrt{-3}]$$

where the map f sends an element $\alpha(k^*)^3$ to the curve $C_{A,\alpha}$. We denote by $S(A)$ the Selmer group $S^{(\sqrt{-3})}(E_A/k)$ (which is defined to be the kernel of the map g in the diagram above) and by $C(A)$ the kernel of the map f. Obviously $C(A)$ is a subgroup of $S(A)$; we can write them explicitly

$$S(A) = \{\alpha(k^*)^3 : C_{A,\alpha} \text{ has } k_\pi\text{-points for any prime } \pi \in k, \ \alpha \in k^*\},$$
$$C(A) = \{\alpha(k^*)^3 : C_{A,\alpha} \text{ has } k\text{-points}, \ \alpha \in k^*\}.$$

Observe also that $C(A)$ is isomorphic to $E_A(k)/\sqrt{-3}E_A(k)$ (this follows immediately from the diagram above).

In the following two lemmas we will determine the structure of $S(A)$ when A satisfies certain conditions.

Lemma 2. — *Let $\rho \in k$ be a prime above 3, and let $a, b, c \in k$ such that a, b, c are congruent to 1 modulo ρ^2. If the projective curve*

$$E: \ aX^3 + b\omega Y^3 + c\omega^2 Z^3 = 0$$

has a point over k_ρ, then $abc \equiv 1 \pmod{\rho^3}$.

Proof. — Let (x, y, z) be a k_ρ-point of the curve E and suppose that

$$\min\{\mathrm{val}_\rho(x), \mathrm{val}_\rho(y), \mathrm{val}_\rho(z)\} = 0.$$

Say

$$\begin{aligned} x &= x_0 + x_1\rho + x_2\rho^2 + \ldots, \\ y &= y_0 + y_1\rho + y_2\rho^2 + \ldots, \\ z &= z_0 + z_1\rho + z_2\rho^2 + \ldots \end{aligned}$$

with $(x_0, y_0, z_0) \neq (0, 0, 0)$.

From the condition $ax^3 + b\omega y^3 + c\omega^2 z^3 = 0$ it follows that $x_0^3 \equiv y_0^3 \equiv z_0^3 \pmod{\rho^2}$. Therefore we may assume $(x_0, y_0, z_0) = (1, 1, 1)$ and hence we get

$$(x^3, y^3, z^3) \equiv (1, 1, 1) \pmod{\rho^3}.$$

Let $a \equiv 1 + a_2\rho^2 \pmod{\rho^3}$, $b \equiv 1 + b_2\rho^2 \pmod{\rho^3}$ and $c \equiv 1 + c_2\rho^2 \pmod{\rho^3}$. Then from $a + b\omega + c\omega^2 \equiv 0 \pmod{\rho^3}$ it follows $a_2 + b_2 + c_2 \equiv 0 \pmod{\rho}$ and therefore $abc \equiv 1 + (a_2 + b_2 + c_2)\rho^2 \equiv 1 \pmod{\rho^3}$. □

Lemma 3. — *Let $A = p_1 p_2 N(\pi)$ where $p_1, p_2 \equiv 2 \pmod 3$ are integer primes, $\pi \equiv 1 \pmod 3$ is a prime in $\mathbb{Z}[\omega]$ such that $A \not\equiv 1 \pmod 9$ and $\left(\frac{\pi}{p_1}\right)_3 = \left(\frac{\pi}{p_2}\right)_3 \neq 1$. Then $S(A)$ is isomorphic to $\mathbb{Z}/3 \times \mathbb{Z}/3$ as an abelian group.*

Proof. — We will prove that $S(A)$ is generated by the elements $A(k^*)^3$ and $p_1 p_2^2 (k^*)^3$.

The curve $C_{A,A}$ has a k-point, namely $(1,0,1)$; hence $A(k^*)^3 \in S(A)$. Suppose now $\alpha \neq A$. Note that, in order to determine for which elements α the coset $\alpha(k^*)^3$ belongs to $S(A)$, it is sufficient to test the elements α where α is a cube-free integer in $\mathbb{Z}[\omega]$ such that

(i) α is composed of primes dividing A;
(ii) we can fix a prime in $\{p_1, p_2, \pi, \bar\pi\}$, say $\bar\pi$, such that $\bar\pi$ does not divide α.

Indeed:

(i) suppose α contains a prime factor ρ which does not divide A; it is then easy to verify that $C_{A,\alpha}$ contains no k_ρ-points. To get a contradiction, we suppose $C_{A,\alpha}$ contains a k_ρ-point (X, Y, Z); so we have

$$\alpha^2 X^3 + Y^3 = \alpha A Z^3.$$

Considering the ρ-adic valuation of the right hand side and of the left hand side, and noting $\operatorname{val}_\rho(\alpha^2 X^3) \neq \operatorname{val}_\rho(Y^3)$, we get

$$\operatorname{val}_\rho(\alpha A Z^3) = \operatorname{val}_\rho(\alpha^2 X^3 + Y^3) = \min\{\operatorname{val}_\rho(\alpha^2 X^3), \operatorname{val}_\rho(Y^3)\}.$$

But this is impossible since, on the one hand, $\operatorname{val}_\rho(\alpha A Z^3) = i + 3j$, where $i \in \{1, 2\}$ is defined to be the ρ-adic valuation of α and j is an integer, and on the other hand

$$\min\{\operatorname{val}_\rho(\alpha^2 X^3), \operatorname{val}_\rho(Y^3)\} = \min\{2i + 3h, 3l\}$$

with h, l integers;

(ii) suppose $\bar\pi^j$ divides α (with $j = 1, 2$); then, since $S(A)$ is a group and $A(k^*)^3$ belongs to $S(A)$, instead of α we may consider the cube-free integer representative of the coset $(\alpha/A^j)(k^*)^3$.

As a result, we may assume $\alpha = \omega^m p_1^{n_1} p_2^{n_2} \pi^n$, where $m, n_1, n_2, n \in \{0, 1, 2\}$. In fact, it is not difficult to prove that if $m \neq 0$ then α does not belong to $S(A)$. Indeed, to get a contradiction suppose that $m \neq 0$ and that $C_{A,\alpha}$ contains a k_ρ-point, where $\rho \in k$ is a prime above 3. Then from Lemma 2 it follows that $A \equiv 1 \pmod{\rho^3}$ and hence $A \equiv 1 \pmod 9$ which contradicts our hypotheses. Therefore we may assume $m = 0$.

Let $\rho \in \mathbb{Z}[\omega]$ be a prime. If ρ does not divide $3A$ then $C_{A,\alpha}$ contains points over k_ρ for any α. Indeed, a smooth curve of genus 1 always contains points over any finite field (see [2]); moreover if ρ does not divide $3A$ then $C_{A,\alpha}$ is non-singular over $\mathbb{Z}[\omega]/\rho$ and therefore it contains a non-singular point over $\mathbb{Z}[\omega]/\rho$ which, by Hensel's lemma, can be lifted to a point over k_ρ.

Suppose now that ρ divides $3A$; hence we have to consider the two cases $\rho \mid A$ and $\rho \mid 3$.

(1) Let ρ divide A, so ρ belongs to $\{p_1, p_2, \pi, \bar{\pi}\}$. We can consider three further subcases.

(1a) Let ρ divide α exactly; then the projective curve
$$\alpha X^3 + \alpha^{-1} Y^3 = AZ^3$$
can also be written as
$$\frac{\alpha^2}{\rho^2}(\rho X)^3 + \rho Y^3 = \frac{\alpha A}{\rho^2}(\rho Z)^3.$$
Hence, considering the transformation $\rho X \to X$, $\rho Z \to Z$ and reducing modulo ρ, we get the curve
$$X^3 - \frac{A}{\alpha} Z^3 = 0$$
which contains a smooth point over $\mathbb{F}_\rho$ if and only if $\frac{A}{\alpha} \in (\mathbb{F}_\rho^*)^3$, i.e. if and only if $\left(\frac{A\alpha^{-1}}{\rho}\right)_3 = 1$. By Hensel's lemma, we can lift this smooth point over $\mathbb{F}_\rho$ to a point over k_ρ.

(1b) Suppose now ρ^2 divides exactly α. Similarly, considering a suitable transformation and reducing modulo ρ, we get the curve
$$Y^3 - \frac{A\alpha}{\rho^3} Z^3 = 0$$
which contains a smooth point over $\mathbb{F}_\rho$ if and only if $\frac{A\alpha}{\rho^3} \in (\mathbb{F}_\rho^*)^3$, i.e. if and only if $\left(\frac{A\alpha\rho^{-3}}{\rho}\right)_3 = 1$.

(1c) Similarly, if ρ does not divide α, we get the curve
$$\alpha^2 X^3 + Y^3 = 0$$
which contains a smooth point over $\mathbb{F}_\rho$ if and only if $\alpha^2 \in (\mathbb{F}_\rho^*)^3$, i.e. $\left(\frac{\alpha}{\rho}\right)_3 = 1$.

Suppose $\rho = p_j$ with $j = 1, 2$.
If p_j divides exactly α, then by case (1a) we must have $\left(\frac{A\alpha^{-1}}{p_j}\right)_3 = 1$ and hence

$$\left(\frac{\alpha p_j^{-1}}{p_j}\right)_3 = 1; \text{ indeed } (*)$$

$$\left(\frac{A\alpha^{-1}}{p_j}\right)_3 = \left(\frac{p_h}{p_j}\right)_3 \left(\frac{p}{p_j}\right)_3 \left(\frac{\alpha^{-1} p_j}{p_j}\right)_3 = \left(\frac{\alpha p_j^{-1}}{p_j}\right)_3^{-1}$$

where $h \in \{1, 2\}$, $h \neq j$ and $p = N(\pi)$.

If p_j^2 divides exactly α, then by case (1b) we must have $\left(\frac{A\alpha p_j^{-3}}{p_j}\right)_3 = 1$ and therefore $\left(\frac{\alpha p_j^{-2}}{p_j}\right)_3 = 1$, since

$$\left(\frac{A\alpha p_j^{-3}}{p_j}\right)_3 = \left(\frac{p_h}{p_j}\right)_3 \left(\frac{p}{p_j}\right)_3 \left(\frac{\alpha p_j^{-2}}{p_j}\right)_3 = \left(\frac{\alpha p_j^{-2}}{p_j}\right)_3,$$

where again $h \in \{1, 2\}$ and $h \neq j$.

If p_j does not divide α, then by case (1c) we must have $\left(\frac{\alpha}{p_j}\right)_3 = 1$.

Finally, we obtain the two following conditions which must be satisfied in order to have $\alpha(k^*)^3 \in S(A)$:

$$\left(\frac{p_2^{n_2} \pi^n}{p_1}\right)_3 = 1 \text{ and } \left(\frac{p_1^{n_1} \pi^n}{p_2}\right)_3 = 1.$$

Since

$$\left(\frac{p_2^{n_2} \pi^n}{p_1}\right)_3 = \left(\frac{p_2}{p_1}\right)_3^{n_2} \left(\frac{\pi}{p_1}\right)_3^n = \left(\frac{\pi}{p_1}\right)_3^n$$

and $\left(\frac{\pi}{p_1}\right)_3 \neq 1$ by hypothesis, the first condition is satisfied if and only if $n = 0$. It is easy to verify that for $n = 0$ the second condition is satisfied as well. Hence $\alpha = p_1^{n_1} p_2^{n_2}$. So, by case (1c), it follows that we must have $\left(\frac{\alpha}{\pi}\right)_3 = 1$; therefore, since by the Cubic Reciprocity Law $\left(\frac{p_1}{\pi}\right)_3 = \left(\frac{p_2}{\pi}\right)_3 \neq 1$, we get the condition $n_1 + n_2 \equiv 0 \pmod{3}$. We may suppose $n_1 = 1$; then $\alpha = p_1 p_2^2$ or, equivalently, $\alpha = p_1 p_2^{-1}$.

(2) Suppose now that ρ divides 3; in other words, since $3 = -\omega^2(1-\omega)^2$, suppose $\rho = 1 - \omega$. Recall that now we have $A = p_1 p_2 p$ and $\alpha = p_1 p_2^{-1}$; therefore the curve

$$\alpha X^3 + \alpha^{-1} Y^3 = AZ^3$$

(*) If $q \neq 3$ is a prime integer with $\#\mathbb{F}_q^* \not\equiv 0 \pmod{3}$ then $\left(\frac{n}{q}\right)_3 = 1$ for any $n \in \mathbb{Z}$ such that $(n, q) = 1$.

is isomorphic to the curve
$$E: \ p_1X^3 + p_2Y^3 = pZ^3.$$

We have to prove that E contains a point over k_ρ. Since $\mathbb{Q}_3$ is contained in k_ρ, it suffices to find a point over $\mathbb{Q}_3$.

By hypothesis $p \equiv 1 \pmod{3}$ and $p_1, p_2 \equiv 2 \pmod{3}$, so let
$$p \equiv 1 + 3b \pmod 9 \quad \text{and} \quad p_j \equiv -1 + 3a_j \pmod 9$$
for $j = 1, 2$; since $A \not\equiv 1 \pmod 9$ by hypothesis, we have $a_1 + a_2 - b \not\equiv 0 \pmod 3$. Thus the integers $a_1, a_2, -b$ cannot be all different modulo 3; wlog we may suppose $a_1 = a_2$. Therefore we have $p_1 p_2^{-1} \equiv 1 \pmod 9$; in other words $p_1 p_2^{-1}$ belongs to $1 + 9\mathbb{Z}_3$ which is contained in $(\mathbb{Z}_3^*)^3$. It follows that E contains the point $(1, -y, 0)$ where $y \in \mathbb{Z}_3$ is a cube root of $p_1 p_2^{-1}$. More precisely, if $p_1 p_2^{-1} \equiv 1 + 9l \pmod{27}$, then we use Hensel's lemma to construct $y \in \mathbb{Z}_3$ such that $y \equiv 1 + 3l \pmod 9$ and $y^3 = p_1 p_2^{-1}$.

Hence E contains a k_ρ-point.

In conclusion, we have proved that $S(A)$ is generated by the two elements of order 3 $A(k^*)^3$ and $p_1 p_2^2(k^*)^3$ and thus it is isomorphic to $\mathbb{Z}/3 \times \mathbb{Z}/3$. □

The following two lemmas allow us to conclude that, under the hypotheses of Lemma 3, the two groups $C(A)$ and $S(A)$ coincide. This provides a local-to-global principle for the curves $C_{A,\alpha}$ when A is as in Lemma 3. As in the statement of Theorem 1, our work here is conditional on the finiteness of the Tate-Shafarevich group $\mathrm{III}(E_A/\mathbb{Q})$.

To get a contradiction, let us suppose that $C(A)$ is strictly included in $S(A)$. Note that $C(A)$ cannot be the trivial group as $A(k^*)^3$ belongs to $C(A)$; then $C(A)$ has order 3. From the exactness of the sequence
$$0 \longrightarrow E_A(k)/\sqrt{-3}E_A(k) \longrightarrow k^*/(k^*)^3 \xrightarrow{f} WC(E_A/k)\left[\sqrt{-3}\right] \longrightarrow 0,$$
it follows that $E_A(k)/\sqrt{-3}E_A(k)$ and $\mathbb{Z}/3$ are isomorphic as abelian groups (recall that $C(A)$ is the kernel of the map f). Hence from Lemma 3 and the exact sequence
$$0 \longrightarrow E_A(k)/\sqrt{-3}E_A(k) \longrightarrow S^{(\sqrt{-3})}(E_A/k) \longrightarrow \mathrm{III}(E_A/k)\left[\sqrt{-3}\right] \longrightarrow 0$$
we deduce that $\mathrm{III}(E_A/k)\left[\sqrt{-3}\right]$ is isomorphic to $\mathbb{Z}/3$ and this is impossible, as we will show in Lemma 5. But first we need one more result.

Lemma 4. — *Let E/L be an elliptic curve over a number field L. Let K be a Galois extension of L of degree n. Let m be a positive integer such that*

$(m, n) = 1$. Then:

$$\text{Ш}(E/L)[m] = \text{Ш}(E/K)[m]^{\text{Gal}(K/L)}.$$

In particular, assuming the finiteness of the group $\text{Ш}(E/L)$, the order of the group $\text{Ш}(E/K)[m]^{\text{Gal}(K/L)}$ must be a square.

Proof. — Let us consider the following commutative diagram with exact rows and columns where the rows are obtained by the *multiplication-by-m* endomorphism and the columns are restriction-inflation sequences:

$$\begin{array}{ccccccccc}
& & & & 0 & & 0 & & \\
& & & & \downarrow & & \downarrow & & \\
& & & & H^1(\text{Gal}(K/L), E(K)[m]) & & H^1(\text{Gal}(K/L), E(K))[m] & & \\
& & & & \downarrow & & \downarrow & & \\
0 & \to & E(L)/mE(L) & \to & H^1(L, E[m]) & \to & H^1(L, E)[m] & \to & 0 \\
& & \downarrow & & \downarrow & & \downarrow & & \\
0 & \to & E(K)/mE(K) & \to & H^1(K, E[m]) & \to & H^1(K, E)[m] & \to & 0.
\end{array}$$

Since $\text{Gal}(K/L)$ has order n, every element of $H^1(\text{Gal}(K/L), E(K))$ has order dividing n (this follows from properties of the restriction and corestriction maps; see [5]). Hence, as m and n are coprime,

$$H^1(\text{Gal}(K/L), E(K))[m] = 0.$$

Thus from the diagram above it follows that $H^1(L, E)[m]$ injects into the group $H^1(K, E)[m]$.

From the exactness of the second row of the diagram we get the exact sequence

$$0 \to E(K)/mE(K)^{\text{Gal}(K/L)} \to H^1(K, E[m])^{\text{Gal}(K/L)} \to$$
$$\to H^1(K, E)[m]^{\text{Gal}(K/L)} \to H^1(\text{Gal}(K/L), E(K)/mE(K)) = 0,$$

where $H^1(\text{Gal}(K/L), E(K)/mE(K))$ is the zero group because it is killed by m and by n which are coprime.

On the other hand, from the exact sequence of low degree terms of the Hochschild-Serre spectral sequence, we get the exact sequence

$$H^1(\text{Gal}(K/L), E(K)[m]) \to H^1(L, E[m]) \xrightarrow{\varphi} H^1(K, E[m])^{\text{Gal}(K/L)} \to$$
$$\to H^2(\text{Gal}(K/L), E(K)[m])$$

where the first and the last term are trivial because, again, they are killed by coprime integers. Hence the map φ is an isomorphism. The following diagram of exact rows and columns summarizes the information we have obtained so far

$$
\begin{array}{ccc}
0 & & 0 \\
\downarrow & & \downarrow \\
H^1(L, E[m]) & \longrightarrow & H^1(L, E)[m] \longrightarrow 0 \\
\downarrow \varphi & & \downarrow \varphi' \\
H^1(K, E[m])^{\mathrm{Gal}(K/L)} & \stackrel{\varphi''}{\longrightarrow} & H^1(K, E)[m]^{\mathrm{Gal}(K/L)} \longrightarrow 0 \\
\downarrow & & \\
0. & &
\end{array}
$$

It is immediate to verify that the injective map φ' is also surjective because of the surjectivity of the maps φ'' and φ. Therefore we obtain

$$H^1(L, E)[m] = H^1(K, E)[m]^{\mathrm{Gal}(K/L)}.$$

Let us consider now a place v of L; since K is a Galois extension of L, for any place w of K over v the degrees of the local extensions K_w/L_v divide n and therefore they are coprime to m. Hence the reasoning above can be applied also to the extensions K_w/L_v and thus we obtain

$$H^1(L_v, E)[m] = H^1(K_w, E)[m]^{\mathrm{Gal}(K_w/L_v)}.$$

Considering the corresponding Tate-Shafarevich groups, we get

$$\mathrm{III}(E/L)[m] = \mathrm{III}(E/K)[m]^{\mathrm{Gal}(K/L)}.$$

Furthermore, assuming the finiteness of $\mathrm{III}(E/L)$, it follows from the existence of the Cassels alternating bilinear pairing on $\mathrm{III}(E/L)$ that the order of $\mathrm{III}(E/L)[m]$ is a perfect square and hence the order of $\mathrm{III}(E/K)[m]^{\mathrm{Gal}(K/L)}$ is a square too. □

Lemma 5. — *If $\mathrm{III}(E_A/\mathbb{Q})$ is finite, then $\mathrm{III}(E_A/k)\left[\sqrt{-3}\right]$ cannot have order 3.*

Proof. — To get a contradiction, assume that $\mathrm{III}(E_A/k)\left[\sqrt{-3}\right]$ and $\mathbb{Z}/3$ are isomorphic as abelian groups.

Let $\tilde{E}_A$ be the quadratic twist of E_A corresponding to the class of -3 in $H^1(\mathbb{Q}, \mathbb{Z}/2) = \mathbb{Q}^*/(\mathbb{Q}^*)^2$; $\tilde{E}_A$ has equation

$$\tilde{E}_A: \ -3Y^2 Z = X^3 - 432 A^2 Z^3$$

and is isomorphic to E_A over k through the map

$$\psi: \quad E_A \longrightarrow \tilde{E}_A$$
$$(X,Y,Z) \longmapsto (12AZ, \frac{36A}{\sqrt{-3}}(X-Y), X+Y).$$

Let us consider the dual isogenies $\varphi_1: E_A \longrightarrow \tilde{E}_A$ and $\varphi_2: \tilde{E}_A \longrightarrow E_A$ given by the compositions $\varphi_1 = \psi \circ \sqrt{-3}$ and $\varphi_2 = -\sqrt{-3} \circ \psi^{-1}$; they are defined over $\mathbb{Q}$ and their composition $\varphi_2 \circ \varphi_1$ gives the *multiplication-by-3* map on E_A.

To obtain a contradiction we have assumed that $\text{III}(E_A/k)\left[\sqrt{-3}\right]$ is isomorphic to $\mathbb{Z}/3$ as an abelian group; this is equivalent to the assumption that $\text{III}(E_A/k)[\varphi_1]$ is isomorphic to $\mathbb{Z}/3$.

Let $\text{Gal}(k/\mathbb{Q}) = \langle \sigma \rangle$. We have two possibilities: either σ acts trivially on $\text{III}(E_A/k)[\varphi_1]$ and therefore $\text{III}(E_A/k)[\varphi]$ and $\mathbb{Z}/3$ are isomorphic as $\text{Gal}(k/\mathbb{Q})$-modules; or σ exchanges the two non-trivial elements of $\text{III}(E_A/k)[\varphi_1]$ and so $\text{III}(E_A/k)[\varphi_1]$ is isomorphic to μ_3.

If $\text{III}(E_A/k)[\varphi_1]$ is isomorphic to $\mathbb{Z}/3$ as a $\text{Gal}(k/\mathbb{Q})$-module then the group $\text{III}(\tilde{E}_A/k)[\varphi_2]$ is isomorphic to μ_3 as a $\text{Gal}(k/\mathbb{Q})$-module and vice versa. Indeed, if $\text{III}(E_A/k)[\varphi_1]$ is composed of the cohomology classes ξ_0, ξ_1, ξ_2, then $\text{III}(\tilde{E}_A/k)[\varphi_2]$ is composed of $\psi\xi_0, \psi\xi_1, \psi\xi_2$; moreover $^\sigma\psi = -\psi$. So, if σ acts trivially on $\text{III}(E_A/k)[\varphi_1]$ then it does not on $\text{III}(\tilde{E}_A/k)[\varphi_2]$ and vice versa.

Suppose that $\text{III}(E_A/k)[\varphi_1] \cong \mathbb{Z}/3$ as a $\text{Gal}(k/\mathbb{Q})$-module and consider the exact sequence of $\text{Gal}(k/\mathbb{Q})$-modules

$$0 \longrightarrow \text{III}(E_A/k)[\varphi_1] \longrightarrow \text{III}(E_A/k)[3] \longrightarrow \text{III}(\tilde{E}_A/k)[\varphi_2]$$

where the first map is the natural inclusion and the second one is induced by φ_1. From this sequence we get the exact sequence

$$0 \longrightarrow \text{III}(E_A/k)[\varphi_1]^{\text{Gal}(k/\mathbb{Q})} \longrightarrow \text{III}(E_A/k)[3]^{\text{Gal}(k/\mathbb{Q})} \longrightarrow \text{III}(\tilde{E}_A/k)[\varphi_2]^{\text{Gal}(k/\mathbb{Q})},$$

where $\text{III}(E_A/k)[\varphi_1]^{\text{Gal}(k/\mathbb{Q})} \cong \mathbb{Z}/3$ and $\text{III}(\tilde{E}_A/k)[\varphi_2]^{\text{Gal}(k/\mathbb{Q})} \cong 0$. If $\text{III}(E_A/k)[\varphi_1] \cong \mu_3$, it is sufficient to consider the sequence

$$0 \longrightarrow \text{III}(\tilde{E}_A/k)[\varphi_2] \longrightarrow \text{III}(E_A/k)[3] \longrightarrow \text{III}(E_A/k)[\varphi_1]$$

instead of that above.

In both cases, we can conclude that $\text{III}(E_A/k)[3]^{\text{Gal}(k/\mathbb{Q})}$ is isomorphic to $\mathbb{Z}/3$. This contradicts Lemma 4 which claims that the order of the group $\text{III}(E_A/k)[3]^{\text{Gal}(k/\mathbb{Q})}$ must be a square. As a result, $\text{III}(E_A/k)\left[\sqrt{-3}\right]$ cannot have order 3. $\square$

In conclusion, we have proved that $C(A) = S(A)$ for $A = p_1 p_2 p$. Moreover, in the proof of Lemma 3 we have shown that $p_1 p_2^2 (k^*)^3$ belongs to $S(A)$; it follows that the curve $p_1 X^3 + p_2 Y^3 = p$ has a k_ρ-point for any prime ρ of k

and hence, by the local-to-global principle, it has a point over the quadratic extension k of $\mathbb{Q}$. Therefore it has a point over $\mathbb{Q}$.

In order to prove Theorem 1, it only remains to be shown that given the prime integers $p_1, p_2, p_3, p_4 \equiv 2 \pmod{3}$ there exists a prime π such that the hypotheses of Lemma 3 are satisfied for each of the triples p_1, p_2, π and p_3, p_4, π.

Lemma 6. — *Let p_1, p_2, p_3, p_4 be prime integers congruent to 2 modulo 3. Then there exists a prime $\pi \in \mathbb{Z}[\omega]$ such that*
(i) $\pi \equiv 1 \pmod{3}$;
(ii) $p_1 p_2 N(\pi)$ and $p_3 p_4 N(\pi)$ are not congruent to 1 modulo 9;
(iii) $\left(\frac{\pi}{p_1}\right)_3 = \left(\frac{\pi}{p_2}\right)_3 = \left(\frac{\pi}{p_3}\right)_3 = \left(\frac{\pi}{p_4}\right)_3 \neq 1$.

Proof. — Let $B \in \{1, 4, 7\}$ such that
$$\begin{cases} p_1 p_2 B \not\equiv 1 \pmod{9} \\ p_3 p_4 B \not\equiv 1 \pmod{9}. \end{cases}$$
Take a prime $\pi \in \mathbb{Z}[\omega]$ such that $N(\pi) \equiv B \pmod{9}$. This condition can be satisfied by taking $\pi \equiv \beta \pmod{9}$, where β is an element of $\mathbb{Z}[\omega]$ congruent to $1, -2$ or $1 + 3\omega$ modulo 9 if $B = 1, 4$ or 7, respectively. Hence we have that $p_i p_j B \not\equiv 1 \pmod 9$ if and only if $p_i p_j N(\pi) \not\equiv 1 \pmod 9$. Therefore conditions (i) and (ii) are satisfied.

As far as condition (iii) is concerned, in order for π to satisfy
$$\left(\frac{\pi}{p_1}\right)_3 = \left(\frac{\pi}{p_2}\right)_3 = \left(\frac{\pi}{p_3}\right)_3 = \left(\frac{\pi}{p_4}\right)_3 \neq 1$$
it is sufficient to take π belonging to a suitable congruence class modulo the product $p_1 p_2 p_3 p_4$. The Chinese Remainder Theorem allows us to determine a suitable residue class γ modulo $9 p_1 p_2 p_3 p_3$ such that, if $\pi \equiv \gamma \pmod{9 p_1 p_2 p_3 p_3}$, then π satisfies the required conditions. The existence of such a prime π is assured by Dirichlet's Theorem. □

Hence Theorem 1 is proved.

We thank Alexei Skorobogatov for his help in the preparation of this note.

References

[1] J.W.S. Cassels, *Arithmetic on curves of genus 1. I. On a conjecture of Selmer*, J. reine angew. Math. **202** (1959), 52–99.

[2] J.W.S. Cassels, *Lectures on Elliptic Curves*, Cambridge University Press, 1991.

[3] R. Heath-Brown, *The solubility of diagonal cubic Diophantine equations*, Proc. London Math. Soc. (3) **79** (1999), 241–259.

[4] J.-P. Serre, *A course in Arithmetic*, New York: Springer-Verlag, 1973.
[5] J.-P. Serre, *Galois cohomology*, Springer-Verlag Berlin Heidelberg, 1997.
[6] J.H. Silverman, *The Arithmetic of Elliptic Curves*, New York: Springer-Verlag, 1986.
[7] H.P.F. Swinnerton-Dyer, *The solubility of diagonal cubic surfaces*, to appear in Ann. Sci. École Norm. Sup.

Rational points on algebraic varieties
(E. PEYRE, Y. TSCHINKEL, ed.), p. 13–35

RATIONAL POINTS ON CUBIC SURFACES

Niklas Broberg

Dept. of Mathematics, Chalmers University of Technology, Göteborg University,
SE–412 96 Göteborg, Sweden • *E-mail* : nibro@math.chalmers.se

Abstract. — Let k be an algebraic number field and $F(x_0, x_1, x_2, x_3)$ a non–singular cubic form with coefficients in k. Suppose that the projective cubic k–surface $X \subset \mathbb{P}^3_k$ given by $F = 0$ contains three coplanar lines defined over k, and let $U(k)$ be the set of those k–rational points on X which do not lie on any line on X. We show that the number of points in $U(k)$, with height at most B, is $O_{F,\varepsilon}(B^{4/3+\varepsilon})$ for any $\varepsilon > 0$.

Introduction

It has been known since the 19th century that there are exactly 27 lines on any non–singular projective cubic surface X. If the form F defining X has coefficients in a field k which is not algebraically closed, then some of the lines may not be defined over k. One of the more interesting problems of diophantine geometry is to count the number $n_X(B)$ of rational points of height at most B on such a surface, and to study the asymptotic growth of this counting function as B tends to infinity. It appears that the lines on X defined over k play the dominant role in determining the behaviour of $n_X(B)$. It is known by

---

2000 *Mathematics Subject Classification*. — 11G35, 11E20.
Key words and phrases. — Cubic surfaces, Rational Points, Quadratic Forms.

a theorem of Schanuel [11] that if Z is such a rational line on X and $n_Z(B)$ is the number of rational points on Z of height not exceeding B, then

$$n_Z(B) = cB^2 + O(B \log B)$$

for some positive constant c which depends on k and the choice of coordinates. In this formula, the field k is a finite extension field of $\mathbb{Q}$, and the height function used in the definition of $n_Z(B)$ is the $[k:\mathbb{Q}]$ power of the absolute height on the set of algebraic points $\mathbb{P}^3(\overline{\mathbb{Q}})$ in projective 3–space.

When it comes to the points outside the lines, very little is actually known. It has been conjectured by Manin (see [3], for example) that if U is the open subset of X given by the complement of the 27 lines, then $n_U(B) = O(B^{1+\varepsilon})$ for any $\varepsilon > 0$. In contrast to this hypothesis, the best general result of this kind seems to be $n_U(B) = O(B^{7/3+\varepsilon})$, due to Pila (see [6]). In some special cases, sharper estimates have been established. In the case where all 27 lines are defined over the ground field, it was shown by Manin and Tschinkel [9] that $n_U(B) = O(B^{5/3+\varepsilon})$. Another, much older, result is due to Hooley [7]. He proved, by means of sieve methods, that $n_U(B) = O(B^{5/3+\varepsilon})$ for the surface $x_0^3 + x_1^3 + x_2^3 + x_3^3 = 0$ over $\mathbb{Q}$. This result is not covered by the theorem of Manin and Tschinkel, since 24 of the lines are not defined over $\mathbb{Q}$. A more elementary proof of Hooley's theorem was given by Wooley [14], and the result was improved and generalized by Heath–Brown [5], who showed that $n_U(B) = O(B^{4/3+\varepsilon})$ for cubic surfaces over $\mathbb{Q}$ containing three coplanar rational lines.

The aim of this paper is to extend Heath–Brown's estimate to cubic surfaces over arbitrary number fields. In order to achieve this we shall combine Heath–Brown's arguments with arguments from algebraic geometry and the arithmetic of ternary quadratic forms over arbitrary number fields. The main theorem of this paper is the following.

Theorem 1. — *Let k be an algebraic number field and $F(x_0, x_1, x_2, x_3)$ a non-singular cubic form with coefficients in k. Suppose that the projective cubic k–surface $X \subset \mathbb{P}_k^3$ given by $F = 0$ contains three coplanar lines defined over k. Then the number of rational points on X, not lying on any line, and with height not exceeding B is $O(B^{4/3+\varepsilon})$.*

The content of the paper is as follows.

In section 1 we fix the notation. We choose normalized absolute values, define height functions, and look briefly at the notion of lattices. We also state some preliminary results, in most cases without proof. One of these results is the adelic version of Minkowski's second theorem about successive minima, due to Bombieri and Vaaler [1].

Section 2, which is the larger part of the paper, is devoted to the arithmetic of ternary quadratic forms. We generalize two theorems of Heath–Brown. Both results give uniform estimates for the number of points of bounded height on a conic. By uniform we mean that the estimates only depend on a few parameters of the form defining the conic.

In section 3 we prove theorem 1. The assumption that X contains three rational coplanar lines makes it possible to define three conic bundle morphisms $f_i : X \to \mathbb{P}^1_k$, one for each line. The general theory of height functions shows that if the height of $x \in X(k)$ is $\leqslant B$, then there is an index i such that $x \in f_i^{-1}(a,b)$ for some $(a,b) \in \mathbb{P}^1(k)$ of height $O(B^{2/3})$. The results from section 2 give estimates for the number of rational points on the fibres $f_i^{-1}(a,b)$ of height $\leqslant B$. By uniformity, the estimates only depend on B and the parameter (a,b). By choosing the most favourable estimate for each (a,b) and summing over all such estimates, we get the required bound $O(B^{4/3+\varepsilon})$ for $n_U(B)$.

Acknowledgements. — I wish to thank my supervisor Per Salberger for his guidance and support during the course of this work.

1. Notations and preliminaries

In this section we fix the notation and state some preliminary results.

For any set A, we denote by $|A|$ or $\#A$ the cardinality of A. If A and B are commensurable abelian subgroups of some group, then $[A : B]$ is the quotient $[A : A \cap B]/[B : A \cap B]$, where $[A : A \cap B]$ and $[B : A \cap B]$ are the index of $A \cap B$ in A and B, respectively.

If f and g are two non–negative functions such that $f \leqslant cg$ for some positive constant c on some common domain of f and g, then we write $f \ll g$ or $f = O(g)$. If the constant c is not absolute, then we may indicate in subscript the parameters on which it depends.

For any integral domain R and elements $x_1, \ldots, x_n$ in the quotient field of R, we let $\langle x_1, \ldots, x_n \rangle$ be the fractional ideal of R generated by $x_1, \ldots, x_n$.

Algebraic number fields. — In general we denote a number field by k, and any of its places by v. If $k = \mathbb{Q}$, then by abuse of notation we write p for the non–archimedean (finite) place corresponding to the prime number p, and ∞ for its archimedean (infinite) place. If w is a place of $\mathbb{Q}$ obtained by restriction of a place v on k, we write $v \mid w$ and $d_v = [k_v : \mathbb{Q}_w]$ for the local degree of the extension $\mathbb{Q}_w \subset k_v$ of local fields. Since the number of infinite places of k will occur frequently we denote this number by s_k.

The normalized absolute value $|\ |_v$ on k is the one which induces the ordinary absolute value on $\mathbb{R}$ if $v \mid \infty$, and the p–adic absolute value $|p|_v = 1/p$ if $v \mid p$. We set $\|x\|_v = |x|_v^{d_v}$, so that we have the product formula $\prod_v \|x\|_v = 1$, for all $x \in k^*$.

We denote by $\mathfrak{o}_k = \mathfrak{o}$ and $\mathfrak{o}_v$ the ring of integers of k and k_v, respectively, and write $N_k(\mathfrak{a})$ for the index $[\mathfrak{o} : \mathfrak{a}]$, where $\mathfrak{a}$ is any fractional ideal of $\mathfrak{o}$. Note that if v is a finite place of k corresponding to the prime ideal $\mathfrak{p}$ of $\mathfrak{o}$ and π is a generator of the maximal ideal of $\mathfrak{o}_v$, then $\|\pi\|_v^{-1} = N_k(\mathfrak{p})$. We also have that $N_k(\langle x \rangle)$ is the absolute value of the norm $N_{k/\mathbb{Q}}(x)$ for $x \in k$. In this case we write $N_k(x)$ for short.

We will not use the notion of measure in the last two sections, but we need it in the formulation of Bombieri and Vaaler's theorem in this section.

If $v \mid \infty$ and $k_v = \mathbb{R}$, then $d\mu_v$ is the ordinary Lebesgue measure on $\mathbb{R}$.

If $v \mid \infty$ and $k_v = \mathbb{C}$, then $d\mu_v$ is the ordinary Lebesgue measure on the complex plane multiplied by 2.

If $v \nmid \infty$, then $d\mu_v$ is the measure normalized so that $\mu_v(\mathfrak{o}_v) = \|\mathfrak{D}_v\|_v^{1/2}$, where $\mathfrak{D}_v$ is the local different of k at v.

The following two results are well–known, in one form or another, and may be found in almost any book on algebraic number theory (see [8, V, §1], for example).

Theorem 2 (Unit theorem). — *Let U_k be the group of units of $\mathfrak{o}$ and W_k the subgroup of U_k consisting of the roots of unity in k. The image of the regulator map*

$$l : U_k \to \mathbb{R}^{s_k}, \quad x \mapsto (\log \|x\|_v)_{v \mid \infty}$$

is a $(s_k - 1)$–dimensional lattice in $\mathbb{R}^{s_k}$ and the kernel is W_k.

Let $\mathbf{r} \in (\mathbb{R}_{>0})^{s_k}$ be a vector with components r_v for $v \mid \infty$, and let $L(\mathbf{r})$ be the set of all $x \in \mathfrak{o}$ such that $\|x\|_v \leqslant r_v$. The size $\|\mathbf{r}\|$ of the vector $\mathbf{r}$ is defined to be the product $\prod_{v \mid \infty} r_v$.

Proposition 1. — *For any $\mathbf{r} \in (\mathbb{R}_{>0})^{s_k}$ we have*

$$\|\mathbf{r}\| \ll_k |L(\mathbf{r})| \ll_k \sup\{1, \|\mathbf{r}\|\}.$$

Height functions. — Let k be a number field and $\mathbb{P}^n(k)$ the set of k–points in projective n–space $\mathbb{P}^n_k$. Since k is equipped with a product formula there is a well–defined height function $H_k : \mathbb{P}^n(k) \to \mathbb{R}_{\geqslant 1}$, given by

$$(x_0, \ldots, x_n) \mapsto \prod_v \sup_{0 \leqslant i \leqslant n} \|x_i\|_v = N_k(\langle x_0, \ldots, x_n \rangle)^{-1} \prod_{v \mid \infty} \sup_{0 \leqslant i \leqslant n} \|x_i\|_v.$$

Note that H_k is not a restriction of the absolute height on $\mathbb{P}^n(\overline{\mathbb{Q}})$, where $\overline{\mathbb{Q}}$ is an algebraic closure of $\mathbb{Q}$. If K is an extension field of k, then $H_K(x) = H_k(x)^{[K:k]}$ for all $x \in \mathbb{P}^n(k)$. We mention this because Northcott's finiteness theorem below is formulated in terms of the absolute height on $\mathbb{P}^n(\overline{\mathbb{Q}})$. For a detailed discussion on heights and for proofs of the following two results see [12].

Theorem 3 (Northcott's finiteness theorem). — *Let n, d, and B be positive integers. Then there are only finitely many points in $\mathbb{P}^n(\overline{\mathbb{Q}})$ of absolute height not exceeding B and of degree less or equal to d.*

Proposition 2. — *Let $\varphi_0, \ldots, \varphi_r$ be homogeneous polynomials of degree m and with coefficients in k. If the polynomials are not simultaneously zero on $\overline{k}$ (an algebraic closure of k), then*

$$\log H_k(\varphi_0(x), \ldots, \varphi_r(x)) = m \log H_k(x) + O(1),$$

for all $x \in \mathbb{P}^n(k)$.

Note. — A special, but important, case of this proposition is when $\varphi : k^{n+1} \to k^{n+1}$ is an automorphism. Then $H_k(x) \ll H_k(\varphi(x)) \ll H_k(x)$ for all $x \in \mathbb{P}^n(k)$.

The next result may also be found in [12, 13.2]. We include a proof since we will refer to it later.

Proposition 3. — *Every point $x \in \mathbb{P}^n(k)$ is representable by homogeneous coordinates $(x_0, \ldots, x_n) \in \mathfrak{o}^{n+1}$ such that $\sup_{0 \leq i \leq n} \|x_i\|_v \ll_{k,n} H_k(x)^{1/s_k}$ for all $v \mid \infty$.*

Proof. — Suppose that $\mathfrak{a}_1, \ldots, \mathfrak{a}_h$ are ideals representing the ideal classes of $\mathfrak{o}$. Then any point $x \in \mathbb{P}^n(k)$ has coordinates $(y_0, \ldots, y_n) \in \mathfrak{o}^{n+1}$ for which $\langle y_0, \ldots, y_n \rangle = \mathfrak{a}_j$ for some j. In particular, if

$$(x_0, \ldots, x_n) = (u^{-1} y_0, \ldots, u^{-1} y_n)$$

for some unit u of $\mathfrak{o}$, then

$$H_k(x) = \frac{1}{N_k(\mathfrak{a}_j)} \prod_{v \mid \infty} \sup_{0 \leq i \leq n} \|x_i\|_v.$$

It is therefore sufficient to find a unit u such that

$$\sup_{0 \leq i \leq n} \|x_i\|_v \ll \prod_{v \mid \infty} \sup_{0 \leq i \leq n} \|x_i\|_v^{1/s_k}$$

for all $v \mid \infty$. Let

$$y_v = \sup_{0 \leq i \leq n} \|y_i\|_v \Big/ \prod_{v \mid \infty} \sup_{0 \leq i \leq n} \|y_i\|_v^{1/s_k},$$

so that the inequalities above may be written as $y_v \ll \|u\|_v$. If $Y \subset (\mathbb{R}_{>0})^{s_k}$ is the locally compact subgroup of elements $y = (y_v)$ which satisfy $\prod_v y_v = 1$, then the image of U_k in Y under $u \mapsto (\|u\|_v)$ is a discrete cocompact multiplicative lattice, by theorem 2. Hence there exists u of the desired type. □

We have not found the next result in the literature so we include a proof. The height $h_k(x)$ of an element $x \in k^*$ is defined to be $H_k(1, x)$.

Proposition 4. — *If d is the degree of k over $\mathbb{Q}$, then the number of units $u \in U_k$ such that $h_k(u) \leq B$ is $\ll_d \sup\{1, (\log B)^{s_k}\}$.*

Note. — The exponent s_k can be reduced to $s_k - 1$, but for our application the actual order is not very significant.

Proof. — First note that $h_k(u) = h_k(u^{-1})$ for all $u \in U_k$, so if $h_k(u) \leq B$ for some $u \in U_k$, then $\sup_{v|\infty} |\log \|u\|_v| \leq \log B$. By theorem 2, it is therefore sufficient to show that there are $\ll_d \sup\{1, (\log B)^{s_k}\}$ lattice points of $l(U_k) \subset \mathbb{R}^{s_k}$ in the box centred at the origin with side length $2 \log B$. From theorem 3 we have that there is a constant $c > 1$, only depending on d, such that $u \in W_k$ whenever $u \in U_k$ and $h_k(u) \leq c$. This implies that any box with side length $\frac{2}{s_k} \log c$ contains at most one of the points of $l(U_k)$. By comparing volumes we thus have that the number of points of $l(U_k)$ in the box $|\log \|u\|_v| \leq \log B$ is $\ll (\log B)^{s_k}/(\log c)^{s_k} \ll_d (\log B)^{s_k}$, providing that $B \gg c$. □

Lattices over number fields. — Let k be a number field of degree d over $\mathbb{Q}$, and let n be a positive integer. One usually says that an o–module in k^n is an o–lattice in k^n if it is finitely generated and contains a basis of k^n over k. It is not hard to see that if Λ is an o–lattice in k^n and v is a finite place of k, then $\Lambda_v = o_v \otimes_o \Lambda$ is a free o_v–module in $(k_v)^n$ such that Λ_v contains a basis for $(k_v)^n$ over k_v. We say that such an o_v–module is an o_v–lattice in $(k_v)^n$. The next result states that an o–lattice is known if it is known locally everywhere. For a proof see, for example, [10, 5.3].

Theorem 4. — *For each finite place v of k, let L_v be an o_v–lattice in $(k_v)^n$ such that $L_v = (o_v)^n$ for almost all v. If $\Lambda = \bigcap_{v \nmid \infty} L_v \cap k^n$, then Λ is the unique o–lattice in k^n such that $\Lambda_v = L_v$ for all v.*

Proposition 5. — *If $L \subset \Lambda$ are o–lattices in k^n, then there is an element $a \in k^*$ such that $\Lambda \subset aL$ and $[L : \Lambda] \ll_k N_k(a)$.*

Proof. — By the invariant factor theorem (see [**10**, 4.14], for example), there are elements $u_1, \ldots, u_n \in \Lambda$, fractional $\mathfrak{o}$–ideals $\mathfrak{a}_1, \ldots, \mathfrak{a}_n$, and $\mathfrak{o}$–ideals $\mathfrak{b}_1, \ldots, \mathfrak{b}_n$ such that

$$\Lambda = \bigoplus_{i=1}^n \mathfrak{a}_i u_i \quad \text{and} \quad L = \bigoplus_{i=1}^n \mathfrak{b}_i \mathfrak{a}_i u_i.$$

Then, if we choose $b \in \mathfrak{b}_1 \cdots \mathfrak{b}_n$ with $N_k(b) \ll_k \prod_{i=1}^n N_k(\mathfrak{b}_i) = [\Lambda : L]$, and put $a = b^{-1}$, we have $\Lambda \subset aL$ and $[L : \Lambda] \ll_k N_k(a)$. □

Next we formulate Bombieri and Vaaler's adelic version of Minkowski's second theorem [**1**]. Later, when it is needed, we will use a slightly reformulated version of this result. We give this version as a corollary of the theorem. However, before we can formulate the result we have to make some definitions.

For each finite place v of k, let L_v be an $\mathfrak{o}_v$–lattice in $(k_v)^n$ such that $L_v = (\mathfrak{o}_v)^n$ for almost all v. For each infinite place v of k, let S_v be a nonempty, open, convex, symmetric, bounded subset of $(k_v)^n$. By symmetric we mean that $S_v = -S_v$. Then

$$\Gamma = \prod_{v|\infty} S_v \times \prod_{v\nmid\infty} L_v$$

is a subset of $(k_\mathbb{A})^n$, the n-fold product of adeles over k. In fact, Γ is an open neighbourhood of 0 and the closure of Γ is compact. Thus the volume

$$\mathrm{Vol}(\Gamma) = \prod_{v|\infty} \mu_v^n(S_v) \prod_{v\nmid\infty} \mu_v^n(L_v)$$

exists as a finite number.

Let Λ be the unique $\mathfrak{o}$–lattice such that $\Lambda_v = L_v$ for all $v \nmid \infty$. The i:th successive minimum λ_i for $S = \prod_{v|\infty} S_v$, with respect to Λ is defined to be the infimum of all positive reals λ such that $\Lambda \cap \lambda S$ contains i linearly independent vectors. It is obvious that $\lambda_1 \leqslant \cdots \leqslant \lambda_n$. It is also evident from the definition that there exist n linearly independent vectors $\mathbf{u}_i \in \Lambda$ such that $\mathbf{u}_i \notin \lambda_i S_v$ for some v, but $\{\mathbf{u}_1, \ldots, \mathbf{u}_i\} \subset \lambda_i \overline{S}$, where $\overline{S} = \prod_{v|\infty} \overline{S}_v$ is the closure of S. We will use this observation later. For a proof of the following theorem see [**1**].

Theorem 5. — *The successive minima $\lambda_1 \leqslant \cdots \leqslant \lambda_n$ satisfy the inequality*

$$(\lambda_1 \cdots \lambda_n)^d \mathrm{Vol}(\Gamma) \leqslant 2^{dn}.$$

Corollary. — *The successive minima satisfy the relation*

$$(\lambda_1 \cdots \lambda_n)^d \prod_{v|\infty} \mathrm{Vol}(S_v) \ll_{k,n} [\mathfrak{o}^n : \Lambda].$$

Proof. — All we have to do is to calculate the volume Vol(Γ). By definition the product $\prod_{v|\infty} \mu_v^n(S_v)$ is the product of the volumes Vol(S_v) multiplied by 2^{d-s_k} ($d - s_k$ being the number of complex places of k). By additivity and translation invariance of μ_v^n we obtain

$$\mu_v^n(\Lambda_v) = [\Lambda_v : \Lambda_v \cap (\mathfrak{o}_v)^n] \mu_v^n(\Lambda_v \cap (\mathfrak{o}_v)^n) = [\Lambda_v : (\mathfrak{o}_v)^n] \mu_v^n((\mathfrak{o}_v)^n).$$

But then we are done since

$$\prod_{v \nmid \infty} [(\mathfrak{o}_v)^n : \Lambda_v] = [\mathfrak{o}^n : \Lambda],$$

and

$$\prod_{v \nmid \infty} \mu_v(\mathfrak{o}_v) = \prod_{v \nmid \infty} \|\mathcal{D}_v\|_v^{1/2} = |\Delta_k|^{-1/2},$$

where Δ_k is the discriminant of k. □

2. Ternary quadratic forms

In this section we estimate the number of rational points of bounded height on conics over number fields. The main estimates are uniform in the sense that the implied constants only depend on a few invariants of the coefficients of the quadratic form defining the conic.

The ground field k is assumed to be fixed from now on, so any implicitly given constants may depend on k, even if this is not stated explicitly. Before we begin our discussion, we also want to stress that most of the arguments in this and the next section are generalizations of Heath-Brown's arguments in the case $k = \mathbb{Q}$ (see [5]).

Let M be any invertible $n \times n$-matrix with entries in k. We define $\Delta(M)$ and $\Delta_0(M)$ to be the fractional ideal of $\mathfrak{o}$ generated by the determinant of M and the fractional ideal generated by the $(n-1) \times (n-1)$-minors of M, respectively.

Lemma 1. — $\Delta(MN) = \Delta(M) \Delta(N)$ and $\Delta_0(MN) \subset \Delta_0(M) \Delta_0(N)$.

Proof. — The equality $\Delta(MN) = \Delta(M) \Delta(N)$ is well-known, and the inclusion $\Delta_0(MN) \subset \Delta_0(M) \Delta_0(N)$ follows from the relation $\widetilde{MN} = \widetilde{N}\widetilde{M}$ between the cofactor matrices of M, N, and MN. □

The main result of this section is the following theorem.

Theorem 6. — Let q be a ternary quadratic form with matrix $M \in M_3(\mathfrak{o})$, and let $\mathbf{r}_1, \mathbf{r}_2, \mathbf{r}_3 \in (\mathbb{R}_{>0})^{s_k}$ be given. If q is non-singular, then there are

$$\ll_\varepsilon \left(1 + \sqrt{\frac{\|\mathbf{r}_1\| \|\mathbf{r}_2\| \|\mathbf{r}_3\| N_k(\Delta_0(M))^2}{N_k(\Delta(M))}}\right) N_k(\Delta(M))^\varepsilon$$

points $(x_1, x_2, x_3) \in \mathbb{P}^2(k)$ on the conic $q = 0$ such that $x_i \in L(\mathbf{r}_i)$.

Note. — By proposition 3 we know that for any point $x \in \mathbb{P}^2(k)$ with $H_k(x) \leqslant B$, we can find homogeneous coordinates $(x_1, x_2, x_3) \in \mathfrak{o}^3$ such that $\|x_i\|_v \ll B^{1/s_k}$ for all $v \mid \infty$. The theorem then says that there are $O_q(B^{3/2})$ points on the conic $q = 0$ of height at most B. Intuitively this estimate is not what one expects (q is of degree 2). One can of course do better if one does not require uniformity in q, as the following proposition shows.

Proposition 6. — If q is a non-singular quadratic form, then there are $O_q(B)$ points $x \in \mathbb{P}^2(k)$ on the conic $q(x) = 0$ such that $H_k(x) \leqslant B$.

Proof. — If q is isotropic, then q is "k–equivalent" to $\alpha(x_0^2 - x_1 x_2)$ for some $\alpha \in k^*$. But $(y_0, y_1) \mapsto (y_0 y_1, y_0^2, y_1^2)$ is a parametrization of the solutions of $x_0^2 - x_1 x_2 = 0$, and $H_k(y_0 y_1, y_0^2, y_1^2) = H_k(y_0, y_1)^2$ for all $(y_0, y_1) \in \mathbb{P}^1(k)$, so by Schanuel's theorem,

$$\# \left\{ x \in \mathbb{P}^2(k) : q(x) = 0, H_k(x) \leqslant B \right\} \ll_q$$
$$\# \left\{ y \in \mathbb{P}^1(k) : H_k(y) \leqslant B^{1/2} \right\} \ll B.$$

$\square$

The proof of the theorem will be given in several steps. In lemma 4 we look at the equation $q = 0$ at each finite place of k and find that each solution $(x_1, x_2, x_3) \in \mathfrak{o}^3$ must belong to one of not too many $\mathfrak{o}$–lattices in k^3. We also find lower bounds for the indices of the lattices in $\mathfrak{o}^3$. By using the theory of successive minima we then use this information to reduce the proof to a problem of counting the number of points on a conic in a "bounded domain" of $\mathbb{P}^2(k)$. We obtain a solution to this problem in lemma 5. First, however, we formulate some minor results which are included for completeness and which we need in the proof of lemma 4.

For an ideal $\mathfrak{a}$ of $\mathfrak{o}$, let $\rho(\mathfrak{a})$ be the number of prime ideals containing $\mathfrak{a}$ and $\tau(\mathfrak{a})$ the number of ideals containing $\mathfrak{a}$. Recall that an arithmetical function $f : \mathbb{Z}_{>0} \to \mathbb{C}$ is said to be multiplicative if $f(mn) = f(m)f(n)$ whenever $\gcd(m, n) = 1$. For a proof of the following result see, for example, [13, I. §5.1].

Theorem 7. — *Let f be a multiplicative function. If $\lim_{p^\nu \to \infty} f(p^\nu) = 0$, where p^ν are prime powers, then $\lim_{n \to \infty} f(n) = 0$.*

Corollary. — *If c is a positive number, then $c^{\rho(\mathfrak{a})} \tau(\mathfrak{a}) \ll_{c,\varepsilon} N_k(\mathfrak{a})^\varepsilon$ for all ideals $\mathfrak{a}$ of $\mathfrak{o}$.*

Proof. — Let $\rho(n)$ and $\tau(n)$ be the number of primes dividing $n \in \mathbb{Z}_{>0}$ and the number of divisors of n, respectively. Then we have $\rho(\mathfrak{a}) \leqslant d\,\rho(N_k(\mathfrak{a}))$ and $\tau(\mathfrak{a}) \leqslant \tau(N_k(\mathfrak{a}))^d$ for all ideals $\mathfrak{a}$ of $\mathfrak{o}$, where d is the degree of k over $\mathbb{Q}$. Thus $c^{\rho(\mathfrak{a})}\tau(\mathfrak{a}) \leqslant c^{d\rho(N_k(\mathfrak{a}))}\tau(N_k(\mathfrak{a}))^d \ll_{c,\varepsilon} N_k(\mathfrak{a})^\varepsilon$, since $f(x) = (c^{\rho(x)}\tau(x))^d/x^\varepsilon$ is a multiplicative function and $f(p^\nu) = (c(\nu+1))^d p^{-\nu\varepsilon} \to 0$ as $p^\nu \to \infty$. □

If M, N, and T are square matrices of the same size, with entries in some ring R, and T is unimodular with $T^t M T = N$, then we say that M and N are R-equivalent. This means that M and N only differs by a finite number of simultaneous elementary row- and column-operations. The next lemma says that over a discrete valuation ring any symmetric matrix is almost equivalent to a diagonal matrix.

Lemma 2. — *Let v be a discrete valuation on some field, and let R be the corresponding valuation ring. Then for any symmetric matrix $M \in \mathrm{M}_n(R)$ there exists an element $x \in R$ and matrices $P, D \in \mathrm{M}_n(R)$ such that D is diagonal, $P^t D P = xM$, and $v(\det P) = v(x) \leqslant (n-1)v(2)$.*

Proof. — Let n be a positive integer and M any symmetric $n \times n$-matrix with entries in R. It is sufficient to show that we can find an element $x \in R$ and a matrix $P \in \mathrm{M}_n(R)$ such that $v(\det P) = v(x) \leqslant v(2)$ and $P^t D P = xM$, where D is some matrix with zeros in its first row and column, except perhaps on the diagonal. The lemma then follows by induction on the size of the matrices.

Since the existence of the objects x and P is trivially verifiable when $n = 1$ or $M = 0$, we may assume that $n \geqslant 2$ and that at least one of the entries of M is a unit in R. Then M is equivalent to a matrix with its first row equal to $(a, u, 0, \ldots, 0)$ or $(u, 0, 0, \ldots, 0)$, where u is a unit and a is some element of R such that $0 \leqslant v(a) \leqslant v(2)$. If $v(2) = 0$, then a matrix of the first shape is obviously equivalent to a matrix of the second shape. If, on the other hand, $v(2) > 0$, we can use the identity

$$\begin{pmatrix} \begin{smallmatrix} a & u \\ 0 & 1 \end{smallmatrix} & 0 \\ 0 & I \end{pmatrix}^t \left[\begin{pmatrix} 1 & 0 \\ 0 & aM' \end{pmatrix} - \begin{pmatrix} \begin{smallmatrix} 0 & 0 \\ 0 & u^2 \end{smallmatrix} & 0 \\ 0 & 0 \end{pmatrix} \right] \begin{pmatrix} \begin{smallmatrix} a & u \\ 0 & 1 \end{smallmatrix} & 0 \\ 0 & I \end{pmatrix} = a \begin{pmatrix} a & \mathbf{x} \\ \mathbf{x}^t & M' \end{pmatrix},$$

where M' is a symmetric $(n-1) \times (n-1)$-matrix and $\mathbf{x} = (u, 0, \ldots, 0)$. This establishes the existence of the objects x and P. □

Before the next lemma we just recall that if v is a finite place of k lying over a prime number p, then $\|p\|_v = \|\pi\|_v^r$, where π is a generator of the maximal ideal of $\mathfrak{o}_v$ and r is the ramification index of $\mathbb{Q}_p \subset k_v$.

Lemma 3. — *For a finite place v of k and any integer $n \geqslant 1$, let R_n be the ring $\mathfrak{o}_v/\langle \pi^n \rangle$. If $\varepsilon \in R_n^*$, then the number of solutions of the equation $x^2 = \varepsilon$ is less or equal to $2 \|2\|_v^{-1}$.*

Proof. — Since $(x/y)^2 = 1$ if $x^2 = y^2 = \varepsilon$ for $\varepsilon \in R_n^*$, we only need to study the case $\varepsilon = 1$. Let r be the least integer such that $2 \notin \langle \pi^{r+1} \rangle$. Then either $x+1 \notin \langle \pi^{r+1} \rangle$ or $x-1 \notin \langle \pi^{r+1} \rangle$ for $x \in \mathfrak{o}_v$. Hence, if $x^2 - 1 = (x+1)(x-1) \in \langle \pi^n \rangle$, we must have $x+1 \in \langle \pi^{n-r} \rangle$ or $x-1 \in \langle \pi^{n-r} \rangle$. There are thus at most $2 |\langle \pi^{n-r} \rangle/\langle \pi^n \rangle| = 2 |(R_1)^r|$ solutions of $x^2 = \varepsilon$ in R_n. Moreover, $r = 0$ if $\|2\|_v = 1$, and r is the ramification index of $\mathbb{Q}_2 \subset k_v$ if $\|2\|_v < 1$. □

We now have enough information to prove the first main lemma in the proof of theorem 6.

Lemma 4. — *Let q be non-singular ternary quadratic form with matrix $M \in M_3(\mathfrak{o})$. For each finite place v of k, let π_v be a generator of the maximal ideal of $\mathfrak{o}_v$, and let a_v and b_v be the non-negative integers defined by $\|\Delta(M)\|_v = \|\pi_v\|_v^{a_v}$ and $\|\Delta_0(M)\|_v = \|\pi_v\|_v^{b_v}$, respectively.*
(a) *If $a_v > 0$ and $q(\mathbf{x}) = 0$ for some $\mathbf{x} \in (\mathfrak{o}_v)^3$, then $\mathbf{x}$ belongs to at least one of at most $O(a_v)$ $\mathfrak{o}_v$–lattices Λ in $(k_v)^3$, each satisfying*

$$[(\mathfrak{o}_v)^3 : \Lambda] \geqslant \|2\|_v^8 \, \|\Delta(M)\|_v^{-1} \, \|\Delta_0(M)\|_v^2 \, .$$

Moreover, the implied constant in $O(a_v)$, let us call it c, depends only on k and may be chosen to be the same for all v.
(b) *If $q(\mathbf{x}) = 0$ for some $\mathbf{x} \in \mathfrak{o}^3$, then $\mathbf{x}$ belongs to at least one of at most $O(N_k(\Delta(M))^\varepsilon)$ $\mathfrak{o}$–lattices Λ in k^3, each satisfying*

$$[\mathfrak{o}^3 : \Lambda] \gg N_k(\Delta(M)) \, N_k(\Delta_0(M))^{-2} \, .$$

Proof. — It is (b) that really interests us, so assume for the moment that we have a proof of (a). If $q(\mathbf{x}) = 0$ for some $\mathbf{x} \in \mathfrak{o}^3$, then $\mathbf{x}$ belongs to an $\mathfrak{o}$–lattice Λ in k^3 with $\Lambda_v = (\mathfrak{o}_v)^3$ if $a_v = 0$, and Λ_v equal to one of the $\mathfrak{o}_v$–lattices from (a) if $a_v > 0$. This is a consequence of theorem 4. It follows immediately from the local nature of index that $[\mathfrak{o}^3 : \Lambda] \gg N_k(\Delta(M)) \, N_k(\Delta_0(M))^{-2}$. Moreover, there are at most $\prod_{a_v > 0} c a_v$ such lattices Λ, and

$$\prod_{a_v > 0} c a_v \leqslant c^{\rho(\Delta(M))} \tau(\Delta(M)) \ll N_k(\Delta(M))^\varepsilon$$

by the corollary of theorem 7. It is thus sufficient to find a proof of (a).

Let v be a finite place of k such that $a_v > 0$. To keep the notations as simple as possible we skip the indices and write π, a, and b for π_v, a_v, and b_v, respectively. By lemma 2 there is an element $x \in o_v$ such that $x\,q(\mathbf{x})$ may be diagonalized over the ring o_v, using a matrix P with $\|\det P\|_v = \|x\|_v \geq \|2\|_v^2$. Let D be the matrix of this diagonalized form

$$Q(y_1, y_2, y_3) = \varepsilon_1 \pi^{\alpha_1} y_1^2 + \varepsilon_2 \pi^{\alpha_2} y_2^2 + \varepsilon_3 \pi^{\alpha_3} y_3^2,$$

where $\alpha_1 \geq \alpha_2 \geq \alpha_3 \geq 0$ are integers and ε_1, ε_2, ε_3 are units. From the identity $\Delta(xM) = \Delta(P)^2 \Delta(D)$ and the inclusion $\Delta_0(xM) \subset \Delta_0(P)^2 \Delta_0(D)$, we have $\|x\|_v \|\pi\|_v^a = \|\pi\|_v^{\alpha_1 + \alpha_2 + \alpha_3}$ and $\|x\|_v^2 \|\pi\|_v^b \leq \|\pi\|_v^{\alpha_2 + \alpha_3}$, respectively. If we combine these two relations and use the fact $\|x\|_v \geq \|2\|_v^2$, we get

$$\|\pi\|_v^{-\alpha_1 + \alpha_2 + \alpha_3} \geq \|2\|_v^8 \|\Delta(P)\|_v^{-1} \|\pi\|_v^{2b-a}.$$

This inequality makes the hypothesis of the lemma more understandable. All we have to prove is that any solution $\mathbf{y} \in (o_v)^3$ of $Q = 0$ belongs to at least one of at most $O(a)$ lattices, each with index $\geq \|\pi\|_v^{-\alpha_1 + \alpha_2 + \alpha_3}$ in $(o_v)^3$.

Now suppose that $Q(y_1, y_2, y_3) = 0$ for some $(y_1, y_2, y_3) \in (o_v)^3$. Then we find that

(2.1) $$\varepsilon_2 \pi^{\alpha_2 - \alpha_3} y_2^2 + \varepsilon_3 y_3^2 \in \langle \pi^{\alpha_1 - \alpha_3} \rangle.$$

When α_2 and α_3 have opposite parities this implies that $y_2^2 \in \langle \pi^{\alpha_1 - \alpha_2} \rangle$ and $y_3^2 \in \langle \pi^{\alpha_1 - \alpha_3} \rangle$. It follows that (y_1, y_2, y_3) belongs to the lattice

$$\Lambda = o_v \times \langle \pi^{[\frac{\alpha_1 - \alpha_2 + 1}{2}]} \rangle \times \langle \pi^{[\frac{\alpha_1 - \alpha_3 + 1}{2}]} \rangle.$$

Since $[(o_v)^3 : \Lambda] \geq \|\pi\|_v^{-\alpha_1 + \alpha_2 + \alpha_3}$, this completes the proof in this case.

The case in which $\alpha_2 - \alpha_3 = 2h$, for some h, needs slightly more work. We can assume that $y_2^2 \notin \langle \pi^{\alpha_1 - \alpha_2} \rangle$ and $y_3^2 \notin \langle \pi^{\alpha_1 - \alpha_3} \rangle$. Otherwise (y_1, y_2, y_3) would belong to the lattice Λ and we would not get anything new. By studying (2.1) we see that $y_2 = u_2 \pi^n$ and $y_3 = u_3 \pi^{n+h}$ for some non-negative integer $n < [\frac{\alpha_1 - \alpha_2 + 1}{2}]$ and some units u_2 and u_3, which satisfy $\varepsilon_2 u_2^2 + \varepsilon_3 u_3^2 \in \langle \pi^{\alpha_1 - \alpha_2 - 2n} \rangle$. By lemma 3 there are at most $2 \|2\|_v^{-1}$ solutions of the equation $r^2 + \varepsilon_2/\varepsilon_3 = 0$ in the ring $o_v/\langle \pi^{\alpha_1 - \alpha_2 - 2n} \rangle$. If $r \in o_v$ represents the same solution as u_2/u_3, then $r\,\pi^h y_2 - y_3 \in \langle \pi^{\alpha_1 - \alpha_2 + h - n} \rangle$, so

$$\begin{pmatrix} y_1 \\ y_2 \\ y_3 \end{pmatrix} = \begin{pmatrix} 1 & 0 & 0 \\ 1 & \pi^n & 0 \\ 1 & r\pi^{h+n} & \pi^{\alpha_1 - \alpha_2 + h - n} \end{pmatrix} \begin{pmatrix} z_1 \\ z_2 \\ z_3 \end{pmatrix}$$

for some $(z_1, z_2, z_3) \in (\mathfrak{o}_v)^3$. Thus (y_1, y_2, y_3) belongs to at least one of at most $2\,\|2\|_v^{-1}\,[\frac{\alpha_1-\alpha_2+1}{2}]$ lattices, each with index

$$\|\pi\|_v^{\alpha_2-\alpha_1-h} \geqslant \|\pi\|_v^{-\alpha_1+\alpha_2+\alpha_3}$$

in $(\mathfrak{o}_v)^3$.

To complete the proof we have to show that $(2 + \alpha_1 - \alpha_2)\|2\|_v^{-1} + 1 \leqslant ca$ for some constant c which only depends on k, but this is obviously true. □

The next result is the second tool in the proof of the theorem. The significance of this result is that the bounds are completely independent of the form involved.

Lemma 5. — *Let f be a ternary form of degree d with no linear factor, and let $\mathbf{r}_1, \mathbf{r}_2, \mathbf{r}_3 \in (\mathbb{R}_{>0})^{sk}$ be given. Then there are*

$$\ll_d 1 + \sqrt{\|\mathbf{r}_1\|\,\|\mathbf{r}_2\|\,\|\mathbf{r}_3\|}$$

points $(x_1, x_2, x_3) \in \mathbb{P}^2(k)$ on the curve $f = 0$ such that $x_i \in L(\mathbf{r}_i)$.

Proof. — Let r_{iv} be the v:th component of the vector $\mathbf{r}_i$, and set $R_i = \|\mathbf{r}_i\|$. We begin by showing that any point $(x_1, x_2, x_3) \in \mathbb{P}^2(k)$ with $x_i \in L(\mathbf{r}_i)$ lies on a line $a_1 x_1 + a_2 x_2 + a_3 x_3 = 0$, where the coefficients $a_i \in \mathfrak{o}$ are not all zero and $\|a_i\|_v \ll r_{iv}^{-1} \sqrt{r_{1v} r_{2v} r_{3v}}$. This is a simple application of the box principle. By proposition 1 there are $\gg Y^{3sk} \sqrt{R_1 R_2 R_3}$ sets of coefficients with $\|a\|_i \leqslant Y r_{iv}^{-1} \sqrt{r_{1v} r_{2v} r_{3v}}$. The corresponding values $a_1 x_1 + a_2 x_2 + a_3 x_3$ satisfy

$$\|a_1 x_1 + a_2 x_2 + a_3 x_3\|_v \ll Y \sqrt{r_{1v} r_{2v} r_{3v}},$$

so by the same proposition there are $\ll \sup\{1, Y^{sk} \sqrt{R_1 R_2 R_3}\}$ such values. Hence, if Y is sufficiently large, then two such values must agree, and this for a Y independent of R_1, R_2, R_3 (for the case $R_1 R_2 R_3 < 1$ see below). Since f does not have any linear factors, each of the above lines has at most d points in common with $f = 0$. Moreover, the number of lines is

$$\ll \prod_{i=1}^{3} \sup\{1, R_i^{-1}\sqrt{R_1 R_2 R_3}\} \ll \sqrt{R_1 R_2 R_3},$$

if $R_i^{-1}\sqrt{R_1 R_2 R_3} \geqslant 1$ for all i. If this last condition is not satisfied, then

$$R_i R_j = \inf\{R_1 R_2, R_1 R_3, R_2 R_3\} \leqslant \sqrt{R_1 R_2 R_3}.$$

Since each pair of elements $x_i \in L(\mathbf{r}_i)$, $x_j \in L(\mathbf{r}_j)$ produces at most d solutions of $f = 0$, and the number of such pairs is

$$\ll \sup\{1, R_i\} \sup\{1, R_j\},$$

we can use the estimate $O(R_i R_j)$ in this case. Of course we have to assume that $R_i \geq 1$ and $R_j \geq 1$, but $L(\mathbf{r}_i) = \{0\}$ or $L(\mathbf{r}_j) = \{0\}$ in any other case, and then there are at most d solutions. □

To prove theorem 6 we now use the theory of successive minima to combine these two last lemmas. Recall that q is a non–singular ternary quadratic form with matrix $M \in \mathrm{M}_3(\mathfrak{o})$, and $\mathbf{r}_1, \mathbf{r}_2, \mathbf{r}_3 \in (\mathbb{R}_{>0})^{s_k}$ are vectors with components r_{iv} for $v \mid \infty$. In order to have the same notation as in section 1, we define the nonempty, open, convex, symmetric, bounded subsets S_v of $(k_v)^3$ by

$$S_v = \left\{ (x_1, x_2, x_3) \in (k_v)^3 \,:\, |x_i| < r_{iv}^{1/d_v} \right\},$$

and put $S = \prod_{v\mid\infty} S_v$. Obviously $\mathfrak{o}^3 \cap \overline{S} = L(\mathbf{r}_1) \times L(\mathbf{r}_2) \times L(\mathbf{r}_3)$, where $\overline{S} = \prod_{v\mid\infty} \overline{S}_v$ is the closure of S. If d is the degree of k over $\mathbb{Q}$ and $\lambda_1 \leq \lambda_2 \leq \lambda_3$ are the successive minima of S with respect to one of the lattices Λ from lemma 4, then

$$(\lambda_1 \lambda_2 \lambda_3)^d \prod_{v\mid\infty} \mathrm{Vol}(S_v) \ll [\mathfrak{o}^3 : \Lambda],$$

by the corollary of theorem 5. Moreover,

$$\prod_{v\mid\infty} \mathrm{Vol}(S_v) \gg \|\mathbf{r}_1\| \|\mathbf{r}_2\| \|\mathbf{r}_3\|$$

by the definition of the sets S_v. We mentioned in the preparations for theorem 5 that one can find a basis $\mathbf{u}_1, \mathbf{u}_2, \mathbf{u}_3$ of k^3 over k such that $\mathbf{u}_i \in \lambda_i \overline{S}$, but $\mathbf{u}_i \notin \lambda_i S_v$ for some v. If u_{ij} is the j:th component of $\mathbf{u}_i$, then these conditions imply that $\|u_{ij}\|_v \leq \lambda_i^{d_v} r_{jv}$. Thus, if $(x_1, x_2, x_3) = y_1 \mathbf{u}_1 + y_2 \mathbf{u}_2 + y_3 \mathbf{u}_3 \in \overline{S}$ for some $(y_1, y_2, y_3) \in k^3$, and U is the matrix with columns $\mathbf{u}_i$, then

$$\|y_1\|_v = \frac{1}{\|\det U\|_v} \left\| \det \begin{pmatrix} x_1 & u_{21} & u_{31} \\ x_2 & u_{22} & u_{32} \\ x_3 & u_{23} & u_{33} \end{pmatrix} \right\|_v \ll \frac{r_{1v} r_{2v} r_{3v} (\lambda_2 \lambda_3)^{d_v}}{\|\det U\|_v}$$

by Cramer's rule. There are of course analogous estimates for $\|y_2\|_v$ and $\|y_3\|_v$. Note that $\mathbf{u}_1, \mathbf{u}_2, \mathbf{u}_3$ does not constitute a basis for Λ over $\mathfrak{o}$. Presumably the lattice Λ is not even free. However, if L is the free $\mathfrak{o}$–lattice with $\mathbf{u}_1, \mathbf{u}_2, \mathbf{u}_3$ as generators, then by proposition 5 we have $L \subset \Lambda \subset aL$ for some $a \in k^*$ such that $N_k(a) \gg [L : \Lambda]$. Thus, any element $(x_1, x_2, x_3) \in \Lambda \cap \overline{S}$ may be written as $y_1(a\mathbf{u}_1) + y_2(a\mathbf{u}_2) + y_3(a\mathbf{u}_3)$ for some $(y_1, y_2, y_3) \in \mathfrak{o}^3$. This shows that if q' is the non–singular quadratic form with matrix $U^t M U$, then any solution $(x_1, x_2, x_3) \in \Lambda \cap \overline{S}$ of $q = 0$ gives a solution $(y_1, y_2, y_3) \in \mathfrak{o}^3$ of $q' = 0$ which

satisfies
$$\|y_i\|_v \ll \frac{r_{1v}r_{2v}r_{3v}}{\|a\|_v \|\det U\|_v} \left(\frac{\lambda_1\lambda_2\lambda_3}{\lambda_i}\right)^{d_v} = t_{iv}.$$
Moreover, there are $\ll 1 + \sqrt{\prod_{i,v} t_{iv}}$ such solutions by lemma 5, and one gets
$$\prod_{i,v} t_{iv} = \left(\frac{\|r_1\|\,\|r_2\|\,\|r_3\|}{N_k(a)N_k(\det U)}\right)^3 (\lambda_1\lambda_2\lambda_3)^{2d} \ll$$
$$\left(\frac{\|r_1\|\,\|r_2\|\,\|r_3\|}{[L:\Lambda][\mathfrak{o}^3:L]}\right)^3 \left(\frac{[\mathfrak{o}^3:\Lambda]}{\|r_1\|\,\|r_2\|\,\|r_3\|}\right)^2 = \frac{\|r_1\|\,\|r_2\|\,\|r_3\|}{[\mathfrak{o}^3:\Lambda]}$$
from the definitions and estimates above. Referring to lemma 4, this completes the proof of the theorem.

Next we prove a result that we can use when the determinant $N_k(\Delta(M))$ of the form q is small. In that case the estimate from theorem 6 is obviously not very good.

Proposition 7. — *Let q be a non-singular ternary quadratic form such that the binary form $q(0, x_2, x_3)$ is also non-singular. Let $\mathbf{r} \in (\mathbb{R}_{>0})^{s_k}$ and $R \geq 1$ be given, and denote by $\|q\|$ the supremum of $\|\sigma\|_v$, where σ ranges over the coefficients of q and v over the infinite places of k. Then there are*
$$\ll \sup\{1, \|\mathbf{r}\|\,(\|\bar{\mathbf{r}}\|\,\|q\|\,R)^\varepsilon\}$$
points $(x_1, x_2, x_3) \in \mathbb{P}^2(k)$ on the conic $q = 0$, satisfying $(x_1, x_2, x_3) \in L(\mathbf{r}) \times \mathfrak{o} \times \mathfrak{o}$ and $\sup_{v\mid\infty} \|x_i\|_v \leq R$ for $i = 2, 3$. Here $\bar{\mathbf{r}} \in (\mathbb{R}_{>0})^{s_k}$ is the vector with the v:th component equal to r_v if $r_v \geq 1$ and 1 otherwise. As before, r_v denotes the v:th component of $\mathbf{r}$.

Proof. — Since $q(0, x_2, x_3)$ is non-singular there is an invertible matrix $P \in M_3(k)$ with first row $(1, 0, 0)$ such that $q(\mathbf{x}) = d(P\mathbf{x})$ for some diagonal form d. Moreover, we may choose P so that the entries p_{ij} satisfy
$$\sup\{\|p_{ij}\|_v, \|p_{ij}\|_v^{-1}\} \leq \|q\|^A$$
for all $v \mid \infty$ and some fixed exponent A. The equation $q(x_1, x_2, x_3) = 0$ then becomes
(2.2) $$\alpha L_1(x_1, x_2, x_3)^2 + \beta L_2(x_1, x_2, x_3)^2 = \gamma x_1^2$$
with non-zero coefficients α, β, γ, and linear forms L_i such that $L_1(0, x_2, x_3)$ and $L_2(0, x_2, x_3)$ are linearly independent. By multiplying the equation with a suitable factor, we can assume that all coefficients are integers in k and satisfy $\sup_{v\mid\infty} \|\sigma\|_v \leq \|q\|^A$, possibly with a new constant value of A. By the same

argument, we may also assume that α is a square. The left hand side of (2.2) then factorizes over the field $K = k(\sqrt{\beta})$, and we have

(2.3) $$L_3(x_1, x_2, x_3) L_4(x_1, x_2, x_3) = \gamma x_1^2$$

for some linear forms L_i with coefficients in $\mathfrak{o}_K$.

Now suppose that $(x_1, x_2, x_3) \in L(\mathbf{r}) \times \mathfrak{o} \times \mathfrak{o}$ satisfies (2.3), and that we have $\sup_{v \mid \infty} \|x_i\| \leqslant R$ for $i = 2, 3$. If $\|\mathbf{r}\| \leqslant 1$, then there are at most two such solutions, considered as elements of $\mathbb{P}^2(k)$, since in that case $L(\mathbf{r})$ only consists of the zero element. If $\|\mathbf{r}\| \geqslant 1$, on the other hand, the cardinality of $L(\mathbf{r})$ is $\ll \|\mathbf{r}\|$, and in this case we show that there are $\ll (\|\overline{\mathbf{r}}\| \|q\| R)^\varepsilon$ possible pairs x_2, x_3 for each x_1. That is, we show that there are $\ll (\|\overline{\mathbf{r}}\| \|q\| R)^\varepsilon$ possible factorizations of the element γy_1^2 in the ring $\mathfrak{o}_K$ with certain conditions on the factors induced by the conditions on x_1, x_2, x_3 stated above.

Let $y_1 = x_1$, $y_2 = L_3(x_1, x_2, x_3)$, and $y_3 = L_4(x_1, x_2, x_3)$ so that we may write (2.3) as $y_2 y_3 = \gamma y_1^2$. If w is an infinite place of K, lying over the place v of k, then

$$\|y_i\|_w \ll (\|q\|^A \sup\{r_v, R\})^{d_w/d_v} \ll \|q\|^{2A} \sup\{r_v^2, R^2\} = t_w$$

for $i = 2, 3$. This follows from the conditions on x_1, x_2, x_3 and the conditions on the coefficients of the forms L_3 and L_4. Remember that $d_v = [k_v : \mathbb{Q}_v]$ is the local degree of k at v. We have seen earlier that there are $\ll N_k(\gamma y_1^2)^\varepsilon$ ideals of $\mathfrak{o}$ which contain the element γy_1^2 ($c = 1$ in the corollary of theorem 7). Since a prime of $\mathfrak{o}$ can split in at most two primes of $\mathfrak{o}_K$, there are thus $\ll N_k(\gamma y_1^2)^\varepsilon \ll (\|q\| \|\mathbf{r}\|)^\varepsilon$ ideals of $\mathfrak{o}_K$ containing γy_1^2. To complete the proof, then, it is sufficient to show that there are $\ll (\|\overline{\mathbf{r}}\| \|q\| R)^\varepsilon$ possible generators of any principal ideal $\langle y \rangle \subset \mathfrak{o}_K$ such that $\|y\|_w \ll t_w$. But if uy is another such generator for some unit u of $\mathfrak{o}_K$, then

$$h_K(u) \ll \prod_{w \mid \infty} t_w / \|y\|_w \ll (\|\overline{\mathbf{r}}\| \|q\| R)^{2As_K},$$

and by proposition 4 there are $\ll (\|\overline{\mathbf{r}}\| \|q\| R)^\varepsilon$ such units. □

Note. — In contrary to theorem 6, this proposition gives the expected estimate $O(B^{1+\varepsilon})$ for the number of points of height $\leqslant B$ on the conic $q = 0$.

3. Proof of the main theorem

We have now come to the proof of theorem 1. We will begin by redefining the objects of interest and formulate the hypothesis once again. As in the previous section we assume that the ground field k is fixed.

Let F be a non-singular quaternary cubic form such that the k-surface $X \subset \mathbb{P}_k^3$ given by $F = 0$ contains three coplanar lines defined over the ground field, and let $U \subset X$ be the complement of all the lines on X. We define $n(B)$ to be the number of rational points in U with height not exceeding B (note that $n(B)$ is finite by theorem 3). Our claim is that $n(B) = O(B^{4/3+\varepsilon})$.

To show this we will proceed in several steps. Clearly $n(B)$ depends on the choice of coordinates. But it follows from proposition 2 that the validity of the statement $n(B) \ll B^{4/3+\varepsilon}$ is independent of the choice of coordinates. Proposition 2 says that $H_k(\varphi(x)) \ll_\varphi H_k(x)$ if φ is an automorphism of k^4. The first thing we do, therefore, is to find appropriate coordinates in which to study F. By an initial linear change of variables we may assume that the plane containing the three lines is $x_0 = 0$. Then $F = L_1 L_2 L_3 - x_0 Q$ for some non-zero quadratic form Q and some non-zero linear forms $L_i(x_1, x_2, x_3)$. After re-scaling both the variables and the form F appropriately we may moreover assume that Q and the L_i have coefficients in $\mathfrak{o}$.

The reduction which is about to follow is the most important part of our proof, and it is the reason why we assume that X contains three coplanar rational lines. The reduction is a generalization of an argument of Wooley and Heath-Brown. From the shape of F we see that the projections

$$f_1(x_0, x_1, x_2, x_3) = \begin{cases} (x_0, L_1) & \text{if } (x_0, L_1) \neq (0, 0) \\ (L_2 L_3, Q) & \text{otherwise,} \end{cases}$$

$$f_2(x_0, x_1, x_2, x_3) = \begin{cases} (x_0, L_2) & \text{if } (x_0, L_2) \neq (0, 0) \\ (L_1 L_3, Q) & \text{otherwise,} \end{cases}$$

$$f_3(x_0, x_1, x_2, x_3) = \begin{cases} (x_0, L_3) & \text{if } (x_0, L_3) \neq (0, 0) \\ (L_1 L_2, Q) & \text{otherwise} \end{cases}$$

define three conic bundle morphisms $f_i : X \to \mathbb{P}_k^1$.

Lemma 6. — *For all $x \in X(k)$,*

$$H_1(x) H_2(x) H_3(x) \ll_F H(x)^2,$$

where $H_i(x) = H_k(f_i(x))$, and $H(x) = H_k(x)$.

Proof. — Let $f : X(k) \to \mathbb{P}^1(k) \times \mathbb{P}^1(k) \times \mathbb{P}^1(k)$ be the morphism given by (f_1, f_2, f_3), and let $\psi : \mathbb{P}^1(k) \times \mathbb{P}^1(k) \times \mathbb{P}^1(k) \to \mathbb{P}^7(k)$ be the trilinear Segre embedding. Then from the equation $L_1 L_2 L_3 = x_0 Q$ one sees that $\psi f(x_0, x_1, x_2, x_3)$ is given by

$$(y_0, y_1, y_2, y_3, y_4, y_5, y_6, y_7) = (x_0^2, x_0 L_3, x_0 L_2, L_2 L_3, x_0 L_1, L_1 L_3, L_1 L_2, Q).$$

Hence,
$$H_1(x)H_2(x)H_3(x) = \prod_v \sup_{0 \leq i \leq 7} \|y_i\|_v \ll \prod_v \sup_{0 \leq i,j \leq 3} \|x_i x_j\|_v = H(x)^2$$
for all $x = (x_0, x_1, x_2, x_3) \in \mathbb{P}^3(k)$. □

Note. — The converse $H(x)^2 \ll H_1(x)H_2(x)H_3(x)$ is also true. The assertion that $2 \log H = \log H_1 + \log H_2 + \log H_3 + O(1)$ on $X(k)$ is in fact a special case of a standard result in the theory of heights (cf. e.g. [12, 2.8]).

According to this lemma, we can choose a positive constant c_1 such that $H_1(x)H_2(x)H_3(x) \leq c_1^3 H(x)^2$ on $X(k)$. If $n_i(B)$ is the number of points $x \in U(k)$ such that $H(x) \leq B$ and $H_i(x) \leq c_1 B^{2/3}$, then
$$n(B) \leq n_1(B) + n_2(B) + n_3(B).$$
It is thus sufficient to show that $n_i(B) \ll B^{4/3+\varepsilon}$ for $i = 1, 2, 3$. We will concentrate on the proof of the statement $n_1(B) \ll B^{4/3+\varepsilon}$. In this particular case, it is convenient to have $L_1 = x_1$ and $L_2 = x_2$. Of course, this change of variables will affect the counting function $n_1(B)$, but as pointed out above, the validity of the statement $n_1(B) \ll B^{4/3+\varepsilon}$ is independent of the choice of coordinates.

The next step in the proof is to look at the fibres of $f_1 : X \to \mathbb{P}^1_k$ in U. If we define $n_1(a, b, B)$ to be the number of rational points in $f_1^{-1}(a,b) \cap U$ with height not exceeding B, then we have

(3.1)
$$n_1(B) \leq \sum_{\substack{(a,b) \in \mathbb{P}^1(k) \\ H(a,b) \leq c_1 B^{2/3}}} n_1(a, b, B).$$

From the definition of F we see that $(ay_1, by_1, y_2, y_3) \in f_1^{-1}(a,b) \cap U$ only if $y_1 \neq 0$ and $q(y_1, y_2, y_3; a, b) = 0$, where
$$q(y_1, y_2, y_3; a, b) = 2 b y_2 L_3(b y_1, y_2, y_3) - 2 a Q(a y_1, b y_1, y_2, y_3).$$
In order to estimate the counting functions $n_1(a, b, B)$ in the sum (3.1), we shall now apply the results from the previous section to the quadratic forms $q(y_1, y_2, y_3; a, b)$. However, the results in the previous section were formulated for quadratic forms, defined over the ring of integers of k, and points in projective space, with conditions on their integral coordinates. Before we can continue, we therefore have to look more closely at the definition of $n_1(a, b, B)$ and make some adjustments to be able to apply these results.

Let $\mathfrak{a}_1, \ldots, \mathfrak{a}_h$ be ideals representing the ideal classes of $\mathfrak{o}$. By the proof of proposition 3, we can choose coordinates $(a, b) \in \mathfrak{o}^2$ for each point in $\mathbb{P}^1(k)$ such that $\langle a, b \rangle = \mathfrak{a}_j$ for some j and $\sup\{\|a\|_v, \|b\|_v\} \ll H(a,b)^{1/s_k}$ for all $v \mid \infty$. We let A be the set of all such coordinates and $A(B)$ be the subset consisting of the elements (a,b) with $H(a,b) \leqslant c_1 B^{2/3}$. By the same proposition, each point on the fibres of f_1 has coordinates $(ay_1, by_1, y_2, y_3) \in \mathfrak{o}^4$, where $(a,b) \in A$ and

(3.2) $$\sup\{\|ay_1\|_v, \|by_1\|_v, \|y_2\|_v, \|y_3\|_v\} \ll H(ay_1, by_1, y_2, y_3)^{1/s_k}.$$

A choice of (y_1, y_2, y_3) under these restrictions does not guarantee that $y_1 \in \mathfrak{o}$. However, by multiplying (y_1, y_2, y_3) with a suitable factor, say, a generator of the principal ideal $\prod_{i=1}^{h} \mathfrak{a}_i^2$, and by changing the implied constant of (3.2) accordingly, we may assume that $(y_1, y_2, y_3) \in \mathfrak{o}^3$. If we introduce the symbol c_2 for this modified implied constant, then we see that $n_1(a, b, B)$ is bounded from above by the number of points $(y_1, y_2, y_3) \in \mathbb{P}^2(k)$ satisfying $y_i \in \mathfrak{o}$, $(ay_1, by_1, y_2, y_3) \in U(k)$, and

$$\sup\{\|ay_1\|_v, \|by_1\|_v, \|y_2\|_v, \|y_3\|_v\} \leqslant c_2 B^{1/s_k}.$$

In fact, we redefine $n_1(a, b, B)$ to be this cardinality.

Let $\mathbf{r} \in (\mathbb{R}_{>0})^{s_k}$ be a vector with components r_v. Then by the definition of the set $L(\mathbf{r}) \subset \mathfrak{o}$ we have that an integer x of k belongs to $L(\mathbf{r})$ if and only if $\|x\|_v \leqslant r_v$ for all $v \mid \infty$. In agreement with the observations above, we now see that we can define $n_1(a, b, B)$ to be the number of points $(y_1, y_2, y_3) \in \mathbb{P}^2(k)$ on the conic $q(y_1, y_2, y_3; a, b) = 0$, not lying on any line on X, and such that $y_i \in L(\mathbf{r}_i)$, where $\mathbf{r}_1$ is the vector with components $c_2 B^{1/s_k}/\sup\{\|a\|_v, \|b\|_v\}$ and $\mathbf{r}_2 = \mathbf{r}_3$ the vectors with components $c_2 B^{1/s_k}$. With this definition of $n_1(a, b, B)$, we are now in a position to use the results from the previous section to handle the sum (3.1). Note that with our new definitions we may write this sum as

$$\sum_{(a,b) \in A(B)} n_1(a, b, B).$$

Also note that we only have to sum over those $(a,b) \in A(B)$ for which the quadratic form $q(y_1, y_2, y_3; a, b)$ is non–singular. By definition, $n_1(a,b,B) = 0$ whenever the components of $f_1^{-1}(a,b)$ are lines on X.

On the hypothesis that $q(y_1, y_2, y_3; a, b)$ is non–singular, theorem 6 gives

(3.3) $$n_1(a, b, B) \ll \left(1 + \frac{B^{3/2} N_k(\Delta_0(a,b))}{H(a,b)^{1/2} N_k(\Delta(a,b))^{1/2}}\right) N_k(\Delta(a,b))^\varepsilon,$$

where $\Delta(a,b)$ is the determinant of $q(y_1, y_2, y_3; a, b)$ and $\Delta_0(a,b)$ the ideal generated by all the 2×2–minors of the matrix of $q(y_1, y_2, y_3; a, b)$. If, in addition,

$q(0, y_2, y_3; a, b)$ is non–singular, proposition 7 gives
$$(3.4) \qquad n_1(a, b, B) \ll B^{1+\varepsilon}/H(a, b).$$
In order for these estimates to be useful we need the following two lemmas.

Lemma 7. — $N_k(\Delta_0(a, b)) \ll_F 1$ for all $(a, b) \in A$.

Proof. — Let $M(a, b)$ be the matrix of $q(y_1, y_2, y_3; a, b)$. The ij:th minor of $M(a, b)$ will be a certain integral form $M_{ij}(a, b)$. If these forms were to have a common factor $(\beta a - \alpha b)$ over some algebraic closure of k, then the rank of $q(y_1, y_2, y_3; \alpha, \beta)$ would be at most one. But $f_1^{-1}(\alpha, \beta)$ is not allowed to be a double line since X is non–singular. Thus the forms $M_{ij}(a, b)$ do not have a common factor. According to Hilbert's Nullstellensatz, there is a positive integer n and polynomials f_{ij}, g_{ij} such that $a^n = \sum f_{ij} M_{ij}$ and $b^n = \sum g_{ij} M_{ij}$. Then by clearing denominators we see that $\gamma a^n, \gamma b^n \in \Delta_0(a, b)$ for some $\gamma \in \mathfrak{o}$. Hence, $N_k(\Delta_0(a, b)) \ll N_k(\langle a^n, b^n \rangle) = N_k(\mathfrak{a}_j)^n$ for some $j = 1, 2, \ldots, h$. □

The lemma tells us that we can forget about the factor $\Delta_0(a, b)$ in (3.3).

Lemma 8. — $q(0, y_2, y_3; a, b)$ is non–singular for almost all $(a, b) \in A$.

Proof. — By definition
$$q(0, y_2, y_3; a, b) = 2 b y_2 L_3(0, y_2, y_3) - 2 a Q(0, 0, y_2, y_3).$$
If $L_3(0, y_2, y_3) = \mu y_2 + \nu y_3$ and $Q(0, 0, y_2, y_3) = \alpha y_2^2 + \beta y_2 y_3 + \gamma y_3^2$ for some α, β, γ, μ, $\nu \in \mathfrak{o}$, then
$$q(0, y_2, y_3; a, b) = 2 (b\mu - a\alpha) y_2^2 + 2 (a\beta - b\nu) y_2 y_3 - 2 a \gamma y_3^2.$$
Hence, $q(0, y_2, y_3; a, b)$ is singular precisely when
$$(4\alpha\gamma - \beta^2) a^2 + 2 (\beta\nu - 2\gamma\mu) a b - \nu^2 b^2 = 0.$$
It is straightforward to check that X would be singular if this form were to vanish identically. Thus $q(0, y_2, y_3; a, b)$ is singular for at most two pairs $(a, b) \in A$. □

From this last lemma, we see that if $q(0, y_2, y_3; a, b)$ is singular, then we may use the estimate $n_1(a, b, B) \ll B$ given by proposition 6. It thus remains to sum $n_1(a, b, B)$ over those $(a, b) \in A(B)$ for which both $q(y_1, y_2, y_3; a, b)$ and $q(0, y_2, y_3; a, b)$ are non–singular. In this sum, the constant term of (3.3) gives the contribution $O(B^{4/3+\varepsilon})$. Simply because the cardinality of $A(B)$ is $O(B^{4/3+\varepsilon})$. We therefore only have to account for the second term

$$(3.5) \qquad B^{3/2+\varepsilon} H(a, b)^{-1/2} N_k(\Delta(a, b))^{-1/2}$$

in the contribution to $n_1(a, b, B)$.

We divide the ranges of $H(a,b)$ and $N_k(\Delta(a,b))$ into intervals $(R, 2R]$ and $(S, 2S]$, respectively. Since $\Delta(a,b)$ is a form of degree 5, the next lemma states that there are $O_F(S^{1/5} R^{1+\varepsilon})$ elements $(a,b) \in A$ which satisfy $H(a,b) \leqslant 2R$ and $N_k(\Delta(a,b)) \leqslant 2S$. The fact that $\Delta(a,b)$ is of degree 5 also implies that $N_k(\Delta(a,b)) \ll H(a,b)^5 \ll B^{10/3}$.

If $S \leqslant B^{5/3}$, then

$$\sum_{\substack{H(a,b)\in(R,2R] \\ N_k(\Delta(a,b))\in(S,2S]}} n_1(a,b,B) \ll S^{1/5} R^{1+\varepsilon} B^{1+\varepsilon} R^{-1} \ll B^{4/3+\varepsilon}$$

by (3.4), and if $S \geqslant B^{5/3}$, then

$$\sum_{\substack{H(a,b)\in(R,2R] \\ N_k(\Delta(a,b))\in(S,2S]}} n_1(a,b,B) \ll S^{1/5} R^{1+\varepsilon} B^{3/2} R^{-1/2} S^{-1/2+\varepsilon} \ll B^{4/3+\varepsilon}$$

by (3.5). This completes the proof of the theorem, on summing over appropriate values of R and S.

Lemma 9. — *Let $G(x,y)$ be a form of degree n with coefficients in $\mathfrak{o}$, and let $R, S \geqslant 1$ be given. Then there are $O_G(S^{1/n} R^{1+\varepsilon})$ elements $(x,y) \in \mathfrak{o}^2$ such that $0 < N_k(G(x,y)) \leqslant S$ and $\sup\{\|x\|_v, \|y\|_v\} \leqslant R^{1/s_k}$ for all $v \mid \infty$.*

Proof. — First we note that we may assume that $G(x,y)$ is irreducible over k. If not, $G(x,y) = G_1(x,y) G_2(x,y)$ for some forms of positive degrees n_1 and n_2, respectively. If $N_k(G(x,y)) \leqslant S$ for some $(x,y) \in \mathfrak{o}^2$, then either $N_k(G_1(x,y)) \leqslant S^{n_1/n}$ or $N_k(G_2(x,y)) \leqslant S^{n_2/n}$. On the assumption that the lemma is valid for all forms of degree less than n, we thus have that it is valid for all reducible forms of degree n.

Now assume that $G(x,y)$ is irreducible and let $\sigma_v : k \to \mathbb{C}$ be embeddings such that $|x|_v = |\sigma_v(x)|$ for all $x \in k$. Then there are non–zero complex numbers μ_{vi} and ν_{vi} such that

$$\sigma_v(G(x,y)) = \prod_{i=1}^n (\mu_{vi}\, \sigma_v(x) - \nu_{vi}\, \sigma_v(y)).$$

If $(x,y) \in \mathfrak{o}^2$ satisfies $N_k(G(x,y)) \leqslant S$, then

(3.6) $$\prod_{v \mid \infty} |\mu_{vi}\, \sigma_v(x) - \nu_{vi}\, \sigma_v(y)|^{d_v} \leqslant S^{1/n}$$

for some i. It is thus sufficient to show that there are $O(S^{1/n} R^{1+\varepsilon})$ elements $(x,y) \in \mathfrak{o}^2$ satisfying the assumptions of the lemma and (3.6) for a fixed i. Note that there are $O(R)$ elements $y \in \mathfrak{o}$ such that $\sup \|y\|_v \leqslant R^{1/s_k}$. We therefore assume that y is fixed and show that there are $O(S^{1/n}R^\varepsilon)$ elements $x \in \mathfrak{o}$ such that (x,y) have the required properties. In order to accomplish this, we choose one place w of k and divide the ranges of $|\mu_{vi}\,\sigma_v(x) - \nu_{vi}\,\sigma_v(y)|^{d_v}$ for the places $v \neq w$ into intervals $(T_v, 2T_v]$. Note that

$$R^{1/s_k - n} \ll |\mu_{vi}\,\sigma_v(x) - \nu_{vi}\,\sigma_v(y)|^{d_v} \ll R^{1/s_k}$$

for all $(x,y) \in \mathfrak{o}^2$ satisfying the assumptions of the lemma. We get the lower bounds from the above bounds and the fact $1 \leqslant N_k(G(x,y))$.

Now, if $|\mu_{vi}\,\sigma_v(x) - \nu_{vi}\,\sigma_v(y)|^{d_v} \in (T_v, 2T_v]$ and (3.6) holds, then

$$|\mu_{wi}\,\sigma_w(x) - \nu_{wi}\,\sigma_w(y)|^{d_w} \ll S^{1/n} / \prod_{v \neq w} T_v.$$

Since $\|x - x'\|_v \ll Y/|\mu_{vi}|^{d_v}$ whenever

$$|\mu_{vi}\,\sigma_v(x) - \nu_{vi}\,\sigma_v(y)|^{d_v} \leqslant Y \quad \text{and} \quad |\mu_{vi}\,\sigma_v(x') - \nu_{vi}\,\sigma_v(y)|^{d_v} \leqslant Y,$$

proposition 1 gives the estimate $O_G(S^{1/n})$ for the number of elements $x \in \mathfrak{o}$ which satisfies the above conditions. By summing over all the intervals $(T_v, 2T_v]$, we get the required estimate $O_G(S^{1/n}R^\varepsilon)$, and this completes the proof. □

References

[1] E. Bombieri and J. Vaaler, *On Siegel's lemma*, Invent. Math. **73** (1983), 11–32.

[2] J. W. S. Cassels, *An Introduction to the Geometry of Numbers*. Springer–Verlag, 1959.

[3] J. Franke, Yu. I. Manin, and Yu. Tschinkel, *Rational points of bounded height on Fano varieties*, Invent. Math. **95** (1989), 421–435.

[4] R. Hartshorne, *Algebraic Geometry*, Springer–Verlag, 1977.

[5] D. R. Heath-Brown, *The density of rational points on cubic surfaces*, Acta Arith. **79(1)** (1997), 17–30.

[6] D. R. Heath-Brown, *Counting rational points on cubic surfaces*, in E. Peyre, editor, Nombre et Répartition de points de hauteur bornée, Astérisque **251** (1998), Société mathématique de France, 13–29.

[7] C. Hooley, *On the numbers that are representable as the sum of two cubes*, Angew. Math. **314** (1980), 146–173.

[8] S. Lang, *Algebraic Number Theory*, Springer–Verlag, second edition, 1994.

[9] Yu. I. Manin and Yu. Tschinkel, *Points of bounded height on del Pezzo surfaces*, Compositio Math. **85** (1993), 315–332.

[10] I. Reiner, *Maximal Orders*. Academic Press Inc., 1975.

[11] S. H. Schanuel, *Heights in number fields*, Bull. Soc. Math. Fr. **107** (1979), 433–449.

[12] J.-P. Serre, *Lectures on the Mordell-Weil Theorem.* Vieweg, 1989.

[13] G. Tenenbaum, *Introduction to analytic and probabalistic number theory*, Cambridge University Press, 1995.

[14] T. D. Wooley, *Sums of two cubes*, Internat. Math. Res. Notices **4** (1985), 181–185.

Rational points on algebraic varieties
(E. Peyre, Y. Tschinkel, ed.), p. 37–70

TORSEURS ARITHMÉTIQUES ET ESPACES FIBRÉS

Antoine Chambert-Loir
Institut de mathématiques de Jussieu, Boite 247, 4, place Jussieu, F-75252
Paris Cedex 05 • *E-mail* : chambert@math.jussieu.fr

Yuri Tschinkel
Department of Mathematics, Princeton University, Princeton, NJ, 08544
E-mail : ytschink@math.princeton.edu

Abstract. — We study the compatibility of Manin's conjecture with natural geometric constructions, like fibrations induced from torsors under linear algebraic groups. The main problem it to understand the variation of metrics from fiber to fiber. For this we introduce the notions of "arithmetic torsors", "adelic torsion" and "Arakelov L-functions". We discuss concrete examples, like horospherical varieties and equivariant compactifications of semi-abelian varieties. These techniques are applied to prove "going up" and "descent" theorems for height zeta functions on such fibrations.

Introduction

Cet article est le premier d'une série dont le thème principal est l'étude des hauteurs sur certaines variétés algébriques sur un corps de nombres. On voudrait notamment comprendre la distribution des points rationnels de hauteur bornée.

Précisément, soient X une variété algébrique projective lisse sur un corps de nombres F, $\mathscr{L}$ un fibré en droites sur X et $H_{\mathscr{L}} : X(\overline{F}) \to \mathbf{R}_+^*$ une fonction

hauteur (exponentielle) pour $\mathscr{L}$. Si U est un ouvert de Zariski de X, on cherche à estimer le nombre
$$N_U(\mathscr{L}, H) = \#\{x \in U(F)\,;\, H_{\mathscr{L}}(x) \leqslant H\}$$
lorsque H tend vers $+\infty$. L'étude de nombreux exemples a montré que l'on peut s'attendre à un équivalent de la forme
$$(*) \quad N_U(\mathscr{L}, H) = \Theta(\mathscr{L}) H^{a(\mathscr{L})} (\log H)^{b(\mathscr{L})-1}(1 + o(1)), \quad H \to +\infty$$
pour un ouvert U convenable et lorsque par exemple $\mathscr{L}$ et ω_X^{-1} (fibré anticanonique) sont amples. On a en effet un résultat de ce genre lorsque X est une variété de drapeaux [12], une intersection complète lisse de bas degré (méthode du cercle), une variété torique [5], une variété horosphérique [23], etc. On dispose de plus d'une description conjecturale assez précise des constantes $a(\mathscr{L})$ et $b(\mathscr{L})$ en termes du cône des diviseurs effectifs [2] ainsi que de la constante $\Theta(\mathscr{L})$ ([16], [6]).

En fait, on étudie plutôt la *fonction zêta des hauteurs*, définie par la série de Dirichlet
$$Z_U(\mathscr{L}, s) = \sum_{x \in U(F)} H_{\mathscr{L}}(x)^{-s}$$
à laquelle on applique des théorèmes taubériens standard. Sur cette série, on peut se poser les questions suivantes : domaine de convergence, prolongement méromorphe, ordre du premier pôle, terme principal, sans oublier la croissance dans les bandes verticales à gauche du premier pôle. Cela permet de proposer des conjectures de précision variable.

Il est naturel de vouloir tester la compatibilité de cette conjecture avec les constructions usuelles de la géométrie algébrique. Par exemple, on n'arrive pas à démontrer cette conjecture pour un éclatement X' d'une variété X pour laquelle cette conjecture est connue. Même pour un éclatement de 4 points dans le plan projectif, on n'a pas de résultat complet !

Dans cet article, nous considérons certaines fibrations localement triviales construites de la façon suivante. Soient G un groupe algébrique linéaire sur F agissant sur une variété projective lisse X, B une variété projective lisse sur F et T un G-torseur sur B localement trivial pour la topologie de Zariski. Ces données définissent une variété algébrique projective Y munie d'un morphisme $Y \to B$ dont les fibres sont isomorphes à X. On donne au §2.7 de nombreux exemples « concrets » de variétés algébriques provenant d'une telle construction.

Le cœur du problème est de comprendre le comportement de la fonction hauteur lorsqu'on passe d'une fibre à l'autre, comportement vraiment non trivial bien qu'elles soient toutes isomorphes.

Pour définir et étudier de façon systématique les fonctions hauteurs sur Y, on est amené à dégager de nouvelles notions dans l'esprit de la géométrie d'Arakelov. Apparaissent notamment les notions de G-torseur arithmétique au §1.1.3, ainsi que la définition de la *fonction L d'Arakelov* attachée à un tel torseur arithmétique et à une fonction sur le groupe adélique $G(\mathbf{A}_F)$ invariante par $G(F)$ et par un sous-groupe compact convenable (§1.4). Elles généralisent les notions usuelles de fibré inversible métrisé ainsi que la fonction zêta des hauteurs introduits par S. Arakelov [1].

Ceci fait, on peut voir que les fonctions hauteurs d'une fibre Y_b de la projection $Y \to B$ diffèrent de la fonction hauteur sur X par ce que nous appelons *torsion adélique*, dans laquelle on retrouve explicitement la classe d'isomorphisme du G-torseur arithmétique T_b sur F (§2.4).

Dans un deuxième article, nous appliquerons ces considérations générales au cas d'une fibration en variétés toriques provenant d'un torseur sous un tore pour l'ouvert U défini par le tore. Le principe de l'étude généralise [23] et est le suivant. On construit les hauteurs à l'aide d'un prolongement du torseur géométrique en un torseur arithmétique, ce qui correspond en l'occurence au choix de métriques hermitiennes sur certains fibrés en droites. On écrit ensuite la fonction zêta comme la somme des fonctions zêta des fibres

$$Z_U(\mathscr{L}, s) = \sum_{b \in B(F)} \sum_{x \in U_b(F)} H_{\mathscr{L}}(x)^{-s} = \sum_{b \in B(F)} Z_{U_b}(\mathscr{L}|_{U_b}, s).$$

Chaque U_b est isomorphe au tore et on peut récrire la fonction zêta des hauteurs de U_b à l'aide de la formule de Poisson adélique. De cette façon, la fonction zêta de U apparaît comme une intégrale sur certains caractères du tore adélique de la fonction L d'Arakelov d'un torseur arithmétique sur B.

Cette expression nous permettra d'établir un théorème de montée : supposons que B vérifie une conjecture, alors Y la vérifie. Bien sûr, la méthode reprend les outils utilisés dans la démonstration de ces conjectures pour les variétés toriques ([5, 3, 4]).

Alors que le présent article contient des considérations générales de « théorie d'Arakelov équivariante » dont on peut espérer qu'elles seront utiles dans d'autres contextes, le deuxième verra intervenir des outils de théorie analytique des nombres (formule de Poisson, théorème des résidus, estimations, etc.).

Remerciements. — Nous remercions J.-B. Bost pour d'utiles discussions. Pendant la préparation de cet article, le second auteur[*] était invité à l'I.H.E.S. et à Jussieu ; il est reconnaissant envers ces institutions pour leur hospitalité.

Notations et conventions

Si $\mathscr{X}$ est un schéma, on désigne par $\mathrm{QCoh}(\mathscr{X})$ et $\mathrm{Fib}_d(\mathscr{X})$ les catégories des faisceaux quasi-cohérents (resp. des faisceaux localement libres de rang d) sur $\mathscr{X}$. On note $\mathrm{Pic}(\mathscr{X})$ le groupe des classes d'isomorphisme de faisceaux inversibles sur $\mathscr{X}$. Si $\mathscr{F}$ est un faisceau localement libre sur $\mathscr{X}$, on note $\mathbf{V}(\mathscr{F}) = \mathrm{Spec\,Sym}\,\mathscr{F}$ et $\mathbf{P}(\mathscr{F}) = \mathrm{Proj\,Sym}\,\mathscr{F}$ les fibrés vectoriels et projectifs associés à $\mathscr{F}$.

On note $\widehat{\mathrm{Fib}}_d(\mathscr{X})$ la catégorie des fibrés vectoriels hermitiens sur $\mathscr{X}$ (c'est-à-dire des faisceaux localement libres de rang d munis d'une métrique hermitienne continue sur $\mathscr{X}(\mathbf{C})$ et invariante par la conjugaison complexe). On note $\widehat{\mathrm{Pic}}(\mathscr{X})$ le groupe des classes d'isomorphisme de fibrés en droites hermitiens sur $\mathscr{X}$.

Si $\mathscr{X}$ est un S-schéma, et si $\sigma \in S(\mathbf{C})$, on désigne par $\mathscr{X}_\sigma$ le $\mathbf{C}$-schéma $\mathscr{X} \times_\sigma \mathbf{C}$. Cette notation servira lorsque S est le spectre d'un localisé de l'anneau des entiers d'un corps de nombres F, de sorte que σ n'est autre qu'un plongement de F dans $\mathbf{C}$.

Si G est un schéma en groupes sur S, $X^*(G)$ désigne le groupe des S-homomorphismes $G \to \mathbf{G}_m$ (caractères algébriques).

Si $\mathscr{X}/S$ est lisse, le faisceau canonique de $\mathscr{X}/S$, noté $\omega_{\mathscr{X}/S}$ est la puissance extérieure maximale de $\Omega^1_{\mathscr{X}/S}$.

1. Torseurs arithmétiques

1.1. Définitions

Rappelons la définition d'un torseur en géométrie algébrique.

Définition 1.1.1. — Soient S un schéma, $\mathscr{B}$ un S-schéma et G un S-schéma en groupes plat et localement de présentation finie.

---

[*]partially supported by the N.S.A.

Un *G-torseur* sur un $\mathscr{B}$ est un $\mathscr{B}$-schéma $\pi : \mathscr{T} \to \mathscr{B}$ fidèlement plat et localement de présentation finie muni d'une action de G au-dessus de $\mathscr{B}$, $m : G \times_S \mathscr{T} \to \mathscr{T}$, de sorte que le morphisme

$$(m, p_2) : G \times_S \mathscr{T} \to \mathscr{T} \times_\mathscr{B} \mathscr{T}$$

soit un isomorphisme. On le suppose de plus *localement trivial pour la topologie de Zariski*.

On note $\mathrm{H}^1(\mathscr{B}, G)$ l'ensemble des classes d'isomorphisme de G-torseurs sur $\mathscr{B}$.

SITUATION 1.1.2. — *Supposons que S est le spectre de l'anneau des entiers d'un corps de nombres F et que G est un S-schéma en groupes linéaire connexe plat et de présentation finie. Fixons pour tout plongement complexe de F, $\sigma \in S(\mathbf{C})$, un sous-groupe compact maximal K_σ de $G(\mathbf{C})$ et notons $\mathbf{K}_\infty$ la collection $(K_\sigma)_\sigma$. On suppose que pour deux plongements complexes conjugués, les sous-groupes compacts maximaux correspondants sont échangés par la conjugaison complexe.*

Définition 1.1.3. — On appelle $(G, \mathbf{K}_\infty)$-*torseur arithmétique* sur $\mathscr{B}$ la donnée d'un G-torseur $\mathscr{T}$ sur $\mathscr{B}$ ainsi que pour tout $\sigma \in S(\mathbf{C})$, d'une section du $K_\sigma \backslash G_\sigma(\mathbf{C})$-fibré sur $\mathscr{B}_\sigma(\mathbf{C})$ quotient à $\mathscr{T}_\sigma(\mathbf{C})$ par l'action de K_σ. On suppose de plus que pour deux plongements complexes conjugués, les sections sont échangées par la conjugaison complexe.

On note $\widehat{\mathrm{H}}^1(\mathscr{B}, (G, \mathbf{K}_\infty))$ l'ensemble des classes d'isomorphisme de $(G, \mathbf{K}_\infty)$-torseurs arithmétiques sur $\mathscr{B}$.

On note aussi $\widehat{\mathrm{H}}^0(\mathscr{B}, (G, \mathbf{K}_\infty))$ l'ensemble des sections $g \in \mathrm{H}^0(\mathscr{B}, G)$ telles que pour toute place à l'infini σ, g définisse une section $\mathscr{B}_\sigma(\mathbf{C}) \to \mathbf{K}_\sigma$.

Remarque 1.1.4. — Se donner une section du $K_\sigma \backslash G_\sigma(\mathbf{C})$-fibré associé à $\mathscr{T}_\sigma(\mathbf{C})$ sur $\mathscr{B}_\sigma(\mathbf{C})$ revient à fixer dans un recouvrement ouvert (U_i) pour la topologie complexe les fonctions de transition $g_{ij} \in \Gamma(U_i \cap U_j, G)$ à valeurs dans K_σ. Il en existe car $G_\sigma(\mathbf{C})$ est homéomorphe au produit de K_σ par un $\mathbf{R}$-espace vectoriel de dimension finie, cf. par exemple [7].

D'autre part, on choisit dans cet article de supposer la section continue. Dans certaines situations, il pourrait être judicieux de la supposer indéfiniment différentiable.

La dépendance de cette notion en les sous-groupes maximaux fixés est la suivante : toute famille $(x_\sigma) \in \prod_\sigma G_\sigma(\mathbf{C})$ telle que $K'_\sigma = x_\sigma K_\sigma x_\sigma^{-1}$ détermine une bijection canonique
$$\widehat{\mathrm{H}}^1(\mathscr{B},(G,\mathbf{K}_\infty)) \simeq \widehat{\mathrm{H}}^1(\mathscr{B},(G,\mathbf{K}'_\infty)).$$
(Rappelons que deux sous-groupes compacts maximaux sont conjugués.)

1.1.5. Variante adélique. — Il existe une variante adélique des considérations précédentes qui supprime en apparence la référence à un modèle sur $\operatorname{Spec} \mathfrak{o}_F$. En effet, si $\mathscr{B}$ est propre sur $\operatorname{Spec} \mathfrak{o}_F$, remarquons que pour toute place finie de F, un G-torseur arithmétique sur $\mathscr{B}$ induit une section du morphisme $G(\mathfrak{o}_v) \backslash \mathscr{T}(F_v) \to \mathscr{B}(F_v)$.

Définition 1.1.6. — Soit G_F un F-schéma en groupes de type fini et fixons un sous-groupe compact maximal[1] $\mathbf{K} = \prod_v K_v$ du groupe adélique $G(\mathbf{A}_F)$. Soit $\mathscr{B}_F$ un F-schéma propre.

On appelle $(G_F,\mathbf{K})$-*torseur adélique* sur $\mathscr{B}_F$ la donnée d'un G_F-torseur $\mathscr{T}_F \to \mathscr{B}_F$, ainsi que pour toute place v de F, d'une section continue de $K_v \backslash \mathscr{T}_F(F_v) \to \mathscr{B}_F(F_v)$. On suppose de plus qu'il existe un ouvert non vide U de $\operatorname{Spec} \mathfrak{o}_F$, un U-schéma en groupes plat et de présentation fini G, un U-schéma $\mathscr{B}$ propre, plat et de type fini, ainsi qu'un G-torseur $\mathscr{T} \to \mathscr{B}$ qui prolongent respectivement G_F, $\mathscr{B}_F$ et $\mathscr{T}_F$ et vérifiant : pour toute place finie v de F dominant U, $G(\mathfrak{o}_v) = K_v$ et la section continue de $K_v \backslash \mathscr{T}_F(F_v) \to \mathscr{B}_F(F_v)$ est celle fournie par le modèle $\mathscr{T} \to \mathscr{B}$.

On note $\overline{\mathrm{H}}^1(\mathscr{B}_F,(G_F,\mathbf{K}))$ l'ensemble des classes d'isomorphisme de $(G_F,\mathbf{K})$-torseurs adéliques sur $\mathscr{B}_F$.

Bien sûr, si $\mathscr{B}$ est un $\mathfrak{o}_F$-schéma propre et G un $\mathfrak{o}_F$-schéma en groupes plat et de présentation finie, tout $(G,\mathbf{K}_\infty)$-torseur arithmétique sur $\mathscr{B}$ définit un $(G_F,\mathbf{K})$-torseur adélique où $\mathbf{K}$ est le compact adélique $\prod_{v \text{ finie}} G(\mathfrak{o}_v) \prod_\sigma K_\sigma$.

1.1.7. Exemples. — a) Quand $G = \operatorname{GL}(d)$, le torseur $\mathscr{T}$ correspond naturellement à la donnée d'un fibré vectoriel $\mathscr{E}$ de rang d sur $\mathscr{B}$ par la formule $\mathscr{T} = \mathbf{Isom}(\mathscr{O}_\mathscr{B}^n, \mathscr{E})$. Si l'on choisit $K_\sigma = \mathrm{U}(d)$, une section du $\mathrm{U}(d,\mathbf{C}) \backslash \operatorname{GL}(d,\mathbf{C})$-fibré associé correspond à une métrique hermitienne (continue) sur $\mathscr{E}$. Ainsi, les

---
[1] Cela signifiera pour nous que les K_v sont des sous-groupes compacts ouverts aux places finies, et maximaux aux places infinies.

$(\mathrm{GL}(d), \mathrm{U}(d))$-torseurs arithmétiques sont en bijection naturelle avec les fibrés vectoriels hermitiens.

b) En particulier, lorsque $G = \mathbf{G}_m$, la famille des sous-groupes compacts maximaux $\mathbf{K}_\infty$ est canoniquement définie (ce qui permet de les omettre dans la notation) et $\widehat{\mathrm{H}}^1(\mathscr{B}, \mathbf{G}_m) = \widehat{\mathrm{Pic}}(\mathscr{B})$, le groupe des classes d'isomorphisme de fibrés en droites sur $\mathscr{B}$ munis d'une métrique hermitienne continue compatible à la conjugaison complexe. Les $\mathbf{G}_m$-torseurs adéliques s'identifient de même aux fibrés inversibles munis d'une métrique adélique. Nous rappelons cette théorie au paragraphe 1.3

c) Dans ce texte, nous ne considérons que des G-torseurs localement triviaux pour la topologie de Zariski. Néanmoins, lorsque G/S est un S-schéma abélien, un exemple de G-torseur localement trivial pour la topologie étale sur $\mathscr{B}$ est fourni par un schéma abélien $\mathscr{A}/\mathscr{B}$ obtenu par torsion de G/S, c'est-à-dire tel qu'il existe un revêtement étale $\mathscr{B}' \to \mathscr{B}$ de sorte que $\mathscr{A} \times_\mathscr{B} \mathscr{B}'$ soit isomorphe à $G \times_S \mathscr{B}'$ (famille de schémas abéliens à module constant). De tels exemples devraient bien sûr faire partie d'une étude plus générale de la géométrie d'Arakelov des torseurs que nous reportons à une occasion ultérieure.

1.2. Propriétés

Les ensembles de classes d'isomorphisme de $(G, \mathbf{K}_\infty)$-torseurs arithmétiques vérifient un certain nombre de propriétés formelles, dont les analogues algébriques sont bien connus. Leur démonstration est standard et laissée au lecteur.

PROPOSITION 1.2.1. — *L'oubli de la structure arithmétique induit une application*
$$\widehat{\mathrm{H}}^1(\mathscr{B}, (G, \mathbf{K}_\infty)) \to \mathrm{H}^1(\mathscr{B}, G).$$
On a aussi une suite exacte d'ensembles pointés :
$$1 \to \widehat{\mathrm{H}}^0(\mathscr{B}, (G, \mathbf{K}_\infty)) \to \mathrm{H}^0(\mathscr{B}, G) \to \left(\bigoplus_\sigma \Gamma(\mathscr{B}_\sigma(\mathbf{C}), K_\sigma \backslash G_\sigma(\mathbf{C}))\right)^{F_\infty} \to$$
$$\to \widehat{\mathrm{H}}^1(\mathscr{B}, (G, \mathbf{K}_\infty)) \to \mathrm{H}^1(\mathscr{B}, G) \to 1.$$

(F_∞ désigne la conjugaison complexe et $(\cdot)^{F_\infty}$ la partie invariante par la conjugaison complexe.)

Remarque 1.2.2. — Lorsque $G = \mathbf{G}_m$, en identifiant $\mathbf{G}_m(\mathbf{C})/K$ à $\mathbf{R}_+^*$, nous retrouvons la suite exacte bien connue pour $\widehat{\mathrm{Pic}}$ et Pic (cf. [**14**], 3.3.5 ou 3.4.2).

D'autre part, on devrait pouvoir interpréter cette suite exacte à l'aide de la *mapping cylinder category* introduite par S. Lichtenbaum dans son étude des valeurs spéciales des fonctions zêta des corps de nombres (exposé à Paris 6, 1998). En effet, cette catégorie est (? !) la catégorie des faisceaux en groupes abéliens sur, disons Spec $\mathbf{Z} \cup \{\infty\}$.

PROPOSITION 1.2.3. — *Supposons que le groupe G est commutatif. Alors, les sous-groupes compacts maximaux sont uniques et l'ensemble $\widehat{\mathrm{H}}^1(\mathscr{B}, (G, \mathbf{K}_\infty))$ hérite d'une structure de groupe abélien compatible avec la structure de groupe abélien sur $\mathrm{H}^1(\mathscr{B}, G)$. Dans ce cas, la suite exacte 1.2.1 est une suite exacte de groupes abéliens.*

PROPOSITION 1.2.4. — (Changement de base) *Tout morphisme de S-schémas $\mathscr{B}' \to \mathscr{B}$ induit un foncteur des $(G, \mathbf{K}_\infty)$-torseurs arithmétiques sur $\mathscr{B}$ vers les $(G, \mathbf{K}_\infty)$-torseurs arithmétiques sur $\mathscr{B}'$, compatible à l'oubli des structures arithmétiques et aux classes d'isomorphisme.*

(Changement du corps de base) *Si F' est une extension de F, $S' = \mathrm{Spec}\, \mathfrak{o}_{F'}$ et si on choisit pour tout plongement complexe σ' de F' $K_{\sigma'} = K_{\sigma'|F}$, on dispose d'un foncteur des $(G, \mathbf{K}_\infty)$-torseurs arithmétiques sur $\mathscr{B}$ vers les $(G \times_S S', \mathbf{K}_\infty)$-torseurs arithmétiques sur $\mathscr{B} \times_S S'$, compatible à l'oubli des structures arithmétiques et aux classes d'isomorphisme.*

PROPOSITION 1.2.5. — (Changement de groupe) *Si $p : G \to G'$ est un morphisme de S-schémas en groupes et que les sous-groupes compacts maximaux $\mathbf{K}_\infty$ et $\mathbf{K}'_\infty$ sont choisis de sorte que pour tout plongement complexe σ, tel que $p(K'_\sigma) \subset K'_\sigma$, il y a un foncteur des $(G, \mathbf{K}_\infty)$-torseurs arithmétiques vers les $(G', \mathbf{K}'_\infty)$-torseurs arithmétiques, compatible à l'oubli des structures arithmétiques et aux classes d'isomorphisme.*

(Suite exacte courte) *Soit*

$$1 \to G'' \xrightarrow{\iota} G \xrightarrow{p} G' \to 1$$

une suite exacte de S-schémas en groupes. Soient $\mathbf{K}_\infty$, $\mathbf{K}'_\infty$ et $\mathbf{K}''_\infty$ des familles de sous-groupes compacts maximaux pour G, G' et G'' aux places archimédiennes choisis de sorte que $K''_\sigma = \iota^{-1}(K_\sigma)$ et $p(K_\sigma) = K'_\sigma$ pour toute place σ.

Si p admet localement une section (comme S-schéma), alors on a une suite exacte courte canonique d'ensembles pointés :

$$1 \to \widehat{\mathrm{H}}^0(\mathscr{B},(G'',\mathbf{K}''_\infty)) \xrightarrow{\iota} \widehat{\mathrm{H}}^0(\mathscr{B},(G,\mathbf{K}_\infty)) \xrightarrow{p} \widehat{\mathrm{H}}^0(\mathscr{B},(G',\mathbf{K}'_\infty)) \xrightarrow{\delta}$$
$$\to \widehat{\mathrm{H}}^1(\mathscr{B},(G'',\mathbf{K}''_\infty)) \xrightarrow{\iota} \widehat{\mathrm{H}}^1(\mathscr{B},(G,\mathbf{K}_\infty)) \xrightarrow{p} \widehat{\mathrm{H}}^1(\mathscr{B},(G',\mathbf{K}'_\infty)).$$

Sur $\mathrm{Spec}\,\mathfrak{o}_F$, l'ensemble des classes d'isomorphisme de $(G,\mathbf{K}_\infty)$-torseurs arithmétiques a une description très simple, similaire à la description classique des classes d'isomorphisme de G-torseurs sur une courbe projective sur un corps fini. Cela généralise la description analogue du groupe $\widehat{\mathrm{Pic}}(\mathrm{Spec}\,\mathfrak{o}_F)$ (cf. [14], 3.4.3, p. 131, où le groupe correspondant est noté $\widehat{\mathrm{CH}}^1(\mathrm{Spec}\,\mathfrak{o}_F)$).

PROPOSITION 1.2.6. — *On a des isomorphismes canoniques*
$$\widehat{\mathrm{H}}^1(\mathrm{Spec}\,\mathfrak{o}_F,(G,\mathbf{K}_\infty)) \simeq G(F)\backslash G(\mathbf{A}_F)/\mathbf{K}_G,$$
où $\mathbf{K}_G$ *désigne le produit* $\prod_{v\ \mathrm{finie}} G(\mathfrak{o}_v) \prod_{\sigma\ \mathrm{infinie}} K_\sigma$.

De même, pour un sous-groupe compact maximal $\mathbf{K}$ *de* $G(\mathbf{A}_F)$, *on a un isomorphisme canonique*
$$\overline{\mathrm{H}}^1(\mathrm{Spec}\,F,(G_F,\mathbf{K})) \simeq G(F)\backslash G(\mathbf{A}_F)/\mathbf{K}.$$

Démonstration. — Soit $\widehat{\mathscr{T}}$ un $(G,\mathbf{K})$-torseur arithmétique sur $\mathrm{Spec}(\mathfrak{o}_F)$, localement trivial pour la topologie de Zariski. Commençons par fixer un section $\tau_F \in \mathscr{T}(F)$. Si v est une place finie de F, comme $\mathrm{H}^1(\mathrm{Spec}\,\mathfrak{o}_v, G) = 0$, il existe une section $\tau_v \in \mathscr{T}(\mathfrak{o}_v)$, unique modulo l'action de $G(\mathfrak{o}_v)$. Cette section se relie à τ_F par un élément bien défini $g_v \in G(F_v)/G(\mathfrak{o}_v)$ tel que $g_v^{-1} \cdot \tau_F = \tau_v$. Comme τ_F s'étend en une section de $\mathscr{T}$ sur un ouvert de $\mathrm{Spec}\,\mathfrak{o}_F$, on a $g_v \in \mathbf{K}_v$ pour presque toute place v. D'autre part, si σ est une place infinie, la section de $K_\sigma\backslash\mathscr{T}(\mathbf{C})$ donnée par la structure de $(G,\mathbf{K}_\infty)$-torseur arithmétique est de la forme $K_\sigma g_\sigma^{-1}\tau_F$, pour un unique $g_\sigma \in G(\mathbf{C})/K_\sigma$. On a ainsi défini un élement $\mathbf{g}$ dans $G(\mathbf{A}_F)/\mathbf{K}_G$. Il dépend de la section τ_F, mais si on choisit une autre section, elle sera de la forme $g_F\tau_F$, ce qui revient à changer l'élément $\mathbf{g}$ par $g_F^{-1}\mathbf{g}$. Nous avons donc attaché au $(G,\mathbf{K}_\infty)$-torseur arithmétique $\widehat{\mathscr{T}}$ un élément dans $G(F)\backslash G(\mathbf{A}_F)/\mathbf{K}_G$ qui visiblement ne dépend que de la classe d'isomorphisme de $\widehat{\mathscr{T}}$.

Pour la bijection réciproque, on choisit un représentant de $\mathbf{g}$ appartenant à $G(F)\backslash G(\mathbf{A}_F)/\mathbf{K}_G$ où pour toute place finie v, $g_v \in G(F)$, et où presque tous les g_v valent 1. Soit alors U le plus grand ouvert de $\operatorname{Spec} \mathfrak{o}_F$ tel que pour toute place finie v, $g_v \in G(U)$; si v est une place finie qui ne domine pas U, soit $U_v = U \cup \{v\}$. On définit un G-torseur $\mathscr{T}$ sur $\operatorname{Spec} \mathfrak{o}_F$ comme isomorphe à G sur U et sur chaque U_v, les isomorphismes de transition étant fixés par l'isomorphisme entre $\mathscr{T}|_U = G|_U$ et $\mathscr{T}|_{U_v} \times U = G|_U$ induit par la multiplication à gauche par g_v^{-1}. On munit ce G-torseur de la K_σ-classe à gauche $K_\sigma g_\sigma^{-1}$ dans la trivialisation canonique sur l'ouvert U qui contient $\operatorname{Spec} F$, d'où un $(G, \mathbf{K}_\infty)$-torseur arithmétique sur $\operatorname{Spec} \mathfrak{o}_F$.

On laisse au lecteur le soin de vérifier plus en détail que la classe d'isomorphisme du $(G, \mathbf{K}_\infty)$-torseur arithmétique ainsi construit est indépendante du représentant choisi, et que cela définit effectivement la bijection réciproque voulue.

La variante adélique $\overline{\mathrm{H}}^1(\operatorname{Spec} F, (G_F, \mathbf{K}))$ se traite de même (et plus facilement car on n'a pas de torseur à construire !). □

Remarque 1.2.7. — On aurait aussi pu construire le G-torseur $\mathscr{T}$ associé à un point adélique (g_v) en décrétant que les sections de $\mathscr{T}$ sur un ouvert U de $\operatorname{Spec} \mathfrak{o}_F$ sont les $\gamma \in G(F)$ tels que pour toute place finie v dominant U, $\gamma g_v \in G(\mathfrak{o}_v)$.

1.3. Métriques adéliques

Pour la commodité du lecteur, nous rappelons la théorie des métriques adéliques sur les fibrés en droites. C'est un cas particulier bien connu des constructions précédentes lorsque le groupe est $\mathbf{G}_m$, mais l'exposer nous permettra de fixer quelques notations.

Définition 1.3.1. — Soient F un corps valué, X un schéma de type fini sur F et $\mathscr{L}$ un fibré en droites sur X. Une métrique sur $\mathscr{L}$ est une application continue $\mathbf{V}(\mathscr{L}^\vee)(F) \to \mathbf{R}_+$ de sorte que pour tout $x \in X(F)$, la restriction de cette application à la fibre en x (identifiée naturellement à F) soit une norme.

Soient F un corps de nombres, X un schéma projectif sur F et $\mathscr{L}$ un fibré en droites sur X. La donnée d'un schéma projectif et plat $\tilde{X}$ sur le spectre $S = \operatorname{Spec} \mathfrak{o}_F$ de l'anneau des entiers de F dont la fibre générique est X définit pour toute place non-archimédienne v de F une métrique sur le fibré en droites $\mathscr{L} \otimes F_v$ sur $X \times F_v$.

Définition 1.3.2. — On appelle *métrique adélique* sur $\mathscr{L}$ toute collection de métriques $(\|\cdot\|_v)_v$ sur $\mathscr{L} \otimes F_v$ pour toutes les places v de F qui est obtenue de cette façon pour presque toutes les places (non-archimédiennes) de F.

On note $\overline{\mathrm{Pic}}(X) = \overline{\mathrm{H}}^1(X, \mathbf{G}_m)$ le groupe des classes d'isomorphisme de fibrés en droites sur X munis de métriques adéliques.

Donnons nous une métrique adélique sur $\mathscr{L}$. Tout morphisme $f : Y \to X$ de F-schémas projectifs fournit par image réciproque une métrique adélique sur $f^*\mathscr{L}$. Si Y n'est pas projective, on obtient tout de même de la sorte une collection de métriques pour toutes les places de F.

Définition 1.3.3. — Si $\overline{\mathscr{L}} = (\mathscr{L}, (\|\cdot\|_v)_v)$ est un fibré en droites sur X muni d'une métrique adélique, on appelle *fonction hauteur* (exponentielle) associée à $\overline{\mathscr{L}}$ la fonction

$$H(\overline{\mathscr{L}}; \cdot) : X(F) \to \mathbf{R}_+, \quad x \mapsto \prod_v \|\mathsf{s}\|_v (x)^{-1},$$

s étant une section non nulle arbitraire de $\mathscr{L}|_x \simeq F$.

Si s est une section globale non nulle de $\mathscr{L}$, on définit une *fonction hauteur* (exponentielle) *sur les points adéliques de* X en posant

$$H(\overline{\mathscr{L}}, \mathsf{s}; \cdot) : X(\mathbf{A}_F) \setminus |\mathrm{div}(s)| \to \mathbf{R}_+, \quad \mathbf{x} = (x_v)_v \mapsto \prod_v \|\mathsf{s}\|_v (x_v)^{-1}.$$

(Dans les deux cas, le produit converge en effet car il n'y a qu'un nombre fini de termes différents de 1.) D'autre part, elle est multiplicative en le fibré en droites (resp. en la section), ce qui permettra de l'étendre aux groupes de Picard tensorisés par $\mathbf{C}$.

Comme on a un isomorphisme canonique $\overline{\mathrm{Pic}}(\mathrm{Spec}\, F) = \widehat{\mathrm{Pic}}(\mathrm{Spec}\, \mathfrak{o}_F)$, on remarque que

$$H(\overline{\mathscr{L}}; x) = \exp(\widehat{\deg}\, \overline{\mathscr{L}}|_x)$$

où $\widehat{\deg} : \widehat{\mathrm{Pic}}(\mathrm{Spec}\, \mathfrak{o}_F) \to \mathbf{R}$ est l'homomorphisme « degré arithmétique » défini dans [14], 3.4.3, p. 131. Par l'isomorphisme de *loc. cit.*,

$$\widehat{\mathrm{Pic}}(\mathrm{Spec}\, \mathfrak{o}_F) \to F^\times \backslash \mathbf{A}_F^\times / K,$$

$\exp \circ \widehat{\deg}$ correspond à l'inverse de la norme.

Définition 1.3.4. — Soit X une variété sur F. Si $U \subset X$ est un ouvert de Zariski, la fonction zêta des hauteurs de U est la fonction sur $\overline{\mathrm{Pic}}(X)_\mathbf{C}$ (le groupe

abélien des fibrés inversibles sur X munis d'une métrique adélique tensorisé par $\mathbf{C}$) à valeurs dans $\mathbf{C}$ qui associe à $\overline{\mathscr{L}}$ la somme,
$$Z_U(\overline{\mathscr{L}}) = \sum_{x \in U(F)} H(\overline{\mathscr{L}}; x)^{-1},$$
quand elle existe.

Remarque 1.3.5. — La convergence absolue de la série ne dépend que de la partie réelle de $\overline{\mathscr{L}}$ dans $\operatorname{Pic}(X)_{\mathbf{R}}$ (on peut comparer deux métriques adéliques). De plus, l'ensemble des $\mathscr{L} \in \operatorname{Pic}(X)_{\mathbf{R}}$ pour lesquels la série converge est une partie convexe (inégalité arithmético-géométrique). Enfin, si $\mathscr{L}$ est ample, alors $Z_U(s\overline{\mathscr{L}})$ converge pour $\Re(s)$ assez grand et définit une fonction analytique de s, notée $Z_U(\overline{\mathscr{L}}, s)$ dans l'introduction.

Les considérations analogues sont évidemment valables pour le groupe de Picard–Arakelov $\widehat{\operatorname{Pic}}(\mathscr{X})$ d'un modèle propre et plat $\mathscr{X}$ de X sur $\operatorname{Spec} \mathfrak{o}_F$.

Exemple 1.3.6. — Lorsque X est une variété torique, $\operatorname{Pic}(X)_{\mathbf{R}}$ est un espace vectoriel de dimension finie et il y a des métriques canoniques sur les fibrés en droites sur X (cf. [3]), d'où un homomorphisme canonique $\operatorname{Pic}(X)_{\mathbf{C}} \to \overline{\operatorname{Pic}}(X)_{\mathbf{C}}$. Batyrev et Tschinkel ont montré dans [5] que la série définissant la fonction zêta des hauteurs du tore converge dès que $\mathscr{L} \otimes \omega_X$ est dans l'intérieur du cône effectif $\Lambda_{\mathrm{eff}}^{\circ}(X) \subset \operatorname{Pic}(X)_{\mathbf{R}}$, le fibré en droites $\mathscr{L}$ étant muni de sa métrique adélique canonique. Elle définit même une fonction holomorphe dans le tube sur ce cône.

1.4. Fonctions L d'Arakelov

On se place dans la situation 1.1.2. Soient $\mathscr{B}$ un schéma propre et fidèlement plat sur $S = \operatorname{Spec} \mathfrak{o}_F$ et $\widehat{\mathscr{T}}$ un $(G, \mathbf{K}_\infty)$-torseur arithmétique sur $\mathscr{B}$.

Pour tout $b \in \mathscr{B}(F)$, il existe une unique section $\varepsilon_b : \operatorname{Spec} \mathfrak{o}_F \to \mathscr{B}$ qui prolonge b. On dispose ainsi d'un $(G, \mathbf{K}_\infty)$-torseur arithmétique $\varepsilon_b^* \widehat{\mathscr{T}}$ sur $\operatorname{Spec} \mathfrak{o}_F$ que l'on notera $\widehat{\mathscr{T}}|_b$. En particulier, si Φ est une fonction à valeurs complexes sur
$$G(F) \backslash G(\mathbf{A}_F) / \mathbf{K}_G \simeq \widehat{\mathrm{H}}^1(\operatorname{Spec} \mathfrak{o}_F, (G, \mathbf{K}_\infty)),$$
la composition
$$\widehat{\mathrm{H}}^1(\mathscr{B}, (G, \mathbf{K}_\infty)) \xrightarrow{\varepsilon_b} \widehat{\mathrm{H}}^1(\operatorname{Spec} \mathfrak{o}_F, (G, \mathbf{K}_\infty)) \xrightarrow{\Phi} \mathbf{C}$$
définit un nombre complexe $\Phi(\widehat{\mathscr{T}}|_b)$.

Définition 1.4.1. — Soient Φ une fonction sur $G(F)\backslash G(\mathbf{A}_F)/\mathbf{K}_G$ et U une partie de $\mathscr{B}(F)$. On appelle *fonction L d'Arakelov* l'expression
$$L(\widehat{\mathscr{T}},U,\Phi) = \sum_{b\in U\subset \mathscr{B}(F)} \Phi(\widehat{\mathscr{T}}|_b),$$
quand la série converge (absolument).

1.4.2. Exemple. — Soit $\widehat{\mathscr{L}} \in \widehat{\mathrm{Pic}}(\mathscr{B})$ identifié au $\mathbf{G}_m$-torseur arithmétique qu'il définit. Si U est l'ensemble des points rationnels d'un ouvert de $\mathscr{B}$, la fonction L d'Arakelov $L(\widehat{\mathscr{L}},U,\|\cdot\|^s)$ définie au § 1.4 ($\|\cdot\|$ désigne la norme adélique) n'est autre que la fonction zêta d'Arakelov $Z_U(\mathscr{L},s)$, introduite par Arakelov et largement étudiée depuis.

En revanche, lorsque χ est un quasi-caractère arbitraire de $\widehat{\mathrm{Pic}}(\mathrm{Spec}\,\mathfrak{o}_F)$ (pour la topologie adélique), on obtient un nouvel invariant $L(\widehat{\mathscr{L}},U,\chi)$ dont l'importance apparaîtra dans l'étude de la fonction zêta des hauteurs (usuelle) d'un espace fibré.

On peut trouver une trace de ces fonctions L d'Arakelov dans l'estimation par Schanuel [21] du nombre de points de hauteur donnée dans l'espace projectif sur un corps de nombres F. En effet, Schanuel donne dans *loc. cit.* une estimation du nombre de points $(x_0 : \ldots : x_n) \in \mathbf{P}^n(F)$ de hauteur $\leqslant B$ et tels que l'idéal fractionnaire $(x_0,\ldots,x_n)$ soit dans une classe d'idéaux fixée. Implicitement, cela revient à considérer, χ étant un caractère de Dirichlet sur $\mathrm{Pic}(\mathfrak{o}_F)$, la fonction L d'Arakelov pour la fonction $\chi\,\|\cdot\|^s$. Le fait que chaque classe d'idéaux « contienne » autant de points de hauteur donnée devrait impliquer l'holomorphie de cette fonction en $s = n+1$ si χ n'est pas le caractère trivial.

Remarque 1.4.3. — Bien entendu, on définit de la même façon une fonction L d'Arakelov, $L(\overline{\mathscr{T}},U,\Phi)$ attachée à un torseur adélique $\overline{\mathscr{T}}$ sur $\mathscr{B}$ (sur F) et à une fonction Φ sur $G(F)\backslash G(\mathbf{A}_F)/\mathbf{K}$.

1.4.4. Fonctions θ et ζ. — Dans la suite de cette section, on suppose pour simplifier que $F = \mathbf{Q}$. Un $\mathrm{GL}(d)$-torseur arithmétique $\widehat{E}$ sur $\mathrm{Spec}\,\mathbf{Z}$ (pour le choix du sous-groupe compact maximal $U(d)$) n'est autre qu'un $\mathbf{Z}$-module libre de rang d muni d'une norme euclidienne, auquel on sait attacher (au moins) deux invariants :
$$\theta(\widehat{E},t) = \sum_{e\in \widehat{E}} \exp(-\pi t\,\|e\|^2) \quad \text{et} \quad \zeta(\widehat{E},s) = \sum_{e\in \widehat{E}\setminus\{0\}} \frac{1}{\|e\|^s}.$$

(Ces séries convergent respectivement pour $\Re(t) > 0$ et $\Re(s) > d$.) Comme il est bien connu, la formule de Poisson standard implique l'équation fonctionnelle

$$\theta(\widehat{E}, t) = \frac{1}{t^{d/2} \operatorname{vol}(\widehat{E})} \theta(\widehat{E}^\vee, 1/t)$$

où $\operatorname{vol}(\widehat{E}) = \exp(-\widehat{\deg} \widehat{E})$ est le covolume du réseau $\widehat{E}$ dans $\widehat{E} \otimes_{\mathbf{Z}} \mathbf{R} \simeq \mathbf{R}^d$, $\widehat{E}^\vee$ désigne le réseau dual (muni de la norme euclidienne duale) et où la détermination de $t^{d/2}$ est usuelle pour $t > 0$. Il est aussi bien connu comment utiliser cette équation pour en déduire que la fonction définie par

$$\Lambda(\widehat{E}, s) = \sqrt{\operatorname{vol}(\widehat{E})} \zeta(\widehat{E}, s) \pi^{-s/2} \Gamma(s/2)$$

possède un prolongement méromorphe à $\mathbf{C}$, avec des pôles simples en $s = 0$ et $s = d$ de résidus respectivement $-2\sqrt{\operatorname{vol}(\widehat{E})}$ et $2/\sqrt{\operatorname{vol}(\widehat{E})}$ et vérifie l'équation fonctionnelle

$$\Lambda(\widehat{E}, s) = \Lambda(\widehat{E}^\vee, d - s).$$

Sur un corps de nombres quelconque, il faudrait tenir compte de la différente, comme dans l'article récent de van der Geer et Schoof [**13**]. Selon ces mêmes auteurs, l'invariant $\theta(\widehat{E}, 1)$ mesure l'*effectivité* du fibré vectoriel hermitien $\widehat{E}$. Ils interprètent en particulier l'équation fonctionnelle de la fonction θ comme une formule de Riemann–Roch.

1.4.5. Exemples exotiques de fonctions L. — Soit maintenant $\widehat{\mathscr{E}} \in \widehat{\operatorname{Fib}}_d(\mathscr{B})$. On peut définir des fonctions L d'Arakelov (pour une partie $U \subset \mathscr{B}(F)$ fixée)

$$\Theta(\widehat{\mathscr{E}}, s) = L(\widehat{\mathscr{E}}, U, \theta(\cdot, 1) \operatorname{vol}(\cdot)^s) = \sum_{b \in U \subset \mathscr{B}(F)} \theta(\widehat{\mathscr{E}}|_b, 1) \operatorname{vol}(\widehat{\mathscr{E}}|_b)^s$$

et

$$Z(\widehat{\mathscr{E}}, s) = L(\widehat{\mathscr{E}}, U, \zeta(\cdot, ds) \operatorname{vol}(\cdot)^s) = \sum_{b \in U \subset \mathscr{B}(F)} \zeta(\widehat{\mathscr{E}}|_b, ds) \operatorname{vol}(\widehat{\mathscr{E}}|_b)^s$$

et l'on a les égalités, où chacun des membres converge absolument quand l'autre converge absolument,

$$\Theta(\widehat{\mathscr{E}}, s) = \Theta(\widehat{\mathscr{E}}^\vee, 1 - s) \quad \text{et} \quad Z(\widehat{\mathscr{E}}, s) = Z(\widehat{\mathscr{E}}^\vee, 1 - s).$$

Par exemple, pour $\mathscr{B} = \mathbf{P}^1_{\mathbf{Z}}$ et $\widehat{\mathscr{E}} = \mathscr{O}_{\mathbf{P}}(1)$ avec la métrique « max. des coordonnées », on a

$$\Theta(\widehat{\mathscr{E}}, s) = \sum_{N \geqslant 1} 2(1 + 2\varphi(N)) \theta(N^2) N^{1-s},$$

expression qui converge pour $\Re(s) > 3$ et dans laquelle θ désigne la fonction thêta de Riemann.

2. Espaces fibrés

2.1. Constructions

SITUATION 2.1.1. — *Soient S un schéma, G un S-schéma en groupes linéaire et plat, dont on suppose pour simplifier les fibres géométriquement connexes $f : \mathscr{X} \to S$ un S-schéma plat (quasi-compact et quasi-séparé), muni d'une action de G/S. Soient aussi $g : \mathscr{B} \to S$ un S-schéma plat ainsi qu'un G-torseur $\mathscr{T} \to \mathscr{B}$ localement trivial pour la topologie de Zariski.*

CONSTRUCTION 2.1.2. — *On définit un S-schéma $\mathscr{Y}$, muni d'un morphisme $\pi : \mathscr{Y} \to \mathscr{B}$ localement isomorphe à $\mathscr{X}$ sur $\mathscr{B}$, par le changement de groupe structural $G \to \mathrm{Aut}_S(\mathscr{X})$.*

En effet, soit $(U_i)_{i \in I}$ un recouvrement ouvert de $\mathscr{B}$ tel qu'il existe une trivialisation $\varphi_i : G \times_S U_i \xrightarrow{\sim} \mathscr{T}|_{U_i}$. Si $i, j \in I$, soit $g_{ij} \in \Gamma(U_i \cap U_j, G)$ l'unique section telle que $\varphi_i = g_{ij}\varphi_j$ sur $U_i \cap U_j$. En particulier, les g_{ij} donnent un cocycle dont la classe dans $\mathrm{H}^1(\mathscr{B}, G)$ représente la classe d'isomorphisme du G-torseur $\mathscr{T}$. Posons $\mathscr{Y}_i = \mathscr{X} \times_S U_i$; alors, g_{ij} agit sur $\mathscr{X} \times_S (U_i \cap U_j)$ et induit un isomorphisme
$$\varphi_{ij} : \mathscr{Y}_j|_{U_i \cap U_j} \simeq \mathscr{Y}_i|_{U_i \cap U_j}$$
que l'on utilise pour recoller les $\mathscr{Y}_i$.

On laisse vérifier que $\mathscr{Y}$ est un $\mathscr{B}$-schéma bien défini, c'est-à-dire qu'il ne dépend pas à isomorphisme canonique près du choix des trivialisations locales que l'on a fait.

LEMME 2.1.3. — *On a $\pi_* \mathscr{O}_\mathscr{Y} = g^* f_* \mathscr{O}_\mathscr{X}$.*

Remarque 2.1.4. — Dans certains cas, $\mathscr{Y}$ hérite d'une action d'un sous-groupe de G, notamment quand G est commutatif.

CONSTRUCTION 2.1.5. — *Il résulte de la construction précédente une application*
$$\vartheta : \mathrm{Z}^{d,G}(\mathscr{X}) \to \mathrm{Z}^d(\mathscr{Y})$$
des cycles G-invariants de codimension d sur $\mathscr{X}$ dans les cycles de codimension d sur $\mathscr{Y}$.

Définition 2.1.6. — Une *G-linéarisation* d'un faisceau quasi-cohérent $\mathscr{F}$ sur $\mathscr{X}$ est une action de G sur $\mathbf{V}(\mathscr{F})$ qui relève l'action de G sur $\mathscr{X}$.

Un morphisme (resp. le produit tensoriel, le dual, la somme directe, le faisceau des homomorphismes, des extensions, etc.) de faisceaux quasi-cohérents G-linéarisés est défini naturellement. On note $\operatorname{QCoh}^G(\mathscr{X})$ (resp. $\operatorname{Fib}_d^G(\mathscr{X})$, resp. $\operatorname{Pic}^G(\mathscr{X})$) la catégorie des faisceaux quasi-cohérents (resp. de fibrés vectoriels de rang d, resp. des classes d'isomorphisme de fibrés inversibles) G-linéarisés sur $\mathscr{X}$.

CONSTRUCTION 2.1.7. — *On construit un foncteur*
$$\vartheta : \operatorname{QCoh}^G(\mathscr{X}) \to \operatorname{QCoh}(\mathscr{Y})$$
qui est compatible avec les opérations standard sur les faisceaux quasi-cohérents.

Soit $\mathscr{F}$ un faisceau quasi-cohérent G-linéarisé sur $\mathscr{X}$. Reprenons les notations de la construction 2.1.2 de $\mathscr{Y}$. Posons $\mathscr{F}_i$ le faisceau quasi-cohérent sur $\mathscr{Y}_i = \mathscr{X} \times_S U_i$ image réciproque de $\mathscr{F}$ par la première projection. Grâce à la G-linéarisation sur $\mathscr{F}$, les g_{ij} induisent des isomorphismes
$$\varphi_{ij}^* \mathscr{F}_j \big|_{\mathscr{X} \times (U_i \cap U_j)} \simeq \mathscr{F}_i \big|_{\mathscr{X} \times (U_i \cap U_j)}$$
qui fournissent par recollement un faisceau quasi-cohérent sur $\mathscr{Y}$.

On laisse vérifier que ce foncteur est bien défini, c'est-à-dire, est indépendant des choix que l'on a fait.

Si $\mathscr{F}$ est un fibré vectoriel G-linéarisé de rang d sur $\mathscr{X}$, il est clair que le faisceau obtenu sur $\mathscr{Y}$ est aussi un fibré vectoriel de rang d.

On laisse vérifier que cette application est compatible aux opérations standard, et en particulier qu'elle descend en une application sur les classes d'isomorphisme.

Un cas particulier des constructions précédentes est obtenu lorsque $\mathscr{X} = S$, auquel cas $\mathscr{Y} = \mathscr{B}$. On notera $\eta_{\mathscr{T}}$ l'application qui en résulte des faisceaux quasi-cohérents sur S avec action de G/S vers les faisceaux quasi-cohérents sur $\mathscr{B}$. Bien sûr, $\eta_{\mathscr{T}} : \operatorname{Rep}_d(G) \to \operatorname{Fib}_d(\mathscr{B})$ n'est autre que l'application usuelle de changement de groupe structural (passage d'un G-torseur à un $\operatorname{GL}(d)$-torseur).

PROPOSITION 2.1.8. — *Le faisceau $\Omega^1_{\mathscr{X}/S}$ est muni d'une linéarisation canonique de G. Par la construction 2.1.7, on obtient le faisceau $\Omega^1_{\mathscr{X}/\mathscr{B}}$.*

Supposons en particulier que $\mathscr{X}$ et $\mathscr{B}$ sont lisses sur S ; le faisceau canonique sur $\mathscr{X}/S$ est alors automatiquement G-linéarisé et on a un isomorphisme

$$\omega_{\mathscr{Y}/S} \simeq \vartheta(\omega_{\mathscr{X}/S}) \otimes \pi^*\omega_{\mathscr{B}/S}.$$

Démonstration. — Si (U_i) est un recouvrement ouvert de $\mathscr{B}$ avec des isomorphismes $(\varphi_i, \pi) : \pi^{-1}(U_i) \simeq \mathscr{X} \times_S U_i$ comme dans la construction 2.1.2, on a un isomorphisme naturel

$$\Omega^1_{\mathscr{Y}/\mathscr{B}}|_{\pi^{-1}(U_i)} = \Omega^1_{\pi^{-1}(U_i)/U_i} \simeq \varphi_i^* \Omega^1_{\mathscr{X}/S}$$

qui se recollent précisément comme dans la construction 2.1.7.

Dans le cas où $\mathscr{X}/S$ et $\mathscr{B}/S$ sont lisses, la suite exacte

$$0 \to \Omega^1_{\mathscr{Y}/\mathscr{B}} \to \Omega^1_{\mathscr{Y}/S} \to \Omega^1_{\mathscr{B}/S} \to 0$$

implique que

$$\omega_{\mathscr{Y}/S} \simeq \det \Omega^1_{\mathscr{Y}/\mathscr{B}} \otimes \pi^*\omega_{\mathscr{B}/S} \simeq \vartheta(\omega_{\mathscr{X}/\mathscr{B}}) \otimes \pi^*\omega_{\mathscr{B}/S}. \quad \square$$

LEMME 2.1.9. — *Si $\mathscr{F} \in \mathrm{QCoh}^G(\mathscr{X})$, $f_*\mathscr{F}$ est muni d'une action naturelle de G et $\pi_*\vartheta(\mathscr{F})$ est canoniquement isomorphe à $\eta_{\mathscr{T}}(f_*\mathscr{F})$.*

Démonstration. — Laissée au lecteur. $\quad\square$

PROPOSITION 2.1.10. — *Soient $(\lambda, \alpha) \in \mathrm{Pic}^G(\mathscr{X}) \times \mathrm{Pic}(\mathscr{B})$. Le fibré en droites $\vartheta(\lambda) \otimes \pi^*\alpha$ sur $\mathscr{Y}$ est effectif si et seulement si le fibré vectoriel sur $\mathscr{B}$*

$$\eta_{\mathscr{T}}(f_*\lambda) \otimes \alpha$$

est effectif. Cela implique que λ est effectif.

Démonstration. — On a

$$\pi_*(\vartheta(\lambda) \otimes \pi^*\alpha) = \pi_*(\vartheta(\lambda)) \otimes \alpha = \eta_{\mathscr{T}}(f_*\lambda) \otimes \alpha$$

d'après le lemme 2.1.9. $\quad\square$

Notons ι le morphisme de groupes naturel $X^*(G) \to \mathrm{Pic}^G(\mathscr{X})$ qui associe à un caractère χ le fibré trivial muni de la linéarisation telle que G agit par χ sur le second facteur de $\mathscr{X} \times_S \mathbf{A}^1_S$.

PROPOSITION 2.1.11. — *Pour tout caractère χ, il existe un isomorphisme canonique de faisceaux inversibles*

$$\vartheta(\iota(\chi)) \simeq \pi^*\eta_{\mathscr{T}}(\chi).$$

Démonstration. — Soit (U_i) un recouvrement ouvert de $\mathscr{B}$ avec des isomorphismes $(\varphi_i, \pi) : \pi^{-1}(U_i) \simeq \mathscr{X} \times_S U_i$; notons $g_{ij} \in G(U_i \cap U_J)$ tel que $\varphi_i = g_{ij} \cdot \varphi_j : \pi^{-1}(U_i \cap U_j) \to \mathscr{X}$. Alors, le fibré en droites $\vartheta(\iota(\chi))$ est obtenu en recollant $\mathbf{A}^1 \times \mathscr{X} \times U_i$ et $\mathbf{A}^1 \times \mathscr{X} \times U_j$ par le morphisme $(t, x, u) \mapsto (\chi(g_{ij})t, g_{ij} \cdot x, u)$.

D'autre part, $\eta_\mathscr{T}(\chi)$ est un fibré en droite sur $\mathscr{B}$ obtenu en recollant $\mathbf{A}^1 \times U_i$ et $\mathbf{A}^1 \times U_j$ par $(t, u) \mapsto (\chi(g_{ij})t, u)$. □

2.2. Groupe de Picard

Dans ce paragraphe, on suppose que S est le spectre d'un corps F de caractéristique 0. On cherche à exprimer le groupe de Picard de $\mathscr{Y}$ en fonction de ceux de $\mathscr{X}$ et $\mathscr{B}$. Pour cela, on se place sous les hypothèses suivantes :

2.2.1. Hypothèses sur $\mathscr{X}$. — On suppose que

(1) $\mathscr{X}$ est propre, lisse, géométriquement intègre ;
(2) $\mathrm{H}^1(\mathscr{X}, \mathscr{O}_\mathscr{X}) = 0$;
(3) $\mathscr{X}(F)$ est non vide ;
(4) tout fibré en droites sur $\mathscr{X}$ est G-linéarisable, et de même après toute extension algébrique de F ;
(5) $\mathrm{Pic}(\mathscr{X}_{\overline{F}})$ est sans torsion.

Remarque 2.2.2. — Ces hypothèses concernant $\mathscr{X}$ sont vérifiées lorsque $\mathscr{X}$ est une variété torique projective déployée sur F, ou bien un espace de drapeaux généralisé pour un groupe algébrique déployé sur F.

Elles entraînent que les groupes de Picard et de Néron-Séveri de $\mathscr{X}_{\overline{F}}$ coïncident (voir la preuve du lemme 2.2.3 plus bas). En particulier, $\mathrm{Pic}(\mathscr{X}_{\overline{F}})$ est sous ces hypothèses un **Z**-module libre de rang fini.

D'autre part, il est prouvé dans [15], Cor. 1.6, p. 35, que sous l'hypothèse (i), tout fibré en droites sur $\mathscr{X}$ admet une puissance G-linéarisable. (Rappelons que G est connexe.) Le lecteur qui désirerait s'affranchir de cette hypothèse vérifiera que de nombreux résultats de la suite de ce texte restent vrais, au moins après tensorisation par **Q**.

LEMME 2.2.3. — *Si les hypothèses 2.2.1 sont satisfaites, on a les deux assertions :*

— $\mathrm{H}^0(\mathscr{X}, \mathscr{O}_\mathscr{X}) = F$;

— *pour tout F-schéma connexe U possédant un point F-rationnel, l'homomorphisme naturel*
$$\mathrm{Pic}(\mathscr{X}) \times \mathrm{Pic}(U) \to \mathrm{Pic}(\mathscr{X} \times_F U)$$
est un isomorphisme.

Démonstration. — La première proposition découle de la factorisation de Stein. Pour la seconde, on a d'après [**8**, 8.1/4] une suite exacte
$$0 \to \mathrm{Pic}(U) \to \mathrm{Pic}(\mathscr{X} \times_F U) \to \mathrm{Pic}_{\mathscr{X}/F}(U) \to 0.$$
En particulier, $\mathrm{Pic}(\mathscr{X}) = \mathrm{Pic}_{\mathscr{X}/F}(F)$. La nullité de $\mathrm{H}^1(\mathscr{X}, \mathscr{O}_{\mathscr{X}})$ implique que $\mathrm{Pic}_{\mathscr{X}/F}$ est de dimension 0, donc que sa composante neutre $\mathrm{Pic}^0_{\mathscr{X}/F} = 0$ puisque F est de caractéristique nulle. Ainsi, $\mathrm{Pic}_{\mathscr{X}/F}$ est discret. Alors, tout point u de $U(F)$ définit un homomorphisme $u^* : \mathrm{Pic}_{\mathscr{X}/F}(U) \to \mathrm{Pic}_{\mathscr{X}/F}(F)$ qui par connexité est nécessairement l'inverse de l'homomorphisme naturel $\mathrm{Pic}_{\mathscr{X}/F}(F) \to \mathrm{Pic}_{\mathscr{X}/F}(U)$. $\square$

THÉORÈME 2.2.4. — *Si ι désigne le morphisme de groupes $X^*(G) \to \mathrm{Pic}^G(\mathscr{X})$ introduit au paragraphe précédent, considérons l'homomorphisme*
$$\mathrm{Pic}^G(\mathscr{X}) \oplus \mathrm{Pic}(\mathscr{B}) \to \mathrm{Pic}(\mathscr{Y}), \quad (\lambda, \alpha) \mapsto \vartheta(\lambda) \otimes \pi^* \alpha.$$
Si les hypothèses 2.2.1 sont satisfaites et si $\mathscr{B}(F)$ est Zariski-dense dans $\mathscr{B}$, alors la suite
$$0 \to X^*(G) \xrightarrow{(\iota, -\eta_{\vartheta})} \mathrm{Pic}^G(\mathscr{X}) \oplus \mathrm{Pic}(\mathscr{B}) \xrightarrow{\vartheta \otimes \pi^*} \mathrm{Pic}(\mathscr{Y}) \to 0$$
est exacte.

Démonstration. — Si $\iota(\chi)$ est trivial dans $\mathrm{Pic}^G(\mathscr{X})$, il résulte de ce que
$$\mathrm{H}^0(\mathscr{X}, \mathscr{O}_{\mathscr{X}}) = F$$
que χ est nécessairement le caractère trivial. En particulier, le premier homomorphisme est injectif.

La proposition 2.1.11 implique que la composition des deux premiers homomorphismes est nulle.

Si λ est un fibré en droites G-linéarisé sur $\mathscr{X}$ et α est un fibré en droites sur $\mathscr{B}$, $\vartheta(\lambda) \otimes \pi^* \alpha$ est un fibré en droites sur $\mathscr{Y}$ dont la classe d'isomorphisme ne dépend que des classes d'isomorphismes de λ dans $\mathrm{Pic}^G(\mathscr{X})$ et α dans $\mathrm{Pic}(\mathscr{B})$.

Supposons qu'elle soit triviale. Soit b un point F-rationnel de $\mathscr{B}$. En restreignant $\vartheta(\lambda) \otimes \pi^* \alpha$ à $\pi^{-1}(b)$, la construction 2.1.5 de $\vartheta(\lambda)$ implique que λ est trivial. La G-linéarisation de λ est ainsi donnée par un caractère χ de G

et $\lambda = \iota(\chi)$. D'après la proposition 2.1.11, on a $\vartheta(\lambda) = \pi^*\eta_{\mathscr{T}}(\chi)$. Par suite, $\pi^*\alpha \simeq \pi^*\eta_{\mathscr{T}}(\chi)^{-1}$, ce qui prouve l'exactitude au milieu.

Montrons alors que la dernière flèche est surjective. Soit $\mathscr{L}$ un fibré en droites sur $\mathscr{Y}$. On peut recouvrir $\mathscr{B}$ par des ouverts connexes non vides U_i assez petits de sorte que
$$\pi^{-1}(U_i) \simeq \mathscr{X} \times_F U_i.$$
La restriction de $\mathscr{L}$ à $\pi^{-1}(U_i)$ fournit alors pour tout i un élément de
$$\mathrm{Pic}(\mathscr{X} \times_F U_i) = \mathrm{Pic}(\mathscr{X}) \times \mathrm{Pic}(U_i)$$
puisque chaque U_i a un point F-rationnel. On en déduit d'abord pour tout i un élément de $\mathrm{Pic}(\mathscr{X})$ qui, comme on le voit en les restreignant à $U_i \cap U_j$, ne dépend pas de i. Notons le λ. Finalement, il existe un faisceau inversible $\alpha_i \in \mathrm{Pic}(U_i)$ tel que la restriction de $\mathscr{L}$ à $\pi^{-1}(U_i) \simeq \mathscr{X} \times_F U_i$ est isomorphe à $p_1^*\lambda \otimes p_2^*\alpha_i$. Quitte à raffiner le recouvrement (U_i), on peut de plus supposer que $\alpha_i \simeq \mathscr{O}_{U_i}$.

Choisissons une G-linéarisation sur λ. On constate que la restriction de $\mathscr{L} \otimes \vartheta(\lambda)^{-1}$ à $\pi^{-1}(U_i)$ est triviale. Si l'on choisit des trivialisations on obtient en les comparant sur $\pi^{-1}(U_i \cap U_j)$ un élément de
$$\Gamma(\pi^{-1}(U_i \cap U_j), \mathscr{O}_{\mathscr{Y}}^\times) = \Gamma(U_i \cap U_j, \mathscr{O}_{\mathscr{B}}^\times)$$
car $\mathrm{H}^0(\mathscr{X}, \mathscr{O}_{\mathscr{X}}) = F$. Ces éléments définissent un 2-cocycle de Čech sur $\mathscr{B}$ à valeurs dans le faisceau $\mathscr{O}_{\mathscr{B}}^\times$, d'où un fibré en droites $\alpha \in \mathrm{Pic}(\mathscr{B})$ tel que
$$\mathscr{L} \otimes \vartheta(\lambda)^{-1} \simeq \pi^*\alpha.$$
Autrement dit, $\mathscr{L}$ appartient à l'image de l'homomorphisme $\vartheta \otimes \pi^*$.

Le théorème est ainsi démontré. □

COROLLAIRE 2.2.5. — *Supposons vérifiées les hypothèses 2.2.1 et supposons que $\mathscr{B}(F)$ est Zariski-dense dans $\mathscr{B}$. On dispose alors de suites exactes de $\mathbf{Z}[\mathrm{Gal}(\overline{F}/F)]$-modules :*

(2.2.6) $\qquad 0 \to X^*(G_{\overline{F}}) \to \mathrm{Pic}^G(\mathscr{X}_{\overline{F}}) \to \mathrm{Pic}(\mathscr{X}_{\overline{F}}) \to 0$

(2.2.7) $\qquad 0 \to X^*(G_{\overline{F}}) \to \mathrm{Pic}^G(\mathscr{X}_{\overline{F}}) \oplus \mathrm{Pic}(\mathscr{B}_{\overline{F}}) \to \mathrm{Pic}(\mathscr{Y}_{\overline{F}}) \to 0$

(2.2.8) $\qquad 0 \to \mathrm{Pic}(\mathscr{B}_{\overline{F}}) \xrightarrow{\pi^*} \mathrm{Pic}(\mathscr{Y}_{\overline{F}}) \to \mathrm{Pic}(\mathscr{X}_{\overline{F}}) \to 0.$

Démonstration. — Il suffit d'appliquer le théorème 2.2.4 sur $\overline{F}$, et de constater que la suite exacte obtenue est $\mathrm{Gal}(\overline{F}/F)$-équivariante. □

Théorème 2.2.9. — *Supposons vérifiées les hypothèses 2.2.1, que $\mathscr{B}(F)$ est Zariski-dense dans $\mathscr{B}$, et supposons de plus que G est un groupe algébrique F-résoluble*[(2)], *un fibré en droites sur $\mathscr{Y}$ est alors effectif si et seulement s'il s'écrit comme l'image d'un couple $(\lambda, \alpha) \in \mathrm{Pic}^G(\mathscr{X}) \times \mathrm{Pic}(\mathscr{B})$ où λ et α sont effectifs.*

Démonstration. — Soient $\lambda \in \mathrm{Pic}^G(\mathscr{X})$ et $\alpha \in \mathrm{Pic}(\mathscr{B})$ effectifs. On veut montrer que $\vartheta(\lambda) \otimes \pi^*\alpha$ est effectif. Il suffit de prouver que $\vartheta(\lambda)$ est effectif, et pour cela, il suffit de prouver qu'il existe un diviseur de Cartier G-invariant D sur $\mathscr{X}$ tel que l'on ait un isomorphisme de fibrés en droites G-linéarisés, $\lambda \simeq \mathscr{O}(D)$. Autrement dit, il faut montrer que la représentation de G sur $f_*\lambda$ admet une F-droite stable, ce qu'implique le théorème de point fixe de Borel puisque G est F-résoluble.

Soit maintenant $\mathscr{L}$ un fibré en droites effectif sur $\mathscr{Y}$. Comme G est connexe et $\mathrm{Pic}(G) = 0$, la démonstration de la proposition 1.5, p. 34, de [**15**] implique que tout fibré inversible sur $\mathscr{X}$ est G-linéarisable. Le théorème 2.2.4 implique donc qu'il existe $\lambda \in \mathrm{Pic}^G(\mathscr{X})$ et $\alpha \in \mathrm{Pic}(\mathscr{B})$ tels que $\mathscr{L} = \vartheta(\lambda) \otimes \pi^*\alpha$. D'après la proposition 2.1.10, $\eta_{\mathscr{T}}(f_*\lambda) \otimes \alpha$ est effectif. Comme G est F-résoluble, toute représentation linéaire de G est extension successive de représentations de dimension 1. Cela implique que $\eta_{\mathscr{T}}(f_*\lambda)$ est extension successive de fibrés en droites ; notons les λ_i. Alors, $\eta_{\mathscr{T}}(f_*\lambda) \otimes \alpha$ est extension des $\lambda_i \otimes \alpha$, et l'effectivité de $\mathscr{L}$ implique que l'un au moins des $\lambda_i \otimes \alpha$ est effectif.

Or, λ_i est associé à un caractère χ_i de G; si on remplace λ par le fibré en droite G-linéarisé $\lambda \otimes \iota(\chi_i)^{-1}$ où l'action a été divisée par χ_i, on représente ainsi $\mathscr{L}$ sous la forme

$$\mathscr{L} \simeq \vartheta(\lambda \otimes \iota(\chi_i)^{-1}) \otimes (\lambda_i \otimes \alpha),$$

ce qui conclut la démonstration, $\lambda \otimes \iota(\chi_i)^{-1}$ étant isomorphe à λ comme fibré en droites, donc effectif. □

2.3. Métriques hermitiennes

Dans ce paragraphe, nous étendons la construction 2.1.7 en supposant que S est le spectre d'un corps de nombres et en faisant intervenir des métriques hermitiennes.

[(2)] Cela signifie que G est extension itérée de $\mathbf{G}_m$ et $\mathbf{G}_a$, autrement dit, que G est résoluble et déployé sur F.

Définition 2.3.1. — Soit G un groupe de Lie connexe sur $\mathbf{C}$; fixons un sous-groupe compact maximal K de G. Soit X une variété analytique complexe munie d'une action de G.

Si $\mathscr{E}$ est un fibré vectoriel complexe G-linéarisé sur X, on dit qu'une métrique hermitienne est K-*invariante* si l'action de K sur $\mathbf{V}(\mathscr{E}) \times X$ est isométrique.

On remarquera que les constructions usuelles (tensorielles) de fibrés hermitiens préservent la K-invariance des métriques hermitiennes.

Remarque 2.3.2. — Avec les notations de la définition précédente, tout fibré vectoriel sur X admet une métrique hermitienne K-invariante : si $\|\cdot\|_0$ est une métrique hermitienne sur $\mathscr{E}$, on peut en effet choisir une mesure de Haar sur K et poser pour toute section $\mathbf{s}$,
$$\|\mathbf{s}\|^2(x) = \int_K \|k \cdot \mathbf{s}\|(x)^2\, dk.$$

Rappelons l'énoncé de la situation 1.1.2 :

SITUATION. — *Supposons que S est le spectre de l'anneau des entiers d'un corps de nombres F et que G est un S-schéma en groupes linéaire connexe. Fixons pour tout plongement complexe de F $\sigma \in S(\mathbf{C})$ un sous-groupe compact maximal K_σ de $G(\mathbf{C})$ et notons $\mathbf{K}_\infty$ la collection $(K_\sigma)_\sigma$.*

Définition 2.3.3. — Supposons que G agit sur un S-schéma plat $\mathscr{X}$. On appelle *fibré vectoriel hermitien $(G, \mathbf{K}_\infty)$-linéarisé* un fibré vectoriel $\mathscr{E}$ sur $\mathscr{X}$ muni d'une G-linéarisation et, pour tout $\sigma \in S(\mathbf{C})$, d'une métrique hermitienne sur le fibré vectoriel $\mathscr{E} \otimes_\sigma \mathbf{C}$ sur $\mathscr{X}(\mathbf{C})$ qui est K_σ-invariante.

On note $\widehat{\mathrm{Fib}}_d^{G,\mathbf{K}_\infty}(\mathscr{X})$ la catégorie des fibrés vectoriels hermitiens $(G, \mathbf{K}_\infty)$-linéarisés de rang d sur $\mathscr{X}$. Si $d = 1$, on notera $\widehat{\mathrm{Pic}}^{G,\mathbf{K}_\infty}(\mathscr{X})$ le groupe des classes d'isomorphisme de fibrés vectoriels hermitiens de rang 1 $(G, \mathbf{K}_\infty)$-linéarisés sur $\mathscr{X}$.

SITUATION 2.3.4. — *Plaçons-nous dans la situation 1.1.2. Soit $f : \mathscr{X} \to S$ un S-schéma plat, muni d'une action de G/S. Soient aussi $g : \mathscr{B} \to S$ un S-schéma plat ainsi qu'un $(G, \mathbf{K})$-torseur arithmétique $\widehat{\mathscr{T}}$ sur $\mathscr{B}$ (voir la définition 1.1.3).*

CONSTRUCTION 2.3.5. — *Le foncteur $\vartheta : \mathrm{Fib}_d^G(\mathscr{X}) \to \mathrm{Fib}_d(\mathscr{Y})$ s'étend en un foncteur*
$$\vartheta : \widehat{\mathrm{Fib}}_d^{G,\mathbf{K}_\infty}(\mathscr{X}) \to \widehat{\mathrm{Fib}}_d(\mathscr{Y})$$

qui est compatible avec les opérations tensorielles standard sur les fibrés vectoriels hermitiens $(G, \mathbf{K}_\infty)$-linéarisés (resp. les fibrés vectoriels hermitiens).

Soit $\mathscr{F}$ un fibré vectoriel hermitien $(G, \mathbf{K}_\infty)$-linéarisé sur $\mathscr{X}$. Soit $\sigma \in S(\mathbf{C})$. De manière analogue à ce qu'on a fait dans la construction 2.1.5, choisissons un recouvrement ouvert (U_i) de $\mathscr{B}_\sigma(\mathbf{C})$ pour la topologie complexe de sorte que la restriction du torseur $\mathscr{T}$ à U_i est triviale et qu'il existe des trivialisations dont les fonctions de transistions associés $g_{ij} \in \Gamma(U_i \cap U_j, G)$ soient à valeurs dans K_σ. Le choix de telles trivialisations induit des isomorphismes

$$\pi^{-1}(U_i) \simeq \mathscr{X}(\mathbf{C}) \times U_i, \qquad \vartheta(\mathscr{F})|_{\pi^{-1}(U_i)} \simeq p_1^* \mathscr{F}.$$

Pour tout i, on a ainsi une métrique hermitienne naturelle sur $\vartheta(\mathscr{F})|_{\pi^{-1}(U_i)}$ par image réciproque de la métrique hermitienne sur $\mathscr{F}$. Comme $g_{ij} \in K_\sigma$ et comme la métrique hermitienne sur $\mathscr{F}$ est K_σ-invariante, les métriques hermitiennes sur $\vartheta(\mathscr{F})|_{U_i \cap U_j}$ induites par U_i et par U_j coïncident, d'où une métrique hermitienne bien définie sur $\vartheta(\mathscr{F})$.

Enfin, la proposition 2.1.11 admet une généralisation avec métriques hermitiennes :

PROPOSITION 2.3.6. — *Pour tout caractère $\chi \in X^*(G)$, l'isomorphisme canonique de la proposition 2.1.11 est une isométrie.*

Démonstration. — Si l'on reproduit la démonstration de la proposition 2.1.11 pour un recouvrement ouvert pour la topologie complexe (les g_{ij} étant donc dans le sous-groupe compact maximal), chacun des fibrés est défini par recollement de la même manière, et les métriques sur ces fibrés sont définies de sorte que cette identification soit une isométrie. Il en résulte que l'isomorphisme de cette proposition, qui consistait en l'application évidente sur les ouverts $\mathscr{X} \times U_i$ est une isométrie. □

2.4. Torsion des métriques adéliques

Plaçons nous alors dans la situation 2.3.4, toujours avec $S = \operatorname{Spec} \mathfrak{o}_F$. Soit $\mathscr{L}$ un fibré en droites hermitien $(G, \mathbf{K}_\infty)$-linéarisé sur $\mathscr{X}$. La restriction de $\mathscr{L}$ à $\mathscr{X}_F$ est ainsi munie d'une métrique adélique naturelle.

PROPOSITION-DÉFINITION 2.4.1. — *Soit $\mathbf{g} = (g_v)_v \in G(\mathbf{A}_F)$. On définit une métrique adélique sur $\mathscr{L}$, appelée métrique adélique tordue par $\mathbf{g}$ en posant*

pour toute place v de F, tout point $x \in \mathscr{X}(F_v)$ et toute section $\mathsf{s} \in \mathscr{L}_x$,
$$\|\mathsf{s}\|'_v(x) = \|g_v \cdot \mathsf{s}\|_v(g_v \cdot x).$$

Démonstration. — Il est clair que pour toute place v, on a défini une métrique v-adique. L'ensemble des places non-archimédiennes v telles que $g_v \in G(\mathfrak{o}_v)$ est par définition de complémentaire fini. Pour ces places, $\|\mathsf{s}\|'_v(x) = \|\mathsf{s}\|_v(x)$ car g_v étant un automorphisme de $\mathscr{L}$ sur $\operatorname{Spec} \mathfrak{o}_v$, la section $g_v \cdot s$ est entière en $g_v \cdot x$ si et seulement si la section s est entière en x. Ainsi, hors d'un nombre fini de places, la nouvelle collection de métriques v-adiques est définie par un modèle entier. Elle définit donc une métrique adélique. □

Remarquons que $G(\mathbf{A}_F)$ n'agit en fait qu'à travers $G(\mathbf{A}_F)/\mathbf{K}_G$.

Exemple 2.4.2. — Soit E un F-espace vectoriel de dimension finie et notons $\mathbf{P}$ l'espace projectif des droites de E. Faisons agir $\operatorname{GL}(E)$ de manière naturelle sur $\mathbf{P}$. Le faisceau $\mathscr{O}_\mathbf{P}(1)$ possède une $\operatorname{GL}(E)$-linéarisation naturelle dès qu'on a remarqué qu'une section de $\mathscr{O}_\mathbf{P}(-1)$ en un point $\mathbf{x} \in \mathbf{P}$ correspond à un point de la droite $D_\mathbf{x}$ définie par $\mathbf{x}$. De manière explicite, l'espace vectoriel des sections globales de $\mathscr{O}(1)$ sur $\mathbf{P}$ s'identifie au dual E^* de E sur lequel la $\operatorname{GL}(E)$-linéarisation sur $\mathscr{O}(1)$ induit la représentation contragrédiente $\varphi \mapsto \varphi \circ g^{-1}$.

Supposons que E est muni d'une métrique adélique. On a alors une métrique adélique sur $\mathscr{O}_\mathbf{P}(1)$ par la formule
$$\|\varphi\|(\mathbf{x}) = \frac{|\varphi(e)|}{\|e\|_v}, \quad \varphi \in E^*, \quad e \in D_\mathbf{x} \setminus \{0\}.$$

Il résulte de la formule du produit que la hauteur exponentielle d'un point $\mathbf{x} \in \mathbf{P}(F)$ est donnée par la formule
$$H(\mathbf{x}) = \prod_v \|e\|_v, \quad e \in D_\mathbf{x} \setminus \{0\}.$$

Soit alors $(g_v)_v \in \operatorname{GL}(E)(\mathbf{A}_F)$. La métrique v-adique tordue par g_v sur $\mathscr{O}_\mathbf{P}(1)$ est ainsi donnée par
$$\|\varphi\|'(\mathbf{x}) = \frac{|\varphi(e)|}{\|g_v \cdot e\|_v}, \quad \varphi \in E^*, \quad e \in D_\mathbf{x} \setminus \{0\}.$$

Autrement dit, la hauteur exponentielle tordue de $\mathbf{x} \in \mathbf{P}(F)$ est définie par l'expression
$$H'(\mathbf{x}) = \prod_v \|g_v \cdot e\|_v, \quad e \in D_\mathbf{x} \setminus \{0\}.$$

Cette formule était donnée comme définition de la hauteur tordue par Roy et Thunder dans [19].

Dans certains cas, on peut comparer la métrique adélique initiale sur $\mathscr{L}$ et la métrique adélique tordue.

PROPOSITION 2.4.3. — *Supposons que* s *est une section globale de $\mathscr{L}$ sur $\mathscr{X}_F$ dont le diviseur est G-invariant. Il existe alors un unique caractère $\chi \in X^*(G)$ F-rationnel (le poids de* s*) tel que pour tout $g \in G$, $g \cdot \mathsf{s} = \chi(g)\mathsf{s}$.*

Soit $\mathbf{g} \in G(\mathbf{A}_F)$, et considérons $\overline{\mathscr{L}'}$ la métrique adélique tordue par $\mathbf{g}$. Si $x \in X(F)$ n'appartient pas au diviseur de s*, on a la formule*

$$H(\overline{\mathscr{L}'};x) = \prod_v |\chi(g_v)|_v^{-1} H(\overline{\mathscr{L}}, \mathsf{s}, \mathbf{g}\cdot x).$$

Démonstration. — Comme le diviseur de s est G-invariant, il existe pour tout $g \in G$ un élément $\chi(g) \neq 0$ tel que $g \cdot \mathsf{s} = \chi(g)\mathsf{s}$. Il est alors clair que $g \mapsto \chi(g)$ définit un caractère F-rationnel (algébrique) de g.

D'autre part, on a pour toute place v de F,

$$\|\mathsf{s}\|'_v(x) = \|g_v \cdot \mathsf{s}\|_v(g_v x) = \|\chi(g_v)\mathsf{s}\|_v(g_v x) = |\chi(g_v)|_v \|\mathsf{s}\|_v(g_v x).$$

La proposition en découle en prenant le produit. □

Remarque 2.4.4. — Bien sûr, dans l'énoncé précédent, il suffit de supposer que la section s est propre pour les éléments g_v. En particulier, si G' est un sous-groupe de G tel que div(s) est invariant par G', on aura une formule du même type pour les métriques adéliques tordue par un élément de $G'(\mathbf{A}_F)$.

Remarque 2.4.5 (Choix des sections). — La formule précédente permet en fait de comparer la restriction à $G(\mathbf{A}_F)\mathscr{X}(F)$ des hauteurs sur les points adéliques associées à deux sections s_1 et s_2 de poids respectivement χ_1 et χ_2. En effet, si $\mathbf{x} = \mathbf{g} \cdot x \in G(\mathbf{A}_F)\mathscr{X}(F)$, on a, $\overline{\mathscr{L}'}$ désignant la métrique adélique tordue par $\mathbf{g}$,

$$H(\overline{\mathscr{L}}, \mathsf{s}_1; \mathbf{x}) = \prod_v |\chi_1(g_v)|_v H(\overline{\mathscr{L}'};x) = \prod_v |\chi_1\chi_2^{-1}(g_v)|_v H(\overline{\mathscr{L}}, \mathsf{s}_2; \mathbf{x}).$$

Appliquée à des sections de même poids χ, cela permet d'étendre les fonctions $H(\overline{\mathscr{L}}, \mathsf{s}; \cdot)$ au complémentaire dans $G(\mathbf{A}_F)\mathscr{X}(F)$ de l'intersection des diviseurs des sections de poids χ.

Remarque 2.4.6. — Lorsque $\mathscr{X}$ est une variété torique, compactification équivariante lisse d'un tore G, tout fibré en droites effectif $\mathscr{L}$ qui est G-linéarisé possède une unique droite F-rationnelle de sections pour lesquelles G agit par le caractère trivial. On peut utiliser cette section pour définir une hauteur sur les points adéliques du complémentaire de son diviseur, donc en particulier sur $G(\mathbf{A}_F)$.

Expliquons maintenant comment la torsion des métriques adéliques intervient dans nos constructions. Nous allons préciser un peu la situation 2.3.4 en faisant désormais l'hypothèse suivante :

SITUATION 2.4.7. — *Nous faisons les hypothèses contenues dans la situation 2.3.4. En particulier, S est le spectre de l'anneau des entiers de corps de nombres F. De plus, supposons que $\mathscr{X}$ et $\mathscr{B}$ sont propres sur S.*

Soit b un point F-rationnel de $\mathscr{B}$. Comme $\mathscr{B}$ est propre sur S, il en résulte une unique section $\varepsilon_b : S \to \mathscr{B}$ qui prolonge b. Toute trivialisation du G_F-torseur $G_F \simeq \mathscr{T}|_b$ sur $\operatorname{Spec} F$ (il en existe car c'est un torseur pour la topologie de Zariski) induit un isomorphisme $\mathscr{X}_F \simeq \mathscr{Y}|_b$. Fixons un tel isomorphisme φ. Si $\lambda \in \operatorname{Pic}^G(\mathscr{X})$, $\varphi^*\vartheta(\lambda)$ est un fibré en droite sur $\mathscr{X}_F$ canoniquement isomorphe à λ. En revanche, les métriques (adéliques) sont en général distinctes.

Soit v une place finie de F, notons $\mathfrak{o}_v$ le complété de l'anneau local de $\mathfrak{o}_F$ en v. Soit $\varepsilon_v : \operatorname{Spec} \mathfrak{o}_v \to \mathscr{B}$ la restriction de ε_b à $\operatorname{Spec} \mathfrak{o}_v$. Alors, $\varepsilon_v^*\mathscr{T}$ est un $G \otimes \mathfrak{o}_v$-torseur sur $\operatorname{Spec} \mathfrak{o}_v$, et est donc trivialisable. Ainsi, $\varepsilon_v^*\mathscr{Y}$ est isomorphe à $\mathscr{X} \otimes \mathfrak{o}_v$. Fixons un isomorphisme φ_v induit par une trivialisation du torseur. Il existe par définition $g_v \in G(F_v)$ tel que

$$\varphi = \varphi_v \circ [g_v], \quad \mathscr{X} \otimes F_v \to \mathscr{Y}|_b \otimes F_v,$$

$[g_v]$ désignant l'automorphisme de $\mathscr{X} \otimes F_v$ défini par g_v. La définition de la métrique v-adique associée à un modèle montre que φ_v est une isométrie. Ainsi, en tant que fibré inversible métrisé sur $\mathscr{X} \otimes F_v$, $\varphi^*(\vartheta(\lambda))$ est isomorphe (isométrique) à $[g_v]^*\lambda$.

Soit maintenant v une place à l'infini. Comme on s'était fixé une trivialisation du $G(\mathbf{C})/K_v$-fibré sur $\mathscr{B}(\mathbf{C})$, on dispose d'un isomorphisme φ_v bien défini modulo K_v qui par définition ne modifie pas les métriques. La comparaison entre φ et φ_v se fait comme précédemment par un élément $g_v \in G(\mathbf{C})$.

Il en résulte le théorème :

THÉORÈME 2.4.8. — *Soit* $\mathbf{g} = (g_v)_v \in G(\mathbf{A}_F)$ *l'élément du groupe adélique que nous venons d'introduire. Il représente la classe de la restriction à b du $(G, \mathbf{K}_\infty)$-torseur arithmétique $\widehat{\mathscr{T}}$ dans l'isomorphisme de la proposition 1.2.6. De plus, la métrique adélique image réciproque sur $\varphi^*\vartheta(\lambda)$ s'identifie à la métrique adélique tordue par $\mathbf{g}$ sur λ.*

2.5. Nombres de Tamagawa

Commençons par rappeler la définition, due à Peyre (cf. [16] et [18]) des nombres de Tamagawa associés à une métrique adélique sur le faisceau anticanonique.

2.5.1. *Hypothèses.* — Soit X une variété propre, lisse et géométriquement intègre sur F telle que $\mathrm{H}^1(X, \mathscr{O}_X) = \mathrm{H}^2(X, \mathscr{O}_X) = 0$ et que $X(F)$ soit Zariski-dense dans X. Sous ces conditions, $\mathrm{Pic}(X_{\overline{F}})_\mathbf{Q}$ est un $\mathbf{Q}$-espace vectoriel de dimension finie.

2.5.2. *Définition.* — Munissons le fibré canonique ω_X d'une métrique adélique. Pour toute place v de F, une construction classique de Weil fournit une mesure $\mu_{X,v}$ sur $X(F_v)$ à partir de la métrique v-adique sur ω_X. Notons $L_v(s, \mathrm{Pic}(X_{\overline{F}}))$ le facteur local en v de la fonction L d'Artin de la représentation de $\mathrm{Gal}(\overline{F}/F)$ sur $\mathrm{Pic}(X_{\overline{F}})_\mathbf{Q}$. Le théorème de Weil sur la mesure de $X(F_v)$ pour $\mu_{X,v}$ et le théorème de Deligne sur les conjectures de Weil concernant le nombre de points rationnels des variétés sur les corps finis ont la conséquence suivante : il existe un ensemble fini Σ de places de F, contenant les places archimédiennes, tel que

$$\prod_{v \in \Sigma} \mu_{X,v} \times \prod_{v \notin \Sigma} \left(L_v^{-1}(1, \mathrm{Pic}(X_{\overline{F}}))\mu_{X,v}\right)$$

définisse une mesure $\mu_{X,\Sigma}$ sur $X(\mathbf{A}_F)$ pour laquelle $X(\mathbf{A}_F)$ a un volume fini.

Soit $L_\Sigma(s, \mathrm{Pic}(X_{\overline{F}})) = \prod_{v \notin \Sigma} L_v(s, \mathrm{Pic}(X_{\overline{F}}))$ la fonction L partielle associée à $\mathrm{Pic}(X_{\overline{F}})$. Le produit eulérien converge en effet pour $\Re(s) > 1$ et L_Σ a un pôle en $s = 1$ d'ordre la dimension t des invariants sous $\mathrm{Gal}(\overline{F}/F)$ de $\mathrm{Pic}(X_{\overline{F}})_\mathbf{Q}$. Notons

$$L_\Sigma^*(1, \mathrm{Pic}(X_{\overline{F}})) = \lim_{s \to 1}(s-1)^r L_\Sigma(s, \mathrm{Pic}(X_{\overline{F}})).$$

On définit alors le nombre de Tamagawa de X (associé à la métrique adélique choisie sur ω_X) par
$$\tau(X) = L_\Sigma^*(s, \operatorname{Pic}(X_{\overline{F}})) \int_{\overline{X(F)}} \mu_{X,\Sigma}.$$
Il est facile de vérifier qu'il ne dépend pas de l'ensemble fini de places Σ choisi.

Nous aurons à utiliser le lemme suivant.

LEMME 2.5.3. — *Supposons réalisées les hypothèses 2.5.1. Soit U un ouvert non vide de X. Notons $\overline{U(F)}$ l'adhérence de $U(F)$ dans $\prod_v U(F_v)$ pour la topologie produit (qui est la topologie induite sur $\prod_v U(F_v)$ par la topologie adélique de $X(\mathbf{A}_F)$). Alors, on a l'égalité*
$$\int_{\overline{U(F)}} \mu_{X,\Sigma} = \int_{\overline{X(F)}} \mu_{X,\Sigma}.$$

Démonstration. — Tout point $x = (x_v) \in \prod_v U(F_v)$ possède par définition un voisinage (pour la topologie induite) contenu dans $\prod_v U(F_v)$. Par suite, si une suite $(x^{(n)})$ de points de $X(F)$ converge vers x, à partir d'un certain rang, $x^{(n)}$ appartient à $U(F_v)$ pour toute place v, et donc $x^{(n)} \in U(F)$. Cela montre que $\overline{U(F)} = \overline{X(F)} \cap \prod_v U(F_v)$. Ainsi, le complémentaire de $\overline{U(F)}$ dans $\overline{X(F)}$ est contenu dans $X(\mathbf{A}_F) \setminus \prod_v U(F_v)$, donc dans la réunion
$$\bigcup_v (X \setminus U)(F_v) \prod_{w \neq v} X(F_w).$$
La définition de la mesure $\mu_{X,v}$ implique que $(X \setminus U)(F_v)$ est de mesure nulle pour $\mu_{X,v}$. On voit donc que $\overline{X(F)} \setminus \overline{U(F)}$ est réunion dénombrable d'ensembles de mesure nulle pour la mesure de Tamagawa sur $X(\mathbf{A}_F)$, donc est de mesure nulle. $\square$

On se place maintenant dans la situation 2.3.4, S étant le spectre $\operatorname{Spec} \mathfrak{o}_F$ de l'anneau des entiers d'un corps de nombres F.

LEMME 2.5.4. — *Si $\mathscr{X}_F$ et $\mathscr{B}_F$ satisfont les hypothèses 2.5.1 nécessaires pour la définition des nombres de Tamagawa, $\mathscr{Y}_F$ les satisfait aussi.*

Démonstration. — Que $\mathscr{Y}_F$ soit lisse, propre et géométrique intègre est clair. D'autre part, les points rationnels de $\mathscr{Y}_F$ sont denses dans chaque fibre au-dessus d'un point rationnel de $\mathscr{B}_F$, lesquels sont supposés denses dans $\mathscr{B}_F$.

Comme $\mathscr{Y}_F \to \mathscr{B}_F$ est propre, un argument élémentaire de platitude puis de dimension implique que les points rationnels de $\mathscr{Y}_F$ sont Zariski-denses.

D'autre part, les hypothèses sur $\mathscr{X}_F$ impliquent que $R^0\pi_*\mathscr{O}_{\mathscr{Y}_F} = \mathscr{O}_{\mathscr{B}_F}$ et que
$$R^1\pi_*\mathscr{O}_{\mathscr{Y}_F} = R^2\pi_*\mathscr{O}_{\mathscr{Y}_F} = 0.$$
La suite spectrale des foncteurs composés implique que $\mathrm{H}^q(\mathscr{O}_{\mathscr{Y}_F})$ est un quotient de $\bigoplus_{i+j=q}\mathrm{H}^i(\mathscr{B}_F, R^j\pi_*\mathscr{O}_{\mathscr{Y}_F})$. Si $j=1$ ou si $j=2$, on a $\mathrm{H}^i(R^j\pi_*\mathscr{O}_{\mathscr{Y}_F}) = 0$ puisque $R^j\pi_*\mathscr{O}_{\mathscr{Y}_F} = 0$. Si $j=0$ et $i \in \{1,2\}$, $\mathrm{H}^i(R^0\pi_*\mathscr{O}_{\mathscr{Y}_F}) = \mathrm{H}^i(\mathscr{O}_{\mathscr{B}_F}) = 0$ en vertu des hypothèses faites sur $\mathscr{B}_F$. $\square$

Supposons donc que $\mathscr{X}_F$ et $\mathscr{B}_F$ satisfont ces hypothèses 2.5.1. Le faisceau canonique sur $\mathscr{Y}$ admet d'après la proposition 2.1.8 une décomposition
$$\omega_{\mathscr{Y}} = \vartheta(\omega_{\mathscr{X}}/S) \otimes \pi^*\omega_{\mathscr{B}/S}.$$

Choisissons une structure de fibré en droite hermitien $(G, \mathbf{K}_\infty)$ linéarisé sur $\omega_{\mathscr{X}/S}$ compatible à la linéarisation canonique sur $\omega_{\mathscr{X}/S}$ (autrement dit, pour toute place archimédienne σ, une métrique hermitienne K_σ-invariante sur $\mathscr{X} \times_\sigma \mathbf{C}$). Choisissons aussi une métrique hermitienne sur $\omega_{\mathscr{B}/S}$. Il en résulte une métrique hermitienne canonique sur $\omega_{\mathscr{Y}/S}$ par la construction 2.3.5. Le fait de disposer d'un modèle sur $\mathfrak{o}_F$ induit de plus des métriques v-adiques au places finies, d'où des métriques adéliques sur $\omega_{\mathscr{X}_F}$, sur $\omega_{\mathscr{B}_F}$ et sur $\omega_{\mathscr{Y}_F}$.

Théorème 2.5.5. — *Muni de ces métriques adéliques, on a l'égalité*
$$\tau(\mathscr{Y}_F) = \tau(\mathscr{X}_F)\tau(\mathscr{B}_F).$$

Démonstration. — Soit U un ouvert de Zariski non vide de $\mathscr{B}_F$ tel que $\mathscr{T}|_U \simeq G \times_S U$. Notons $V = \pi^{-1}(U) \subset \mathscr{Y}_F$, de sorte que V est un ouvert non vide de $\mathscr{Y}|_F$ isomorphe à $\mathscr{X}_F \times U$, et que dans cette décomposition, la mesure

(2.5.6) $$\mu_{\mathscr{Y},v}|_{\pi^{-1}(U)} = \mu_{\mathscr{X},v} \otimes \mu_{\mathscr{B},v}|_U.$$

Pour toute place v de F, il résulte du corollaire au théorème 2.2.4 la relation entre facteurs locaux

(2.5.7) $$L_v(s, \mathrm{Pic}(\mathscr{Y}_{\overline{F}})) = L_v(s, \mathrm{Pic}(\mathscr{X}_{\overline{F}}))L_v(s, \mathrm{Pic}(\mathscr{B}_{\overline{F}})).$$

Alors, les équations (2.5.6) et (2.5.7) impliquent que la restriction de la mesure de Tamagawa de $\mathscr{X}(\mathbf{A}_F)$ à $\prod_v V(F_v)$ s'écrit comme le produit
$$\mu_{\mathscr{Y},\Sigma}|_{\prod_v V(F_v)} = \mu_{\mathscr{X},\Sigma} \otimes \mu_{\mathscr{B},\Sigma}|_{\prod_v U(F_v)}.$$

Or, si $\overline{U(F)}$ est l'adhérence de $U(F)$ dans le produit $\prod_v U(F_v)$, l'adhérence de $V(F)$ dans $\prod_v V(F_v)$ s'identifie à $\overline{\mathscr{X}(F)} \times \overline{U(F)}$. Intégrons $\mu_{\mathscr{Y},\Sigma}$ sur $\overline{V(F)}$; en utilisant le lemme 2.5.3, on obtient

$$\int_{\overline{\mathscr{Y}(F)}} \mu_{\mathscr{Y},\Sigma} = \int_{\overline{\mathscr{X}(F)}} \mu_{\mathscr{X},\Sigma} \times \int_{\overline{\mathscr{B}(F)}} \mu_{\mathscr{B},\Sigma}.$$

L'équation (2.5.7) implique aussi que pour $\Re(s) > 1$,

$$L_\Sigma(s, \mathrm{Pic}(\mathscr{Y}_{\overline{F}})) = L_\Sigma(s, \mathrm{Pic}(\mathscr{X}_{\overline{F}})) L_\Sigma(s, \mathrm{Pic}(\mathscr{B}_{\overline{F}})).$$

Par suite, l'ordre du pôle en $s = 1$ pour la fonction L_Σ de $\mathscr{Y}$ est la somme des ordes des pôles pour $\mathscr{X}$ et $\mathscr{B}$, et donc

$$L_\Sigma^*(1, \mathrm{Pic}(\mathscr{Y}_{\overline{F}})) = L_\Sigma^*(1, \mathrm{Pic}(\mathscr{X}_{\overline{F}})) L_\Sigma^*(1, \mathrm{Pic}(\mathscr{B}_{\overline{F}})).$$

Le théorème est donc démontré. □

2.6. Torseurs trivialisants

Le paragraphe 2.4 a montré que le phénomène de torsion des métriques adéliques intervient naturellement dans nos constructions. Cependant, la hauteur tordue n'est facile à calculer que lorsqu'il existe des sections propres pour l'action du groupe. L'existence de sections canoniques permet comme on l'a vu de disposer d'une fonction hauteur sur les points adéliques.

Les torseurs trivialisants que nous introduisons ici ont pour fonction de fournir — au prix d'un changement de variété — d'une droite canonique de sections.

Dans ce paragraphe, nous nous plaçons sur un corps F. Supposons que $\mathrm{Pic}^G(\mathscr{X}) \simeq \mathrm{Pic}(\mathscr{X}) \times X^*(G)$ est un groupe de type fini.

Soit H un groupe algébrique sur F, $\mathscr{X}_1 \to \mathscr{X}$ un H-torseur qui induise par fonctorialité covariante des torseurs un isomorphisme $X^*(H) \to \mathrm{Pic}(\mathscr{X})$. On suppose de plus que $\mathscr{X}_1$ est muni d'une action de G qui relève l'action de G sur $\mathscr{X}$ et qui commute à l'action de H. On peut construire un tel $\mathscr{X}_1$ en fixant $\lambda_1, \ldots, \lambda_h$ des fibrés inversibles G-linéarisés sur $\mathscr{X}$ dont les classes forment une base de $\mathrm{Pic}(\mathscr{X})$. On pose alors $\mathscr{X}_1 = \prod_{i=1}^h (\mathbf{V}(\lambda_i^\vee) \setminus \{0\})$ et $H = \mathbf{G}_m^h$.

Soit T le plus grand quotient de G tel que l'application naturelle $X^*(T) \to X^*(G)$ est un isomorphisme. (C'est le quotient de G par l'intersection des noyaux des caractères de G). On pose $\widetilde{\mathscr{X}} = \widetilde{\mathscr{X}_1} \times T$ et $\pi : \widetilde{\mathscr{X}} \to \mathscr{X}$ la composition de la première projection de de la projection $\widetilde{\mathscr{X}_1} \to \mathscr{X}$. C'est un $H \times T$-torseur muni d'une action de G (diagonale).

Exemple 2.6.1. — Supposons que $\mathscr{X} = P\backslash G$ est un espace de drapeaux généralisé pour un groupe algébrique simplement connexe semi-simple G sur F. On a $\mathrm{Pic}(\mathscr{X}) \simeq X^*(P)$ et $G \to \mathscr{X}$ est un P-torseur qui induit un isomorphisme $X^*(P) \simeq \mathrm{Pic}(\mathscr{X})$. De plus, $T = \{1\}$. Ainsi, on peut prendre $\widetilde{\mathscr{X}} = G$.

Exemple 2.6.2. — Lorsque le groupe G est trivial, on retrouve les torseurs universels introduits dans le contexte des hauteurs par Salberger et Peyre (cf. [**20**], [**18**]).

FAIT 2.6.3. — *Si $\lambda \in \mathrm{Pic}^G(\mathscr{X})$, $\pi^*\lambda$ admet une F-droite canonique de sections G-invariantes.*

Remarque 2.6.4. — L'isomorphisme canonique
$$\mathrm{Pic}^G(\mathscr{X}) \simeq X^*(H \times T) = X^*(H) \times X^*(G)$$
admet une réciproque qu'il est facile d'expliciter. En effet, soient χ_H et χ_G deux caractères de H et G respectivement. On définit un fibré inversible G-linéarisé sur $\mathscr{X}$ comme suit : on quotiente $\widetilde{\mathscr{X}} \times \mathbf{A}^1 = \widetilde{\mathscr{X}_1} \times T \times \mathbf{A}^1$ par l'action de H donnée par
$$h \cdot (\widetilde{x}, t, u) = (h \cdot \widetilde{x}, t, \chi_H(h)^{-1} u), \quad h \in H, \quad (\widetilde{x}, t, u) \in \widetilde{\mathscr{X}_1} \times T \times \mathbf{A}^1$$
et la G-linéarisation provient de l'action de G sur $\widetilde{\mathscr{X}} \times \mathbf{A}^1$ fournie par
$$(\widetilde{x}, t, u) \cdot g = (g \cdot \widetilde{x}, g \cdot t, \chi_G^{-1}(g) u), \quad g \in G, \quad (\widetilde{x}, t, u) \in \widetilde{\mathscr{X}_1} \times T \times \mathbf{A}^1.$$
Par la construction 2.1.2, on obtient ainsi un F-schéma $\widetilde{\mathscr{Y}}$ avec une projection $\widetilde{\mathscr{Y}} \to \mathscr{Y}$. Supposons que $\mathscr{Y}$ provient de la situation 2.3.4, on dispose en particulier de fibrés inversibles sur $\mathscr{Y}_F$ munis de métriques adéliques associés aux fibrés inversibles $(G, \mathbf{K})$-linéarisés sur $\mathscr{X}$. En particulier, on obtient sur $\widetilde{\mathscr{Y}}$ des fibrés inversibles avec métriques adéliques. Le fait nouveau est que l'on dispose d'une hauteur sur les points adéliques de $\widetilde{\mathscr{Y}}$ associée à ces fibrés inversibles. En effet, une fois remontés à $\widetilde{\mathscr{Y}}$, ces fibrés inversibles possèdent une droite de sections F-rationnelle canonique.

2.7. Exemples

2.7.1. Action d'un tore. — Pour les applications auxquelles notre deuxième article sera consacré, on considère l'action d'un tore T.

Un tel tore peut agir non seulement sur des variétés toriques, mais aussi sur des variétés de drapeaux généralisées $P\backslash G$, via un morphisme $T \to G$.

Dans le cas des variétés toriques sur un corps de nombres F, on dispose de modèles canoniques sur $\operatorname{Spec} \mathfrak{o}_F$ (si le tore est déployé), et de métriques hermitiennes à l'infini canoniques sur les fibrés en droites. Pour tout plongement σ de F dans $\mathbf{C}$, les points complexes $T(\mathbf{C})$ du tore admettent un unique sous-groupe compact maximal K_σ, et les métriques hermitiennes introduites sont automatiquement K_σ-invariantes. On obtient ainsi des fibrés hermitiens $(T, \mathbf{K})$-linéarisés (cf. par exemple [3]).

Dans le cas des variétés de drapeaux $P \backslash G$, une fois fixé des sous-groupes compacts maximaux de G aux places à l'infini, il est aussi possible de munir les fibrés en droites P-linéarisés de métriques hermitiennes invariantes pour ces sous-groupes compacts maximaux et donc pour le sous-groupe compact maximal de $T(\mathbf{C})$. Aux places finies, les métriques v-adiques qu'on obtient admettent une description analogue en termes de la décomposition d'Iwasawa (cf. [12]).

D'autre part, un T-torseur sur un F-schéma $\mathscr{B}$, du moins quand le tore est déployé, à la donnée d'un morphisme $X^*(T) \to \operatorname{Pic}(\mathscr{B})$, et donc, une fois fixé une base de $X^*(T)$, à des fibrés en droites $\lambda_1, \ldots, \lambda_t \in \operatorname{Pic}(\mathscr{B})$. (On a noté $t = \dim T$.) La trivialisation des T/K_σ-torseurs correspond, ainsi qu'on l'a dit après la définition 1.1.3 d'un T-torseur arithmétique, à une métrique hermitienne sur les fibrés en droites λ_i.

Dans le cas où T agit sur une variété torique, on obtient alors par la construction 2.3.4 une famille de variétés toriques sur $\mathscr{B}$. On peut notamment compactifier ainsi une variété semi-abélienne $\mathscr{T} \to \mathscr{B}$ et construire sur la compactification $\mathscr{Y}$ des fonctions hauteurs canoniques. Dans ce cas, les λ_i sont des fibrés en droites algébriquement équivalent à 0 sur une variété abélienne $\mathscr{B}$. Si on a pris soin de les munir, ainsi que tous les fibrés en droites sur $\mathscr{B}$, de leur métrique adélique canonique, pour laquelle le théorème du cube est une isométrie, on obtient sur $\mathscr{Y}$ les hauteurs canoniques, au sens de la hauteur de Néron–Tate. (Dans ce cas particulier, cf. [9] où l'on trouvera cette construction dans un esprit analogue, et [11], où est donnée une construction « à la Tate » de ces hauteurs canoniques, due à M. Waldschmidt).

Dans le cas où T agit sur une variété de drapeaux généralisée, on obtient la variété de drapeaux (généralisée) d'un fibré vectoriel sur $\mathscr{B}$ construit naturellement à partir des λ_i. Ce cas était étudié (lorsque la base est aussi une variété de drapeaux) dans la thèse de M. Strauch ([22]).

2.7.2. Variétés de drapeaux. — Tout fibré vectoriel sur $\mathscr{B}$ donne lieu à des variétés de drapeaux généralisées. Dans ce cas, le groupe G est le groupe linéaire $\mathrm{GL}(d)$, $\mathscr{X}$ est une variété $P\backslash G$. On identifie en effet un fibré vectoriel de rang n sur $\mathscr{B}$ à un $\mathrm{GL}(d)$-torseur. Si l'on choisit comme sous-groupe compact à l'infini le groupe unitaire $\mathrm{U}(d)$, la trivialisation à l'infini du G/K-fibré correspond à une métrique hermitienne sur le fibré vectoriel.

Il est à noter que cette situation se retrouve, mais dans l'autre sens, dans le calcul du comportement de la fonction zêta des hauteurs d'une puissance symétrique d'une courbe $\mathscr{C}$ de genre $g \geqslant 2$. Dans ce cas en effet, si $d > 2g - 2$, $\mathrm{Sym}^d \mathscr{C}$ est un fibré projectif au-dessus de la jacobienne de $\mathscr{C}$ associé à un fibré vectoriel de rang $d + 1 - g$.

2.7.3. Action d'un groupe vectoriel. — Dans [10] et [9], on étudie des compactifications d'extensions vectorielles de variétés abéliennes. Expliquons comment ce travail s'insère dans les constructions de cet article lorsque, pour simplifier les notations, on prend $G = \mathbf{G}_a$.

Un $\mathbf{G}_a$-torseur sur $\mathscr{B}$ correspond à une extension de $\mathscr{O}_{\mathscr{B}}$ par lui-même, soit un fibré vectoriel $\mathscr{E}$ de rang 2 sur $\mathscr{B}$. La trivialisation du $\mathbf{G}_a$-torseur à l'infini correspond à un scindage $\mathscr{C}^\infty$ de l'extension sur $\mathscr{B}(\mathbf{C})$. D'autre part, $\mathbf{G}_a$ agit naturellement sur $\mathbf{P}^1$ (via son plongement dans $\mathrm{GL}(2)$, $a \mapsto \left(\begin{smallmatrix}1 & a \\ 0 & 1\end{smallmatrix}\right)$). On obtient ainsi une compactification du $\mathbf{G}_a$-torseur en une famille de droites projectives sur $\mathscr{B}$.

Références

[1] S. Ju. Arakelov, *Theory of intersections on the arithmetic surface*, Proceedings of the International Congress of Mathematicians (Vancouver, 1974), 1974, 405–408.

[2] V. V. Batyrev & Yu. I. Manin, *Sur le nombre de points rationnels de hauteur bornée des variétés algébriques*, Math. Ann. **286** (1990), 27–43.

[3] V. V. Batyrev & Yu. Tschinkel, *Rational points on bounded height on compactifications of anisotropic tori*, Internat. Math. Res. Notices **12** (1995), 591–635.

[4] _____, *Height zeta functions of toric varieties*, Journal Math. Sciences **82** (1996), no. 1, 3220–3239.

[5] _____, *Manin's conjecture for toric varieties*, J. Algebraic Geometry **7** (1998), no. 1, 15–53.

[6] _____, *Tamagawa numbers of polarized algebraic varieties*, in Nombre et répartition des points de hauteur bornée [**17**], 299–340.

[7] A. Borel, *Sous-groupes compacts maximaux des groupes de Lie (d'après Cartan, Iwasawa et Mostow)*, Séminaire Bourbaki 1950/51, Exp. 33.

[8] S. Bosch, W. Lütkebohmert & M. Raynaud, *Néron models*, Ergeb., no. 21, Springer Verlag, 1990.

[9] A. Chambert-Loir, *Extensions vectorielles, périodes et hauteurs*, Thèse, Univ. P. et M. Curie, Paris, 1995.

[10] _____, *Extension universelle d'une variété abélienne et hauteurs des points de torsion*, Compositio Math. **103** (1996), 243–267.

[11] P. Cohen, *Heights of torsion points on commutative group varieties*, Proc. London Math. Soc. **52** (1986), 427–444.

[12] J. Franke, Yu. I. Manin & Yu. Tschinkel, *Rational points of bounded height on Fano varieties*, Invent. Math. **95** (1989), no. 2, 421–435.

[13] G. van der Geer & R. Schoof, *Effectivity of Arakelov divisors and the Theta divisor of a number field*, Tech. report, math.AG/9802121, 1998.

[14] H. Gillet & C. Soulé, *Arithmetic intersection theory*, Publ. Math. Inst. Hautes Études Sci. **72** (1990), 94–174.

[15] D. Mumford, J. Fogarty & F. Kirwan, *Geometric invariant theory*, Ergeb., no. 34, Springer Verlag, 1994.

[16] E. Peyre, *Hauteurs et mesures de Tamagawa sur les variétés de Fano*, Duke Math. J. **79** (1995), 101–218.

[17] _____ (éd.), *Nombre et répartition des points de hauteur bornée*, Astérisque, no. 251, 1998.

[18] _____, *Terme principal de la fonction zêta des hauteurs et torseurs universels*, in Nombre et répartition des points de hauteur bornée [**17**], 259–298.

[19] D. Roy & J. L. Thunder, *An absolute Siegel's lemma*, J. Reine Angew. Math. **476** (1996), 1–26.

[20] P. Salberger, *Tamagawa measures on universal torsors and points of bounded height on Fano varieties*, in Nombre et répartition des points de hauteur bornée [**17**], 91–258.

[21] S. Schanuel, *Heights in number fields*, Bull. Soc. Math. France **107** (1979), 433–449.

[22] M. Strauch, *Thèse*, Universität Bonn, 1997.

[23] M. Strauch & Yu. Tschinkel, *Height zeta functions of toric bundles over flag varieties*, Selecta Math. (N.S.) **5** (1999), no. 3, 325–396.

Rational points on algebraic varieties
(E. PEYRE, Y. TSCHINKEL, ed.), p. 71–115

FONCTIONS ZÊTA DES HAUTEURS DES ESPACES FIBRÉS

Antoine Chambert-Loir
Institut de mathématiques de Jussieu, Boite 247, 4, place Jussieu, F-75252 Paris Cedex 05 • *E-mail* : chambert@math.jussieu.fr

Yuri Tschinkel
Department of Mathematics, Princeton University, Princeton, NJ, 08544
E-mail : ytschink@math.princeton.edu

Abstract. — In this paper we study the compatibility of Manin's conjectures concerning asymptotics of rational points on algebraic varieties with certain natural geometric constructions. More precisely, we consider locally trivial fibrations constructed from torsors under linear algebraic groups. The main problem is to understand the behaviour of the height function as one passes from fiber to fiber - a difficult problem, even though all fibers are isomorphic. We will be mostly interested in fibrations induced from torsors under split tori. Asymptotic properties follow from analytic properties of height zeta functions. Under reasonable assumptions on the analytic behaviour of the height zeta function for the base we establish analytic properties of the height zeta function of the total space.

Introduction

Cet article est le deuxième d'une série consacrée à l'étude des hauteurs sur certaines variétés algébriques sur un corps de nombres, notamment en ce qui concerne la distribution des points rationnels de hauteur bornée.

Précisément, soient X une variété algébrique projective lisse sur un corps de nombres F, $\mathscr{L}$ un fibré en droites sur X et $H_{\mathscr{L}} : X(\overline{F}) \to \mathbf{R}_+^*$ une fonction

hauteur (exponentielle) pour $\mathscr{L}$. Si U est un ouvert de Zariski de X, on cherche à estimer le nombre
$$N_U(\mathscr{L}, H) = \#\{x \in U(F)\,;\, H_{\mathscr{L}}(x) \leqslant H\}$$
lorsque H tend vers $+\infty$. L'étude de nombreux exemples a montré que l'on peut s'attendre à un équivalent de la forme

(∗) $\quad N_U(\mathscr{L}, H) = \Theta(\mathscr{L}) H^{a(\mathscr{L})} (\log H)^{b(\mathscr{L})-1} (1 + o(1)), \quad H \to +\infty$

pour un ouvert U convenable et lorsque par exemple $\mathscr{L}$ et ω_X^{-1} (fibré anticanonique) sont amples. On a en effet un résultat de ce genre lorsque X est une variété de drapeaux [11], une intersection complète lisse de bas degré (méthode du cercle), une variété torique [4], une variété horosphérique [19], une compactification équivariante d'un groupe vectoriel [10], etc. On dispose de plus d'une description conjecturale assez précise des constantes $a(\mathscr{L})$ et $b(\mathscr{L})$ en termes du cône des diviseurs effectifs [1] ainsi que de la constante $\Theta(\mathscr{L})$ ([15], [5]).

En fait, on étudie plutôt la *fonction zêta des hauteurs*, définie par la série de Dirichlet
$$Z_U(\mathscr{L}, s) = \sum_{x \in U(F)} H_{\mathscr{L}}(x)^{-s}$$
à laquelle on applique des théorèmes taubériens standard. Sur cette série, on peut se poser les questions suivantes : domaine de convergence, prolongement méromorphe, ordre du premier pôle, terme principal, sans oublier la croissance dans les bandes verticales à gauche du premier pôle. Cela permet de proposer des conjectures de précision variable.

Dans cet article, nous considérons certaines fibrations localement triviales construites de la façon suivante. Soient G un groupe algébrique linéaire sur F agissant sur une variété projective lisse X, B une variété projective lisse sur F et T un G-torseur sur B localement trivial pour la topologie de Zariski. Ces données définissent une variété algébrique projective Y munie d'un morphisme $Y \to B$ dont les fibres sont isomorphes à X. Le cœur du problème est de comprendre le comportement de la fonction hauteur lorsqu'on passe d'une fibre à l'autre, comportement vraiment non trivial bien qu'elles soient toutes isomorphes.

Dans notre premier article (*Torseurs arithmétiques et espaces fibrés*, [9]), nous avons exposé en détail la construction de hauteurs sur de telles variétés. Dans celui-ci, nous appliquons ces considérations générales au cas d'une fibration en variétés toriques provenant d'un torseur sous un *tore déployé*, pour l'ouvert U défini par le tore. Nous avons construit les hauteurs à l'aide d'un prolongement du torseur géométrique en un torseur arithmétique, ce qui correspond en l'occurence au choix de métriques hermitiennes sur certains fibrés en droites. Écrivons la fonction zêta comme la somme des fonctions zêta des fibres

$$Z_U(\mathscr{L},s) = \sum_{b \in B(F)} \sum_{x \in U_b(F)} H_{\mathscr{L}}(x)^{-s} = \sum_{b \in B(F)} Z_{U_b}(\mathscr{L}|_{U_b},s).$$

Chaque U_b est isomorphe au tore et on peut exprimer la fonction zêta des hauteurs de U_b à l'aide de la formule de Poisson adélique. De cette façon, la fonction zêta de U apparaît comme une intégrale sur certains caractères du tore adélique de la fonction L d'Arakelov du torseur arithmétique sur B.

Ainsi, nous pouvons démontrer des *théorèmes de montée* : supposons que B vérifie une conjecture, alors Y la vérifie. Bien sûr, la méthode reprend les outils utilisés dans la démonstration de ces conjectures pour les variétés toriques ([**4, 2, 3**]).

Par exemple, nous démontrerons au § 5.1 l'holomorphie de la fonction $Z_U(\mathscr{L},s)$ pour $\mathrm{Re}(s) > a(\mathscr{L})$ sous la seule hypothèse de la convergence de la fonction zêta des hauteurs analogue sur B ; cela implique que pour tout $\varepsilon > 0$, le nombre de points rationnels de hauteur $H_{\mathscr{L}}$ inférieure à H est $O(H^{a(\mathscr{L})+\varepsilon})$. Ensuite, sous des hypothèses raisonnables concernant B, nous établissons un prolongement méromorphe de cette fonction zêta à gauche de $a(\mathscr{L})$ et nous démontrons que l'ordre du pôle est inférieur ou égal à $b(\mathscr{L})$; cela précise la majoration du nombre de points en $O(H^{a(\mathscr{L})}(\log H)^{b(\mathscr{L})-1})$. Enfin, lorsque $\mathscr{L} = \omega_Y^{-1}$, nous démontrons que le pôle est effectivement d'ordre $b(\mathscr{L})$ d'où une estimation de la forme $(*)$ et nous identifions la constante $\Theta(\mathscr{L})$, établissant ainsi la conjecture de Manin raffinée par Peyre. Pour un fibré en droites quelconque, la preuve de la conjecture de Batyrev–Manin [**1**] avec son raffinement par Batyrev–Tschinkel [**5**] est ramenée à la détermination exacte de l'ordre du pôle, c'est-à-dire à la non-annulation d'une certaine constante. Dans le cas des variétés toriques ou des variétés horosphériques, l'utilisation de « fibrations $\mathscr{L}$-primitives » dans [**3**] et [**19**] a permis d'établir cette conjecture. Moyennant des hypothèses sur B, cette méthode devrait s'étendre au sujet de notre étude.

Notre méthode impose de disposer de majorations de la fonction zêta des hauteurs (pour B) dans les bandes verticales à gauche du premier pôle ; nous avons ainsi tâché d'obtenir de telles majorations pour la variété Y. Il est en outre bien connu que cela entraîne un développement asymptotique assez précis pour le nombre de points de hauteur bornée, cf. le théorème taubérien donné en appendice. Quelques cas de variétés toriques sur $\mathbf{Q}$ avaient en effet attiré l'attention des spécialistes de théorie analytique des nombres (voir notamment les articles de É. Fouvry et R. de la Bretèche dans [16], ainsi que [6]). Notre méthode établit un tel développement pour les variétés toriques lisses, les variétés horosphériques, etc. sur tout corps de nombres.

La démonstration de l'existence d'un prolongement méromorphe de la fonction zêta des hauteurs pour les variétés toriques ou pour les variétés horosphériques faisait intervenir un théorème technique d'analyse complexe à plusieurs variables dont la démonstration se trouve dans [4], [3] et [19]. En vue d'obtenir les majorations exigées dans les bandes verticales, nous sommes obligés d'en préciser la preuve; ceci est l'objet du § 3.

Dans les § 4 et § 5 se situe l'étude de la fonction zêta des hauteurs d'une variété torique et d'une fibration en variétés toriques. Pour les variétés toriques, nous améliorons le terme d'erreur à la suite de [4, 18, 8]. Le théorème de montée pour les fibrations généralise le résultat principal de [19].

Notations et conventions

Si $\mathscr{X}$ est un schéma, on note $\operatorname{Pic}(\mathscr{X})$ le groupe des classes d'isomorphisme de faisceaux inversibles sur $\mathscr{X}$. Si $\mathscr{F}$ est un faisceau quasi-cohérent sur $\mathscr{X}$, on note $\mathbf{V}(\mathscr{F}) = \operatorname{Spec} \operatorname{Sym} \mathscr{F}$ et $\mathbf{P}(\mathscr{F}) = \operatorname{Proj} \operatorname{Sym} \mathscr{F}$ les fibrés vectoriels et projectifs associés à $\mathscr{F}$.

On note $\widehat{\operatorname{Pic}}(\mathscr{X})$ le groupe des classes d'isomorphisme de fibrés en droites hermitiens sur $\mathscr{X}$ (c'est-à-dire des fibrés en droites munis d'une métrique hermitienne continue sur $\mathscr{X}(\mathbf{C})$ et invariante par la conjugaison complexe)..

Si $\mathscr{X}$ est un S-schéma, et si $\sigma \in S(\mathbf{C})$, on désigne par $\mathscr{X}_\sigma$ le $\mathbf{C}$-schéma $\mathscr{X} \times_\sigma \mathbf{C}$. Cette notation servira lorsque S est le spectre d'un localisé de l'anneau des entiers d'un corps de nombres F, de sorte que σ n'est autre qu'un plongement de F dans $\mathbf{C}$.

Si G est un schéma en groupes sur S, $X^*(G)$ désigne le groupe des S-homomorphismes $G \to \mathbf{G}_m$ (caractères algébriques).

Si $\mathscr{X}/S$ est lisse, le faisceau canonique de $\mathscr{X}/S$, noté $\omega_{\mathscr{X}/S}$, est la puissance extérieure maximale de $\Omega^1_{\mathscr{X}/S}$.

Enfin, cet article commence au paragraphe 3. Les références aux paragraphes 1 et 2 renvoient ainsi à l'article précédent [9].

3. Fonctions holomorphes dans un tube

Le but de ce paragraphe est de prouver un théorème d'analyse sur le prolongement méromorphe de certaines intégrales et leur estimation dans des bandes verticales. Ce théorème généralise un énoncé analogue de [4, 19]. La présentation en est un peu différente et le formalisme que nous introduisons permet de contrôler la croissance des fonctions obtenues. Ce contrôle est nécessaire pour utiliser des théorèmes taubériens précis et améliorer ainsi le développement asymptotique du nombre de points rationnels de hauteur bornée.

Les résultats de ce paragraphe n'interviennent que dans la preuve des théorèmes 4.4.6 et 5.2.5.

3.1. Énoncé du théorème

Soit V un $\mathbf{R}$-espace vectoriel réel de dimension finie muni d'une mesure de Lebesgue dv et d'une norme $\|\cdot\|$. On dispose alors d'une mesure canonique dv^* sur le dual V^*. Notons $V_{\mathbf{C}} = V \otimes_{\mathbf{R}} \mathbf{C}$ le complexifié de V. On appelle *tube* toute partie connexe de $V_{\mathbf{C}}$ de la forme $\Omega + iV$ où Ω est une partie connexe de V ; on le notera $\mathsf{T}(\Omega)$.

Soit enfin M un sous-espace vectoriel de V muni d'une mesure de Lebesgue dm.

Définition 3.1.1. — Une *classe de contrôle* $\mathscr{D}$ est la donnée pour tout couple $M \subset V$ de $\mathbf{R}$-espaces vectoriels de dimension finie d'un ensemble $\mathscr{D}(M,V)$ de fonctions mesurables $\kappa : V \to \mathbf{R}_+$ dites $\mathscr{D}(M,V)$-*contrôlantes* vérifiant les propriétés suivantes :

(a) si κ_1 et κ_2 sont deux fonctions de $\mathscr{D}(M,V)$, λ_1 et λ_2 deux réels positifs, et si κ est une fonction mesurable $V \to \mathbf{R}_+$ telle que $\kappa \leqslant \lambda_1 \kappa_1 + \lambda_2 \kappa_2$, alors $\kappa \in \mathscr{D}(M,V)$;

(b) Si $\kappa \in \mathscr{D}(M,V)$ et si K est un compact de V, la fonction qui à v associe $\sup_{u \in K} \kappa(v+u)$ appartient à $\mathscr{D}(M,V)$;

(c) si $\kappa \in \mathscr{D}(M,V)$, pour tout $v \in M \setminus 0$, $\kappa(tv)$ tend vers 0 lorsque t tend vers $+\infty$;

(d) si $\kappa \in \mathscr{D}(M,V)$, pour tout sous-espace $M_1 \subset M$, la fonction M_1-invariante
$$\kappa_{M_1} : v \mapsto \int_{M_1} \kappa(v+m_1)\,dm_1$$
est finie et appartient à $\mathscr{D}(M/M_1, V/M_1)$;

(e) si $\kappa \in \mathscr{D}(M,V)$, pour tout sous-espace $M_1 \subset M$ et tout projecteur $p : V \to V$ de noyau M_1, la fonction M_1-invariante $\kappa \circ p$ appartient à l'ensemble $\mathscr{D}(M/M_1, V/M_1)$.

3.1.2. Il existe une classe de contrôle $\mathscr{D}^{\max}$ contenant toutes les classes de contrôles : l'ensemble $\mathscr{D}^{\max}(M,V)$ est défini par récurrence sur la dimension de M par les trois conditions (a, c, e) dans la définition 3.1.1. La dernière condition est alors automatique.

Dans la suite, on fixe une classe de contrôle $\mathscr{D}$, et on abrège l'expression $\mathscr{D}(M,V)$-*contrôlante* en M-*contrôlante*.

Définition 3.1.3. — *Une fonction $f : \mathsf{T}(\Omega) \to \mathbf{C}$ sur un tube est dite M-contrôlée s'il existe une fonction M-contrôlante κ telle que pour tout compact $K \subset \mathsf{T}(\Omega)$, il existe un réel $c(K)$ de sorte que l'inégalité*
$$|f(z+iv)| \leqslant c(K)\kappa(v)$$
soit vérifiée pour tout $z \in K$ et tout $v \in V$.

3.1.4. Considérons une fonction sur un tube, $f : \mathsf{T}(\Omega) \to \mathbf{C}$. Soit M un sous-espace vectoriel de V, muni d'une mesure de Lebesgue dm. On considère la projection $\pi : V \to V' = V/M$ et on munit V' de la mesure de Lebesgue quotient. On pose, quand cela a un sens,

(3.1.5) $$\mathscr{S}_M(f)(z) = \frac{1}{(2\pi)^{\dim M}} \int_M f(z+im)\,dm, \qquad z \in \mathsf{T}(\Omega).$$

Lemme 3.1.6. — *Soit $\Omega \subset V$ et $f : \mathsf{T}(\Omega) \to \mathbf{C}$ une fonction holomorphe M-contrôlée. Soit M' un sous-espace vectoriel de M et Ω' l'image de Ω par la projection $V \to V/M'$. Alors, l'intégrale qui définit $\mathscr{S}_{M'}(f)$ converge en tout $z \in \mathsf{T}(\Omega)$ et définit une fonction holomorphe M/M'-contrôlée sur $\mathsf{T}(\Omega')$.*

Démonstration. — Comme f est M-contrôlée, il existe une fonction κ de l'ensemble $\mathscr{D}(M,V)$ et, pour tout compact $K \subset \mathsf{T}(\Omega)$, un réel $c(K) > 0$ de sorte que pour tout $v \in V$ et tout $z \in K$, on ait $|f(z+iv)| \leqslant c(K)\kappa(v)$. La condition (3.1.1, d) des classes de contrôles jointe au théorème de convergence dominée de Lebesgue implique que l'intégrale qui définit $\mathscr{S}_{M'}(f)$ converge et que la somme est une fonction holomorphe sur $\mathsf{T}(\Omega)$. Par construction, cette fonction est iM'-invariante. Comme elle est analytique, elle est donc invariante par M' et définit ainsi une fonction holomorphe sur $\mathsf{T}(\Omega')$. De plus, si π désigne la projection $V \to V/M'$, pour tout $z \in K$ et tout $v \in V$, on a

$$|\mathscr{S}_{M'}(f)(\pi(z)+i\pi(v))| \leqslant c(K) \int_{M'} \kappa(v+m')\,dm' = c(K)\kappa'(\pi(v))$$

où κ' appartient par définition à $\mathscr{D}(M/M', V/M')$. Tout compact de $\mathsf{T}(\Omega')$ étant de la forme $\pi(K)$ pour un compact K de $\mathsf{T}(\Omega)$, le lemme est ainsi démontré. $\square$

3.1.7. Fonction caractéristique d'un cône. — Soit Λ un cône convexe polyédral ouvert de V. La fonction caractéristique de Λ est la fonction sur $\mathsf{T}(\Lambda)$ définie par l'intégrale convergente

(3.1.8) $$\mathsf{X}_\Lambda(z) = \int_{\Lambda^*} e^{-\langle z,v^*\rangle}\,dv^*,$$

où $\Lambda^* \subset V^*$ est le cône dual de Λ, V^* étant muni de la mesure de Lebesgue dv^* duale de la mesure dv.

Si Λ est simplicial, c'est-à-dire qu'il existe $n = \dim V$ formes linéaires indépendantes $\ell_1, \ldots, \ell_n$ telles que $v \in \Lambda$ si et seulement si $\ell_j(v) > 0$ pour tout j, alors

$$\mathsf{X}_\Lambda(z) = \|d\ell_1 \wedge \cdots \wedge d\ell_n\| \prod_{j=1}^n \frac{1}{\ell_j(z)}.$$

(On a noté $\|d\ell_1 \wedge \cdots \wedge d\ell_n\|$ le volume du parallélépipède fondamental dans V^* de base les ℓ_j.) Dans le cas général, toute triangulation de Λ^* par des cônes simpliciaux permet d'exprimer X_Λ sous la forme d'une somme de fractions rationnelles de ce type. Elle se prolonge ainsi en une fonction rationnelle sur $\mathsf{T}(V)$ dont les pôles sont exactement les hyperplans de $V_\mathbf{C}$ définis par les équations des faces de Λ. Elle est de plus strictement positive sur Λ.

Une autre façon de construire un cône est de s'en donner des générateurs, autrement dit de l'écrire comme *quotient* d'un cône simplicial. À ce titre, on a la proposition suivante.

PROPOSITION 3.1.9. — *Soit Λ un cône polyédral convexe ouvert de V dont l'adhérence $\overline{\Lambda}$ ne contient pas de droite. Soit M un sous-espace vectoriel de V tel que $\overline{\Lambda} \cap M = \{0\}$. On note π la projection $V \to V' = V/M$.*

La restriction à $\mathsf{T}(\Lambda)$ *de la fonction* X_Λ *est M-contrôlée (pour la classe* $\mathscr{D}^{\max}(M,V)$). *L'intégrale qui définit* $\mathscr{S}_M(\mathsf{X}_\Lambda)$ *converge donc absolument et pour tout* $z \in \mathsf{T}(\Lambda)$, *on a*

$$\mathscr{S}_M(\mathsf{X}_\Lambda)(z) = \mathsf{X}_{\Lambda'}(\pi(z)).$$

Remarque 3.1.10. — Les hypothèses impliquent que $\overline{\Lambda}'$ ne contient pas de droite. En effet, s'il existait un vecteur non nul de $\overline{\Lambda}' \cap -\overline{\Lambda}'$, il existerait deux vecteurs v_1 et v_2 de $\overline{\Lambda}$ tels que $v_1 + v_2 \in M$ mais $v_1 \notin M$. Comme $\overline{\Lambda} \cap M = \{0\}$, $v_1 = -v_2$ ce qui contredit l'hypothèse que $\overline{\Lambda}$ ne contient pas de droite.

Démonstration. — La preuve est une adaptation des paragraphes 7.1 et 7.2 de [**19**]. Soit (e_i) une famille minimale de générateurs de Λ. Chaque face de Λ^* dont la dimension est $\dim V - 1$ engendre un sous-espace vectoriel qui est l'orthogonal d'un des e_i.

Comme $M \cap \overline{\Lambda} = \{0\}$, il existe une forme linéaire $\ell \in V^*$ qui est nulle sur M mais qui n'appartient à aucune face de Λ^* ; posons $H = \ker \ell$. Soit H' un supplémentaire de $\mathbf{R}\ell$ dans V^*. Si $\varphi \in V^*$ et $t \in \mathbf{R}$ sont tels que $\varphi + t\ell \in \Lambda^*$, on doit avoir pour tout générateur e_j de Λ l'inégalité $\varphi(e_j) + t\ell(e_j) > 0$, soit (rappelons que $\ell(e_j)$ n'est pas nul), $t > -\varphi(e_j)/\ell(e_j)$ quand $\ell(e_j) > 0$ et $t < -\varphi(e_j)/\ell(e_j)$ quand $\ell(e_j) < 0$. Soit alors $I(\varphi) =]h_1(\varphi), h_2(\varphi)[$ l'intervalle de $\mathbf{R}$ défini par ces inégalités. (Si tous les $\ell(e_j)$ sont positifs, c'est-à-dire $\ell \in \Lambda^*$, on a $h_1 \equiv -\infty$, tandis que s'ils sont tous négatifs, $h_2 \equiv +\infty$.) Les fonctions h_1 et h_2 sont linéaires par morceaux par rapport à un éventail de H' qu'on peut supposer complet et régulier (voir par exemple [**12**] pour la définition, ou [**2**]).

Alors, si $v \in \mathsf{T}(\Lambda)$ et $m \in H$, on a

$$\begin{aligned}\mathsf{X}_\Lambda(v+im) &= \int_{V^*} \mathbf{1}_{\Lambda^*}(\varphi) e^{-\langle v+im,\varphi\rangle}\, d\varphi \\ &= \int_{H'}\int_{\mathbf{R}} \mathbf{1}_{\Lambda^*}(\varphi+t\ell) e^{-\langle v+im,\varphi\rangle} e^{-t\langle v,\ell\rangle}\, dt\, d\varphi \\ &= \int_{H'} \int_{h_1(\varphi)}^{h_2(\varphi)} e^{-\langle v+im,\varphi\rangle} e^{-t\langle v,\ell\rangle}\, dt\, d\varphi \\ &= \int_{H'} e^{-\langle v+im,\varphi\rangle} \frac{e^{-h_1(\varphi)\langle v,\ell\rangle} - e^{-h_2(\varphi)\langle v,\ell\rangle}}{\langle v,\ell\rangle}\, d\varphi\end{aligned}$$

de sorte que la fonction $H \to \mathbf{C}$ telle que $m \mapsto \mathsf{X}_\Lambda(v+im)$ est (à une constante multiplicative près) la différence des transformées de Fourier des fonctions

$$H' \to \mathbf{C}, \quad \varphi \mapsto e^{-\langle v, \varphi + h_j(\varphi)\ell\rangle}$$

pour $j = 1$ et 2.

Comme $v \in \mathsf{T}(\Lambda)$ et $\varphi + h_j(\varphi)\ell$ appartient au bord de Λ^*, $\langle v, \varphi + h_j(\varphi)\ell\rangle$ est de partie réelle strictement positive, à moins que $\varphi = 0$. Soit K un compact de $\mathsf{T}(\Lambda)$. Il résulte alors des estimations des transformées de Fourier de fonctions linéaires par morceaux et positives (voir [2], proposition 2.3.2, p. 614, et aussi *infra*, prop. 4.2.4) une majoration de la fonction

$$f_{\Lambda,K}(m) := \sum_{v \in K} |\mathsf{X}_\Lambda(v+im)|$$

de la forme

$$f_{\Lambda,K}(m) \leqslant c(K) \sum_\alpha \prod_{j=1}^{\dim H} \frac{1}{(1+|\langle m, \ell_{\alpha,j}\rangle|)^{1+1/\dim H}},$$

où pour tout α, la famille $(\ell_{\alpha,j})_j$ est une base de H^*. D'après le lemme 3.1.11 ci-dessous, la fonction $f_{\Lambda,K}$ appartient à $\mathscr{D}^{\max}(M, V)$.

La fonction $m \mapsto \mathsf{X}_\Lambda(v+im)$ est donc absolument intégrable sur M. C'est la transformée de Fourier de la fonction $\varphi \mapsto \mathbf{1}_{\Lambda^*}(\varphi)e^{-\langle v,\varphi\rangle}$ dont il est facile de voir qu'elle est intégrable sur tout sous-espace et donc aussi $M^\perp$. La formule de Poisson s'applique (après un léger argument de régularisation) et s'écrit

$$\int_M \mathsf{X}_\Lambda(v+im)\, dm = (2\pi)^{\dim M} \int_{\Lambda^* \cap M^\perp} e^{-\langle v,\varphi\rangle}\, d\varphi.$$

Or, l'application $V \to V'$ identifie $(V')^*$ à $M^\perp$, et $\Lambda^* \cap M^\perp$ à $(\Lambda')^*$. Ainsi, on obtient

$$\mathscr{S}_M(\mathsf{X}_\Lambda)(v) = \int_{(\Lambda')^*} e^{-\langle \pi(v),\varphi\rangle}\, d\varphi = \mathsf{X}_{\Lambda'}(\pi(v)).$$

$\square$

LEMME 3.1.11. — *Soit V un $\mathbf{R}$-espace vectoriel de dimension d, $(\ell_1, \ldots, \ell_d)$ une base de V^* et f la fonction $v \mapsto \prod_{j=1}^d (1+|\ell_j(v)|)^{-1-1/d}$. Alors, $f \in \mathscr{D}^{\max}(V,V)$.*

Démonstration. — Soit M un sous-espace vectoriel de V de dimension m. Quitte à réordonner les indices, on peut supposer que M est l'image d'une application linéaire $\mathbf{R}^m \to \mathbf{R}^d = V$ de la forme

$$t = (t_1, \ldots, t_m) \mapsto (t_1, \ldots, t_m, \varphi_{m+1}(t), \ldots, \varphi_d(t)).$$

Si on réalise V/M par son supplémentaire $\{0\}^m \times \mathbf{R}^{d-m}$, la fonction $f_M : v \mapsto \int_M f(v+m)\, dm$ est donnée par l'intégrale

$$\int_{\mathbf{R}^m} \frac{1}{(1+|t_1|)^{1+1/d}} \cdots \frac{1}{(1+|t_m|)^{1+1/d}} \prod_{j=m+1}^{d} \frac{1}{(1+|v_j + \varphi_j(t)|)^{1+1/d}}\, dt_1 \ldots dt_m.$$

Elle est dominée par l'intégrale convergente

$$\int_{\mathbf{R}^m} \frac{1}{(1+|t_1|)^{1+1/d}} \cdots \frac{1}{(1+|t_m|)^{1+1/d}}\, dt_1 \ldots dt_m$$

et le théorème de convergence dominée implique alors que pour tout vecteur $v = (0, \ldots, 0, v_{m+1}, \ldots, v_d)$ distinct de 0,

$$\lim_{s \to +\infty} f_M(sv) = 0.$$

Le lemme est ainsi démontré. □

Définition 3.1.12. — Soient C un ouvert convexe de V ayant 0 pour point adhérent et Λ un cône polyédral ouvert contenant C.

Soit $\Phi \subset V^*$ une famille de formes linéaires deux à deux non proportionnelles définissant les faces de Λ.

On note $\mathscr{H}_M(\Lambda; C)$ l'ensemble des fonctions holomorphes $f : \mathsf{T}(C) \to \mathbf{C}$ telles qu'il existe un voisinage convexe B de 0 dans V de sorte que la fonction g définie par

$$g(z) = f(z) \prod_{\varphi \in \Phi} \frac{\varphi(z)}{1 + \varphi(z)}$$

admet un prolongement holomorphe M-contrôlé dans $\mathsf{T}(B)$.

Par le théorème d'extension de Bochner (voir par exemple [13]), une telle fonction s'étend en une fonction holomorphe sur le tube de base l'enveloppe convexe C' de $B \cup C$. En particulier, il n'aurait pas été restrictif de prendre pour C l'intersection du cône Λ avec un voisinage convexe de 0 dans V.

On constate aussi que f est nécessairement M-contrôlée dans $\mathsf{T}(C)$. Enfin, il est facile de vérifier que $\mathscr{H}_M(\Lambda;C)$ ne dépend pas du choix des formes linéaires qui définissent les faces de Λ.

3.1.13. Si Λ est un cône polyédral et si M est un sous-espace vectoriel de V tel que l'image de $\overline{\Lambda}$ dans V/M ne contient pas de droite, la proposition 3.1.9 implique donc que la fonction X_Λ appartient à l'espace $\mathscr{H}_M(\Lambda;\Lambda)$ défini par la classe de contrôle $\mathscr{D}^{\max}$.

Le théorème principal de cette section est le suivant.

THÉORÈME 3.1.14. — *Soit $M \subset V$ un sous-espace vectoriel muni d'une mesure de Lebesgue.*

Soit C l'intersection de Λ avec un voisinage convexe de 0 et soit f un élément de $\mathscr{H}_M(\Lambda;C)$. Soit M' un sous-espace vectoriel de M, π la projection $V \to V' = V/M'$, $\Lambda' = \pi(\Lambda)$ et $C' = \pi(C)$.

Alors, la fonction $\mathscr{S}_{M'}(f)$ appartient à $\mathscr{H}_{M/M'}(\Lambda';C')$.

Si de plus l'adhérence du cône Λ' ne contient pas de droite et si pour tout $z \in \Lambda$,
$$\lim_{s \to 0^+} \frac{f(sz)}{\mathsf{X}_\Lambda(sz)} = 1,$$

alors pour tout $z' \in \Lambda'$,
$$\lim_{s \to 0^+} \frac{\mathscr{S}_{M'}(f)(sz')}{\mathsf{X}_{\Lambda'}(sz')} = 1.$$

COROLLAIRE 3.1.15. — *Supposons de plus que f est la restriction à $\Lambda \cap C$ d'une fonction holomorphe M-contrôlée sur Λ. Alors, la fonction $\mathscr{S}_M(f)$ sur V' est méromorphe dans un voisinage convexe de Λ', ses pôles étant simples définis par les faces (de codimension 1) de Λ'.*

3.2. Démonstration du théorème

D'après le lemme 3.1.6, la fonction $\mathscr{S}_{M'}(f)$ est holomorphe et M/M'-contrôlée sur $\mathsf{T}(C')$. Le but est de montrer qu'elle y est la restriction d'une fonction méromorphe dont on contrôle les pôles et la croissance. La démonstration est fondée sur l'application successive du théorème des résidus pour obtenir le prolongement méromorphe. La définition des classes de contrôle est faite pour assurer l'intégrabilité ultérieure de chacun des termes obtenus.

Par récurrence, il suffit de démontrer le résultat lorsque $\dim M' = 1$. Soit m_0 un générateur de M'. Munissons la droite $\mathbf{R}m_0$ de la mesure de Lebesgue $d\rho$. Soit $\Phi \subset V^*$ une famille de formes linéaires deux à deux non proportionnelles positives sur Λ et dont les noyaux sont les faces de Λ.

Soit B un ouvert convexe et symétrique par rapport à l'origine, assez petit de sorte que pour tout $\varphi \in \Phi$ et tout $v \in B$, $|\varphi(v)| < 1$ et que la fonction

$$g(z) = f(z) \prod_{\varphi \in \Phi} \frac{\varphi(z)}{1+\varphi(z)}$$

admette un prolongement holomorphe M-contrôlé sur $\mathsf{T}(B)$. L'intégrale à étudier est

$$\int_{-\infty}^{+\infty} g(z+itm_0) \prod_{\varphi \in \Phi} \frac{1+\varphi(z+itm_0)}{\varphi(z+itm_0)} \, dt.$$

On veut déplacer la droite d'intégration vers la gauche. Fixons $\tau > 0$ tel que $2\tau m_0 \in B$. Ainsi, si $\operatorname{Re}(z) \in \frac{1}{2}B$, $z + (u+it)m_0$ appartient à $\mathsf{T}(B)$ pour tout $u \in [-\tau; 0]$ et tout $t \in \mathbf{R}$.

Notons Φ^+, Φ^- et Φ^0 les ensembles des $\varphi \in \Phi$ tels que respectivement $\varphi(m_0) > 0$, $\varphi(m_0) < 0$ et $\varphi(m_0) = 0$. Soit $B_1 \subset \frac{1}{2}B$ l'ensemble des $v \in \frac{1}{2}B$ tels que pour tout $\varphi \in \Phi^+$, $|\varphi(v)| < \frac{\tau}{2}\varphi(m_0)$.

Dans la bande $-\tau \leqslant s \leqslant 0$, les pôles de la fonction holomorphe

$$s \mapsto g(z+sm_0) \prod_{\varphi \in \Phi} \frac{1+\varphi(z+sm_0)}{\varphi(z+sm_0)}$$

sont ainsi donnés par

$$s_\varphi(z) = -\frac{\varphi(z)}{\varphi(m_0)}, \quad \varphi \in \Phi^+.$$

Le pôle $s = s_\varphi(z)$ est simple si et seulement si pour tout $\psi \in \Phi^+$ tel que $\psi \neq \varphi$,

$$\varphi(z)\psi(m_0) - \psi(z)\varphi(m_0) \neq 0.$$

Comme φ et ψ sont non proportionnelles, $\psi(m_0)\varphi - \varphi(m_0)\psi$ est une forme linéaire non nulle ; notons $B_1^\dagger \subset B_1$ le complémentaire des hyperplans qu'elles définissent lorsque $\varphi \neq \psi$ parcourent les éléments de Φ^+.

Si $z \in \mathsf{T}(B_1^\dagger)$ et si $T > \max\{|\operatorname{Im}(s_\varphi(z))| \, ; \, \varphi \in \Phi^+\}$, la formule des résidus pour le contour délimité par le rectangle $-\tau \leqslant \operatorname{Re}(s) \leqslant 0$, $-T \leqslant \operatorname{Im}(s) \leqslant T$

s'écrit

$$\int_{-T}^{T} g(z+itm_0) \prod_{\varphi \in \Phi} \frac{1+\varphi(z+itm_0)}{\varphi(z+itm_0)} dt$$

$$= \sum_{\varphi \in \Phi^+} \frac{2i\pi}{\varphi(m_0)} g(z+s_\varphi(z)m_0) \prod_{\psi \neq \varphi} \frac{1+\psi(z+s_\varphi(z)m_0)}{\psi(z+s_\varphi(z)m_0)}$$

$$+ \int_{-T}^{T} g(z-\tau m_0 + itm_0) \prod_{\varphi \in \Phi} \frac{1+\varphi(z-\tau m_0 + itm_0)}{\varphi(z-\tau m_0 + itm_0)} dt$$

$$+ \int_{0}^{-\tau} g(z+sm_0 + iTm_0) \prod_{\varphi \in \Phi} \frac{1+\varphi(z+sm_0 + iTm_0)}{\varphi(z+sm_0 + iTm_0)} ds$$

$$+ \int_{-\tau}^{0} g(z+sm_0 - iTm_0) \prod_{\varphi \in \Phi} \frac{1+\varphi(z+sm_0 - iTm_0)}{\varphi(z+sm_0 - iTm_0)} ds.$$

Lorsque $T \to +\infty$, l'hypothèse que g est M-contrôlée et l'axiome (3.1.1,c) des classes de contrôles impliquent que ces deux dernières intégrales (sur les segments horizontaux du rectangle) tendent vers 0. De même, l'axiome (3.1.1,d) assure la convergence des deux premières intégrales vers les intégrales correspondantes de $-\infty$ à $+\infty$.

Par suite, si $z \in \mathsf{T}(B_1^\dagger \cap \Lambda)$, on a

$$(3.2.1) \quad \mathscr{S}_{\mathbf{R}m_0}(f)(z) = \sum_{\varphi \in \Phi^+} g(z+s_\varphi(z)m_0) \prod_{\psi \neq \varphi} \frac{1+\psi(z+s_\varphi(z)m_0)}{\psi(z+s_\varphi(z)m_0)}$$

$$+ \prod_{\varphi \in \Phi^0} \frac{1+\varphi(z)}{\varphi(z)} \int_{-\infty}^{\infty} g(z-\tau m_0 + itm_0) \prod_{\varphi \in \Phi \setminus \Phi^0} \frac{1+\varphi(z-\tau m_0 + itm_0)}{\varphi(z-\tau m_0 + itm_0)} dt.$$

Il résulte alors des axiomes (3.1.1,e) et (3.1.1,d) des classes de contrôles que la fonction

$$(3.2.2) \quad z \mapsto \mathscr{S}_{\mathbf{R}m_0}(f)(z) \prod_{\varphi \in \Phi^0} \frac{\varphi(z)}{1+\varphi(z)} \prod_{\varphi \in \Phi^+} \prod_{\psi \notin \Phi^0 \cup \{\varphi\}} \frac{\varphi(s+s_\psi(z)m_0)}{1+\psi(s+s_\varphi(z)m_0)}$$

définie sur $\mathsf{T}(B_1^\dagger \cap \Lambda)$ s'étend en une fonction holomorphe M/M'-contrôlée sur $\mathsf{T}(\pi(B_1^\dagger))$. En particulier, $\mathscr{S}_{\mathbf{R}m_0}(f)$ se prolonge méromorphiquement à $\mathsf{T}(B_1^\dagger)$ et les pôles de $\mathscr{S}_{\mathbf{R}m_0}(f)$ sont donnés par une famille finie de formes linéaires. Le lemme suivant les interprète géométriquement.

LEMME 3.2.3. — *Les faces de Λ' sont les noyaux des formes linéaires deux à deux non proportionnelles sur $V/\mathbf{R}m_0$ $\varphi \in \Phi^0$ et $\varphi - \frac{\varphi(m_0)}{\psi(m_0)}\psi$ pour $\varphi \in \Phi^+$ et $\psi \in \Phi^-$.*
De plus, si φ et $\psi \in \Phi^+$, le noyau de $\varphi - \frac{\varphi(m_0)}{\psi(m_0)}\psi$ rencontre Λ'.

Démonstration. — Un vecteur $x \in V$ appartient à Λ si et seulement si $\varphi(x) > 0$ pour tout $\varphi \in \Phi$. Par suite, $\pi(x) \in \Lambda'$ si et seulement si il existe $\alpha \in \mathbf{R}$ tel que $\varphi(x - \alpha m_0) > 0$ pour tout $\varphi \in \Phi$. Si $\varphi \in \Phi^0$, cette condition est exactement $\varphi(x) > 0$. Pour les autres φ, elle devient

$$\max_{\varphi \in \Phi^-} \frac{\varphi(x)}{\varphi(m_0)} < \alpha < \min_{\varphi \in \Phi^+} \frac{\varphi(x)}{\varphi(m_0)}$$

d'où la première partie du lemme.

Pour la seconde, soit φ et ψ deux éléments distincts de Φ^+. Si le noyau de $\varphi - \frac{\varphi(m_0)}{\psi(m_0)}\psi$ ne recontre pas Λ', quitte à permuter φ et ψ, on a

$$\frac{\varphi(v)}{\varphi(m_0)} > \frac{\psi(v)}{\psi(m_0)}$$

pour tout $v \in \Lambda$ et cela contredit le fait que φ et ψ définissent deux faces distinctes de Λ. □

On sait que $\mathscr{S}_{\mathbf{R}m_0}(f)$ est holomorphe sur $\mathsf{T}(\Lambda')$. Il résulte du lemme que les formes linéaires $\psi + s_\varphi(z)\varphi$ avec $\varphi \in \Phi^+$ et $\psi \notin \Phi^0 \cup \{\varphi\}$ sont des pôles apparents dès que $\psi \in \Phi^+$. Les autres correspondent aux faces de Λ'!

Autrement dit, nous avons déjà prouvé que $\mathscr{S}_{\mathbf{R}m_0}(f)$ est la restriction à $\mathsf{T}(\pi(B_1))$ d'une fonction méromorphe dont les pôles (simples) sont donnés par les faces de Λ'. Montrons comment contrôler la croissance de $\mathscr{S}_{\mathbf{R}m_0}(f)$ dans les bandes verticales.

LEMME 3.2.4. — *Soit V un espace vectoriel, M un sous-espace vectoriel, B un voisinage de 0 dans V. Soit h une fonction holomorphe sur $\mathsf{T}(B)$ et soit ℓ une forme linéaire sur V. Si la fonction $z \mapsto h(z)\frac{\ell(z)}{1+\ell(z)}$ est M-contrôlée, h est M-contrôlée.*

Démonstration. — Il faut montrer que h est M-contrôlée dans un voisinage de tout point de B. Soit donc $x_0 \in B$ et K un voisinage compact de x_0 contenu dans B. Soit $\kappa \in \mathscr{D}(M,V)$ telle que pour tout $x \in K$ et tout $y \in V$,

$$\left| h(x+iy) \frac{\ell(x+iy)}{1+\ell(x+iy)} \right| \leqslant \kappa(y).$$

Supposons d'abord que $\ell(x_0) \neq 0$. Si $\rho = |\ell(x_0)|/2 > 0$, il existe un voisiange compact $K_1 \subset K$ de x_0 où $|\ell| \geqslant \rho$. Alors, pour tout $x \in K_1$ et tout $y \in V$, on a

$$|h(x+iy)| \leqslant \kappa(y)\frac{1+|\ell(x+iy)|}{\ell(x+iy)} \leqslant \frac{1+\rho}{\rho}\kappa(y),$$

ce qui prouve que h est M-contrôlée dans K_1.

Si $\ell(x_0) = 0$, soit $u \in V$ tel que $\ell(u) = 1$, K_1 un voisinage compact de x_0 assez petit et $\rho > 0$ tels que pour tout $t \in [-1;1]$ et tout $x \in K_1$, $x + t\rho u \in K$. La fonction $s \mapsto h(x+iy+s\rho u)$ est une fonction holomorphe sur le disque unité fermé $|s| \leqslant 1$. D'après le principe du maximum, on a donc pour tout $x+iy \in \mathsf{T}(K_1)$,

$$|h(x+iy)| \leqslant \sup_{|s|\leqslant 1} |h(x+iy+s\rho u)| = \sup_{|s|=1} |h(x+iy+s\rho u)|$$
$$\leqslant \frac{1+\rho}{\rho} \sup_{|s|\leqslant 1} \kappa(y+su).$$

L'axiome (3.1.1,b) assure alors l'existence d'une fonction $\kappa_1 \in \mathscr{D}(M,V)$ telle que pour tout $x+iy \in \mathsf{T}(K_1)$,

$$|h(x+iy)| \leqslant \kappa_1(y).$$

La fonction h est donc M-contrôlée dans un voisinage de x_0. □

Il reste à démontrer que si pour tout $z \in \Lambda$, $\lim_{t \to 0^+} f(tz)/\mathsf{X}_\Lambda(tz) = 1$, alors

$$\lim_{t \to 0^+} \mathscr{S}_{\mathbf{R}m_0}(f)(tz')/\mathsf{X}_{\Lambda'}(tz') = 1.$$

Comme $\mathsf{X}_\Lambda(tz) = t^{-\dim V}\mathsf{X}_\Lambda(z)$, l'hypothèse $f(tz) \sim \mathsf{X}_\Lambda(tz)$ se récrit

$$\lim_{t \to 0} t^{\dim V - \#\Phi} g(tz) = \mathsf{X}_\Lambda(z).$$

D'autre part, la formule (3.2.1) donne

$$t^{-1+\dim V} \mathscr{S}_{\mathbf{R}m_0}(f)(tz)$$
$$= t^{-1+\dim V} \sum_{\varphi \in \Phi^+} g(tz + s_\varphi(tz)m_0) \prod_{\psi \neq \varphi} \frac{1 + \psi(tz + s_\varphi(tz)m_0)}{\psi(tz + s_\varphi(tz)m_0)}$$
$$+ t^{-1+\dim V} \prod_{\varphi \in \Phi^0} \frac{1 + \varphi(tz)}{\varphi(tz)} \times$$
$$\times \int_{-\infty}^{\infty} g(tz - \tau m_0 + itm_0) \prod_{\varphi \in \Phi \setminus \Phi^0} \frac{1 + \varphi(tz - \tau m_0 + itm_0)}{\varphi(tz - \tau m_0 + itm_0)} dt$$
$$= \sum_{\varphi \in \Phi^+} t^{\dim V - \#\Phi} g(t(z + s_\varphi(z)m_0)) \prod_{\psi \neq \varphi} \frac{1 + t\psi(z + s_\varphi(z)m_0)}{\psi(z + s_\varphi(z)m_0)}$$
$$+ t^{-1+\dim V - \#\Phi^0} \prod_{\varphi \in \Phi^0} \frac{1 + t\varphi(tz)}{\varphi(z)} \times$$
$$\times \int_{-\infty}^{\infty} g(tz - \tau m_0 + itm_0) \prod_{\varphi \in \Phi \setminus \Phi^0} \frac{1 + \varphi(tz - \tau m_0 + itm_0)}{\varphi(tz - \tau m_0 + itm_0)} dt.$$

Un vecteur non nul de V ne peut appartenir qu'à au plus $\dim V - 1$ faces de Λ et seuls les générateurs de Λ appartiennent à $\dim V - 1$ faces. Comme m_0 est supposé n'être pas un générateur de Λ, $\#\Phi^0 \leq \dim V - 2$. Lorsque t tend vers 0, on a donc

$$\lim t^{-1+\dim V} \mathscr{S}_{\mathbf{R}m_0}(f)(tz) = \sum_{\varphi \in \Phi^+} X_\Lambda(z + s_\varphi(z)m_0) \prod_{\psi \neq \varphi} \frac{1}{\psi(z + s_\varphi(z)m_0)}$$

où le second membre ne dépend plus de f. Comme on peut appliquer cette formule à $f = X_\Lambda$, on obtient donc

$$\lim t^{1-\dim V}(\mathscr{S}_{\mathbf{R}m_0}(f))(tz) = \lim t^{1-\dim V}(\mathscr{S}_{\mathbf{R}m_0}(X_\Lambda))(tz)$$
$$= \lim t^{1-\dim V} X_{\Lambda'}(tz) = X_{\Lambda'}(z).$$

Le théorème est ainsi démontré.

Remarque 3.2.5. — La démonstration s'adapte sans peine lorsque f dépend uniformément de paramètres supplémentaires.

4. Variétés toriques

Dans ce paragraphe, nous montrons comment les raffinements analytiques du paragraphe 3 permettent de préciser le développement asymptotique obtenu par Batyrev–Tschinkel dans [4] pour la fonction zêta des hauteurs d'une variété torique. Les résultats techniques que nous rappelons à l'occasion seront réutilisés au paragraphe suivant, lorsque nous traiterons le cas d'une fibration en variétés toriques.

4.1. Préliminaires

4.1.1. Rappels adéliques. — Notons $S = \operatorname{Spec} \mathfrak{o}_F$ le spectre de l'anneau des entiers de F. Si v est une place de F, on définit la norme $\|\cdot\|_v$ sur F_v de la manière habituelle, comme le *module* associé à une mesure de Haar additive sur F_v. En particulier, si π_v est une uniformisante en une place finie v, $\|\pi_v\|_v$ est l'inverse du cardinal du corps résiduel en v.

Soit G un tore déployé de dimension d sur S. Désignons par $\mathbf{K}_\infty$ la collection de ses sous-groupes compacts maximaux aux places à l'infini et $\mathbf{K}_G = \prod_{v \nmid \infty} G(\mathfrak{o}_v) \prod_{v \mid \infty} \mathbf{K}_v \subset G(\mathbf{A}_F)$. Il nous faut faire quelques rappels sur la structure du groupe $\mathscr{A}_G$ des caractères de $G(F)\backslash G(\mathbf{A}_F)/\mathbf{K}_G$. On a un homomorphisme de noyau fini $\mathscr{A}_G \to \bigoplus_{v \mid \infty} X^*(G)_{\mathbf{R}}$, $\chi \mapsto \chi_\infty$ obtenu en associant à un caractère adélique son *type à l'infini*, c'est-à-dire sa restriction au sous-groupe de $G(\mathbf{A})$ dont les composantes aux places finies sont triviales. En choisissant une norme sur $X^*(G)_{\mathbf{R}}$, on obtient ainsi une « norme » $\chi \mapsto \|\chi_\infty\|$ sur $\mathscr{A}_G$.

Il existe enfin un homomorphisme $X^*(G)_{\mathbf{R}} \to \mathscr{A}_G$, tel que l'image du caractère algébrique $\chi \in X^*(G)$ est le caractère adélique $\mathbf{g} \mapsto |\chi(\mathbf{g})|^i$ dont le type à l'infini s'identifie à χ sur chaque composante.

Le quotient $\mathscr{A}_G/X^*(G)_{\mathbf{R}}$ est un $\mathbf{Z}$-module de type fini et de rang $(\rho - 1)d$ (où $\rho = r_1 + r_2$, r_1 et r_2 désignant comme d'habitude les nombres de places réelles et complexes) et l'on peut fixer une décomposition $\mathscr{A}_G = X^*(G)_{\mathbf{R}} \oplus \mathscr{U}_G$, par exemple à l'aide d'un scindage de la suite exacte

$$1 \to \mathbf{G}_m(\mathbf{A}_F)^1 \to \mathbf{G}_m(\mathbf{A}_F) \xrightarrow{|\cdot|} \mathbf{R}^* \to 1.$$

(Rappelons que G est supposé déployé.)

4.1.2. Rappels sur les variétés toriques. — Notons $M = X^*(G)_{\mathbf{R}}$, c'est un espace vectoriel sur $\mathbf{R}$ de dimension finie d. Considérons une compactification équivariante $\mathscr{X}$ de G, lisse sur S. D'après la théorie des variétés toriques (cf.

par exemple [14], [12]), $\mathscr{X}$ est définie par un éventail complet et régulier Σ de $N := \operatorname{Hom}(M, \mathbf{R})$ formé de cônes convexes simpliciaux rationnels. Il existe ainsi une famille (minimale) $(e_j)_{j \in J}$ de vecteurs de N telle que tout cône $\sigma \in \Sigma$ soit engendré par une sous-famille $(e_j)_{j \in J_\sigma}$ de cardinal $\dim \operatorname{vect}(\sigma)$. On note $\Sigma(d)$ l'ensemble des cônes de Σ de dimension d.

L'espace vectoriel $\operatorname{PL}(\Sigma)$ des fonctions continues $N \to \mathbf{R}$ dont la restriction à chaque cône de Σ est linéaire est un espace vectoriel de dimension finie sur $\mathbf{R}$, d'ailleurs égale à $\#J$; munissons le d'une norme arbitraire. L'espace vectoriel $\operatorname{Pic}^G(\mathscr{X}_F)_\mathbf{R}$ est isomorphe à $\operatorname{PL}(\Sigma)$; il possède une base canonique formée des fibrés en droites G-linéarisés associés aux diviseurs G-invariants sur $\mathscr{X}_F$. À chaque e_j correspond un tel diviseur D_j ; à un diviseur G-invariant $D = \sum_j \lambda_j D_j$ correspond l'unique fonction $\varphi \in \operatorname{PL}(\Sigma)$ telle que $\varphi(e_j) = \lambda_j$. Dans cette description, le cône des diviseurs effectifs correspond simplement l'ensemble des éléments de $\operatorname{Pic}^G(\mathscr{X}_F)$ dont les coordonnées (λ_j) vérifient $\lambda_j \geqslant 0$ pour tout j. Plus généralement, on notera Λ_t l'ensemble des éléments de $\operatorname{Pic}^G(\mathscr{X}_F)$ tels que $\lambda_j > t$ pour tout j ; le cône ouvert Λ_0 est aussi noté $\operatorname{PL}^+(\Sigma)$ et encore $\Lambda^0_{\text{eff}}(\mathscr{X}_F)$.

Cette base (D_j) de $\operatorname{Pic}^G(\mathscr{X}_F)$ et l'homomorphisme canonique $\iota : X^*(G) \to \operatorname{Pic}^G(\mathscr{X})$ induisent des sous-groupes à un paramètre $\mathbf{G}_m \to G$, d'où, pour tout caractère $\chi \in \mathscr{A}_G$, des caractères χ_j de $\mathbf{G}_m(F) \backslash \mathbf{G}_m(\mathbf{A}_F) / \mathbf{K}_{\mathbf{G}_m}$, autrement dit des caractères de Hecke.

Les fibrés en droites sur $\mathscr{X}_F$ seront systématiquement munis de leur métrique adélique canonique introduite notamment dans [2]. Cela nous fournit un homomorphisme canonique $\operatorname{Pic}(\mathscr{X}_F) \to \widehat{\operatorname{Pic}}(\mathscr{X})$ qui induit un homomorphisme

$$(4.1.3) \qquad \operatorname{Pic}^G(\mathscr{X}_F) \to \widehat{\operatorname{Pic}}^{G,\mathbf{K}}(\mathscr{X}).$$

On vérifie aisément, par exemple sur les formules données dans [2], que les sous-groupes compacts maximaux aux places archimédiennes agissent de manière isométrique. De plus, le choix d'une G-linéarisation fournit une unique F-droite de sections ne s'annulant pas sur G, donc en particulier une fonction hauteur sur les points adéliques de $\mathscr{X}_F$ comme dans la définition 1.3.3. Cette fonction s'étend en une application « bilinéaire »

$$H : \operatorname{PL}(\Sigma)_\mathbf{C} \times G(\mathbf{A}_F) \to \mathbf{C}^*.$$

(On a identifié $\operatorname{Pic}^G(\mathscr{X}_F)_\mathbf{C}$ et $\operatorname{PL}(\Sigma)_\mathbf{C}$.)

LEMME 4.1.4. — *Soit* $m \in X^*(G)$ *et notons* $\chi_m \in \mathscr{A}_G$ *le caractère adélique qu'il définit. On a alors*
$$\chi_m(\mathbf{g}) = H(\iota(m), \mathbf{g})^{-i}.$$

Démonstration. — Par définition, $\iota(m)$ est le fibré en droite trivial sur $\mathscr{X}$ muni de la G-linéarisation dans laquelle G agit par multiplication par le caractère algébrique m. Ainsi, la droite de sections rationnelles G-invariante et ne s'annulant pas sur G est engendrée par le caractère m vu comme fonction rationnelle sur $\mathscr{X}$. La définition de H implique que
$$H(\iota(m), \mathbf{g}) = \prod_v \|m(g_v)\|^{-1} = \|m(\mathbf{g})\|^{-1}.$$
Or,
$$\chi_m(\mathbf{g}) = \|m(\mathbf{g})\|^i = H(\iota(m), \mathbf{g})^{-i}.$$
□

4.1.5. Mesures. — Pour toute place v de F, on fixe une mesure de Haar dx_v sur F_v. On suppose que pour presque toute place finie v, la mesure du sous-groupe compact $\mathfrak{o}_v$ est égale à 1. Alors, $dx = \prod_v dx_v$ est une mesure de Haar sur le groupe localement compact $\mathbf{A}_F$. On en déduit pour tout v une mesure de Haar $\mu'_{\mathbf{G}_m, v} = \|x_v\|_v^{-1} dx_v$ sur F_v^*. Pour presque toute place finie v, la mesure de $\mathfrak{o}_v^*$ est égale à $1 - q_v^{-1}$; définissons ainsi, si v est une place finie, $\mu_{\mathbf{G}_m, v} = (1 - q_v^{-1})^{-1} \mu'_{\mathbf{G}_m, v}$. On munit alors $\mathbf{A}_F^*$ de la mesure
$$\prod_v \mu_{\mathbf{G}_m, v} = \prod_{v \nmid \infty} (1 - q_v^{-1})^{-1} \|x_v\|^{-1} dx_v \times \prod_{v \mid \infty} \|x_v\|^{-1} dx_v.$$
Remarquons que $\zeta_{F,v}(1) = (1 - q_v^{-1})^{-1}$ est le facteur local en la place finie v de la fonction zêta de Dedekind du corps F.

Tout $\mathfrak{o}_F$-isomorphisme $G \simeq \mathbf{G}_m^d$ induit alors des mesures de Haar $\mu'_{G,v}$ et $\mu_{G,v} = \zeta_{F,v}(1)^d \mu'_{G,v}$ sur $G(F_v)$ pour toute place v de F, indépendantes de l'isomorphisme. On en déduit aussi une mesure de Haar $\prod \mu_{G,v}$ sur $G(\mathbf{A}_F)$.

D'autre part, le fibré canonique sur $\mathscr{X}$ est métrisé. Peyre a montré dans [15] comment en déduire une mesure sur $\mathscr{X}(\mathbf{A}_F)$. Pour toute place v, on dispose d'une mesure $\mu'_{\mathscr{X}, v}$ sur $\mathscr{X}(F_v)$ définie par la formule
$$\mu'_{\mathscr{X}, v} = \|d\xi_1 \wedge \cdots \wedge d\xi_d\|_v^{-1} d\xi_1 \ldots d\xi_d$$

si $(\xi_1, \ldots, \xi_d)$ est un système arbitraire de coordonnées locales sur $\mathscr{X}(F_v)$. Si l'on restreint la mesure $\mu'_{\mathscr{X},v}$ à $G(F_v)$, on obtient donc

(4.1.6) $$H_v(-\rho, x)\mu'_{G,v},$$

ρ désignant la fonction de PL(Σ) telle que pour tout j, $e_j \mapsto 1$ (ρ correspond à la classe anticanonique).

Pour presque toute place finie v, on a alors

$$\mu'_{\mathscr{X},v}(\mathscr{X}(F_v)) = q_v^{-d} \# \mathscr{X}(k_v).$$

La décomposition cellulaire des variétés toriques (point n'est besoin ici d'invoquer le théorème de Deligne sur les conjectures de Weil) implique alors que

$$\#\mathscr{X}(k_v) = q_v^d + \operatorname{rang}(\operatorname{Pic} \mathscr{X}_F) q_v^{d-1} + O(q_v^{d-2}).$$

Par suite, le produit infini

$$\prod_{v \nmid \infty} \mu'_v(\mathscr{X}(F_v)) \zeta_{F,v}(1)^{-\operatorname{rang}(\operatorname{Pic} \mathscr{X}_F)}$$

est convergent. Définissons une mesure $\mu_{\mathscr{X},v}$ sur $\mathscr{X}(F_v)$ par

$$\mu_{\mathscr{X},v} = \zeta_{F,v}(1)^{-\operatorname{rang} \operatorname{Pic} \mathscr{X}_F} \mu'_{\mathscr{X},v}$$

si v est finie et $\mu_{\mathscr{X},v} = \mu'_{\mathscr{X},v}$ si v est archimédienne. Ainsi, le produit infini $\prod_v \mu_{\mathscr{X},v}$ converge et définit une mesure, dite *mesure de Tamagawa* sur $\mathscr{X}(\mathbf{A}_F)$. Le nombre de Tamagawa de $\mathscr{X}(\mathbf{A}_F)$ est alors définie par

(4.1.7) $$\tau(\mathscr{X}) = \mu(\mathbf{A}_F/F)^{-d} \operatorname{res}_{s=1} \zeta_F(s)^{\operatorname{rang}(\operatorname{Pic} \mathscr{X}_F)} \mu_{\mathscr{X}}(\mathbf{A}_F).$$

Remarque 4.1.8. — La différence de formulation avec la définition que donne Peyre dans [15] n'est qu'apparente. Peyre a choisi la mesure sur F_v de la façon suivante : si v est une place finie, $dx_v(\mathfrak{o}_v) = 1$, si v est une place réelle, dx_v est la mesure de Lebesgue usuelle sur $\mathbf{R}$ et si v est une place complexe, dx_v est le double de la mesure usuelle sur $\mathbf{C}$. Le volume de $\mathbf{A}_F/F$ est alors égal à $\Delta_F^{1/2}$.

4.2. Transformations de Fourier

On s'intéresse à la transformée de Fourier de la fonction $g \mapsto H(-\lambda, g)$ sur le groupe abélien localement compact $G(\mathbf{A}_F)$. Rappelons qu'on a noté Λ_1 l'ensemble des $\lambda \in \operatorname{PL}(\Sigma)$ tels que $\lambda_j > 1$ pour tout j. Alors, si $\lambda \in \mathsf{T}(\Lambda_1)$, la fonction $g \mapsto H(-\lambda, g)$ est intégrable (cf. [19], § 3.4), si bien que la transformée de Fourier existe pour tout $\lambda \in \mathsf{T}(\Lambda_1)$. Elle se décompose par construction en un produit $\check{H} = \check{H}_f \times \check{H}_\infty$, où

$$\check{H}_f = (\operatorname{res}_{s=1} \zeta_F(s))^{-d} \prod_{v \nmid \infty} (1 - q_v^{-1})^{-d} \check{H}_v$$

et $\check{H}_\infty = \prod_{v|\infty} \check{H}_v$ sont les produits des intégrales locales (renormalisées) aux places finies et archimédiennes. (Les transformées de Fourier locales existent même dès que pour tout j, $\operatorname{Re}(\lambda_j) > 0$.)

LEMME 4.2.1. — *Soit $\Lambda_{2/3} \subset \operatorname{PL}(\Sigma)$ la partie convexe définie par $\lambda_j > 2/3$ pour tout j. Il existe une fonction*

$$c_f : \mathsf{T}(\Lambda_{2/3}) \times \mathscr{A}_G \to \mathbf{C}, \quad (\lambda, \chi) \mapsto c_f(\lambda, \chi),$$

holomorphe en λ telle que $\log |c_f|$ est bornée et telle que le produit des transformées de Fourier locales aux places non archimédiennes s'écrive, pour tout $\chi \in \mathscr{A}_G$ et tout $\lambda \in \mathsf{T}(\Lambda)$

$$\check{H}_f(-\lambda, \chi) = c_f(\lambda, \chi) \prod_j L(\lambda_j, \chi_j).$$

Démonstration. — Si χ est fixé, c'est la proposition 2.2.6 de [2]. Le fait que $\log |c_f|$ soit borné indépendamment de χ se déduit immédiatement de la preuve dans *loc. cit.* □

COROLLAIRE 4.2.2. — *La fonction $\check{H}_f$ se prolonge en une fonction méromorphe pour $\lambda \in \mathsf{T}(\Lambda_{2/3})$. Plus précisément, le produit $\prod_j (\lambda_j - 1) \check{H}_f(-\lambda, \chi)$ se prolonge en une fonction holomorphe dans $\mathsf{T}(\Lambda_0)$ et*

$$\lim_{\lambda \to (1, \ldots, 1)} \prod_j (\lambda_j - 1) \check{H}_f(-\lambda, \chi) = 0$$

si et seulement si $\chi \neq \mathbf{1}$.

Comme conséquence facile de l'estimation par Rademacher des valeurs des fonctions L de Hecke pour les caractères non ramifiés, estimation qui repose sur le principe de Phragmén–Lindelöf, on obtient la majoration suivante :

COROLLAIRE 4.2.3. — *Pour tout $\varepsilon > 0$, il existe $0 < \delta < 1/3$ et un réel c_ε tels que si $\operatorname{Re}(\lambda_j) > 1 - \delta$,*

$$\prod_j \frac{|\lambda_j - 1|}{|\lambda_j|} \check{H}_f(-\lambda, \chi) \leqslant c_\varepsilon \bigl(1 + \|\operatorname{Im}(\lambda)\|\bigr)^\varepsilon \bigl(1 + \|\chi\|\bigr)^\varepsilon.$$

Passons maintenant aux places archimédiennes. La proposition suivante précise la proposition 2.3.2 de [2].

PROPOSITION 4.2.4. — *Pour tout compact $K \subset \Lambda_{2/3} \subset \mathrm{PL}(\Sigma)_{\mathbf{R}}$, il existe un réel c_K telle que pour tout $\varphi \in \mathsf{T}(K)$ et tout $m \in M$, on ait la majoration*

$$|\mathscr{F}(m)| \leqslant c_K \frac{1}{1+\|m\|} \sum_{\sigma \in \Sigma(d)} \frac{1+\|\varphi\|_\sigma}{\prod_{j \in J_\sigma}(1+|\langle e_j, m\rangle|)}.$$

COROLLAIRE 4.2.5. — *Désignons par $\widetilde{\Sigma}$ l'éventail $\prod_{v\mid\infty} \Sigma$ dans $\widetilde{N} = \prod_{v\mid\infty} N$. Si $\varphi \in \mathrm{PL}(\Sigma)$, désignons par $\widetilde{\varphi}$ la fonction $\widetilde{N} \to \mathbf{R}$ définie par $(n_v)_v \mapsto \sum \varphi(n_v)$. Pour tout compact K de $\mathrm{PL}(\Sigma)$ contenu dans $\Lambda_{2/3}$, il existe une constante c_K telle que pour tout $\varphi \in \mathsf{T}(K)$ et tout $\chi \in \mathscr{A}_G$ décomposé sous la forme $\chi = im + \chi_u \in iM \oplus \mathscr{U}_G$, on ait*

$$\check{H}_\infty(\varphi, \chi) \leqslant \frac{c_K}{1+\|\chi\|} \sum_{\widetilde{\sigma} \in \widetilde{\Sigma}} \frac{1+\|\mathrm{Im}\,\widetilde{\varphi}\|_{\widetilde{\sigma}}}{\prod_{e \in \widetilde{\sigma}}(1+|\langle e, \mathrm{Im}\,\widetilde{\varphi}|_{\widetilde{\sigma}} + \widetilde{m}\rangle|)}.$$

Démonstration. — Si l'on note $\widetilde{m} = (m_v)_v$ la décomposition de χ à l'infini, on remarque que

$$\check{H}_\infty(\varphi, \chi) = \prod_{v\mid\infty} \check{H}_v(\varphi, \chi) = \prod_{v\mid\infty} \mathscr{F}(\varphi, m_v) = \mathscr{F}(\widetilde{\varphi}, \widetilde{m}).$$

Il suffit alors d'appliquer la proposition précédente. □

Preuve de la proposition 4.2.4. — Il faut estimer

$$\mathscr{F}(m) = \int_N \exp(-\varphi(v) - i\langle v, m\rangle)\, dv.$$

Soit $\sigma \in \Sigma$ un cône de base $(e_1, \ldots, e_d)$. Si $|\det(e_j)|$ désigne la mesure du parallélotope de base les e_j, on a

$$\int_\sigma \exp(-\varphi(v) - i\langle v, m\rangle)\, dv$$

$$= \int_{\mathbf{R}_+^d} \prod_{j=1}^d \exp\bigl(-t_j(\varphi(e_j) + i\langle e_j, m\rangle)\bigr)\, |\det(e_j)| \prod dt_j$$

$$= c(\sigma) \prod_{j=1}^d \frac{1}{\varphi(e_j) + i\langle e_j, m\rangle}.$$

Ainsi, on a

(4.2.6) $$\mathscr{F}(m) = \sum_\sigma c(\sigma) \prod_{e\in\sigma} \frac{1}{\varphi(e) + i\langle e,m\rangle}.$$

D'autre part, supposons que $m_j \neq 0$, on peut intégrer par parties et écrire

$$\mathscr{F}(m) = \int_N \frac{1}{im_j}\left(-\frac{\partial\varphi}{\partial v_j}\right)\exp(-\varphi(v) - i\langle v,m\rangle)\,dv$$

$$-im_j\mathscr{F}(m) = \int_N \left(\frac{\partial\varphi}{\partial v_j}\right)\exp(-\varphi(v) - i\langle v,m\rangle)\,dv$$

$$= \sum_\sigma \left.\frac{\partial\varphi}{\partial v_j}\right|_\sigma \int_\sigma \exp(-\varphi(v) - i\langle v,m\rangle)\,dv$$

(4.2.7) $$= \sum_\sigma c(\sigma)\left.\frac{\partial\varphi}{\partial v_j}\right|_\sigma \prod_{e\in\sigma}\frac{1}{\varphi(e) + i\langle e,m\rangle}.$$

En combinant les égalités (4.2.6) et (4.2.7) pour tous les indices j tels que $m_j \neq 0$, on obtient une majoration

$$|\mathscr{F}(m)| \leqslant \frac{1}{1 + \|m\|}\sum_\sigma c(\sigma)\frac{1 + \|\varphi\|_\sigma}{\prod_{e\in\sigma}|\varphi(e) + i\langle e,m\rangle|}.$$

Finalement, comme $\varphi \in \mathsf{T}(K)$, on a une estimation

$$|\varphi(e) + i\langle e,m\rangle| \gg 1 + |\mathrm{Im}(\varphi)(e) + \langle e,m\rangle|$$

et la proposition s'en déduit. □

4.3. Définition d'une classe de contrôle

Soit β un réel strictement positif. Si M et V sont deux $\mathbf{R}$-espaces vectoriels de dimension finie avec $M \subset V$, notons $\mathscr{D}_{\beta,\varepsilon}(M,V)$ le sous-monoïde de $\mathscr{F}(V,\mathbf{R}_+)$ engendré par les fonctions $h : V \to \mathbf{R}_+$ telles que pour tout $\varepsilon > 0$, il existe $c > 0$, $\varepsilon \in {]}0;1{[}$ et une famille (ℓ_j) de formes linéaires sur V vérifiant :
- la famille $(\ell_j|_M)$ forme une base de M^* ;
- pour tout $v \in V$ et tout $m \in M$, on a

(4.3.1) $$h(v + m) \leqslant c\frac{(1 + \|v\|)^\beta}{(1 + \|m\|)^{1-\varepsilon}}\frac{1}{\prod(1 + |\ell_j(v+m)|)}.$$

Notons alors $\mathscr{D}_\beta = \bigcap_{\varepsilon > 0}\mathscr{D}_{\beta,\varepsilon}$.

PROPOSITION 4.3.2. — *Les $\mathscr{D}_\beta(M,V)$ définissent une classe de contrôle au sens de la définition 3.1.1.*

La preuve de cette proposition consiste en une série d'inégalités faciles mais techniques. Nous la repoussons à l'appendice B.

4.4. La fonction zêta des hauteurs et la formule de Poisson

On s'intéresse fonction zêta des hauteurs de $\mathscr{X}$ restreinte à l'ouvert dense formé par le tore G; c'est par définition la série génératrice

$$Z(\lambda) = \sum_{x \in G(F)} H(-\lambda, x),$$

quand elle converge. Des théorèmes taubériens standard (voir l'appendice) permettront de déduire de résultats analytiques sur Z un développement asymptotique du nombre de points de hauteur bornée

$$N(\lambda, H) = \#\{x \in G(F)\,;\, H(\lambda, x) \leqslant H\}.$$

LEMME 4.4.1. — *Lorsque* $\mathrm{Re}(\lambda)$ *décrit un compact de* Λ_1, *la fonction zêta des hauteurs converge uniformémént en* λ. *Plus généralement, la série*

$$\sum_{x \in G(F)} H(-\lambda, x\mathbf{g})$$

converge absolument uniformément lorsque $\mathrm{Re}(\lambda)$ *décrit un compact de* Λ_1 *et* $\mathbf{g}$ *un compact de* $G(\mathbf{A}_F)$.

Démonstration. — Compte tenu d'estimations pour $H(-\lambda, x\mathbf{g})/H(-\lambda, x)$ lorsque $\mathbf{g}$ décrit un compact de $G(\mathbf{A}_F)$, $x \in G(F)$ et $\lambda \in \mathsf{T}(\Lambda_1)$, c'est en fait un corollaire de l'intégrabilité de la fonction $H(-\lambda, \cdot)$ sur $G(\mathbf{A}_F)$. Voir [**4**], Th. 4.2 et aussi [**19**], Prop. 4.3. □

Par conséquent, on peut appliquer la formule sommatoire de Poisson sur le tore adélique $G(\mathbf{A}_F)$ pour le sous-groupe discret $G(F)$. Compte tenu de l'invariance de l'accouplement de hauteurs par le sous-groupe compact maximal $\mathbf{K}_G$ de $G(\mathbf{A}_F)$, on en déduit la formule

$$(4.4.2) \qquad Z(\lambda) = \int_{\mathscr{A}_G} \check{H}(-\lambda, \chi)\, d\chi$$

où $d\chi$ est la mesure de Haar sur le groupe $\mathscr{A}_G$ des caractères unitaires continus sur le groupe $G(F)\backslash G(\mathbf{A}_F)/\mathbf{K}_G$ duale de la mesure de comptage sur $G(F)$.

Rappelons que l'on a décomposé le groupe $\mathscr{A}_G = M \oplus \mathscr{U}_G$, où $\mathscr{U}_G$ est un groupe discret. De plus, si $\chi = m \oplus \chi_u$,

$$\check{H}(-\lambda, \chi) = \check{H}(-\lambda - im, \chi_u)$$

si bien que
$$Z(\lambda) = \int_M \left(\sum_{\chi_u \in \mathscr{U}_G} \check{H}(-\lambda - im, \chi_u) \right) dm$$
où dm est la mesure de Lebesgue sur M telle que $dm\, d\chi_u = d\chi$, $d\chi_u$ étant la mesure de comptage sur $\mathscr{U}_G$.

LEMME 4.4.3. — *Si d^0m est la mesure de Lebesgue sur M définie par le réseau M, on a*
$$dm = (2\pi \operatorname{vol}(\mathbf{A}_F/F) \operatorname{res}_{s=1} \zeta_F(s))^{-d} d^0m.$$

Démonstration. — Par multiplicativité, il suffit de traiter le cas $G = \mathbf{G}_m$ et $d = 1$. Notons $\mathbf{A}_F^1$ le sous-groupe de $\mathbf{A}_F^*$ formé des x tels que $\|x\| = 1$. La suite exacte
$$1 \to \mathbf{A}_F^1/F^* \to \mathbf{A}_F^*/F^* \xrightarrow{\log\|x\|} \mathbf{R} \to 0$$
permet de munir $\mathbf{A}_F^1/F^*$ de la mesure de Haar dx^1 telle que $d^*x = dx^1\, d^0n$. La suite exacte duale
$$1 \to \mathbf{R} \to (\mathbf{A}_F^*/F^*)^* \to (\mathbf{A}_F^1/F^*)^* \to 1$$
et la discrétude du groupe des caractères de $\mathbf{A}_F^1/F^*$ permet de munir $(\mathbf{A}_F^*/F^*)^*$ de le mesure $d^0m \sum$. Avec ces normalisations, la constante devant la formule de Poisson est $(2\pi \operatorname{vol}(\mathbf{A}_F^1/F^*))^{-1}$. Compte tenu des normalisations choisies, le théorème classique selon lequel $\tau(\mathbf{G}_m) = \tau(\mathbf{G}_a) = 1$, cf. par exemple [20], p. 116, devient
$$\operatorname{vol}(\mathbf{A}_F^1/F^*) = \operatorname{vol}(\mathbf{A}_F/F) \operatorname{res}_{s=1} \zeta_F(s),$$
d'où le lemme. □

Soit $\rho = (1,\ldots,1) \in \operatorname{PL}(\Sigma)$. On décale la fonction zêta des hauteurs de ρ : si $\lambda \in \operatorname{PL}(\Sigma)^+$,
$$Z(\rho + \lambda) = \int_M \left(\sum_{\chi_u \in \mathscr{U}_G} \check{H}(-\lambda - \rho - im, \chi_u) \right) dm$$
Soit F la fonction $\operatorname{PL}(\Sigma)^+ \to \mathbf{C}$ définie par la série
$$\lambda \mapsto (\operatorname{vol}(\mathbf{A}_F/F) \operatorname{res}_{s=1} \zeta_F(s))^{-d} \sum_{\chi_u \in \mathscr{U}_G} \check{H}(-1 - \lambda, \chi_u),$$

de sorte que si $\lambda \in \mathrm{PL}(\Sigma)^+$,

(4.4.4) $$Z(\lambda + \rho) = \frac{1}{(2\pi)^d} \int_M F(\lambda + im) \, d^0 m.$$

PROPOSITION 4.4.5. — *Si $\beta > 1$, la fonction F appartient à $\mathscr{H}_M(\mathrm{PL}(\Sigma)^+)$ défini par la classe de contrôle $\mathscr{D}_\beta$ du paragraphe 4.3.*

De plus, pour tout $\lambda \in \mathrm{PL}(\Sigma)^+$,
$$\lim_{s \to 0} \frac{F(s\lambda)}{\mathsf{X}_{\mathrm{PL}(\Sigma)^+}(s\lambda)} = \tau(\mathscr{X}),$$

le nombre de Tamagawa de $\mathscr{X}$.

Démonstration. — On a vu que l'on pouvait écrire
$$\check{H}(-\rho - \lambda, \chi) = c_f(\lambda + \rho, \chi) \check{H}_\infty(-\rho\lambda, \chi) \prod_j L(\lambda_j + 1, \chi_j).$$

Par suite, la fonction
$$\lambda \mapsto \check{H}(-\rho - \lambda, \chi) \prod_j \frac{\lambda_j}{\lambda_j + 1}$$

admet un prolongement holomorphe pour $\mathrm{Re}(\lambda_j) > -1$.

De plus, il résulte des corollaires 4.2.3 et 4.2.5 que pour tout $\varepsilon > 0$, il existe $\delta < 1/3$ tel que si pour tout j on a $\mathrm{Re}(\lambda_j) > -\delta$, alors

$$\left| \check{H}(-\rho - \lambda, \chi) \prod_j \frac{\lambda_j}{\lambda_j + 1} \right|$$
$$\ll \frac{(1 + \|\mathrm{Im}(\lambda)\|)^{1+\varepsilon}}{(1 + \|\chi_\infty\|)^{1-\varepsilon}} \sum_{\tilde{\sigma} \in \tilde{\Sigma}(d)} \frac{1}{\prod_{e \in \tilde{\sigma}_1} (1 + |\langle e, \mathrm{Im}(\lambda)|_{\tilde{\sigma}} + \chi_\infty \rangle|)},$$

formule dans laquelle χ_∞ désigne l'image de χ par l'homomorphisme de noyau fini « type à l'infini » $\mathscr{A}_G \to M_\infty = \bigoplus_{v|\infty} M$. Ainsi, on obtient un prolongement holomorphe de la fonction $\Phi : \lambda \mapsto F(\lambda) \prod_j \lambda_j/(1+\lambda_j)$ pour $\mathrm{Re}(\lambda_j) > -\delta$ si l'on prouve que pour tout $\tilde{\sigma} \in \tilde{\Sigma}(d)$, la série

$$\sum_{\chi_u \in \mathscr{U}_G} \frac{1}{(1 + \|\chi_{u,\infty}\|)^{1-\varepsilon}} \frac{1}{\prod_{e \in \tilde{\sigma}_1} (1 + |\langle e, \mathrm{Im}(\lambda)|_{\tilde{\sigma}} + \chi_{u,\infty} \rangle|)}$$

converge localement uniformément en λ si $\mathrm{Re}(\lambda_j) > -\delta$. Fixons $\tilde{\sigma} \in \tilde{\Sigma}(d)$. Alors, lorsque $e \in \tilde{\sigma}_1$, les formes linéaires $\langle e, \cdot \rangle$ forment une base de M_∞^*. Il est facile de remplacer la sommation sur le sous-groupe discret $\mathcal{U}_{G,\infty}$ par une intégrale sur l'espace vectoriel qu'il engendre, lequel est d'ailleurs un supplémentaire de M envoyé diagonalement dans M_∞. La convergence est alors une conséquence de la proposition B.3.

Pour obtenir l'assertion sur la croissance de F, il faut montrer que si $\beta > 1$, K est un compact de $\mathrm{PL}(\Sigma)^+$, $\lambda \in \mathsf{T}(K)$ et $m \in M$, on a une majoration

$$|\Phi(\lambda + im)| \ll \frac{(1 + \|\mathrm{Im}(\lambda)\|)^\beta}{(1 + \|m\|)^{1-\varepsilon}} \sum_\alpha \prod_k \frac{1}{1 + |\ell_{\alpha,k}(\mathrm{Im}(\lambda) + m)|}$$

où α parcourt un ensemble fini et où pour tout α, $\{\ell_{\alpha,k}\}_k$ est une base de $\mathrm{PL}(\Sigma)^*$. Il nous faut récrire un peu différemment la majoration de $\check{H}$ obtenue ci-dessus en remarquant que si la forme des transformées de Fourier aux places finies fournit le prolongement méromorphe, la convergence de la série provient, elle, des estimations archimédiennes. On écrit ainsi

$$\check{H}(-\rho - \lambda - im, \chi_u)$$
$$= c_f(\rho + \lambda + im, \chi_u) \prod_j L(\lambda_j + 1 + im, \chi_{u,j}) \check{H}_\infty(-\rho - \lambda, \chi_m \chi_u)$$

et donc

$$\left| \check{H}(-\rho - \lambda - im, \chi_u) \prod_j \frac{\lambda_j + im}{1 + \lambda_j + im} \right|$$
$$\ll \frac{(1 + \|\mathrm{Im}(\lambda) + m\|)^\varepsilon (1 + \|\chi_u\|)^\varepsilon}{1 + \|m + \chi_{u,\infty}\|} \sum_{\widetilde{\sigma} \in \widetilde{\Sigma}(d)} \frac{1 + \|\mathrm{Im}(\lambda)\|_{\widetilde{\sigma}}}{\prod_{e \in \widetilde{\sigma}_1}(1 + |\langle e, \mathrm{Im}(\lambda)\rangle_{\widetilde{\sigma}} + m + \chi_\infty\rangle|)}.$$

Par suite,

$$|\Phi(\lambda + im)| \leqslant \sum_{\widetilde{\sigma} \in \widetilde{\Sigma}(d)} (1 + \|\mathrm{Im}(\lambda)\|_{\widetilde{\sigma}})(1 + \|\mathrm{Im}(\lambda) + m\|)^\varepsilon G_{\widetilde{\sigma}}(\mathrm{Im}(\lambda), m)$$

où $\Phi_{\widetilde{\sigma}(\varphi, m)}$ est défini par la série

$$\Phi_{\widetilde{\sigma}}(\varphi, m) = \sum_{\chi_u \in \mathcal{U}_G} \frac{(1 + \|\chi_u\|)^\varepsilon}{1 + \|m + \chi_{u,\infty}\|} \prod_{e \in \widetilde{\sigma}_1} \frac{1}{1 + |\langle e, \widetilde{\varphi}\rangle_{\widetilde{\sigma}} + m + \chi_{u,\infty}\rangle|}.$$

On a la majoration

$$1 + \|\chi_u\| \leqslant 1 + \|m + \chi_{u,\infty}\| + \|m\| \leqslant (1 + \|m + \chi_{u,\infty}\|)(1 + \|m\|)$$

et comme précédemment, on remplace la sommation sur le sous-groupe discret $\mathscr{U}_G$ par l'intégrale sur l'espace vectoriel qu'il engendre. La proposition B.3 fournit alors pour tout $\varepsilon' > \varepsilon$ une estimation

$$G_{\widetilde{\sigma}}(\varphi, m) \ll \frac{1}{(1+\|m\|)^{1-\varepsilon'}} \sum_\alpha \prod_k \frac{1}{1+|\ell_{\alpha,k}(m+\varphi|_{\widetilde{\sigma}})|}$$

où $\{\ell_{\alpha,k}\}_k$ est une base de M^* et $\varphi|_{\widetilde{\sigma}}$ l'élément de M qui coïncide avec $(\varphi, \ldots, \varphi)$ de $\bigoplus_{v\mid\infty} \mathrm{PL}(\Sigma)$ sur le cône $\widetilde{\sigma}$ de l'éventail $\widetilde{\Sigma}$. L'application $\varphi \mapsto \ell_{\alpha,k}(\varphi|_{\widetilde{\sigma}})$ est une forme linéaire $\ell_{\widetilde{\sigma},\alpha,k}$ sur $\mathrm{PL}(\Sigma)$. On a ainsi

$$|G(\lambda + im)| \ll \frac{(1+\|\mathrm{Im}(\lambda)\|)(1+\|\mathrm{Im}(\lambda)+m\|)^\varepsilon}{(1+\|m\|)^{1-\varepsilon-\varepsilon'}} \times$$
$$\times \sum_{\widetilde{\sigma}} \sum_\alpha \prod_k \frac{1}{1+|\ell_{\widetilde{\sigma},\alpha,k}(\mathrm{Im}(\lambda)+m)|}$$
$$\ll \frac{(1+\|\mathrm{Im}(\lambda)\|)^{1+\varepsilon}}{(1+\|m\|)^{1-2\varepsilon-\varepsilon'}} \sum_{\widetilde{\sigma}} \sum_\alpha \prod_k \frac{1}{1+|\ell_{\widetilde{\sigma},\alpha,k}(\mathrm{Im}(\lambda)+m)|}.$$

Comme on peut prendre ε et ε' arbitrairement petits, la contrôlabilité est établie.

Il reste à calculer la limite quand $s \to 0$ par valeurs supérieures de l'expression $F(s\lambda)/\mathsf{X}_{\mathrm{PL}(\Sigma)^+}(s\lambda)$. Le cône $\mathrm{PL}(\Sigma)^+$ est simplicial et

$$\mathsf{X}_{\mathrm{PL}(\Sigma)^+}(\lambda) = \frac{1}{\prod_j \lambda_j}.$$

Ainsi,

$$\frac{F(\lambda)}{\mathsf{X}_{\mathrm{PL}(\Sigma)^+}(\lambda)} = (\mathrm{vol}(\mathbf{A}_F/F)\,\mathrm{res}_{s=1}\,\zeta_F(s))^{-d} \prod_j \lambda_j \sum_{\chi_u \in \mathscr{U}_G} \check{H}(-\lambda-\rho, \chi_u).$$

D'après ce qui précède, la série qui définit F converge uniformément pour $\mathrm{Re}(\lambda_j) > -\delta$; cela permet de permuter sommation et limite, si bien que

$$\lim_{s\to 0^+} \frac{F(s\lambda)}{\mathsf{X}_{\mathrm{PL}(\Sigma)^+}(s\lambda)}$$
$$= (\mathrm{vol}(\mathbf{A}_F/F)\,\mathrm{res}_{s=1}\,\zeta_F(s))^{-d} \sum_{\chi_u \in \mathscr{U}_G} \left(\lim_{s\to 0^+} \check{H}(-s\lambda-\rho, \chi_u) \prod_j (s\lambda_j) \right).$$

En écrivant,
$$\check{H}(-s\lambda-\rho,\chi)\prod_j(s\lambda_j) = c_f(s\lambda,\chi)\prod_j\left(s\lambda_j L(s\lambda_j+1,\chi_j)\right)\check{H}_\infty(s\lambda,\chi),$$

on voit que la limite est nulle si l'un des $\chi_j \neq \mathbf{1}$ (car une des fonctions $L(\cdot,\chi_j)$ n'a pas de pôle en 1, les autres ont au plus un pôle simple). Étudions maintenant le cas $\chi = \mathbf{1}$. Utilisant la formule (4.1.6), il vient

$$\check{H}(-s\lambda-\rho,\mathbf{1})\prod_j \zeta_F(1+\lambda_j s)^{-1}$$
$$= \prod_{v\nmid\infty}\zeta_v(1)^d\prod_j\zeta_v(1+\lambda_j s)^{-1}\int_{G(F_v)} H(-s\lambda-\rho,x)\mu'_{G,v}\times$$
$$\times \prod_{v\mid\infty}\int_{G(F_v)} H(-s\lambda-\rho,x)\mu'_{G,v}$$
$$= \prod_{v\nmid\infty}\zeta_v(1)^d\prod_j\zeta_v(1+\lambda_j s)^{-1}\int_{G(F_v)} H(-s\lambda)\mu'_{\mathscr{X},v}\times$$
$$\times \prod_{v\mid\infty}\int_{G(F_v)} H(-s\lambda)\mu'_{\mathscr{X},v}.$$

C'est un produit eulérien absolument convergent pour $\mathrm{Re}(s) > -\varepsilon$, d'où un prolongement par continuité en $s=0$, de valeur

$$\prod_{v\nmid\infty}\zeta_v(1)^{d-\#J}\mu'_{\mathscr{X},v}(\mathscr{X}(F_v))\prod_{v\mid\infty}\mu'_{\mathscr{X},v}(\mathscr{X}(F_v))$$
$$= \tau(\mathscr{X})\mu(\mathbf{A}_F/F)^d(\mathrm{res}_{s=1}\zeta_F(s))^{-\mathrm{rang}(\mathrm{Pic}\,\mathscr{X}_F)}$$

en vertu de la définition (4.1.7) de la mesure de Tamagawa de $\mathscr{X}(\mathbf{A}_F)$. Ainsi,

$$\lim_{s\to 0}\check{H}(-s\lambda-\rho,\mathbf{1})(\prod_j s\lambda_j)$$
$$= (\mathrm{res}_{s=1}\zeta_F(s))^{\#J}\lim_{s\to 0}\check{H}(-s\lambda-\rho,\mathbf{1})\prod_j\zeta_F(1+s\lambda_j)^{-1}$$
$$= \tau(\mathscr{X})\mu(\mathbf{A}_F/F)^d(\mathrm{res}_{s=1}\zeta_F(s))^d.$$

Finalement, on a donc

$$\lim_{s\to 0} F(\lambda s) \mathsf{X}_{\mathrm{PL}^+(\Sigma)}(\lambda s)^{-1}$$
$$= (\mathrm{vol}(\mathbf{A}_F/F)\,\mathrm{res}_{s=1}\zeta_F(s))^{-d}\mu(\mathbf{A}_F/F)^d(\mathrm{res}_{s=1}\zeta_F(s))^d\tau(\mathscr{X})$$
$$= \tau(\mathscr{X}),$$

ainsi qu'il fallait démontrer. □

L'équation (4.4.4) et le théorème 3.1.14 impliquent alors le théorème suivant.

THÉORÈME 4.4.6. — *La fonction zêta des hauteurs (décalée)*

$$\lambda \mapsto Z(\rho + \lambda)$$

converge localement uniformément sur le tube $\mathsf{T}(\mathrm{PL}(\Sigma)^+)$ *et définit une fonction holomorphe sur* $\mathsf{T}(\Lambda^0_{\mathrm{eff}}(\mathscr{X}_F))$. *Si* $\beta > 1$ *et si* $\mathscr{D}_\beta$ *désigne la classe de contrôle introduite au sous-paragraphe 4.3, elle appartient à l'espace*

$$\mathscr{H}_{\{0\}}(\Lambda^0_{\mathrm{eff}}(\mathscr{X}_F); \Lambda^0_{\mathrm{eff}}(\mathscr{X}_F))$$

(défini en 3.1.12) des fonctions méromorphes $\{0\}$-*contrôlées dont les pôles sont simples et donnés par les faces du cône* $\Lambda^0_{\mathrm{eff}}(\mathscr{X}_F)$.

De plus, pour tout $\lambda \in \Lambda^0_{\mathrm{eff}}(\mathscr{X}_F)$,

$$\lim_{s\to 0} \frac{Z(s\lambda + \rho)}{\mathsf{X}_{\Lambda^0_{\mathrm{eff}}}(s\lambda)} = \tau(\mathscr{X}).$$

En spécialisant la fonction zêta des hauteurs à la droite $\mathbf{C}\rho$ qui correspond au fibré en droite anticanonique, on obtient le corollaire :

COROLLAIRE 4.4.7. — *Si* $\beta > 1$, *il existe* $\varepsilon > 0$, *une fonction* f *holomorphe pour* $\mathrm{Re}(s) \geqslant 1 - \varepsilon$ *telle que*

(i) $f(1) = \tau(\mathscr{X})$;
(ii) *Pour tout* $\sigma \in [1-\varepsilon; 1+\varepsilon]$ *et tout* $\tau \in \mathbf{R}$, $|f(\sigma+i\tau)| \ll (1+|\tau|)^\beta$;
(iii) *Pour tout* $\sigma > 1$ *et tout* $\tau \in \mathbf{R}$, $Z(s\omega) = \left(\frac{s}{s-1}\right)^r f(s)$.

COROLLAIRE 4.4.8. — *Si* r *désigne le rang de* $\mathrm{Pic}(\mathscr{X}_F)$, *il existe un polynôme unitaire* P *de degré* $r-1$ *et un réel* $\varepsilon > 0$ *tels que pour tout* $H > 0$,

$$N(\omega_{\mathscr{X}}^{-1}; H) = \frac{\tau(\mathscr{X})}{(r-1)!} H P(\log H) + O(H^{1-\varepsilon}).$$

Lorsque $F = \mathbf{Q}$ et lorsque la variété torique $\mathscr{X}$ est projective et telle que $\omega_{\mathscr{X}}^{-1}$ est engendré par ses sections globales, ce corollaire avait été démontré précédemment par R. de la Bretèche. Sa méthode est différente ; elle est fondée sur le travail de P. Salberger [18] et une étude fine des sommes de fonctions arithmétiques en plusieurs variables (voir [7, 6] et [8] pour un cas particulier).

5. Application aux fibrations en variétés toriques

5.1. Holomorphie

Soit $\mathscr{B}$ un S-schéma projectif et plat. Soit $\mathscr{T} \to \mathscr{B}$ un G-torseur, et notons $\eta : X^*(G) \to \mathrm{Pic}(\mathscr{B})$ l'homomorphisme de fonctorialité des torseurs. Fixons un relèvement $\widehat{\eta} : X^*(G) \to \widehat{\mathrm{Pic}}(\mathscr{B})$ de cet homomorphisme (c'est-à-dire, un choix de métriques hermitiennes à l'infini sur les images d'une base de $X^*(G)$, prolongés par multiplicativité à l'image de η).

Donnons nous une S-variété torique lisse $\mathscr{X}$, compactification équivariante de G. Soit $\mathscr{Y}$ le S-schéma obtenu par les constructions du § 2.1.

On obtient alors un diagramme canonique, qui provient des propositions 2.1.11, 2.3.6, du théorème 2.2.4 et de l'oubli des métriques hermitiennes :

$$(5.1.1) \quad \begin{array}{ccccccccc} 0 & \longrightarrow & X^*(G) & \longrightarrow & \mathrm{Pic}^G(\mathscr{X}_F) \oplus \mathrm{Pic}(\mathscr{B}_F) & \longrightarrow & \mathrm{Pic}(\mathscr{Y}_F) & \longrightarrow & 0 \\ & & \| & & \uparrow & & \uparrow & & \\ 0 & \longrightarrow & X^*(G) & \longrightarrow & \widehat{\mathrm{Pic}}^{G,\mathbf{K}}(\mathscr{X}) \oplus \widehat{\mathrm{Pic}}(\mathscr{B}) & \longrightarrow & \widehat{\mathrm{Pic}}(\mathscr{Y}) & & \end{array}$$

Le schéma $\mathscr{Y}$ contient $\mathscr{T}$ comme ouvert dense. On s'intéresse à la fonction zêta des hauteurs de $\mathscr{T}$. Lorsque $\lambda \in \mathrm{Pic}^G(\mathscr{X}_F)_{\mathbf{C}}$, notons $\widehat{\lambda}$ l'image de λ par l'homomorphisme (4.1.3). Si de plus $\widehat{\alpha} \in \widehat{\mathrm{Pic}}(\mathscr{B})$, on notera enfin

$$Z(\widehat{\lambda}, \widehat{\alpha}) = Z(\vartheta(\widehat{\lambda}) \otimes \pi^*\widehat{\alpha}; \mathscr{Y}) = \sum_{y \in \mathscr{T}(F)} H(\vartheta(\widehat{\lambda}) \otimes \pi^*\widehat{\alpha}; y)^{-1}.$$

PROPOSITION 5.1.2. — *Soient* $\widehat{\Lambda} \subset \widehat{\mathrm{Pic}}(\mathscr{B})_{\mathbf{R}}$ *une partie convexe telle que* $Z(\widehat{\alpha}; \mathscr{B})$ *converge normalement si la partie réelle de* $\widehat{\alpha} \in \widehat{\mathrm{Pic}}(\mathscr{B})_{\mathbf{C}}$ *appartient à* Λ.

Alors, la fonction zêta des hauteurs de $\mathscr{T}$ *converge absolument pour tout* $(\widehat{\lambda}, \widehat{\alpha})$ *tel que la partie réelle de* $\lambda \otimes \omega_{\mathscr{X}}$ *appartient à* $\Lambda_{\mathrm{eff}}^0(\mathscr{X}_F)$ *et la partie réelle de* α *appartient à* Λ. *La convergence est de plus uniforme si la partie la partie réelle de* $\lambda \otimes \omega_{\mathscr{X}}$ *décrit un compact de* $\Lambda_{\mathrm{eff}}^0(\mathscr{X}_F)$.

Démonstration. — On peut décomposer la fonction zêta des hauteurs de $\mathscr{T}$ en écrivant

(5.1.3) $$Z(\widehat{\lambda}, \widehat{\alpha}) = \sum_{b \in \mathscr{B}(F)} H(\widehat{\alpha}; b)^{-1} Z(\vartheta(\widehat{\lambda}); \mathscr{T}|_b).$$

D'après la remarque 2.4.6, le fibré inversible λ admet une section G-invariante s qui n'a ni pôles ni zéros sur l'ouvert $G \subset \mathscr{X}$. En utilisant cette section, on obtient, en vertu du théorème 2.4.8 et de la proposition 2.4.3 une égalité

(5.1.4) $$Z(\vartheta(\widehat{\lambda}); \mathscr{T}|_b) = \sum_{x \in G(F)} H(\widehat{\lambda}, \mathsf{s}; \mathbf{g}_b \cdot x)^{-1},$$

où $\mathbf{g}_b \in G(\mathbf{A}_F)$ représente la classe du G-torseur arithmétique $\widehat{\mathscr{T}}|_b$. On rappelle que si $\mathbf{x} \in G(\mathbf{A}_F)$, on a une expression de la hauteur en produit de hauteurs locales

$$H(\widehat{\lambda}, \mathsf{s}, \mathbf{x}) = \prod_v \|\mathsf{s}\|_v (x_v)^{-1}.$$

On peut appliquer la formule sommatoire de Poisson sur le tore adélique $G(\mathbf{A}_F)$, d'où, en utilisant l'invariance des hauteurs locales par les sous-groupes compacts maximaux,

(5.1.5) $$Z(\vartheta(\widehat{\lambda}); \mathscr{T}|_b) = \int_{\mathscr{A}_G} \chi^{-1}(\mathbf{g}_b) \check{H}(-\widehat{\lambda}; \chi) \, d\chi$$

où l'intégration est sur le groupe $\mathscr{A}_G$ des caractères (unitaires continus) du groupe localement compact $G(F) \backslash G(\mathbf{A}_F)/K_G$, muni de son unique mesure de Haar $d\chi$ qui permet cette formule.

L'utilisation de la formule de Poisson est justifiée par le fait que les deux membres convergent absolument. La série du membre de gauche est traitée dans [4], Theorem 4.2, lorsque $\mathbf{g}_b = 1$, c'est-à-dire lorsqu'il n'y a pas de torsion. Comme il existe une constante $C(\lambda, \mathbf{g}_b)$ ne dépendant que de $\mathbf{g}_b$ et $\widehat{\lambda}$ telle que

$$\left| H(\widehat{\lambda}, \mathsf{s}; \mathbf{g}_b \cdot x) \right|^{-1} \leqslant C(\lambda, \mathbf{g}_b) \left| H(\widehat{\lambda}, \mathsf{s}; x) \right|^{-1}$$

et comme $H(\widehat{\lambda}, \mathsf{s}; x) = H(\widehat{\lambda}; x)$, la convergence absolue du membre de gauche en résulte. (Voir aussi le lemme 4.4.1.) Quant à l'intégrale du membre de droite, on peut négliger le caractère χ dont la valeur absolue est 1 et on retrouve une intégrale dont la convergence absolue est prouvée dans [4] (preuve du théorème 4.4). Cela prouve aussi que lorsque $\mathrm{Re}(\lambda)$ décrit un compact de $\omega_{\mathscr{X}}^{-1} + \Lambda_{\mathrm{eff}}^0(\mathscr{X}_F)$, la fonction zêta des hauteurs $Z(\vartheta(\widehat{\lambda}); \mathscr{T}|_b)$ de la fibre en $b \in \mathscr{B}(F)$ est bornée indépendamment de b.

En reportant cette majoration dans la décomposition (5.1.3), il en résulte la convergence absolue de la fonction zêta des hauteurs de $\mathscr{T}$ lorsque la partie réelle de $\widehat{\alpha}$ appartient à $\widehat{\Lambda}$ et $\lambda \otimes \omega_{\mathscr{X}}$ appartient à $\Lambda^0_{\text{eff}}(\mathscr{X}_F)$, uniformément lorsque $\lambda \otimes \omega_{\mathscr{X}}$ décrit un compact de ce cône. □

Dans [9], définition 1.4.1, on a défini la notion de fonction L d'Arakelov attachée à un torseur arithmétique et à une fonction sur un espace adélique. Appliquée au $G \times \mathbf{G}_m$-torseur arithmétique sur $\mathscr{B}$ défini par $\widehat{\mathscr{T}} \times_{\mathscr{B}} \widehat{\alpha}$ et à la fonction $\chi^{-1} \cdot \|\cdot\|$, la définition devient

$$L(\widehat{\mathscr{T}} \boxtimes \widehat{\alpha}, \chi^{-1} \boxtimes \|\cdot\|) = \sum_{b \in \mathscr{B}(F)} \chi^{-1}(\mathbf{g}_b) H(\widehat{\alpha}; b)^{-1}.$$

(On a utilisé le fait que $\mathbf{g}_b \in G(F)\backslash G(\mathbf{A}_F)/K_G$ est la classe du G-torseur arithmétique $\mathscr{T}|_b$.)

Un corollaire de la démonstration de la proposition précédente est alors le suivant :

COROLLAIRE 5.1.6. — *Sous les hypothèses de la proposition 5.1.2, on a la formule*

$$Z(\widehat{\lambda}, \widehat{\alpha}) = \int_{\mathscr{A}_G} \check{H}(-\widehat{\lambda}; \chi) L(\widehat{\mathscr{T}} \boxtimes \widehat{\alpha}, \chi^{-1} \boxtimes \|\cdot\|) \, d\chi.$$

Démonstration. — Compte tenu de la majoration établie à la fin de la preuve du théorème précédent et des rappels faits sur les fonctions L d'Arakelov, il suffit de reporter l'équation (5.1.5) dans la formule (5.1.3) et d'échanger les signes somme et intégrale. □

Cette dernière formule est le point de départ pour établir, moyennant des hypothèses supplémentaires sur $\mathscr{B}$, un prolongement méromorphe de la fonction zêta des hauteurs de $\mathscr{T}$.

5.2. Prolongement méromorphe

Fixons une section de l'homomorphisme canonique

$$\widehat{\text{Pic}}(\mathscr{B}) \otimes_{\mathbf{Z}} \mathbf{Q} \to \text{Pic}(\mathscr{B}_F) \otimes \mathbf{Q},$$

autrement dit un choix de fonctions hauteurs compatible au produit tensoriel, ce que Peyre appelle *système de hauteurs* dans [17], 2.2. Concernant

$\mathscr{X}$, on utilise toujours les métriques adéliques canoniques utilisées au paragraphe 4. Ainsi, on écrira λ et α, les chapeaux devenant inutiles. L'application $\widehat{\eta} : X^*(G) \to \widehat{\mathrm{Pic}}(\mathscr{B})$ est supposée être la composée de l'application $\eta : X^*(G) \to \mathrm{Pic}(\mathscr{B}_F)$ donnée par la restriction du torseur à la fibre générique, et de la section $\mathrm{Pic}(\mathscr{B}_F) \otimes \mathbf{Q} \to \widehat{\mathrm{Pic}}(\mathscr{B}) \otimes \mathbf{Q}$ fixée.

Ces restrictions ne sont pas vraiment essentielles mais simplifient beaucoup les notations.

Notons $V_1 = \mathrm{Pic}^G(\mathscr{X}_F)_\mathbf{R}$, $M_1 = X^*(G)_\mathbf{R}$, $n_1 = \dim V_1$ et $V_2 = \mathrm{Pic}(\mathscr{B}_F)_\mathbf{R}$. Soient $\Lambda_1 \subset V_1$ et $\Lambda_2 \subset V_2$ les cônes ouverts, intérieurs des cônes effectifs dans $\mathrm{Pic}^G(\mathscr{X}_F)_\mathbf{R}$ et $\mathrm{Pic}(\mathscr{B}_F)_\mathbf{R}$. L'espace vectoriel V_1 possède une base naturelle, formée des fibrés en droites G-linéarisés associés aux diviseurs G-invariants sur $\mathscr{X} \otimes F$. Dans cette base, le cône Λ_1 est simplement l'ensemble des $(s_1, \ldots, s_{n_1})$ strictement positifs.

On note $\eta : M_1 \to V_2$ l'application linéaire déduite de $\widehat{\eta}$ et $M = (\mathrm{id}, -\eta)(M_1)$, contenu dans $V_1 \times V_2$. Notons $V = V_1 \times V_2$. Les théorèmes 2.2.4 et 2.2.9 identifient $\mathrm{Pic}(\mathscr{Y}_F)_\mathbf{R}$ à V/M, et l'intérieur du cône effectif de $\mathscr{Y}_F$ à l'image de $\Lambda_1 \times \Lambda_2$ par la projection $V \to V/M$. Si $\omega_\mathscr{X}$ est muni de sa G-linéarisation canonique, la proposition 2.1.8 dit que $\omega_\mathscr{Y}$ est l'image du couple $(\omega_\mathscr{X}, \omega_\mathscr{B})$ par cette même projection.

LEMME 5.2.1. — *Avec ces notations, la formule du corollaire du paragraphe précédent peut se récrire* :
$$Z(\lambda + \omega_\mathscr{X}^{-1}, \alpha + \omega_\mathscr{B}^{-1}) = \int_{M_1} f(\lambda + im_1; \alpha - i\eta(m_1))\, dm_1,$$

où la fonction
$$f : \mathsf{T}(\Lambda_1 \times \Lambda_2) \to \mathbf{C}$$

est définie par
$$f(\lambda; \alpha) = \int_{\mathscr{U}_G} \check{H}(-(\lambda + \omega_\mathscr{X}^{-1}); \chi_u) L(\widehat{\mathscr{T}} \boxtimes (\alpha + \omega_\mathscr{B}^{-1}); \chi_u^{-1} \boxtimes \|\cdot\|)\, d\chi_u$$

et dm_1, $d\chi_u$ sont des mesures de Haar sur M_1 et $\mathscr{U}_G$ telles que $d\chi = dm_1\, d\chi_u$ dans la décomposition $\mathscr{A}_G = M_1 \oplus \mathscr{U}_G$ du paragraphe 4.1.1 (cf. aussi le lemme 4.4.3).

On note que $\mathscr{U}_G$ est un groupe discret et que la mesure $d\chi_u$ est donc proportionnelle à la mesure de comptage.

Démonstration. — Si $\chi \in \mathscr{A}_G$ s'écrit (m_1, χ_u) dans $M_1 \oplus \mathscr{U}_G$, on remarque que l'on a les égalités
$$\check{H}(-\lambda; \chi) = \check{H}(-\lambda - \iota(i\, m_1); \chi_u)$$
et
$$\chi^{-1}(\mathbf{g}_b) H(\widehat{\alpha}; b)^{-1} = \chi_u^{-1}(\mathbf{g}_b) H(\widehat{\alpha} - \eta(m_1); b)^{-1}$$
car (lemme 4.1.4)
$$\chi_{m_1}(\mathbf{g}_b) = \exp(i\, \|\cdot\|)([m_1]_* \widehat{\mathscr{T}}|_b) = \exp(i\, \|\widehat{\eta}(m_1)|_b\|) = H(-\widehat{\eta}(m_1); b).$$
On utilise ensuite le théorème de Fubini. □

On utilise enfin les notations du § 3.

HYPOTHÈSES 5.2.2. — *On fait les hypothèses suivantes :*

— *le cône Λ_2 est un cône polyédral (de type fini). Notons (ℓ_j) les formes linéaires définissant ses faces ;*

— *la fonction zêta des hauteurs de $\mathscr{B}$ converge localement normalement pour $\alpha + \omega_\mathscr{B} \in \Lambda_2$;*

— *il existe un voisinage convexe B_2 de l'origine dans V_2 et pour tout caractère $\chi \in \mathscr{A}_G$ une fonction holomorphe $g(\chi; \cdot)$ sur le tube $\mathsf{T}(B_2)$ tels que, si $\mathrm{Re}(\alpha + \omega_\mathscr{B}) \in \Lambda_2$,*
$$L(\widehat{\mathscr{T}} \boxtimes \alpha, \chi^{-1} \boxtimes \|\cdot\|) = \prod_j \frac{\ell_j(\alpha)}{\ell_j(\alpha + \omega_\mathscr{B})} g(\chi; \alpha + \omega_\mathscr{B});$$

— *il existe un réel γ strictement positif tel que pour tout $\varepsilon > 0$, les fonctions $g(\chi; \cdot)$ vérifient une majoration uniforme*
$$|g(\chi; \alpha + \omega_\mathscr{B})| \leqslant C_\varepsilon \bigl(1 + \|\mathrm{Im}(\alpha)\|\bigr)^\gamma \bigl(1 + \|\chi\|\bigr)^\varepsilon,$$
pour un réel $\varepsilon < 1$ et une constante C_ε ;

— *si $\tau(\mathscr{B})$ désigne le nombre de Tamagawa de $\mathscr{B}$, pour tout α appartenant à Λ_2,*
$$\lim_{s \to 0^+} \frac{Z(\mathscr{B}; s\alpha + \omega_\mathscr{B}^{-1})}{\mathsf{X}_{\Lambda_2}(s\alpha)} = \tau(\mathscr{B}) \neq 0.$$

Remarque 5.2.3. — Dans le cas où $\mathscr{B}$ est une variété de drapeaux généralisée, ces hypothèses correspondent à des énoncés sur les séries d'Eisenstein tordues par des caractères de Hecke. Ils sont établis dans [19].

Dans la suite, on travaille avec les classes de contrôle $\mathscr{D}_\beta$ introduites au paragraphe 4.3.

LEMME 5.2.4. — *Sous les hypothèses précédentes, pour tout réel $\beta > 1$, la fonction f appartient à $\mathscr{H}_M(\Lambda_1 \times \Lambda_2)$, pour la classe $\mathscr{D}_{\beta+\gamma}$.*

Démonstration. — Il suffit de reprendre la démonstration de la proposition 4.4.5, d'y insérer les majorations que nous avons supposées et de majorer

$$(1 + \|\operatorname{Im} \lambda\|)^\beta (1 + \|\operatorname{Im} \alpha\|)^\gamma \leqslant (1 + \|\operatorname{Im} \lambda\| + \|\operatorname{Im} \alpha\|)^{\beta+\gamma}.$$

□

Grâce au théorème d'analyse 3.1.14, on en déduit un prolongement méromorphe pour la fonction zêta des hauteurs de $\mathscr{T}$.

THÉORÈME 5.2.5. — *La fonction zêta des hauteurs décalée de $\mathscr{T}$ admet un prolongement méromorphe dans un voisinage de $\mathsf{T}(\Lambda^0_{\mathrm{eff}}(\mathscr{Y}))$ dans $\mathrm{Pic}(\mathscr{Y})_{\mathbf{C}}$. Cette fonction a des pôles simples donnés par les équations des faces de $\Lambda^0_{\mathrm{eff}}(\mathscr{Y})$. De plus, si $\lambda \in \Lambda^0_{\mathrm{eff}}(\mathscr{Y})$,*

$$\lim_{s \to 0^+} \frac{Z(\mathscr{T}; s\lambda + \omega_{\mathscr{Y}}^{-1})}{\mathsf{X}_{\Lambda_{\mathrm{eff}}(\mathscr{Y})}(s\lambda)} = \tau(\mathscr{Y}),$$

le nombre de Tamagawa de $\mathscr{Y}$.

Démonstration. — Le seul point qui n'a pas été rappelé est que le nombre de Tamagawa est $\mathscr{Y}$ est le produit de ceux de $\mathscr{X}$ et $\mathscr{B}$ ([9], théorème 2.5.5). □

COROLLAIRE 5.2.6. — *Il existe $\varepsilon > 0$ et un polynôme P tels que le nombre de points de $\mathscr{T}(F)$ dont la hauteur anticanonique est inférieure ou égale à H vérifie un développement asymptotique*

$$N(H) = HP(\log H) + O(H^{1-\varepsilon})$$

lorsque H tend vers $+\infty$. Le degré de P est égal au rang de $\mathrm{Pic}(\mathscr{Y}_F)$ moins 1 et son coefficient dominant vaut

$$\mathsf{X}_{\Lambda_{\mathrm{eff}}(\mathscr{Y})}(\omega_{\mathscr{Y}}^{-1})\tau(\mathscr{Y}).$$

Appendice A
Un théorème taubérien

Le but de ce paragraphe est de démontrer un théorème taubérien dont la preuve nous a été communiquée par P. Etingof. Ce théorème est certainement bien connu des experts mais que nous n'avons pu le trouver sous cette forme dans la littérature.

THÉORÈME A.1. — *Soient $(\lambda_n)_{n \in \mathbf{N}}$ une suite croissante de réels strictement positifs, $(c_n)_{n \in \mathbf{N}}$ une suite de réels positifs et f la série de Dirichlet*

$$f(s) = \sum_{n=0}^{\infty} c_n \frac{1}{\lambda_n^s}.$$

On fait les hypothèses suivantes :
- *la série définissant f converge dans un demi-plan $\operatorname{Re}(s) > a > 0$;*
- *elle admet un prolongement méromorphe dans un demi-plan $\operatorname{Re}(s) > a - \delta_0 > 0$;*
- *dans ce domaine, elle possède un unique pôle en $s = a$, de multiplicité $b \in \mathbf{N}$. On note $\Theta = \lim_{s \to a} f(s)(s-a)^b > 0$;*
- *enfin, il existe un réel $\kappa > 0$ de sorte que l'on ait pour $\operatorname{Re}(s) > a - \delta_0$ l'estimation,*

$$\left| f(s) \frac{(s-a)^b}{s^b} \right| = O\bigl((1 + \operatorname{Im}(s))^\kappa \bigr).$$

Alors il existe un polynôme unitaire P de degré $b - 1$ tel que pour tout $\delta < \delta_0$, on ait, lorsque X tend vers $+\infty$,

$$N(X) \stackrel{\mathrm{def}}{=} \sum_{\lambda_n \leqslant X} c_n = \frac{\Theta}{a\,(b-1)!} X^a P(\log X) + O(X^{a-\delta}).$$

On introduit pour tout entier $k \geqslant 0$ la fonction

$$\varphi_k(X) = \sum_{\lambda_n \leqslant X} a_n \left(\log(X/\lambda_n) \right)^k,$$

de sorte que $\varphi_0 = N$.

LEMME A.2. — *Sous les hypothèses du théorème A.1, il existe pour tout entier $k > \kappa$ un polynôme Q de degré $b - 1$ et de coefficient dominant $k!\Theta/(a^{k+1}(b-1)!)$ tel que pour tout $\delta < \delta_0$, on ait l'estimation, lorsque X tend vers $+\infty$,*

$$\varphi_k(X) = X^a Q(\log X) + O(X^{a-\delta}).$$

Démonstration. — Soit $a' > a$ arbitraire. On remarque, en vertu de l'intégrale classique

$$\int_{a'+i\mathbf{R}} \lambda^s \frac{ds}{s^{k+1}} = \frac{2i\pi}{k!} \left(\log^+(\lambda) \right)^k, \quad \lambda > 0$$

que l'on a la formule
$$\varphi(X) = \frac{k!}{2i\pi} \int_{a'+i\mathbf{R}} f(s) X^s \frac{ds}{s^{k+1}},$$
l'intégrale étant absolument convergente puisque $\kappa < k$.

On veut décaler le coutour d'intégration vers la droite verticale $\mathrm{Re}(s) = a - \delta$, où δ est un réel arbitraire tel que $0 < \delta < \delta_0$. Dans le rectangle $a - \delta \leqslant \mathrm{Re}(s) \leqslant a'$, $|\mathrm{Im}(s)| \leqslant T$, il y a un unique pôle en $s = a$. Le résidu y vaut
$$\mathrm{Res}_{s=a} f(s) \frac{X^s}{s^{k+1}} = \frac{\Theta}{a^{k+1}(b-1)!} X^a Q(\log X)$$
où Q est un polynôme unitaire de degré $b - 1$. Il en résulte que

$$\frac{1}{2i\pi} \int_{a'-iT}^{a'+iT} f(s) X^s \frac{ds}{s^{k+1}}$$
$$= \frac{1}{2i\pi} \int_{a-\delta-iT}^{a-\delta+iT} f(s) X^s \frac{ds}{s^{k+1}} + I_+ - I_- + \frac{\Theta}{a^{k+1}(b-1)!} X^a Q(\log X),$$

où I_+ et I_- sont les intégrales sur les segments horizontaux (orientés de la gauche vers la droite). Lorsque T tend vers $+\infty$, ces intégrales sont $O(T^{\kappa-k-1} X^{a'})$ et tendent donc vers 0. Les hypothèses sur f et le fait que $k > \kappa$ montrent que $f(s) X^s / s^{k+1}$ est absolument intégrable sur la droite $\mathrm{Re}(s) = a - \delta$, l'intégrale étant majorée par $O(X^{a-\delta})$. Par conséquent, on a
$$\varphi(X) = \Theta \frac{k!}{a^{k+1}(b-1)!} X^a Q(\log X) + O(X^{a-\delta}).$$

Le lemme est ainsi démontré. □

Preuve du théorème. — On va démontrer par récurrence descendante que la conclusion du lemme précédent vaut en fait pour tout entier $k \geqslant 0$. Arrivés à $k = 0$, le théorème sera prouvé. Montrons donc comment passer de $k \geqslant 1$ à $k - 1$.

Pour tout $\eta \in {]0;1[}$, on a facilement l'inégalité
$$\frac{\varphi_k(X(1-\eta)) - \varphi_k(X)}{\log(1-\eta)} \leqslant k\varphi_{k-1}(X) \leqslant \frac{\varphi_k(X(1+\eta)) - \varphi_k(X)}{\log(1+\eta)}.$$

Fixons un réel δ' tel que $0 < \delta' < \delta < \delta_0$. D'après le lemme précédent, il existe un réel C tel que
$$\left| \varphi_k(X) - \frac{k!\Theta}{a^{k+1}(b-1)!} X^a Q(\log X) \right| \leqslant C X^{a-\delta'}.$$

On constate que l'on a alors, si $-1 < u < 1$,

$$\frac{\varphi_k(X(1+u)) - \varphi_k(X)}{\log(1+u)}$$
$$= \frac{k!\Theta}{a^{k+1}(b-1)!} X^a \frac{Q(\log X + \log(1+u))(1+u)^a - Q(\log X)}{\log(1+u)} + R(X),$$

où

$$|R(X)| \leqslant 2CX^{a-\delta'}/|\log(1+u)| = O(X^{a-\delta'}/u)$$

si u tend vers 0 et $X \to +\infty$. Toujours lorsque $X \to +\infty$ et $u \to 0$, on a

$$\frac{Q(\log X + \log(1+u))(1+u)^a - Q(\log X)}{\log(1+u)}$$
$$= Q(\log X) \frac{(1+u)^a - 1}{\log(1+u)} + \sum_{n=1}^{b-1} \frac{1}{n!} Q^{(n)}(\log X) \log(1+u)^{n-1}(1+u)^a$$
$$= Q(\log X)(a + O(u)) + Q'(\log X)(1 + O(u)) + O((\log X)^{b-1}u)$$
$$= (aQ + Q')(\log X) + O((\log X)^{b-1}u).$$

Prenons $u = \pm 1/X^\varepsilon$ où $\varepsilon > 0$ est choisi de sorte que $\delta' + \varepsilon < \delta$. Alors, lorsque $X \to +\infty$, $|R(X)| = O(X^{a-\delta})$ et

$$\frac{Q(\log X + \log(1+u))(1+u)^a - Q(\log X)}{\log(1+u)} = (aQ + Q')(\log X) + O(X^{-\delta}).$$

On a alors un développement

$$\varphi_{k-1}(X) = \frac{1}{k} X^a (aQ + Q')(\log X) + O(X^{a-\delta})$$

Le coefficient dominant de $(aQ + Q')/k$ est égal à $(k-1)!\Theta/(a^k(b-1)!)$ d'où le théorème par récurrence descendante. □

Appendice B
Démonstration de quelques inégalités

Le but de cet appendice est de démontrer les inégalités sous-jacentes à la proposition 4.3.2 qui affirmait l'existence d'une classe de contrôle.

Rappelons les notations.

Soit β un réel strictement positif. Si M et V sont deux **R**-espaces vectoriels de dimension finie avec $M \subset V$, notons $\mathcal{D}_{\beta,\varepsilon}(M,V)$ le sous-monoïde de $\mathcal{F}(V, \mathbf{R}_+)$ engendré par les fonctions $h : V \to \mathbf{R}_+$ telles qu'il existe $c > 0$ et une famille (ℓ_j) de formes linéaires sur V vérifiant :

– la famille $(\ell_j|_M)$ forme une base de M^* ;

– pour tout $v \in V$ et tout $m \in M$, on a

(B.1) $$h(v+m) \leqslant c \frac{(1+\|v\|)^\beta}{(1+\|m\|)^{1-\varepsilon}} \frac{1}{\prod(1+|\ell_j(v+m)|)}.$$

On définit ensuite $\mathscr{D}_\beta(M,V) = \bigcap_{\varepsilon>0} \mathscr{D}_{\beta,\varepsilon}(M,V)$.

Théorème B.2. — *Les $\mathscr{D}_\beta(M,V)$ définissent une classe de contrôle au sens de la définition 3.1.1.*

Démonstration. — Les points (3.1.1,a) et (3.1.1,c) sont clairs. L'axiome (3.1.1,e) est vrai car la famille $(\ell_j \circ p|_M)$ contient une base de $(M/M_1)^*$. L'axiome (3.1.1,b) résulte de l'inégalité

$$\min_{|t|\leqslant 1}(1+|\ell(v+tu+m)|) \geqslant \frac{1}{1+|\ell(u)|}(1+|\ell(v+m)|$$

valable pour tous $v \in V$, $u \in V$ et $m \in M$. Enfin, l'axiome (3.1.1,d), le plus délicat, fait l'objet de la proposition suivante. □

Proposition B.3. — *Soient $M \subset V$, V' un supplémentaire de M dans V, dm une mesure de Lebesgue sur M, (ℓ_j) une base de V^*. Pour tout $\varepsilon' > \varepsilon$, il existe une constante $c_{\varepsilon'}$ et un ensemble $((\ell_{j,\alpha})_j)_\alpha$ de bases de $(V')^*$ tels que pour tous v_1 et $v_2 \in M'$,*

$$\int_M \frac{1}{(1+\|v_1+m\|)^{1-\varepsilon}} \frac{dm}{\prod(1+|\ell_j(v_2+m)|)} \leqslant \frac{c_{\varepsilon'}}{(1+\|v_1\|)^{1-\varepsilon'}} \sum_\alpha \frac{1}{\prod_j(1+|\ell_{j,\alpha}(v_2)|)}.$$

Démonstration. — On raisonne par récurrence sur $\dim M$. Soient $\mathbf{u} \in M$, $M' \subset M$ tels que $M = M' \oplus \mathbf{R}\mathbf{u}$ et fixons une mesure de Lebesgue dm' sur M' telle que $dm' \cdot dt = dm$. Alors,

$$\int_{\mathbf{R}\mathbf{u}} \cdots \ll \int_{\mathbf{R}} \frac{1}{(1+\|v_1+m'\|+|t|)^{1-\varepsilon}} \frac{dt}{\prod_j(1+|\ell_j(v_2+m')t\ell_j(\mathbf{u})|)} dt$$

$$\ll \prod_{j\,;\,\ell_j(\mathbf{u})=0} \frac{1}{1+|\ell_j(v_2+m')|} \times$$

$$\times \int_{\mathbf{R}} \frac{1}{(1+\|v_1+m'\|+|t|)^{1-\varepsilon}} \prod_{j\,;\,\ell_j(\mathbf{u})\neq 0} \frac{1}{1+|\ell_j(v_2+m')+t|} dt$$

et, en appliquant le lemme B.4 ci-dessous,

$$\ll \frac{1+\log(1+\|v_1+m'\|)}{(1+\|v_1+m'\|)^{1-\varepsilon}} \sum_\alpha \prod_j \frac{1}{1+|\ell_{j,\alpha}(v_2+m')|}$$

$$\ll_{\varepsilon'} \frac{1}{(1+\|v_1+m'\|)^{1-\varepsilon'}} \sum_\alpha \frac{1}{\prod_j(1+|\ell_{j,\alpha}(v_2)|)}. \qquad \square$$

Lemme B.4. — *On a une majoration, valable pour tous réels $t_1 \leqslant \cdots \leqslant t_n$ et tout $A \geqslant 0$,*

$$\int_{-\infty}^{\infty} \frac{1}{(1+A+|t|)^{1-\varepsilon}} \prod_{j=1}^{n} \frac{1}{1+|t-t_j|} \, dt \ll \frac{1+\log(1+A)}{(1+A)^{1-\varepsilon}} \sum_{\alpha} \prod_{j=1}^{n-1} \frac{1}{1+|\tau_{\alpha,j}|}$$

où pour tout α et tout j, $\tau_{\alpha,j} = t_{a(\alpha,j)} - t_{b(\alpha,j)}$ de sorte que pour tout α, notant $(e_1, \ldots, e_n)$ la base canonique de $\mathbf{R}^n$, les familles $(e_{\alpha,j} = e_{a(\alpha,j)} - e_{b(\alpha,j)})_j$ sont libres.

Démonstration. — On découpe l'intégrale en $\int_{-\infty}^{t_1}, \int_{t_1}^{t_2}, \ldots, \int_{t_n}^{\infty}$ et on majore chaque terme.

Pour l'intégrale de $-\infty$ à t_1, on a

$$\int_{-\infty}^{t_1} \cdots \leqslant \prod_{j=2}^{n} \frac{1}{1+|t_j-t_1|} \int_{-\infty}^{t_1} \frac{1}{(1+A+|t|)^{1-\varepsilon}} \frac{dt}{1+t_1-t}$$

$$\leqslant \prod_{j=2}^{n} \frac{1}{1+|t_j-t_1|} \int_{0}^{\infty} \frac{1}{(1+A+|t-t_1|)^{1-\varepsilon}} \frac{dt}{1+t}$$

$$\leqslant \prod_{j=2}^{n} \frac{1}{1+|t_j-t_1|} \frac{1+\log(1+A)}{(1+A)^{1-\varepsilon}}$$

d'après le lemme B.5. La dernière intégrale (de t_n à $+\infty$) se traite de même. Enfin,

$$\int_{t_k}^{t_{k+1}} \cdots \leqslant \prod_{j<k} \frac{1}{1+|t_k-t_j|} \prod_{j>k+1} \frac{1}{1+|t_{k+1}-t_j|} \times$$

$$\times \int_{t_k}^{t_{k+1}} \frac{1}{(1+A+|t|)^{1-\varepsilon}} \frac{dt}{(1+t-t_k)(1+t_{k+1}-t)}$$

et cette dernière intégrale s'estime comme suit :

$$\frac{1}{(1+A+|t|)^{1-\varepsilon}} \frac{dt}{(1+t-t_k)(1+t_{k+1}-t)} =$$

$$= \int_{t_k}^{t_{k+1}} \frac{1}{(1+A+|t|)^{1-\varepsilon}} \frac{1}{2+t_{k+1}-t_k} \left(\frac{1}{1+t-t_k} + \frac{1}{1+t_{k+1}-t} \right) dt$$

$$\leqslant \frac{1}{2+t_{k+1}-t_k} \left(\int_{t_k}^{\infty} \frac{1}{(1+A+|t|)^{1-\varepsilon}} \frac{dt}{1+t-t_k} \right.$$

$$\left. + \int_{-\infty}^{t_{k+1}} \frac{1}{(1+A+|t|)^{1-\varepsilon}} \frac{dt}{1+t_{k+1}-t} \right)$$

$$\leqslant \frac{1}{2+t_{k+1}-t_k} \left(\int_{0}^{\infty} \frac{1}{(1+A+|t+t_k|)^{1-\varepsilon}} \frac{dt}{1+t} \right.$$

$$\left. + \int_{0}^{\infty} \frac{1}{(1+A+|t-t_{k+1}|)^{1-\varepsilon}} \frac{dt}{1+t} \right)$$

$$\ll \frac{1}{1+t_{k+1}-t_k} \frac{1+\log(1+A)}{(1+A)^{1-\varepsilon}}$$

en vertu du lemme B.5. □

LEMME B.5. — *On a une majoration, valable pour tout $A \geqslant 1$ et tout $a > 0$,*

$$\int_0^\infty \frac{1}{(A+|t+a|)^\alpha} \frac{dt}{1+t} \ll \frac{1+\log A}{A^\alpha}.$$

Il reste à démontrer ce lemme. Pour cela, on a besoin de deux lemmes supplémentaires !

LEMME B.6. — *Pour tous A et $B \geqslant 1$ et tous $\alpha, \beta > 0$ tels que $\alpha + \beta > 1$,*

$$\int_0^\infty \frac{dt}{(A+t)^\alpha (B+t)^\beta} \ll_{\alpha,\beta} \frac{\min(A,B)}{A^\alpha B^\beta} \times \begin{cases} 1+\log(B/A) & si\ \alpha = 1\ et\ B > A\,; \\ 1+\log(A/B) & si\ \beta = 1\ et\ A > B\,; \\ 1 & sinon. \end{cases}$$

Démonstration. — On ne traite que le cas $A < B$, l'autre étant symétrique et le cas $A = B$ élémentaire. Faisons le changement de variables $A + T = (B-A)e^u$, d'où $B + T = (B-A)(1+e^u)$. Pour $t = 0$, $u = \log A/(B-A)$. Lorsque $t \to +\infty$, $u \to +\infty$. Ainsi, l'intégrale vaut

$$I(A, B; \alpha, \beta) = \frac{1}{(B-A)^{\alpha+\beta-1}} \int_{\log A/(B-A)}^\infty \frac{e^{(1-\alpha)u}}{(1+e^u)^\beta} du.$$

Si $A < B \leqslant 2A$, on majore l'intégrale par

$$I(A, B; \alpha, \beta) \leqslant \frac{1}{(B-A)^{\alpha+\beta-1}} \int_{\log A/(B-A)}^\infty e^{(1-\alpha-\beta)u} du$$
$$\leqslant \frac{1}{(B-A)^{\alpha+\beta-1}} \frac{1}{1-\alpha-\beta} \left(\frac{B-A}{A}\right)^{\alpha+\beta-1}$$
$$\ll \frac{1}{A^{\alpha+\beta-1}} \ll \frac{A}{A^\alpha B^\beta}$$

puisque $1/A \leqslant 2/B$.

Lorsque $B \geqslant 2A$, $\log A/(B-A) \leqslant 0$. On minore $1 + e^u$ par 1 lorsque $u \leqslant 0$ et par e^u lorsque $u \geqslant 0$, d'où les inégalités

$$(B-A)^{\alpha+\beta-1} I(A,B;\alpha,\beta) = \int_{\log A/(B-A)}^{0} + \int_{0}^{\infty}$$

$$\leqslant \int_{0}^{\infty} \frac{e^{(1-\alpha)u}}{(1+e^u)^\beta} du + \int_{\log A/(B-A)}^{0} e^{(1-\alpha)u} du$$

$$\ll 1 + \begin{cases} \log(B-A)/A & \text{si } \alpha = 1; \\ \frac{1}{1-\alpha}\left(1 - \left(\frac{B-A}{A}\right)^{\alpha-1}\right) & \text{si } \alpha \neq 1 \end{cases}$$

$$\ll \begin{cases} 1 + \log(B/A) & \text{si } \alpha = 1; \\ 1 + \left(\frac{B-A}{A}\right)^{\alpha-1} & \text{sinon.} \end{cases}$$

De plus, $\dfrac{1}{B-A} \leqslant \dfrac{2}{B} \leqslant \dfrac{1}{A}$, si bien que

$$I(A,B;\alpha,\beta) \ll \frac{1}{(B-A)^{\alpha+\beta-1}} \times \begin{cases} 1 + \log(B/A) \\ 1 + ((B-A)/A)^{\alpha-1} \end{cases}$$

$$\ll \begin{cases} (1 + \log(B/A))/A^{\alpha-1} B^\beta & \text{si } \alpha = 1; \\ 1/A^{\alpha-1} B^\beta & \text{sinon.} \end{cases}$$

Le lemme est donc démontré. □

LEMME B.7. — *Si $A, B \geqslant 1$, $\alpha \leqslant 1$, on a*

$$\int_0^{B-1} \frac{du}{(A+u)^\alpha (B-u)} \ll_\alpha \frac{1 + \log A}{A^\alpha}.$$

Démonstration. — On fait le changement de variables $A + u = (A+B)(1-t)$, soit $B - u = (A+B)t$. Ainsi, l'intégrale vaut

$$J(A,B;\alpha) = \frac{1}{(A+B)^\alpha} \int_{1/(A+B)}^{B/(A+B)} \frac{du}{(1-u)^\alpha u}.$$

Si $A \geqslant B$, $u \leqslant B/(A+B) \leqslant 1/2$, donc $1 - u \geqslant 1/2$ et l'intégrale vérifie

$$J(A,B;\alpha) \ll \frac{1}{(A+B)^\alpha} \int_{1/(A+B)}^{B/(A+B)} \frac{du}{u} = \frac{\log B}{(A+B)^\alpha} \ll \frac{1 + \log A}{A^\alpha}.$$

Si $A \leqslant B$, on découpe l'intégrale de $1/(A+B)$ à $1/2$ et de $1/2$ à $B/(A+B)$.

$$\int_{1/(A+B)}^{1/2} \frac{du}{(1-u)^\alpha u} \leqslant \int_{1/(A+B)}^{1/2} \frac{du}{u} = \log \frac{A+B}{2}$$

$$\int_{1/2}^{B/(A+B)} \frac{du}{(1-u)^\alpha u} \leqslant \begin{cases} \int_{1/2}^{1}(\dots) & \text{si } \alpha < 1; \\ \log \frac{A+B}{2A} \leqslant \log \frac{A+B}{2} & \text{si } \alpha = 1 \end{cases}$$

Finalement,
$$J(A,B;\alpha) \ll \frac{1+\log(A+B)}{(A+B)^\alpha} \ll \frac{1+\log A}{A^\alpha},$$
ainsi qu'il fallait démontrer. □

Preuve du lemme B.5. — Si $a > 0$, l'intégrale se majore par
$$\int_0^\infty \frac{1}{(A+t)^\alpha} \frac{dt}{1+t} \ll \frac{1+\log A}{A^\alpha}$$
d'après le lemme B.6. Si $a < 0$, on découpe l'intégrale de 0 à $-a$ et de $-a$ à $+\infty$. L'intégrale de 0 à $-a$ vaut
$$\int_0^{-a} \frac{1}{(A-t-a)^\alpha} \frac{dt}{1+t} = \int_0^{-a} \frac{1}{(A+u)^\alpha} \frac{dy}{(1-a)-u} \ll \frac{1+\log A}{A^\alpha}$$
en vertu du lemme B.7, tandis que l'intégrale de $-a$ à $+\infty$ s'estime ainsi :
$$\int_{-a}^\infty \frac{1}{(A+t+a))^\alpha} \frac{dt}{1+t} = \int_0^\infty \frac{1}{(A+u)^\alpha} \frac{du}{1-a+u} \ll \frac{1+\log A}{A^\alpha}$$
en appliquant de nouveau le lemme B.6 et en distinguant suivant que $A \leqslant 1-a$ ou $A \geqslant 1-a$. □

Références

[1] V. V. Batyrev & Yu. I. Manin, *Sur le nombre de points rationnels de hauteur bornée des variétés algébriques*, Math. Ann. **286** (1990), 27–43.

[2] V. V. Batyrev & Yu. Tschinkel, *Rational points on bounded height on compactifications of anisotropic tori*, Internat. Math. Res. Notices **12** (1995), 591–635.

[3] ———, *Height zeta functions of toric varieties*, Journal Math. Sciences **82** (1996), no. 1, 3220–3239.

[4] ———, *Manin's conjecture for toric varieties*, J. Algebraic Geometry **7** (1998), no. 1, 15–53.

[5] ———, *Tamagawa numbers of polarized algebraic varieties*, in Nombre et répartition des points de hauteur bornée [**16**], 299–340.

[6] R. de la Bretèche, *Compter des points d'une variété torique rationnelle*, Prépublication 41, Université Paris Sud (Orsay), 1998.

[7] ———, *Estimations de sommes multiples de fonctions arithmétiques*, Prépublication 42, Université Paris Sud (Orsay), 1998.

[8] ———, *Sur le nombre de points de hauteur bornée d'une certaine surface cubique singulière*, in Nombre et répartition des points de hauteur bornée [**16**], 51–77.

[9] A. Chambert-Loir & Yu. Tschinkel, *Torseurs arithmétiques et espaces fibrés*, E-print, math.NT/9901006, 1999.

[10] ———, *On the distribution of points of bounded height on equivariant compactifications of vector groups*, E-print, math.NT/0005015, 2000.

[11] J. Franke, Yu. I. Manin & Yu. Tschinkel, *Rational points of bounded height on Fano varieties*, Invent. Math. **95** (1989), no. 2, 421–435.

[12] W. Fulton, *Introduction to toric varieties*, Annals of Math. Studies, no. 131, Princeton Univ. Press, 1993.

[13] R. Narasimhan, *Several complex variables*, Chicago Lectures in Mathematics, University of Chicago Press, Chicago, IL, 1995, Reprint of the 1971 original.

[14] T. Oda, *Convex bodies and algebraic geometry*, Ergeb., no. 15, Springer Verlag, 1988.

[15] E. Peyre, *Hauteurs et mesures de Tamagawa sur les variétés de Fano*, Duke Math. J. **79** (1995), 101–218.

[16] _____ (éd.), *Nombre et répartition des points de hauteur bornée*, Astérisque, no. 251, 1998.

[17] _____, *Terme principal de la fonction zêta des hauteurs et torseurs universels*, in Nombre et répartition des points de hauteur bornée [**16**], 259–298.

[18] P. Salberger, *Tamagawa measures on universal torsors and points of bounded height on Fano varieties*, in Nombre et répartition des points de hauteur bornée [**16**], 91–258.

[19] M. Strauch & Yu. Tschinkel, *Height zeta functions of toric bundles over flag varieties*, Selecta Math. (N.S.) **5** (1999), no. 3, 325–396.

[20] A. Weil, *Adeles and algebraic groups*, Progr. Math., no. 23, Birkhäuser, 1982.

Rational points on algebraic varieties
(E. PEYRE, Y. TSCHINKEL, ed.), p. 117–161
Progress in Mathematics, Vol. 199, © 2001 Birkhäuser Verlag Basel/Switzerland

HASSE PRINCIPLE FOR PENCILS OF CURVES OF GENUS ONE WHOSE JACOBIANS HAVE A RATIONAL 2-DIVISION POINT
CLOSE VARIATION ON A PAPER OF BENDER AND SWINNERTON-DYER

Jean-Louis Colliot-Thélène
 C.N.R.S., UMR 8628, Mathématiques, Bâtiment 425, Université de Paris-Sud, F-91405 Orsay, France • *E-mail* : colliot@math.u-psud.fr

Résumé. — Une série d'articles exploite une nouvelle technique qui mène à des conditions suffisantes d'existence et de densité des points rationnels sur certaines surfaces fibrées en courbes de genre un au-dessus de la droite projective. Dans les premiers articles de cette série (plusieurs articles de Swinnerton-Dyer, un article en collaboration de Skorobogatov, Swinnerton-Dyer et l'auteur), la jacobienne de la fibre générique des surfaces considérées a tous ses points d'ordre 2 rationnels. Un article récent de Bender et Swinnerton-Dyer traite de cas où cette jacobienne possède seulement un point d'ordre 2 non trivial (pour que la méthode fonctionne, il semble nécessaire que la jacobienne possède un point de torsion rationnel non trivial). Le présent article est une réécriture de celui de Bender et Swinnerton-Dyer. La principale contribution est une reformulation plus abstraite des hypothèses principales des théorèmes. La première hypothèse est formulée de façon entièrement algébrique (certains groupes de Selmer algébrico-géométriques sont supposés petits) et la seconde hypothèse est simplement : "Il n'y a pas d'obstruction de Brauer-Manin verticale". Comme dans la plupart des articles de cette série, les résultats dépendent de deux conjectures difficiles : l'hypothèse de Schinzel et la finitude des groupes de Tate-Shafarevich.

Let k be a number field and X/k a smooth projective surface equipped with a morphism $f : X \to \mathbf{P}_k^1$ the generic fibre of which is a curve of genus one. Assume that each geometric fibre of f contains a multiplicity one component. If the set $X(k)$ of rational points of X is not empty, one may ask whether it is then Zariski-dense in X. If no k-point is known, one may ask whether the

Brauer-Manin condition (existence of an adelic point $M_v \in \prod_v X(k_v)$ such that $\sum_v \mathrm{inv}_v(A(M_v)) = 0$ for each element A of the Brauer group of X) ensures the existence of (many) rational points.

In [**SwD1**] Swinnerton-Dyer initiated a new technique for studying these problems. The technique was systematized in the joint work [**CSS3**]. Swinnerton-Dyer has since then written a number of papers on the topic: [**SwD2**], [**SwD3**], [**BSwD**] (joint with Bender), [**SwD4**]. Here are some common features of these papers.

(i) They postulate the finiteness of Tate-Shafarevich groups of elliptic curves.

(ii) With the notable exception of the very recent paper [**SwD4**], they build upon Schinzel's Hypothesis (a very likely, but very unreachable conjecture).

(iii) Some assumption is made on the torsion subgroup of the jacobian J/K of the generic fibre of f (here K denotes the function field $k(\mathbf{P}^1)$). In the papers [**SwD1**], [**CSS3**], [**SwD2**], [**SwD3**], one assumes that the entire 2-torsion subgroup $_2J$ of J is rational, i.e. isomorphic to the constant K-group scheme $(\mathbf{Z}/2)_K^2$. In [**BSwD**], the assumption is that $_2J$ contains a copy of the constant group $(\mathbf{Z}/2)_K$. (In the paper [**SwD4**] the 3-torsion satisfies a similar property.)

The paper [**BSwD**] is written in the style of [**SwD2**]. The contribution of the present paper is merely to rewrite most of Sections 1 to 5 of [**BSwD**] in the style of [**CSS3**], i.e. in terms of the Brauer-Manin obstruction (I have not however produced the analogue of Section 4 of [**CSS3**]).

Apart from gaining more practice in the new technique, my immediate motivation for writing the present paper was the hope to produce a more satisfactory statement than Theorem 3 in Section 6 of [**BSwD**]. Namely, under the Schinzel hypothesis and the finiteness of Tate-Shafarevich group, one would hope to prove that the Brauer-Manin obstruction to the Hasse principle is the only obstruction for smooth complete intersections of two quadrics in $\mathbf{P}_k^4$. It has been known for some time (Sansuc and the author, 1986; Harari [**Ha1**], Prop. 5.2.3) that this last statement implies the Hasse principle for smooth complete intersections of two quadrics in $\mathbf{P}_k^n$ for $n \geq 5$ (for $n \geq 8$, the Hasse principle for such intersections of two quadrics is proved in [**CTSaSwD**]). Section 6 of [**BSwD**] contains an interesting geometric construction which one could have tried to combine with the techniques in Harari's papers [**Ha1**] [**Ha2**]. But for the time being I still feel that serious difficulties are in the way.

Theorem A below is a reformulation of Theorem 2 of [**BSwD**]. Theorem B below is a reformulation of Theorem 1 of [**BSwD**]. Conditions (3.a) and (3.b) in the statements of Theorems A and B are purely algebraic renditions of Conditions 7 and 8 of [**BSwD**]. The slightly technical Conditions 5 and 6 of [**BSwD**] (which involved the quadratic reciprocity law, hence had to do with the Brauer-Manin obstruction, but in a rather involved manner) are replaced by the sheer hypothesis that there is no vertical Brauer-Manin obstruction. Out of habit, I have used the one variable, inhomogeneous version of Schinzel's hypothesis. In his papers, Swinnerton-Dyer uses the two variables, homogeneous version of that hypothesis. The latter hypothesis has now been proved in nontrivial cases (Heath-Brown for $x^3 + 2y^3$ over the integers, Heath-Brown and Moroz for any binary cubic form $ax^3 + by^3$ over the integers), hence the homogeneous version should be preferred in the long run. As Swinnerton-Dyer points out, the homogeneous version also has the advantage that the point at infinity on the projective line does not play a special rôle.

The reader of the present paper is supposed to be acquainted with Sections 1 and 2 of the paper [CSS3].

Statement of the Theorems

Let k be a field, char$(k)=0$. Let $c(t), d(t), m(t) \in k[t]$ be polynomials. We assume:

(0.1) *The polynomials $c(t), d(t), m(t)$ are all of even degree,*

$$\deg(d) = 2\deg(c) > 0,$$

$c^2 - d$ *has even positive degree and $d(c^2 - d)$ has no square factor.*

Let $\mathscr{E}$ be the smooth projective surface over k which is the projective regular minimal model over $\mathbf{P}_k^1$ of the surface given in affine 3-space by the affine equation

$$y^2 = (x - c(t))(x^2 - d(t)),$$

the fibration over the affine line $\mathbf{A}_k^1$ being given by $(x, y, t) \mapsto t$.

Given m as above, let $\pi : X(m) \to \mathbf{P}_k^1$ denote the projective regular minimal model over $\mathbf{P}_k^1$ of the surface given in affine 4-space by the system of equations

$$w_1^2 = m(t)(x - c(t)), \quad w_2^2 = m(t)(x^2 - d(t)),$$

the fibration over the affine line $\mathbf{A}_k^1$ being given by $(x, w_1, w_2, t) \mapsto t$. The restrictions on the degrees of d, c and m made above ensure that the fibres at $t = \infty$ of $X(m)/\mathbf{P}_k^1$ and of $\mathscr{E}/\mathbf{P}_k^1$ have a geometrically integral component of multiplicity one; if moreover the degree of $c^2 - d$ is equal to the degree of d, then these fibres are smooth.

Let $U \subset \mathbf{A}_k^1$ be the open set defined by $d(c^2 - d) \neq 0$.

Let $U' \subset \mathbf{A}_k^1$ be the open set defined by $d \neq 0$. Let $\mathfrak{S}' = k[U']^*/k[U']^{*2}$. Let $\mathfrak{S}'_0 \subset \mathfrak{S}'$ be the subgroup of classes of functions whose divisor at infinity is even: these are exactly the classes (modulo squares) of polynomials $m'(t) \in k[t]$

which are of even degree and divide d. Let $\mathscr{M}'$ be the set of closed points M of $\mathbf{A}_k^1$ such that $d(M) = 0$.

Let $U'' \subset \mathbf{A}_k^1$ be the open set defined by $c^2 - d \neq 0$. Let $\mathfrak{S}'' = k[U'']^*/k[U'']^{*2}$. Let $\mathfrak{S}_0'' \subset \mathfrak{S}''$ be the subgroup of classes of functions whose divisor at infinity is even: these are exactly the classes (modulo squares) of polynomials $m''(t) \in k[t]$ which are of even degree and divide $c^2 - d$.

Let $\mathscr{M}''$ be the set of closed points M of $\mathbf{A}_k^1$ such that $(c^2 - d)(M) = 0$.

For each $M \in \mathscr{M} = \mathscr{M}' \cup \mathscr{M}''$, let $r_M(t)$ be the monic irreducible polynomial defining M.

Let
$$\delta' : \mathfrak{S}' \to \oplus_{M \in \mathscr{M}''} k_M^*/k_M^{*2}$$
and
$$\delta'' : \mathfrak{S}'' \to \oplus_{M \in \mathscr{M}'} k_M^*/k_M^{*2}$$
be the map given by evaluation at the relevant points M. Since $d(c^2 - d)$ is squarefree, d and $c^2 - d$ are coprime hence these evaluation maps are well-defined.

These maps induce maps
$$\delta_0' : \mathfrak{S}_0' \to \oplus_{M \in \mathscr{M}''} k_M^*/k_M^{*2}$$
and
$$\delta_0'' : \mathfrak{S}_0'' \to \oplus_{M \in \mathscr{M}'} k_M^*/k_M^{*2},$$
which will sometimes be denoted simply δ' and δ''. Clearly d is in the kernel of the first map, and $c^2 - d$ is in the kernel of the second map.

We refer the reader to [**CTSwD**], Section 3, and [**CT**] for a detailed discussion of the Brauer-Manin obstruction. Given a morphism $\pi : X \to \mathbf{P}_k^1$ with X/k smooth and projective, the vertical subgroup $\mathrm{Br}_{\mathrm{vert}}(X)$ of the Brauer group of X with respect to π is the subgroup of $\mathrm{Br}(X)$ consisting of classes whose restriction to the generic fibre of π comes from $\mathrm{Br}(k(\mathbf{P}^1))$. When $m \in \mathfrak{S}_0''$, each geometric fibre fibre of $\pi : X(m) \to \mathbf{P}^1$ contains a component of multiplicity one. This implies that the quotient $\mathrm{Br}_{\mathrm{vert}}(X(m))/\mathrm{Br}(k)$ is finite (the proof of this fact given in [**Sk**], Cor. 4.5 holds under the mere assumption that the Galois module denoted $\hat{G}$ is torsion free; such is the case when the above multiplicity one condition is fulfilled, as proved in [**Sk**] 3.2.4).

Let Ω be the set of places of the number field k. For $v \in \Omega$, let k_v be the completion of k at v. For any proper k-variety X, let $X(\mathbb{A}_k) = \prod_{v \in \Omega} X(k_v)$ be the space of adèles of X, equipped with the product topology. For any element $\mathscr{A} \in \mathrm{Br}(X)$, the map which sends the adèle $\{P_v\}$ to
$$\sum_{v \in \Omega} \mathrm{inv}_v(\mathscr{A}(P_v)) \in \mathbf{Q}/\mathbf{Z}$$

is a continuous function $\theta_{\mathscr{A}} : X(\mathbb{A}_k) \to \mathbb{Q}/\mathbb{Z}$ with finite image. Given a subset B of $\mathrm{Br}(X)$, we let $X(\mathbb{A}_k)^B \subset X(\mathbb{A}_k)$ denote the closed subset which is the intersection of the kernels $\theta_{\mathscr{A}}^{-1}(0)$ for $\mathscr{A} \in B$. The diagonal inclusion $X(k) \subset X(\mathbb{A}_k)^B$ of the set of k-rational points of X in the set of adèles of X shows that $X(\mathbb{A}_k)^B \neq \emptyset$ is a necessary condition for the existence of a k-rational point on X. For $\pi : X \to \mathbf{P}_k^1$ as above, the group $\mathrm{Br}_{\mathrm{vert}}(X)/\mathrm{Br}(k)$ is finite, hence $X(\mathbb{A}_k)^{\mathrm{Br}_{\mathrm{vert}}(X)}$ is open in $X(\mathbb{A}_k)$.

For any commutative ring A, let $\mathfrak{H}(A) = A^*/A^{*2}$.

Theorem A. — *Let k be a number field. Let $c, d, m \in k[t]$ satisfy (0.1) and let $X = X(m)$.*

(1) Assume that the polynomial m divides $c^2 - d$. The class of m in the group $\mathfrak{H}(k(t)) = k(t)^*/k(t)^{*2}$ then defines a class in $\mathfrak{S}_0''$.

(2) Assume that the class of m in $\mathfrak{S}_0'' \subset \mathfrak{H}(k(t))$ differs from that of 1 and of $c^2 - d$.

(3.a) Assume that the kernel of δ_0' consists of the classes $1, d$.

(3.b) Assume that the kernel of the composite map

$$\mathfrak{S}_0'' \xrightarrow{\delta_0''} \oplus_{M \in \mathscr{M}'} k_M^*/k_M^{*2} \to \oplus_{M \in \mathscr{M}'} k_M^*/(k_M^{*2}, \delta_M''(m)),$$

is spanned by $c^2 - d$ and m.

(4) Assume that the Tate-Shafarevich group of each elliptic curve occurring as a fibre of $\mathscr{E}/\mathbf{P}_k^1$ is finite.

(5) Assume Schinzel's hypothesis.

Let $\mathscr{R} \subset \mathbf{P}^1(k)$ be the set of points $\lambda \in \mathbf{P}^1(k)$ such that the fibre $X_\lambda = \pi^{-1}(\lambda)$ is smooth and has infinitely many k-points. Let $\mathscr{R}_E \subset \mathbf{P}^1(k)$ be the set of points $\lambda \in \mathbf{P}^1(k)$ such that the fibre $\mathscr{E}_\lambda$ of $\mathscr{E}/\mathbf{P}_k^1$ is an elliptic curve over k of rank (defined as $\dim_\mathbb{Q}(\mathscr{E}_\lambda(k) \otimes \mathbb{Q})$) at least equal to one. We have $\mathscr{R} \subset \mathscr{R}_E$. Then:

(a) The closure of $\mathscr{R}$ in $\mathbf{P}^1(\mathbb{A}_k)$ coincides with $\pi(X(\mathbb{A}_k)^{\mathrm{Br}_{\mathrm{vert}}(X)})$.

(b) Assume $X(\mathbb{A}_k)^{\mathrm{Br}_{\mathrm{vert}}(X)} \neq \emptyset$, i.e. assume that there is no vertical Brauer-Manin obstruction to the existence of a rational point (this is certainly the case if $X(k) \neq \emptyset$). Then for any non-empty finite set $S \subset \Omega$, the closure of $\mathscr{R}$ in $\prod_{v \in S} \mathbf{P}^1(k_v)$ contains a non-empty open set. The same therefore holds for $\mathscr{R}_E$. In particular $\mathscr{R}$ and $\mathscr{R}_E$ are Zariski-dense in $\mathbf{P}_k^1$, the set $X(k)$ is Zariski-dense in X and the set $\mathscr{E}(k)$ is Zariski-dense in the surface $\mathscr{E}$.

(c) Assume $X(\mathbb{A}_k)^{\mathrm{Br}_{\mathrm{vert}}(X)} \neq \emptyset$. Then there exists a finite set $S_0 \subset \Omega$ such that for any finite set $S \subset \Omega$ with $S \cap S_0 = \emptyset$, the closure of $\mathscr{R}$ under the diagonal embedding $\mathbf{P}^1(k) \to \prod_{v \in S} \mathbf{P}^1(k_v)$ coincides with $\prod_{v \in S} \pi(X(k_v))$.

Let us comment on the various hypotheses in the Theorem.

Hypothesis (1) ensures in particular that all geometric fibres of the morphism $\pi : X \to \mathbf{P}_k^1$ contain a component of multiplicity one. Some condition of that kind is required, if we hope to conclude that infinitely many k-fibres of π have rational points (see Cor. 2.2, Cor. 2.4, Prop. 4.1, Prop. 4.2 of [**CSS1**]).

If Hypothesis (2) is not satisfied, then the generic fibre of $X/\mathbf{P}_k^1$ has a $k(\mathbf{P}^1)$-rational point; it can be regarded as an elliptic curve isogenous to the generic fibre of $\mathscr{E}/\mathbf{P}_k^1$. In particular X has a k-rational point and the question of (Zariski) density of k-rational points on the surface X is equivalent to that question for the surface $\mathscr{E}$.

By analogy with the work done in Section 4 of [**CSS3**], Hypotheses (3.a) and (3.b) should be slightly stronger than the hypothesis that the 2-primary subgroup of the Brauer group of X is contained in the vertical part of the Brauer group. But we have not done the corresponding (algebraic) work here. As mentioned in [**CSS3**], this would fit in well with the obvious remark that the necessary condition $X(\mathbb{A}_k)^{\mathrm{Br}_{\mathrm{vert}}(X)} \neq \emptyset$ for the existence of a rational point (assumption made in (b) and (c)) is a priori weaker than the equally necessary condition $X(\mathbb{A}_k)^{\mathrm{Br}(X)} \neq \emptyset$ (a condition which does not appear in the Theorem).

Let us now prepare for the statement of Theorem B.

Let k be a field, char(k)=0. Let $\alpha_i, \beta_i, i = 0, \cdots, 4$ be elements of $k[t]$. Consider the variety given in $\mathbf{A}_k^1 \times_k \mathbf{P}_k^3$ by the system of equations

$$(0.2) \quad \begin{aligned} \alpha_0 U_0^2 + \alpha_1 U_1^2 + \alpha_2 U_2^2 + \alpha_3 U_3^2 + 2\alpha_4 U_2 U_3 &= 0, \\ \beta_0 U_0^2 + \beta_1 U_1^2 + \beta_2 U_2^2 + \beta_3 U_3^2 + 2\beta_4 U_2 U_3 &= 0. \end{aligned}$$

This defines a one-parameter family of intersections of two quadrics in $\mathbf{P}_k^3$.

Let $d_{ij} = \alpha_i \beta_j - \alpha_j \beta_i$. These d_{ij} satisfy a number of useful identities. For any subset $\{i, j, k, l\}$ of $\{0, 1, 2, 3, 4\}$, we have the identity

$$(0.3) \quad d_{ij}d_{k\ell} + d_{ik}d_{\ell j} + d_{i\ell}d_{jk} = 0.$$

We have the identities ([**BSwD**], (23))

$$(0.4) \quad \begin{aligned} 4d_{24}^2(d_{04}^2 - d_{02}d_{03}) &= (2d_{04}d_{24} - d_{02}d_{03})^2 - d_{02}^2(d_{23}^2 + 4d_{24}d_{34}) \\ 4d_{24}^2(d_{14}^2 - d_{12}d_{13}) &= (2d_{14}d_{24} - d_{12}d_{23})^2 - d_{12}^2(d_{23}^2 + 4d_{24}d_{34}). \end{aligned}$$

Let

$$\Delta = d_{01}(d_{23}^2 + 4d_{24}d_{34})(d_{04}^2 - d_{02}d_{03})(d_{14}^2 - d_{12}d_{13}).$$

At any point $t \in \mathbf{A}^1$ where Δ does not vanish, the fibre is a smooth, pure intersection of two quadrics in $\mathbf{P}_k^3$. In particular, if Δ does not identically vanish, the generic fibre is a smooth, pure intersection of two quadrics in $\mathbf{P}_k^3$.

We assume:

(0.5) *All the $\alpha_i(t)$ are of the same degree r, all the $\beta_i(t)$ are of the same degree s, and Δ is of degree $r+s$.*

This assumption, which one could slightly relax (see [**BSwD**]), ensures that the surface given by the system (*) admits a regular proper model over $\mathbf{P}_k^1$ whose fibre over $\infty \in \mathbf{P}_k^1$ is smooth.

We also assume:

(0.6) Δ *has no square factor.*

(0.7) $d_{04}^2 - d_{02}d_{03}$ *and d_{02} are coprime; $d_{14}^2 - d_{12}d_{13}$ and d_{12} are coprime.*

Let $U \subset \mathbf{A}_k^1$ be the open set defined by $\Delta \neq 0$. Let $U' \subset \mathbf{A}_k^1$ be the open set defined by $\Delta' = d_{01}(d_{23}^2 + 4d_{24}d_{34}) \neq 0$. Let $\mathfrak{S}' = k[U']^*/k[U']^{*2}$. Let $\mathfrak{S}'_0 \subset \mathfrak{S}'$ be the subgroup of classes of functions whose divisor at infinity is even: these are exactly the classes (modulo squares) of polynomials $r(t) \in k[t]$ which are of even degree and divide Δ'. Let $\mathscr{M}'$ be the set of closed points M of $\mathbf{A}_k^1$ such that $\Delta'(M) = 0$. This set breaks up into the set $\mathscr{M}'_1$ consisting of closed points M such that $d_{01}(M) = 0$ and the set $\mathscr{M}'_2$ consisting of closed points M such that $(d_{23}^2 + 4d_{24}d_{34})(M) = 0$.

Let $U'' \subset \mathbf{A}_k^1$ be the open set defined by

$$\Delta'' = (d_{04}^2 - d_{02}d_{03})(d_{14}^2 - d_{12}d_{13}) \neq 0.$$

Let $\mathfrak{S}'' = k[U'']^*/k[U'']^{*2}$. Let $\mathfrak{S}''_0 \subset \mathfrak{S}''$ be the subgroup of classes of functions whose divisor at infinity is even: these are exactly the classes (modulo squares) of polynomials $r(t) \in k[t]$ which are of even degree and divide Δ''.

Let $\mathscr{M}''$ be the set of closed points M of $\mathbf{A}_k^1$ such that $\Delta''(M) = 0$.

For each $M \in \mathscr{M} = \mathscr{M}' \cup \mathscr{M}''$, let $r_M(t)$ be the monic irreducible polynomial defining M.

Let

$$\delta' : \mathfrak{S}' \to \oplus_{M \in \mathscr{M}''} k_M^*/k_M^{*2}$$

and

$$\delta'' : \mathfrak{S}'' \to \oplus_{M \in \mathscr{M}'} k_M^*/k_M^{*2}$$

be the map given by evaluation at the relevant points M. Since $\Delta = \Delta' . \Delta''$ is squarefree, Δ' and Δ'' are coprime hence these evaluation maps are well-defined.

These maps induce maps

$$\delta'_0 : \mathfrak{S}'_0 \to \oplus_{M \in \mathscr{M}''} k_M^*/k_M^{*2}$$

and

$$\delta''_0 : \mathfrak{S}''_0 \to \oplus_{M \in \mathscr{M}'} k_M^*/k_M^{*2}.$$

Using the formulas (0.4) and the various coprimality assumptions above, one checks that (the class of) $d_{23}^2 + 4d_{24}d_{34} \in \mathfrak{S}'_0$ is in the kernel of δ'_0 and that (the classes of) $d_{04}^2 - d_{02}d_{03}$ and $d_{14}^2 - d_{12}d_{13}$ in $\mathfrak{S}_0''$ are in the kernel of the maps $\delta_{0,M}''$ relative to the points $M \in \mathscr{M}_2'$.

Let
$$c = 4d_{04}d_{14} - 2d_{02}d_{13} - 2d_{03}d_{12}.$$
Let $d = 4d_{01}^2(d_{23}^2 + 4d_{24}d_{34})$. We have
$$(0.8) \qquad c^2 - d = 16(d_{04}^2 - d_{02}d_{03})(d_{14}^2 - d_{12}d_{13}).$$
In particular, at any point $M \in \mathscr{M}_1'$, we have
$$\delta_M''(d_{04}^2 - d_{02}d_{03}) = \delta_M''(d_{14}^2 - d_{12}d_{13}).$$

Theorem B. — *Let k be a number field. Let α_i, β_i be as above, and let X be a smooth projective model of the surface defined by the system (0.2), which we may choose equipped with a k-morphism $X \to \mathbf{P}_k^1$ extending the map to $\mathbf{A}_k^1$ given by the t-coordinate. Let $\mathscr{E}/\mathbf{P}_k^1$ be the associated jacobian fibration.*

(1) Assume that all the $\alpha_i(t)$ are of the same degree r, that all the $\beta_i(t)$ are of the same degree s, and that Δ is of degree $r + s$, and assume that the product
$$\Delta = d_{01}(d_{23}^2 + 4d_{24}d_{34})(d_{04}^2 - d_{02}d_{03})(d_{14}^2 - d_{12}d_{13})$$
has no square factor.

(2) Assume that $d_{04}^2 - d_{02}d_{03}$ and d_{02} are coprime, and that $d_{14}^2 - d_{12}d_{13}$ and d_{12} are coprime.

(3.a) Assume that the kernel of the composite map
$$\mathfrak{S}_0' \xrightarrow{\delta_0'} \oplus_{M \in \mathscr{M}''} k_M^*/k_M^{*2} \to \oplus_{M \in \mathscr{M}''} k_M^*/(k_M^{*2}, \varepsilon_M),$$
where ε_M is the class of $(-d_{01}d_{02})(M)$ if $(d_{04}^2 - d_{02}d_{03})(M) = 0$ and is the class of $(-d_{01}d_{21})(M)$ if $(d_{14}^2 - d_{12}d_{13})(M) = 0$, consists of the classes 1 and $d_{23}^2 + 4d_{24}d_{34}$.

(3.b) Let $c = 4d_{04}d_{14} - 2d_{02}d_{13} - 2d_{03}d_{12}$. Assume that the kernel of the (composite) map
$$\mathfrak{S}_0'' \xrightarrow{\delta_0''} (\oplus_{M \in \mathscr{M}_2'} k_M^*/k_M^{*2}) \oplus (\oplus_{M \in \mathscr{M}_1'} k_M^*/(k_M^{*2}, \delta_M''(d_{14}^2 - d_{12}d_{13}), \delta_M''(-c)))$$
is spanned by the classes $(d_{04}^2 - d_{02}d_{03})$ and $(d_{14}^2 - d_{12}d_{13})$.

(4) Assume that the Tate-Shafarevich group of each elliptic curve occurring as a fibre of $\mathscr{E}'/\mathbf{P}_k^1$ is finite.

(5) Assume Schinzel's hypothesis.

Then we have the same conclusions as in Theorem A, namely:

Let $\mathscr{R} \subset \mathbf{P}^1(k)$ be the set of points $\lambda \in \mathbf{P}^1(k)$ such that the fibre $X_\lambda = \pi^{-1}(\lambda)$ is smooth and has infinitely many k-points. Let $\mathscr{R}_E \subset \mathbf{P}^1(k)$ be the set of points $\lambda \in \mathbf{P}^1(k)$ such that the fibre $\mathscr{E}_\lambda$ of $\mathscr{E}/\mathbf{P}_k^1$ is an elliptic curve over k of rank (defined as $\dim_\mathbf{Q}(\mathscr{E}_\lambda(k) \otimes \mathbf{Q}))$ at least equal to one. We have $\mathscr{R} \subset \mathscr{R}_E$. Then:

(a) *The closure of $\mathscr{R}$ in $\mathbf{P}^1(\mathbb{A}_k)$ coincides with $\pi(X(\mathbb{A}_k)^{\mathrm{Br vert}(X)})$.*

(b) *Assume $X(\mathbb{A}_k)^{\mathrm{Br vert}(X)} \neq \emptyset$, i.e. assume that there is no vertical Brauer-Manin obstruction to the existence of a rational point (this is certainly the case if $X(k) \neq \emptyset$). Then for any non-empty finite set $S \subset \Omega$, the closure of $\mathscr{R}$ in $\prod_{v \in S} \mathbf{P}^1(k_v)$ contains a non-empty open set. The same therefore holds for $\mathscr{R}_E$. In particular $\mathscr{R}$ and $\mathscr{R}_E$ are Zariski-dense in $\mathbf{P}^1_k$, the set $X(k)$ is Zariski-dense in X and the set $\mathscr{E}(k)$ is Zariski-dense in the surface $\mathscr{E}$.*

(c) *Assume $X(\mathbb{A}_k)^{\mathrm{Br vert}(X)} \neq \emptyset$. Then there exists a finite set $S_0 \subset \Omega$ such that for any finite set $S \subset \Omega$ with $S \cap S_0 = \emptyset$, the closure of $\mathscr{R}$ under the diagonal embedding $\mathbf{P}^1(k) \to \prod_{v \in S} \mathbf{P}^1(k_v)$ coincides with $\prod_{v \in S} \pi(X(k_v))$.*

Comment on (3.a). In their (more arithmetic) version (Condition 7 of [BSwD], as applied to their Theorem 1), Bender and Swinnerton-Dyer only have $\oplus_{M \in \mathscr{M}''} k_M^*/k_M^{*2}$ as the target group (i.e. no ε_M), but I think this is an oversight on their part.

Comment on (3.b). To simplify matters, in their (more arithmetic) version (Condition 8 of [BSwD]), Bender and Swinnerton-Dyer replace this by the stronger condition where one only keeps the sum over points in $\mathscr{M}'_2$ in the target group. But they write

$$\oplus_{M \in \mathscr{M}'_2} k_M^*/(k_M^{*2}, \delta''_M(d_{14}^2 - d_{12}d_{13}))$$

whereas the formulas (0.4) above show that for such an M the second class $\delta''_M(d_{14}^2 - d_{12}d_{13})$ is trivial.

1. Selmer groups associated to a degree 2 isogeny

1.1. A statement from global class field theory, and some linear algebra. — Let S be a finite subset of the set Ω of places of the number field k. Consider the vector space $\oplus_{v \in S} k_v^*/k_v^{*2}$ over the finite field $\mathbf{F}_2$. We equip this space with the bilinear form $(\,,\,)_S = \oplus_{v \in S}(\,,\,)_v$, where $(\,,\,)_v$ is the bilinear pairing (the local symbol)

$$H^1(k_v, \mathbf{Z}/2) \times H^1(k_v, \mathbf{Z}/2) \to H^2(k_v, \mathbf{Z}/2) \hookrightarrow \mathbf{Z}/2,$$

where the right hand arrow is an isomorphism provided v is not a complex place. It is a standard fact of local class field theory that the pairing $(\,,\,)_v$, and hence the pairing $(\,,\,)_S$, is nondegenerate. This pairing is skewsymmetric: for any $a, b \in k_v^*$, we have $(a,b)_v + (b,a)_v = 0$. One also has $(a, -a)_v = 0$ for any $a \in k_v^*$, but in general one need not have $(a,a)_v = 0$, i.e. the pairing need not be alternating.

Let $\mathfrak{o}_S$ be the ring of S-integers of k.

Proposition 1.1.1. — *Let S be a finite set of places containing all the archimedean places and all the places over 2. Then the dimension of the $\mathbf{F}_2$-vector space $\mathfrak{o}_S^*/\mathfrak{o}_S^{*2}$ is equal to half the dimension of the $\mathbf{F}_2$-vector space $\oplus_{v \in S} k_v^*/k_v^{*2}$. If moreover S contains a set of generators of the ideal class group of k, then the natural map $\mathfrak{o}_S^*/\mathfrak{o}_S^{*2} \to \oplus_{v \in S} k_v^*/k_v^{*2}$ is injective and its image is a maximal isotropic subspace of $\oplus_{v \in S} k_v^*/k_v^{*2}$.*

This is [CSS3], Prop. 1.1.1. Let S_0 denote a fixed family of places of k satisfying all the conditions of the proposition. Clearly, for any finite set S of places containing S_0, the statement of the proposition holds.

For any place v of k, we let $V_v = H^1(k_v, \mathbf{Z}/2) = \mathfrak{H}(k_v)$. We equip this $\mathbf{F}_2$-vector space with the nondegenerate skewsymmetric bilinear form $(.,.)_v$.

For any finite set S of places containing S_0, and any family of objects $\{*_v\}$, $v \in S$, we shall consistently write $*_S = \oplus_{v \in S} *_v$.

The space V_S comes equipped with the pairing $e_S : V_S \times V_S \to \mathbf{Z}/2$ sum of the pairings $(.,.)_v$. Clearly e_S is skewsymmetric and nondegenerate (but unless $-1 \in k^{*2}$, it need not be alternating).

Because S contains S_0, we have $\mathrm{Pic}(\mathfrak{o}_S) = 0$. Hence the diagonal map

$$H^1(\mathfrak{o}_S, \mathbf{Z}/2) \to \bigoplus_{v \in S} H^1(k_v, \mathbf{Z}/2)$$

may be read

$$\mathfrak{o}_S^*/\mathfrak{o}_S^{*2} \to \bigoplus_{v \in S} k_v^*/k_v^{*2}$$

and it is injective by the previous proposition. By the global reciprocity law, the image I^S of this map is isotropic with respect to the symplectic pairing e_S defined above. By dimension count (using Proposition 1.1.1), we then see that it is maximal isotropic.

For any S, the natural map $H^1(\mathfrak{o}_S, \mathbf{Z}/2) \to H^1(k, \mathbf{Z}/2)$ is injective. For S containing S_0, we shall identify $\mathfrak{o}_S^*/\mathfrak{o}_S^{*2} = H^1(\mathfrak{o}_S, \mathbf{Z}/2)$ with the corresponding subgroup of $k^*/k^{*2} = H^1(k, \mathbf{Z}/2)$, and also with I^S. The elements of I^S may thus be identified with the elements of k^*/k^{*2} whose image in k_v^*/k_v^{*2} belongs to $\mathfrak{o}_v^*/\mathfrak{o}_v^{*2} = H^1(\mathfrak{o}_v, \mathbf{Z}/2)$ for each $v \notin S$.

We shall actually consider two copies of V_S, call them V_S' and V_S'', and consider the bilinear pairing

$$(.,.)_S : V_S' \times V_S'' \to \mathbf{Z}/2.$$

We have $I'^S \subset V_S'$ and $I''^S \subset V_S''$. Note that I'^S is the exact orthogonal of I''^S under the pairing $V_S' \times V_S'' \to \mathbf{Z}/2$. The following Proposition is proved in [BSwD] (Lemma 8 and Corollary). It is linear algebra, but not entirely trivial.

In the present context of degree 2 isogenies, it is the appropriate substitute for Prop. 1.1.2 of [CSS3].

Proposition 1.1.2. — *Let S_0 be as above, and let S be a finite set of places containing S_0. There exist subspaces $K'_v \subset V'_v$, $v \in S$ and $K''_v \subset V''_v$, $v \in S$ such that:*

  (a) *for each place $v \in S$, K'_v is the exact orthogonal of K''_v under the pairing $V'_v \times V''_v \to \mathbf{Z}/2$;*
  (b) *$V'_S = I'^S \oplus K'_S$, where $K'_S = \oplus_{v \in S} K'_v$; and similarly $V''_S = I''^S \oplus K''_S$, where $K''_S = \oplus_{v \in S} K''_v$;*
  (c) *For $v \notin S_0$, one may choose $K'_v = \mathfrak{H}(\mathfrak{o}_v)$ and similarly $K''_v = \mathfrak{H}(\mathfrak{o}_v)$.*

In the sequel, given S_0, we shall take the K'_v and K''_v for $v \in S_0$ as given by the Proposition, and we shall take K'_v and K''_v to be equal to $\mathfrak{H}(\mathfrak{o}_v)$ for $v \notin S_0$.

1.2. Isogenies of degree 2.

— The object of this section, copied directly from Section 3 of [BSwD], is to recall the standard formulae for an elliptic curve with a rational 2-division point, for the associated isogeny, and for multiplication by 2 on such a curve. The notation introduced in this paragraph will be standard throughout the paper. Let k be an arbitrary field of characteristic not 2. An elliptic curve E' defined over k and having a primitive 2-division point P' defined over k can be written in the form

(1.2.1) $$E' : y^2 = (x - c)(x^2 - d)$$

where c, d are in k and $d, (c^2 - d)$ are non-zero, and where P' is $(c, 0)$. If O' is the point at infinity on E', which is the identity element under the group law, then there is an isogeny $\phi' : E' \to E'' = E'/\{O', P'\}$ where E'', whose point at infinity will be denoted by O'', is given by the equation

(1.2.2) $$E'' : y_1^2 = (x_1 + 2c)(x_1^2 + 4(d - c^2)).$$

Explicitly, ϕ' is given by

$$x_1 = ((d - x^2)/(c - x)) - 2c, \quad y_1 = y(x^2 - 2cx + d)/(x - c)^2.$$

There is a dual isogeny $\phi'' : E'' \to E'$, and $\phi'' \circ \phi'$ and $\phi' \circ \phi''$ are the doubling maps on E' and E'' respectively.

The elements of $H^1(k, \mathbf{Z}/2.P') \simeq k^*/k^{*2}$ classify the ϕ'-coverings of E''; the covering corresponding to the class of m' is

(1.2.3) $$v_1^2 = m'(x_1 + 2c), \quad v_2^2 = m'(x_1^2 + 4d - 4c^2)$$

with the obvious two-to-one map to E''. The ϕ'-covering corresponding to P'' is given by $m' = d$. Similarly the ϕ''-coverings of E' are classified by the elements of $H^1(k, \mathbf{Z}/2.P'') \simeq k^*/k^{*2}$, the covering corresponding to the class of m'' being

(1.2.4) $$w_1^2 = m''(x-c), \quad w_2^2 = m''(x^2-d).$$

The ϕ''-covering corresponding to P' is given by $m'' = c^2 - d$.

The composite map $\phi'' \circ \phi' : E' \to E'' \to E'$ is multiplication by 2. There is an obvious commutative diagram

$$\begin{array}{ccccccccc} 0 & \longrightarrow & {}_2E' & \longrightarrow & E' & \xrightarrow{2} & E' & \longrightarrow & 0 \\ & & \downarrow & & \downarrow{\phi'} & & \parallel & & \\ 0 & \longrightarrow & \mathbf{Z}/2 & \longrightarrow & E'' & \xrightarrow{\phi''} & E' & \longrightarrow & 0. \end{array}$$

This diagram induces a map $H^1(k, {}_2E') \to H^1(k, \mathbf{Z}/2)$ which may be interpreted as sending 2-coverings of E' to ϕ''-coverings of E'.

Let us now discuss multiplication by 2 on E'. Let $K = k[t]/(t^2-d)$. The kernel ${}_2E'$ of multiplication by 2 on E' may be identified with the group $R_{K/k}\mu_2$; the point P' defines a natural map $\mathbf{Z}/2 \to R_{K/k}\mu_2$. We have $H^1(k, {}_2E') = H^1(k, R_{K/k}\mu_2) = K^*/K^{*2}$.

From [BSwD], we also extract the following pieces of information.

Let $a + b\sqrt{d} \in K^*$. The 2-covering of E' associated to the class of $a + b\sqrt{d} \in K^*/K^{*2}$ is given in homogeneous form by the intersection of two quadrics

(1.2.5) $$\begin{aligned} Z_2^2 + dZ_3^2 &= (a/(a^2-db^2))Z_1^2 + (ac+bd)Z_0^2, \\ 2Z_2Z_3 &= (b/(a^2-db^2))Z_1^2 + (a+bc)Z_0^2 \end{aligned}$$

in $\mathbf{P}_k^3$. Let us denote this curve by Γ'. The covering map $\Gamma' \to E'$ can be factorized as $\Gamma' \to C'' \to E'$; here C'' is the ϕ''-covering of E' given by (1.2.4), with $m'' = a^2 - db^2$, and the map $\Gamma' \to C''$ is

$$W_1 = Z_1/Z_0, \quad W_2 = (Z_2^2 - dZ_3^2)/Z_0^2.$$

From (1.2.5) it follows that if we are given an elliptic curve which has one nontrivial rational 2-division point then any 2-covering of it can be written as a smooth, complete intersection of two quadrics in $\mathbf{P}_k^3$ given by a system of homogeneous equations

(1.2.6) $$\begin{aligned} \alpha_0 U_0^2 + \alpha_1 U_1^2 + \alpha_2 U_2^2 + \alpha_3 U_3^2 + 2\alpha_4 U_2 U_3 &= 0, \\ \beta_0 U_0^2 + \beta_1 U_1^2 + \beta_2 U_2^2 + \beta_3 U_3^2 + 2\beta_4 U_2 U_3 &= 0. \end{aligned}$$

We refer to [BSwD] for the (simple) proof that conversely any curve of genus 1 defined over k by a system (1.2.6) is a 2-cover of an elliptic curve which has one nontrivial rational 2-division point.

As mentioned in the introduction, letting $d_{ij} = \alpha_i \beta_j - \alpha_j \beta_i$, for any subset $\{i, j, k, l\}$ of $\{0, 1, 2, 3, 4\}$ we have the identity

(1.2.7) $$d_{ij}d_{k\ell} + d_{ik}d_{\ell j} + d_{i\ell}d_{jk} = 0.$$

The system (1.2.6) implies

(1.2.8) $$d_{10}U_0^2 + d_{12}U_2^2 + 2d_{14}U_2U_3 + d_{13}U_3^2 = 0,$$
$$d_{01}U_1^2 + d_{02}U_2^2 + 2d_{04}U_2U_3 + d_{03}U_3^2 = 0.$$

and is in fact equivalent to it, since the assumption that (1.2.6) defines a smooth curve of genus one implies $d_{01} \neq 0 \in k$.

From [BSwD] we reproduce the formulas, some of which have already been mentioned:

$$a = -2(2d_{14}d_{34} + d_{13}d_{23})(d_{14}^2 - d_{12}d_{13}),$$
$$b = -d_{01}^{-1}d_{13}(d_{14}^2 - d_{12}d_{13}),$$
$$c = 4d_{04}d_{14} - 2d_{02}d_{13} - 2d_{03}d_{12},$$
$$d = 4d_{01}^2(d_{23}^2 + 4d_{24}d_{34}),$$
$$c^2 - d = 16(d_{04}^2 - d_{02}d_{03})(d_{14}^2 - d_{12}d_{13}),$$
$$m'' = a^2 - db^2 = 16d_{34}^2(d_{14}^2 - d_{12}d_{13})^3.$$

From these formulas we read the equation $y^2 = (x - c)(x^2 - d)$ of the curve E', we read $K = k[t]/(t^2 - d)$, we read the class $a + b\sqrt{d} \in K^*/K^{*2}$ of the 2-covering Γ' of E' which is isomorphic to the curve (1.2.5), and we also read the curve C'' given by (1.2.4) through which the map $\Gamma' \to E'$ factorizes.

1.3. Local computations. — Let E', E'', ϕ', ϕ'' be as above. We let

$$V' = H^1(k, \{O', P'\}) = H^1(k, \mathbf{Z}/2) = k^*/k^{*2}$$

and

$$V'' = H^1(k, \{O'', P''\}) = H^1(k, \mathbf{Z}/2) = k^*/k^{*2}.$$

The isogeny ϕ' gives rise to a natural embedding $E''(k)/\phi'(E'(k)) \hookrightarrow V'$ and the isogeny ϕ'' gives rise to a natural embedding $E'(k)/\phi''(E''(k)) \hookrightarrow V''$. We let $W' \subset V'$ denote the image of the first embedding and $W'' \subset V''$ be the image of the second. As mentioned above, the group V' classifies the set of ϕ'-coverings of E'' up to isomorphism; the subgroup W' classifies the set of isomorphism classes of such coverings which have a k-point.

In this section k is a local field of characteristic different from 2. If k is nonarchimedean, we denote by A its ring of integers.

As with any pair of dual isogenies, the (nondegenerate) Weil pairing

$$_{\phi'}E' \times {}_{\phi''}E'' \to \mu_2$$

(for a general pair of isogenies, μ_2 should be replaced by a big enough group of roots of unity) induces a pairing

$$V' \times V'' \to H^2(k, \mu_2) \subset \mathrm{Br}(k).$$

Since k is a local field, this pairing is nondegenerate.

Lemma 1.3.1. — *Assume that k is a local field. Under the nondegenerate pairing*

$$V' \times V'' \to H^2(k, \mu_2) \simeq \mathbf{Z}/2$$

the subgroups $W' \subset V'$ and $W'' \subset V''$ are each other's orthogonal.

A proof of this well-known lemma is provided in [**BSwD**] (Lemma 3). Note that once the diagrams written down there have been shown to commute, the result follows from Tate's duality theorem over a nonarchimedean local field, and from Witt's result over the reals. For the commutativity, one may refer to [**CTSa**], Lemme 5.2 et Remarque 5.3. □

We shall need an explicit description of the groups $W' \subset V'$ and $W'' \subset V''$. When k is a nonarchimedean nondyadic local field, and E', hence also E'', has good reduction over the ring A of integers of k, the description is simple: $W' \subset V' = k^*/k^{*2}$ coincides with the subgroup $A^*/A^{*2} \subset k^*/k^{*2}$ and the same result holds for $W'' \subset V''$.

The following general lemma will enable us to compute W' and W'' in the nondyadic case.

Lemma 1.3.2. — *Let A be a discrete valuation ring, K its fraction field, v its valuation, π a uniformizing parameter of A and κ the residue field. Assume $v(2) = 0$. Let $c, d, m \in A$ with $d(c^2 - d) \neq 0$.*

Let $X/\mathrm{Spec}(A)$ be a regular proper integral scheme over $\mathrm{Spec}(A)$ whose generic fibre over $\mathrm{Spec}(K)$ is K-isomorphic to the smooth projective K-curve C given in $\mathbf{P}^3_K$ by the projective system of equations

$$w_1^2 = m(xt - ct^2), \quad w_2^2 = m(x^2 - dt^2).$$

Assume that either $v(d) = 0$ or $v(c^2 - d) = 0$. Assume $v(m) = 0$ or $v(m) = 1$ (one may always reduce to this hypothesis). Then the special fibre of $X/\mathrm{Spec}(A)$, which is a κ-curve, has a geometrically integral component of multiplicity one if and only if c, d, m satisfy one of the conditions:
  (i) $v(d) = 0, v(c^2 - d) = 0$ and $v(m) = 0$.
  (ii) $v(m) = 0$, $v(d) > 0$ is odd and m is a square modulo π.
  (iii) $v(m) = 0$, $v(d) > 0$ is even and either m is a square modulo π or $-mc$ is a square modulo π.
  (iv) $v(c^2 - d) > 0$ is odd.
  (v) $v(c^2 - d) > 0$ is even, $v(m) = 0$.
  (vi) $v(c^2 - d) > 0$ is even, $v(m) = 1$ and $2c$ is a square modulo π.

Proof. — Let us recall the following facts. The property that the special fibre of X/A has a multiplicity one component which is geometrically integral over κ does not depend on the regular proper model X/A of C/K. The genus one curve C actually admits a unique regular proper minimal model over A. If Y/A is some integral (not necessarily regular or proper) model of X/K such that $Y \to \mathrm{Spec}(A)$ is surjective and the special fibre over κ has a geometrically integal component of multiplicity one, then a regular proper model X/A also has this property.

In case (i), the curve $X \subset \mathbf{P}_A^3$ given by the equations in the Lemma is smooth over A, its special fibre is a geometrically integral smooth curve of genus one over κ.

Suppose $v(d) \geq 0, v(c^2 - d) = 0$ and $v(m) = 1$. We claim that in that case any component of the special fibre of X/A has even multiplicity. Let V be a discrete valuation of rank one on the function field L of C, and assume $e = V(\pi) > 0$. From the affine equations $w_1^2 = m(x - c), w_2^2 = m(x^2 - d)$ viewed as identities in L we deduce that both $e + V(x - c)$ and $e + V(x^2 - d)$ are even. Assume e odd. This now implies $V(x - c)$ odd and $V(x^2 - d)$ odd. From $V(x^2 - d)$ odd, and $V(d) = ev(d) \geq 0$ we deduce $V(x) \geq 0$. We now have $V(x - c) \geq 1$, $V(x^2 - d) \geq 1$ hence $V(c^2 - d) > 0$, which implies $v(c^2 - d) > 0$, contradicting our hypothesis.

Assume $v(d) > 0, v(c^2 - d) = 0$ and $v(m) = 0$.

If m is a square modulo π, then the special fibre of the model Y/A given by the system of affine equations $w_1^2 - m(x - c) = 0, w_2^2 - m(x^2 - d) = 0$ breaks up as the union of two smooth irreducible conics defined over the residue field κ.

Assume $d = u\pi^{2n+1}$ with u a unit. Suppose that a regular proper model X/A of the K-curve given by the system of affine equations $w_1^2 = m(x - c)$, $w_2^2 = m(x^2 - d)$ has a multiplicity one component which is geometrically integral over κ. On the function field of the curve, let V denote the valuation associated to the codimension one point of the model defined by that component. The multiplicity one assumption ensures $V(\pi) = 1$. We have $V(x^2 - d) = \inf(2V(x), 2n+1)$. From the second equation, we obtain

$$2V(w_2) = V(m(x^2 - d)) = \inf(2V(x), 2n+1),$$

hence $2V(w_2) = 2V(x) < 2n+1 = V(d)$. The same equation now implies that m is a square modulo π.

For later use, let us consider the special case where $v(d) = 1, v(c^2 - d) = 0$ and $v(m) = 0$. One easily checks that the subscheme of $\mathbf{P}_A^3$ defined by the homogeneous equations

$$w_1^2 = m(xt - ct^2), \quad w_2^2 = m(x^2 - dt^2)$$

is the regular proper minimal model X/A of C/K. The special fibre is the union of two smooth conics meeting transversely in two points. Each conic is defined over $\kappa[t]/(t^2 - m)$. If this extension is a field, they are conjugate. The two points are defined over $\kappa[t]/(t^2 + mc)$.

Assume now that $d = u\pi^{2n}$ with u a unit, and that m is not a square modulo π.

Suppose a regular minimal model (over A) of the K-curve given by the system of affine equations $w_1^2 = m(x - c), w_2^2 = m(x^2 - d)$ has a multiplicity one component which is geometrically integral over κ. On the function field of the curve, let V denote the valuation associated to the codimension one point of the model defined by that component. The multiplicity one assumption ensures $V(\pi) = 1$. If $V(x) \leq 0$, then from the second equation we deduce $V(w_2^2) = v(x^2) \leq 0$ and m is a square modulo π, which we have excluded. Assume $V(x) > 0$. Then from the first equation we deduce that $-mc$ is a square modulo π.

Conversely, assume that $-mc$ is a square modulo π. Make the change of variables $x = \pi^n X$. Then affine equations for C/K read $w_1^2 = m(\pi^n X - c)$, $w_2^2 = m\pi^{2n}(X^2 - u)$. Another change of variables tranforms this system into $w_1^2 = m(\pi^n X - c), w_2^2 = m(X^2 - u)$. Now these equations over A reduce to $w_1^2 = -\overline{mc}, w_2^2 = \overline{m}(X^2 - \overline{u})$ over κ, and provided $-\overline{mc}$ is a square, this decomposes into two smooth geometrically integral conics.

Assume $v(d) = 0, v(c^2 - d) > 0$ and $v(m) = 0$. In this case the special fibre of the model Y/A given by $w_1^2 = m(x - c), w_2^2 = m(x^2 - d)$ is clearly irreducible and of multiplicity one.

Assume $v(d) = 0, v(c^2 - d) > 0$ and $v(m) = 1$. Assume $v(c^2 - d) > 0$ is odd. We claim that here also the special fibre of a suitable model contains a geometrically integral fibre of multiplicity one. Let E be the elliptic curve over K given by the affine equation $y^2 = (x - c)(x^2 - d)$. The curve C is K-birational to the unramified double cover of E given by $w^2 = m(x - c)$. Let σ denote translation by the 2-torsion point P given by $x = c, y = 0$ on E. The divisor of $x - c$ is a double, and the evaluation of the unramified double cover of E given by $x - c = z^2$ at the point P is the class $(c^2 - d) \in K^*/K^{*2}$. This is well-known to imply that in the function field of E there exists a formula $(x - c) \circ \sigma = (c^2 - d).(x - c).g^2$ with g a suitable rational function. Thus C viewed as an unramified double cover of E after pulling-back by σ is birationally given by the set of equations $(x - c) = uz^2$, $y^2 = (x - c)(x^2 - d)$, where $u = m/(c^2 - d) \in K^*$ has even valuation. Multiplying z by a suitable power of the uniformizing parameter reduces to the case where u is a unit, i.e. to the situation $v(m) = 0$.

Assume $v(d) = 0, v(c^2 - d) > 0$ even, and $v(m) = 1$. Note that this implies $v(c) = 0$. We claim that the special fibre has a geometrically integral component of multiplicity one if and only if $2c$ is a square modulo π.

Suppose that there exists such a component, let V be the discrete valuation associated to it. By hypothesis, π is a uniformizing parameter. In the function field of the K-curve, we have $w_1^2 = m(x - c)$ and $w_2^2 = m(x^2 - d)$. We have $V(\pi) = 1$, hence $V(m) = 1$. From the second equality we deduce $V(x) = 0$. From the first equality we conclude $x = c + m^{2r+1}u$, with $r \geq 0$, $V(u) = 0$ and u is a square modulo π. Hence $x^2 = c^2 + 2cm^{2r+1}u + m^{4r+2}u^2$. The second equality now gives

$$w_2^2 = m(c^2 - d + 2cm^{2r+1}u + m^{4r+2}u^2).$$

Since $V(m) = 1$ and $V(c^2 - d)$ is even, this equality implies the inequality $V(2cu.m^{2r+2}) < V(c^2 - d)$, hence $2cu$ is a square modulo π, hence finally $2c$ is a square modulo π.

Conversely, suppose that $2c$ is a square modulo π. By K-birational transformations we transform the system of equations defining C into

$$mw_1^2 = x - c, \quad mw_2^2 = x^2 - d$$

which is K-isomorphic to

$$mw_1^2 = x - c, \quad w_2^2 = 2cw_1^2 + (c^2 - d)/m + mw_1^4.$$

By assumption, $v(m) = 1$ and $v((c^2-d)/m) > 0$ (since it is odd). The reduction modulo π of the above system reads $0 = x - \bar{c}, w_2^2 = 2\bar{c}w_1^2$. Since $2\bar{c}$ is a square, this has a geometrically integral component of multiplicity one. □

Remark 1.3.3. — Lemma 4 in [**BSwD**] is an arithmetic consequence of the above geometric lemma. Suppose A is a complete (or henselian) discrete valuation ring with finite residue field of characteristic not 2. If X/A has a special fibre which contains a multiplicity one geometrically integral component, then one may find on it a zero-cycle of degree one whose support lies in the smooth locus of the special fibre. By Hensel's lemma, one may lift this zero-cycle to a zero-cycle of degree one on C/K. Since C/K is a curve of genus one, this curve admits a K-rational point. Conversely, if C/K admits a rational point, this defines an A-point of the regular proper A-scheme X/A, hence the special fibre of the latter contains a geometrically integral component of multiplicity one with a smooth κ-point (which as a matter of fact does not belong to any other component).

Let k be a p-adic local field, $p \neq 2$, let c, d be in the ring A of integers. Let v denote the valuation of k. The above computation enables us to determine $W'' \subset k^*/k^{*2}$. By duality, this also determines W'.

(i) If $v(d) = 0$ and $v(c^2 - d) = 0$, then $W' = A^*/A^{*2}$ and $W'' = A^*/A^{*2}$.
(ii) If $v(d) = 1$ and $v(c^2 - d) = 0$, then $W' = k^*/k^{*2}$ and $W'' = 1 \subset k^*/k^{*2}$.
(iii) If $v(d) = 0$ and $v(c^2 - d) = 1$, then $W' = 1 \subset k^*/k^{*2}$ and $W'' = k^*/k^{*2}$.

This will be enough to prove Theorem A. But when proving theorem B, we shall also need:

(iv) If $v(d) > 0$ is even and $v(c^2 - d) = 0$ (hence $v(c) = 0$), then $W'' = 1$ or $W'' = A^*/A^{*2}$ depending on whether $-c$ is or is not a square in A^*. In the first case, $W' = k^*/k^{*2}$, in the second case, $W' = A^*/A^{*2}$.

We shall now study, though in less details, the analogue of Lemma 1.3.2 for a curve (1.2.6).

Lemma 1.3.4. — *Let A be a discrete valuation ring, K its fraction field, v its valuation, π be a uniformizing parameter of A and κ the residue field. Assume $v(2) = 0$. Let $\alpha_i, \beta_i, i = 0, \cdots, 3$ be elements of A, and consider the closed subscheme of $\mathbf{P}_A^3$ defined by the system of homogeneous equations*

(1.2.6)
$$\alpha_0 U_0^2 + \alpha_1 U_1^2 + \alpha_2 U_2^2 + \alpha_3 U_3^2 + 2\alpha_4 U_2 U_3 = 0,$$
$$\beta_0 U_0^2 + \beta_1 U_1^2 + \beta_2 U_2^2 + \beta_3 U_3^2 + 2\beta_4 U_2 U_3 = 0.$$

Assume that the generic fibre of this A-scheme is a smooth intersection of two quadrics in $\mathbf{P}_K^3$. Noting as above $d_{ij} = \alpha_i \beta_j - \alpha_j \beta_i \in A$, this amounts to the assumption that the product

$$d_{01}(d_{23}^2 + 4d_{24}d_{34})(d_{04}^2 - d_{02}d_{03})(d_{14}^2 - d_{12}d_{13}) \neq 0 \in A.$$

If the valuation of this product is zero, the above A-scheme is smooth over A, and the special fibre is a smooth intersection of two quadrics in $\mathbf{P}^3_\kappa$.

Assume that the valuation of this product is equal to one (we shall only be interested in this case). Assume moreover that the valuation of $(d_{04}^2 - d_{02}d_{03})d_{02}$ is at most one, and that the valuation of $(d_{14}^2 - d_{12}d_{13})d_{12}$ is at most one.

Let X/A be a regular proper integral scheme over $\mathrm{Spec}(A)$ whose generic fibre over $\mathrm{Spec}(K)$ is K-isomorphic to the curve defined by (1.2.6) over K.

Then the special fibre of $X/\mathrm{Spec}(A)$, which is a κ-curve, has a geometrically integral component of multiplicity one if and only if one of the following conditions is satisfied:

(i) $v(d_{01}) = 1$ and either $d_{04}^2 - d_{02}d_{03}$ or $d_{14}^2 - d_{12}d_{13}$ is a square modulo π (in which case both are squares);
(ii) $v(d_{04}^2 - d_{02}d_{03}) = 1$ and $-d_{01}d_{02}$ is a square modulo π;
(iii) $v(d_{14}^2 - d_{12}d_{13}) = 1$ and $d_{12}d_{01}$ is a square modulo π;
(iv) $v(d_{23}^2 + 4d_{24}d_{34}) = 1$.

Proof. — By assumption, we have

$$v(d_{01}(d_{23}^2 + 4d_{24}d_{34})(d_{04}^2 - d_{02}d_{03})(d_{14}^2 - d_{12}d_{13})) = 1.$$

Thus exactly one of $v(d_{01}), v(d_{04}^2 - d_{02}d_{03}), v(d_{14}^2 - d_{12}d_{13}), v(d_{23}^2 + 4d_{24}d_{34})$ is equal to 1, the other ones are zero.

(i) Assume $v(d_{01}) = v(\alpha_0 \beta_1 - \alpha_1 \beta_0) = 1$. Then each of $v(d_{23}^2 + 4d_{24}d_{34})$, $v(d_{04}^2 - d_{02}d_{03})$, $v(d_{14}^2 - d_{12}d_{13})$ equals zero.

One of $\alpha_0, \beta_0, \alpha_1, \beta_1$ has valuation zero. By the symmetry of the equation, we may assume $v(\alpha_1) = 0$. If we had $v(\alpha_0) > 0$, then we would have $v(\beta_0) > 0$. But $v(\alpha_0) > 0$ and $v(\beta_0) > 0$ imply $v(d_{04}^2 - d_{02}d_{03}) > 0$, which is excluded. Thus we have $v(\alpha_1) = 0$ and $v(\alpha_0) = 0$.

Another (possibly singular) model for (1.2.6) is given by the system

(1.3.1)
$$\alpha_0 U_0^2 + \alpha_1 U_1^2 + \alpha_2 U_2^2 + \alpha_3 U_3^2 + 2\alpha_4 U_2 U_3 = 0,$$
$$d_{10} U_0^2 + d_{12} U_2^2 + d_{13} U_3^2 + 2d_{14} U_2 U_3 = 0.$$

Suppose there exists a multiplicity one, geometrically integral component γ of the special fibre of X/A. Let V be the associated valuation on the function field of our curve. By hypothesis, $V(\pi) = 1$. Suppose $d_{14}^2 - d_{12}d_{13}$ is not a square modulo π. Then it is not a square in the residue field of V either, because the κ-curve γ is geometrically integral. Since $d_{14}^2 - d_{12}d_{13}$ is not a square modulo π, certainly each of d_{12}, d_{13} is a unit with respect to v, hence also with respect to V. In the function field L of X, we have the equality

$$V(-d_{10}(U_0/U_2)^2) = V(d_{12} + d_{13}(U_3/U_2)^2 + 2d_{14}(U_3/U_2)).$$

The left hand side has odd valuation. The right hand side has even valuation, because it is equal to the valuation of a norm of an element in a nontrivial quadratic extension, which is unramified and nonsplit at V (indeed, $d_{14}^2 - d_{12}d_{13}$ is a unit, and not a square in the residue field).

We conclude: if the special fibre of X/A contains a multiplicity one, geometrically integral component, then $d_{14}^2 - d_{12}d_{13}$ is a square mod π.

Conversely, suppose that $d_{14}^2 - d_{12}d_{13}$ is a square mod π. Equations for the special fibre of (1.3.1) are

$$\text{(1.3.2)} \quad \begin{aligned} \tilde{\alpha}_0 U_0^2 + \tilde{\alpha}_1 U_1^2 + \tilde{\alpha}_2 U_2^2 + \tilde{\alpha}_3 U_3^2 + 2\tilde{\alpha}_4 U_2 U_3 &= 0, \\ \tilde{d}_{12} U_2^2 + \tilde{d}_{13} U_3^2 + 2\tilde{d}_{14} U_2 U_3 &= 0. \end{aligned}$$

The discriminant of this last equation is a nonzero square. Hence it breaks up as the product of two nonproportional linear forms, defining two distinct planes.

The 2 by 2 determinants associated to the matrix

$$\begin{pmatrix} \tilde{\alpha}_2 & \tilde{\alpha}_3 & 2\tilde{\alpha}_4 \\ \tilde{d}_{12} & \tilde{d}_{13} & 2\tilde{d}_{14} \end{pmatrix}$$

are $\tilde{\alpha}_1 \tilde{d}_{23}, 2\tilde{\alpha}_1 \tilde{d}_{24}, 2\tilde{\alpha}_1 \tilde{d}_{34}$. One of them is nonzero (recall $v(d_{23}^2 + 4d_{24}d_{34}) = 0$). Thus the form $\tilde{\alpha}_2 U_2^2 + \tilde{\alpha}_3 U_3^2 + 2\tilde{\alpha}_4 U_2 U_3$ does not vanish identically on at least one of the two planes. From $v(\alpha_1) = 0$ and $v(\alpha_0) = 0$ we conclude that the trace of

$$\tilde{\alpha}_0 U_0^2 + \tilde{\alpha}_1 U_1^2 + \tilde{\alpha}_2 U_2^2 + \tilde{\alpha}_3 U_3^2 + 2\tilde{\alpha}_4 U_2 U_3 = 0$$

on that plane is a smooth conic: this produces a geometrically integral component of multiplicity one.

(iii) (the argument for (ii) is entirely similar). Assume $v(d_{14}^2 - d_{12}d_{13}) = 1$. By assumption this implies $v(d_{01}) = 0$ and $v(d_{12}) = 0$. Suppose there exists a multiplicity one, geometrically integral component γ of the special fibre of X/A. Let V be the associated valuation on the function field L of our curve. By hypothesis, $V(\pi) = 1$. Let $u_i = U_i/U_3 \in L$. From the second equation in (1.3.1), we deduce the following equation in the function field of X:

$$u_0^2 - (d_{12}/d_{01})(u_2 + d_{14}/d_{12})^2 = (d_{14}^2 - d_{12}d_{13})/d_{10}d_{12}.$$

If $d_{12}d_{01}$ were not a square in the residue field of v, hence of V, then the V-valuation of the left hand side would be even. But the valuation of the right hand side is odd.

Conversely, assume that $d_{12}d_{01}$ is a square in the residue field of v. As a possibly singular A-model for the K-curve defined by (1.2.6) we may take

$Y \subset \mathbf{P}_A^3$ defined by the system of homogeneous equations
$$d_{10}U_0^2 + d_{12}U_2^2 + 2d_{14}U_2U_3 + d_{13}U_3^2 = 0,$$
$$d_{01}U_1^2 + d_{02}U_2^2 + 2d_{04}U_2U_3 + d_{03}U_3^2 = 0.$$

The fibre of this A-scheme over the residue field κ of A is the intersection of two quadrics in $\mathbf{P}_\kappa^3$. The first of these quadrics decomposes as the product of two planes over κ, each of them being given by an equation $U_0 = \pm l(U_2, U_3)$ where l is a linear form. The second one is the cone over a nonsingular conic in the variables U_1, U_2, U_3 : indeed, $v(d_{01}) = 0$ and $v(d_{04}^2 - d_{02}d_{03}) = 0$. Thus the fibre Y_κ is the union of two distinct smooth conics.

(iv) One considers the possibly singular model $Y \subset \mathbf{P}_A^3$ defined by the system of homogeneous equations
$$d_{10}U_0^2 + d_{12}U_2^2 + 2d_{14}U_2U_3 + d_{13}U_3^2 = 0,$$
$$d_{01}U_1^2 + d_{02}U_2^2 + 2d_{04}U_2U_3 + d_{03}U_3^2 = 0.$$
Under the assumption $v(d_{23}^2 + 4d_{24}d_{34}) = 1$, we have the equalities $v(d_{01}) = 0$, $v(d_{04}^2 - d_{02}d_{03}) = 0$ and $v(d_{14}^2 - d_{12}d_{13}) = 0$. By (1.2.7), we have
$$d_{12}d_{04} - d_{02}d_{14} = d_{01}d_{42},$$
$$d_{12}d_{03} - d_{02}d_{13} = d_{01}d_{32},$$
$$d_{14}d_{03} - d_{04}d_{13} = d_{01}d_{34}.$$

Since the valuation of $d_{23}^2 + 4d_{24}d_{34}$ is one, one at least of d_{42}, d_{34}, d_{32} is a unit in A, and we know that d_{01} is a unit. This implies that the reduction over κ of the two quadratic forms
$$d_{12}U_2^2 + 2d_{14}U_2U_3 + d_{13}U_3^2 \quad \text{and} \quad d_{01}U_0^2 + d_{02}U_2^2 + 2d_{04}U_2U_3 + d_{03}U_3^2$$
are nonproportional. As the reader will check, this implies that the variety in $\mathbf{P}_\kappa^3$ obtained by the reduction of the above system of equations is geometrically irreducible (with possibly one singular point). □

1.4. The Selmer groups as kernels of pairings. — In this section, k is a number field. We follow the model of [CSS3], Section 1.2. We let E', E'', ϕ', ϕ'' be as in Section 1.2. Define $S_{\phi'}$, the ϕ'-Selmer group of E'', to be the set of isomorphism classes of ϕ'-coverings of E'' which are everywhere locally soluble; and similarly for $S_{\phi''}$, the ϕ''-Selmer group of E'.

Let S_0 be as in Section 1.1 and let S be any finite set of places containing S_0.

From Section 1.1 we have the nondegenerate bilinear pairing
$$(.,.)_S : V_S' \times V_S'' \to \mathbf{Z}/2.$$

The group I^S is the image of $\mathfrak{o}_S^*/\mathfrak{o}_S^{*2}$ in V_S, and I'^S, I''^S are its images in V'_S, V''_S respectively. By Proposition 1.1.1, for any S, I'^S and I''^S are each other's orthogonal under the pairing $V'_S \times V''_S \to \mathbf{Z}/2$.

Let K'_v, K''_v, K'_S, K''_S be as in Proposition 1.1.2. Recall that $K'_S \subset V'_S$ and $K''_S \subset V''_S$ are each other's othogonal under the pairing $V'_S \times V''_S \to \mathbf{Z}/2$. For v a place of k, let $W'_v \subset V'_v$ be the image of $E''(k_v)/\phi'(E'(k_v)) \hookrightarrow V'_v$ and similarly let $W''_v \subset V''_v$ be the image of $E'(k_v)/\phi''(E''(k_v)) \hookrightarrow V''_v$.

Let $W'_S = \oplus_{v \in S} W'_v \subset V'_S$ and let $W''_S = \oplus_{v \in S} W''_v \subset V''_S$. By Lemma 1.3.1, W'_S and W''_S are each other's orthogonal under the pairing $V'_S \times V''_S \to \mathbf{Z}/2$.

The following Lemma is the analogue of Prop. 1.2.1 of [**CSS3**].

Lemma 1.4.1. — *Let S_0 be as above and suppose that S contains S_0 and all the primes of bad reduction for E', i.e. c, d and $c^2 - d$ are units at any $v \notin S$. Then $S_{\phi'}$ is isomorphic to each of the following groups:*
(i) *the intersection $I'^S \cap W'_S$;*
(ii) *the left kernel of the map $I'^S \times W''_S \to \mathbf{Z}/2$ induced by $(.,.)_S$;*
(iii) *the left kernel of the map $W'_S \times I''^S \to \mathbf{Z}/2$ induced by $(.,.)_S$.*
A similar result holds for $S_{\phi''}$.

Proof. — If v is not in S then the ϕ'-covering (1.2.3) is soluble in k_v if and only if m' is in $\mathfrak{H}(\mathfrak{o}_v)$; hence $S_{\phi'}$ can be identified with the subgroup of I'^S for which (1.2.3) is soluble at every place of S. This proves (i). Since W'_S and W''_S are orthogonal complements with respect to $(.,.)_S$, we deduce (ii). As for (iii), it follows from the corresponding fact for I'^S and I''^S. □

For any finite set S of places containing S_0, but not necessarily the primes of bad reduction for E' and E'', write

$$\mathbf{I}'^S = I'^S \cap (W'_S + K'_S), \quad \mathbf{W}'_S = W'_S/(W'_S \cap K'_S) = \oplus_{v \in S} W'_v/(W'_v \cap K'_v)$$

and similarly for $\mathbf{I}''^S$ and $\mathbf{W}''_S$. Define $t'_S : V'_S \to I'^S$ to be the projection along K'_S in V'_S, and similarly for t''_S.

The image of t'_S lies in I'^S, and the image of $1 - t'_S$ lies in K'_S, and similarly for t''_S. For $x \in V'_S$ and $y \in V''_S$, we have

$$(t'_S(x), t''_S(y))_S = 0 = (x - t'_S(x), y - t''_S(y))_S,$$

hence

(1.4.1) $\qquad (x,y)_S = (t'_S(x), y)_S + (x, t''_S(y))_S.$

The kernel of t'_S is K'_S, so t'_S induces a map $\mathbf{W}'_S \to I'^S$ whose kernel is trivial and whose image is easily seen to be $\mathbf{I}'^S$; in other words, it induces an

isomorphism
$$\tau'_S : \mathbf{W}'_S \to \mathbf{I}'^S.$$
There is an analogous isomorphism $\tau''_S : \mathbf{W}''_S \to \mathbf{I}''^S$. We shall denote the inverse isomorphisms by σ'_S, σ''_S respectively.

Proposition 1.4.2. — *The pairing $(.,.)_S$ induces pairings*
$$\mathbf{I}'^S \times \mathbf{W}''_S \to \mathbf{Z}/2, \quad \mathbf{W}'_S \times \mathbf{I}''^S \to \mathbf{Z}/2.$$
The action of $\tau'_S \times \sigma''_S$ takes the first pairing into the second.

Proof. — To prove the existence of the first pairing it is enough to show that $\mathbf{I}'^S$ is orthogonal to $W''_S \cap K''_S$ with respect to $(.,.)_S$. But $\mathbf{I}'^S \subset W'_S + K'_S$ which is the orthogonal complement of $W''_S \cap K''_S$. The argument for the second pairing is similar. To prove the last statement, let $\alpha' + \beta'$ be any element of $\mathbf{I}'^S$, where α' is in W'_S and β' in K'_S, and let α'' be any element of W''_S; by abuse of language we also use α'' to denote the corresponding element of $\mathbf{W}''_S$. Then $\sigma'_S(\alpha' + \beta')$ is the class of α', and $\tau''_S \alpha''$ has the form $\alpha'' + \beta''$ for some β'' in K''_S. Since we are in characteristic 2, what we need to prove is
$$(\alpha' + \beta', \alpha'')_S + (\alpha', \alpha'' + \beta'')_S = 0.$$
But $t'_S(\alpha') = \alpha' + \beta'$ because β' is in K'_S and $\alpha' + \beta'$ is in $\mathbf{I}'^S$, and similarly $t''_S(\alpha'') = \alpha'' + \beta''$; the left hand side of the displayed equation is equal to $(\alpha' + \beta', \alpha'' + \beta'')_S$ by (1.4.1) and this vanishes because I'^S is orthogonal to I''^S. □

We shall denote the pairings in Prop. 1.4.2 by $(.,.)'_S$ and $(.,.)''_S$ respectively.

Proposition 1.4.3. — *Suppose that S also contains all the primes of bad reduction for E'. Then the left kernel of either of the pairings in Proposition 1.4.2 is isomorphic to $S_{\phi'}$ and the right kernel to $S_{\phi''}$.*

Proof. — By Lemma 1.4.1, $S_{\phi'}$ can be identified with $I'^S \cap W'_S$ and is therefore contained in $\mathbf{I}'^S$; also it is orthogonal to W''_S and therefore to $\mathbf{W}''_S$. Conversely, any element in $\mathbf{I}'^S$ which is orthogonal to $\mathbf{W}''_S$ must be orthogonal to W''_S and therefore lies in W'_S; since $\mathbf{I}'^S$ lies in I'^S, such an element lies in $I'^S \cap W'_S = S_{\phi'}$.

The left kernel of the second pairing in Prop. 1.4.2 is isomorphic through σ'_S to the left kernel of the first pairing. The proof for the right kernels starts from the second pairing in Prop. 1.4.2 and is similar. □

What we shall actually use is the pairing $\mathbf{W}'_S \times \mathbf{W}''_S \to \mathbf{Z}/2$ given by
$$(x', x'') \mapsto (\tau'_S(x'), x'')'_S = (x', \tau''_S(x''))''_S.$$

The equality here follows from the last sentence of Prop. 1.4.2, and Prop. 1.4.3 asserts that the left kernel of this map is isomorphic to $S_{\phi'}$ and the right kernel to $S_{\phi''}$.

1.5. Small Selmer groups. — In [CSS3], one considers multiplication by 2 on an elliptic curve E/k whose 2-torsion is rational over k and whose 2-Selmer group $S_2(E)$ is of order at most 8. On the Tate-Shafarevich group of E we have the alternating Cassels-Tate pairing. If the Tate-Shararevich group is finite, then the order of the 2-torsion subgroup of $\Sha(E)$ must be a square. Since it is of order at most 2, it has to be trivial and $E(k)/2 \simeq S_2(E)$ which forces any 2-covering of E with points everywhere locally to have a k-rational point. Here we must give an argument taking into account the two isogenies $E' \to E''$ and $E'' \to E'$. This is Lemma 11 of [BSwD], for which I give a slightly revised proof.

Let k be a number field. Let $\phi' : E' \to E''$ be a degree 2 isogeny with kernel $\mathbf{Z}/2.P'$. Let $\phi'' : E'' \to E'$ be the dual isogeny, the kernel of which is $\mathbf{Z}/2.P''$. The composite map $\phi' \circ \phi''$ is multiplication by 2 on E''. Similarly, the composite map $\phi'' \circ \phi'$ is multiplication by 2 on E'.

From the isogenies ϕ' et ϕ'' one obtains exact sequences

$$0 \to E''(k)/\phi'(E'(k)) \to S_{\phi'} \to {}_{\phi'}\Sha(k, E') \to 0$$

and

$$0 \to E'(k)/\phi''(E''(k)) \to S_{\phi''} \to {}_{\phi''}\Sha(k, E'') \to 0,$$

where $S_{\phi'}$ is the Selmer group classifying ϕ'-coverings of E'' which have rational points everywhere locally, $S_{\phi''}$ is the Selmer group classifying ϕ''-coverings of E' which have rational points everywhere locally, and ${}_{\phi'}\Sha(k, E')$ is the kernel of the map $\Sha(k, E') \to \Sha(k, E'')$ induced by ϕ', and similarly for ${}_{\phi''}\Sha(k, E'')$.

Lemma 1.5.1. — *If P' is the unique 2-torsion point of $E'(k)$, then P'' does not belong to $\phi'(E'(k))$. If P'' is the unique 2-torsion point of $E''(k)$, then P' does not belong to $\phi''(E''(k))$.*

Proof. — Assume there exists $M \in E'(k)$ such that $\phi'(M) = P''$. Then we have $2M = \phi'' \circ \phi'(M) = \phi''(P'') = 0$, hence M is a 2-torsion point on E'. Then $M = 0$ or $M = P'$, hence $P'' = \phi'(M) = 0$, which contradicts our hypothesis. □

Proposition 1.5.2. — *Assume that P' is the only 2-torsion point of $E'(k)$ and that P'' is the only 2-torsion point of $E''(k)$. Assume that $S_{\phi'}$ has order 2 and that $S_{\phi''}$ has order at most 4. Then ${}_{\phi'}\Sha(E') = 0$ and $S_2(E')$ has order at most 4. If we moreover assume that the groups $\Sha(E')$ and $\Sha(E'')$ are finite,* then the groups $_{\phi''}\text{III}(E'')$, $_2\text{III}(E')$ and $_2\text{III}(E'')$ are all zero. In particular the curves corresponding to elements in the Selmer groups $S_{\phi'}, S_{\phi''}, S_2(E'), S_2(E'')$ all have rational points. The finitely generated abelian groups $E'(k)$ and $E''(k)$ have (Mordell-Weil) rank r, where $0 \leqslant r \leqslant 1$ and 2^{r+1} is the order of the group $S_{\phi''}$.

Proof. — From Lemma 1.5.1 we have an injection $\mathbf{Z}/2.P'' \hookrightarrow E''(k)/\phi'(E'(k))$. The hypothesis on the order of $S_{\phi'}$ thus implies $\mathbf{Z}/2.P'' = E''(k)/\phi'(E'(k))$ and $_{\phi'}\text{III}(E') = 0$. From the first equality we deduce

$$\phi''(E''(k)) = \phi'' \circ \phi'(E'(k)) = 2E'(k) \subset E'(k).$$

From the commutative diagram

$$\begin{array}{ccccccccc}
0 & \longrightarrow & {}_2E' & \longrightarrow & E' & \stackrel{2}{\longrightarrow} & E' & \longrightarrow & 0 \\
& & \downarrow & & \downarrow{\phi'} & & \| & & \\
0 & \longrightarrow & {}_{\phi''}E'' & \longrightarrow & E'' & \stackrel{\phi''}{\longrightarrow} & E' & \longrightarrow & 0
\end{array}$$

we obtain the commutative diagram of exact sequences

$$\begin{array}{ccccccccc}
0 & \longrightarrow & E'(k)/2E'(k) & \longrightarrow & S_2(E') & \longrightarrow & {}_2\text{III}(E') & \longrightarrow & 0 \\
& & \downarrow & & \downarrow & & \downarrow & & \\
0 & \longrightarrow & E'(k)/\phi''(E''(k)) & \longrightarrow & S_{\phi''} & \longrightarrow & {}_{\phi''}\text{III}(E'') & \longrightarrow & 0,
\end{array}$$

where the middle and right vertical maps are induced by ϕ' and the left vertical map is the natural projection map, which in the case in point is an isomorphism, as noticed earlier on.

The kernel of the map $_2\text{III}(E') \to {}_{\phi''}\text{III}(E'')$ is the group $_{\phi'}\text{III}(E')$, and we have shown that this group vanishes. Thus the left and right vertical maps are one-to-one, hence so is the map $S_2(E') \to S_{\phi''}$, and the order of $S_2(E')$ is at most 4.

According to the lemma, the order of $E'(k)/\phi''(E''(k))$ is at least 2. The group $S_{\phi''}$ has order at most 4. Thus $_{\phi''}\text{III}(E'')$ has order at most 2. Since $_2\text{III}(E') \to {}_{\phi''}\text{III}(E'')$ is one-to-one, we conclude that $_2\text{III}(E')$ has order at most 2. If the group $\text{III}(E')$ is finite, then the (alternating) Cassels-Tate pairing on that group is nondegenerate. This implies that the group $\text{III}(E')$ is a direct sum of groups $(\mathbf{Z}/n)^2$ for various n's, in particular the order of $_2\text{III}(E')$ is a square. Thus $_2\text{III}(E') = 0$ and $E'(k)/2E'(k) \simeq S_2(E')$, this last group being of order 2 or 4.

The isogeny ϕ'' induces a homomorphism $_2\text{III}(E'') \to {}_{\phi'}\text{III}(E')$ whose kernel is $_{\phi''}\text{III}(E'')$, a group of order at most 2. From $_{\phi'}\text{III}(E') = 0$ we now deduce that $_2\text{Ш}(E'')$ is of order at most 2. The hypothesis that $\text{Ш}(E'')$ is finite implies that the order of $_2\text{Ш}(E'')$ is a square. Hence $_2\text{Ш}(E'')=0$, and $_{\phi''}\text{Ш}(E'')=0$. From the commutative diagram above we then conclude that the groups $E'(k)/2E'(k), E'(k)/\phi''(E''(k))$ and $S_{\phi''}$ are isomorphic. The last one has order 2^{r+1}, with $0 \leqslant r \leqslant 1$. Thus $E'(k)$, whose 2-torsion subgroup is of order 2, has Mordell-Weil rank r. Hence also $E''(k)$. □

2. Proof of Theorem A

2.1. Fibres with points everywhere locally. — Let k be a number field and let $c(t), d(t), m(t) \in k[t]$ satisfy the hypothesis of the theorem. Thus all have even degree, $\deg(d) = 2\deg(c) > 0$, $d(c^2 - d)$ has no square factor and m divides $c^2 - d$, hence is coprime with d. We let $r(t) = d(t)(c^2(t) - d(t))$ and we let $r_M(t)$ be the monic irreducible factors of $r(t)$, where M denotes the associated closed point of $\mathbf{A}_k^1$. The polynomial $m(t)$ is the product of a constant $\rho_m \in k^*$ by a product of some of the $r_M(t)$.

We refer the reader to the introduction for the notation. The surface $X = X(m)$ is equipped with the fibration $\pi : X \to \mathbf{P}_k^1$.

Recall that $\mathcal{M}'$ is the set of closed points M such that $d(M) = 0$ and that $\mathcal{M}''$ is the set of closed points where $c^2 - d$ vanishes. We let $r'(t)$ be the product of the $r_M(t)$ for $M \in \mathcal{M}'$ and similarly for $r''(t)$. Note that $r'(t)$ up to a scalar coincides with $d(t)$ and that $r''(t)$ up to a scalar coincides with $c^2(t) - d(t)$.

The assumptions imply (Lemma 1.3.2 for $M \in \mathbf{A}_k^1$, a direct computation for $M = \infty \in \mathbf{P}_k^1$) that for $M \in \mathbf{P}_k^1$, $M \notin \mathcal{M}'$, the fibre X_M contains a component of multiplicity one which is geometrically integral over the residue field k_M (the fibres need not be smooth, in contrast with what happened in [**CSS3**]; however, as is by now well known ([**CSS2**], Lemma 1.2), the above property is just as good for the method to be applied – see Lemma 2.1.1 hereafter). At a point $M \in \mathcal{M}'$, the fibre over k_M when viewed over an algebraic closure of k_M is the union of two smooth conics meeting transversally in two points; each conic is defined over the extension $K_M = k_M[u]/(u^2 - m(M))$.

We let $\text{Br}_{\text{vert}}(X) \subset \text{Br}(X)$ be the vertical Brauer group with respect to the map π, as defined in the introduction. Since the geometric fibres are reduced, the quotient $\text{Br}_{\text{vert}}(X)/\text{Br}(k)$ is finite ([**Sk**], Cor. 4.5). We start with an adèle $\{P_v\} \in X(\mathbf{A}_k) = \prod_{v \in \Omega} X(k_v)$. We assume that for all $\mathscr{A} \in \text{Br}_{\text{vert}}(X)$, we have

$$\sum_{v \in \Omega} \text{inv}_v(\mathscr{A}(P_v)) = 0 \in \mathbf{Q}/\mathbf{Z},$$

in other words we assume that $\{P_v\}$ is in the kernel of all reciprocity maps $\theta_\mathscr{A} : X(\mathbb{A}_k) \to \mathbf{Q}/\mathbf{Z}$ associated to elements $\mathscr{A} \in \mathrm{Br}_{\mathrm{vert}}(X)$ (see [**CT**]).

Let S_1 be a finite set of places of k containing all the archimedean places; for $v \in S_1$, let $\mathfrak{U}_v \subset \mathbf{P}^1(k_v)$ be an open neighbourhood of $\pi(P_v)$. Under the five assumptions of Theorem A, we will find a point $\lambda \in \mathbf{P}^1(k)$, lying in $\mathfrak{U}_v$ for $v \in S_1$, such that the fibre $X_\lambda = \pi^{-1}(\lambda)$ is smooth and has infinitely many rational points. This will prove statement (a) in the Theorem. Statements (b) and (c) then follow easily.

A standard procedure, which uses the finiteness of $\mathrm{Br}_{\mathrm{vert}}(X)/\mathrm{Br}(k)$, and is described in full detail in [**CSS2**] (proof of Theorem 1.1) allows us to assume also that:

(a) For each place $v \in \Omega$, the point P_v lies on a smooth fibre of π and does not lie on the fibre at infinity (i.e. $\pi(P_v)$ lies in $U(k_v)$, for $U \subset \mathbf{A}^1_k$ as in the introduction).

(b) For each place $v \in S_1$, the projection $X(k_v) \to \mathbf{P}^1(k_v)$ admits an analytic section $\sigma_v : \mathfrak{U}_v \to X(k_v)$ over $\mathfrak{U}_v$.

(c) For each real place v, the neighbourhood $\mathfrak{U}_v$ coincides with the open set $t_v > 0$, and at any point of $\mathfrak{U}_v$ the fibre of π is a smooth curve of genus one ; for v complex, $\mathfrak{U}_v$ contains the point at infinity.

The mentioned procedure involves a projective transformation of the projective line $\mathbf{P}^1_k$ sending a point of $\mathbf{A}^1(k)$, with smooth fibre, to the point at infinity. One checks that the procedure does not affect the hypotheses. This is clear for any transformation $t \to t - \alpha$, where α does not belong to the zeroes of $d(c^2 - d)$. Assume, as we may, that $t = 0$ is not a zero of $d(c^2 - d)$. Then the statement is also clear when one sets $T = 1/t$, and one replaces $m(t), c(t), d(t)$ by $M(T) = T^\mu m(1/T)$, $D(T) = T^\delta d(1/T)$, $C(T) = T^\gamma c(1/T)$, where μ, δ, γ are the respective degrees of m, d, c, all even. Note that we then have $\deg(C^2 - D) = \deg(D) = 2\deg(C) > 0$.

From now on, we assume that conditions (a),(b),(c) above are fulfilled, as well as the improved condition (0.1), where the degree of $c^2 - d$ is equal to the degree of d.

Given (the new) m and its associated regular minimal proper model $X = X(m)/\mathbf{P}^1_k$, we shall define a finite set S_0 of 'bad' places of k. As in Proposition 1.1.1, we want it to contain all the archimedean places and all the places above 2, and we want the class group of $\mathfrak{o}_{S_0}$ to be trivial. We also want it to contain the set S_1 given above. For each $v \notin S_0$, we want all polynomials $c(t), d(t), m(t)$ to be $\mathfrak{o}_v$-integral, we want ρ_m to be an $\mathfrak{o}_v$-unit, and we want $r(t)$ to define a closed subscheme of $\mathrm{Spec}(\mathfrak{o}_{S_0}[t])$ which is finite and étale over $\mathfrak{o}_{S_0}$, the ring of S_0-integers of k. This implies in particular that the leading coefficients of $d(t)$ and of $(c^2 - d)(t)$ are units away from S_0. Given any monic irreducible factor $r_M(t)$ of $r(t)$, the closed subscheme $\tilde{M} = \mathrm{Spec}(\mathfrak{o}_{S_0}[t]/r_M(t))$ of $\mathbf{A}^1_{\mathfrak{o}_{S_0}}$, is finite and étale over $\mathrm{Spec}(\mathfrak{o}_{S_0})$. For each $M \in \mathscr{M}'$, we want the trivial or quadratic extension K_M/k_M (see above) to be unramified over the ring of S_0-integers of k_M. For much later use, we want S_0 to contain any finite place of k whose residue characteristic is less than or equal to the product of the degree $[k:\mathbf{Q}]$ and the degree of the polynomial $r(t)$ (this will be used when applying Schinzel's hypothesis). We also want S_0 to be big enough so that the fibration $\pi : X \to \mathbf{P}^1_k$ extends to a fibration $\mathscr{X}/\mathbf{P}^1_{\mathfrak{o}_{S_0}}$ with the following property: the scheme $\mathrm{Spec}(\mathfrak{o}_{S_0}[t]/r'(t)) = \bigcup_{M \in \mathscr{M}'} \tilde{M}$ is the set of points of $\mathbf{P}^1_{\mathfrak{o}_{S_0}}$ whose fibre does not contain a geometrically integral component of multiplicity one, and over any geometric point of this subscheme, the fibre consists of two smooth conics intersecting transversally in two points (that this can be achieved may be seen directly on projective equations, as in the introduction).

Given a finite set S of primes containing S_0, we shall be interested in the affine line $\mathbf{A}^1_{\mathfrak{o}_S}$ over $\mathfrak{o}_S$ and in the open set $\mathscr{U}_{\mathfrak{o}_S} = \mathrm{Spec}(\mathfrak{o}_S[t][1/r])$, and the analogous $\mathscr{U}'_{\mathfrak{o}_S}$ and $\mathscr{U}''_{\mathfrak{o}_S}$.

The group of units of $\mathfrak{o}_S[t][1/r]$ is the direct product of $\mathfrak{o}_S^*$ and the free group generated by the polynomials r_M for $M \in \mathscr{M}$, as one easily checks. Also, one easily checks that the Picard group of $\mathfrak{o}_S[t][1/r]$ vanishes (this uses the vanishing of the Picard group of $\mathfrak{o}_S$). The same results hold for $\mathfrak{o}_S[t][1/r']$) and $\mathfrak{o}_S[t][1/r'']$).

A point $\lambda \in \mathbf{P}^1(k) = \mathbf{P}^1(\mathfrak{o}_{S_0})$ (resp. $\lambda_v \in \mathbf{P}^1(k_v) = \mathbf{P}^1(\mathfrak{o}_v)$ for $v \notin S_0$) defines a section $\tilde{\lambda} : \mathrm{Spec}(\mathfrak{o}_{S_0}) \to \mathbf{P}^1_{\mathfrak{o}_{S_0}}$ (resp. $\tilde{\lambda}_v : \mathrm{Spec}(\mathfrak{o}_v) \to \mathbf{P}^1_{\mathfrak{o}_v}$). To simplify the notation, we shall sometimes write λ where it would be more correct to write λ_v. We shall call such a point *transversal* if the corresponding section $\tilde{\lambda}$ (resp. local section $\tilde{\lambda}_v$) defined over $\mathrm{Spec}(\mathfrak{o}_{S_0})$ (resp. $\mathrm{Spec}(\mathfrak{o}_v)$) transversally intersects with the finite étale $\mathfrak{o}_{S_0}$-scheme (resp. $\mathfrak{o}_v$-scheme) which is the union of all the restrictions over $\mathfrak{o}_{S_0}$ of the $\tilde{M}$'s ($M \in \mathscr{M}$) and of the section at infinity of $\mathbf{P}^1_{\mathfrak{o}_{S_0}}$ (resp. which is the pull-back to $\mathfrak{o}_v$ of this étale scheme). In less geometric terms, this simply means that for each $M \in \mathscr{M}$, the element $r_M(\lambda) \in k^*$ (resp. $r_M(\lambda_v) \in k_v^*$) is either a unit or a uniformizing parameter at each $v \notin S_0$ (resp. at v), and that if $v(\lambda) < 0$, then $v(\lambda) = -1$.

Given any point $\lambda \in \mathbf{A}^1(k)$ such that $d(c^2 - d)$ does not vanish at λ, the fibre X_λ is a smooth curve of genus one over k.

The analogue of Lemma 2.1.1 of [**CSS3**] (see [**BSwD**] Lemma 4) is here:

Lemma 2.1.1. — *Let S_0 be as above. Let $v \notin S_0$ and let $\lambda \in \mathbf{A}^1(k_v)$ be a transversal point. Let $\tilde{\lambda} \subset \mathbf{P}^1(\mathfrak{o}_v)$ be its closure.*

(a) If $\tilde{\lambda}$ does not intersect any of the $\tilde{M}$ for $M \in \mathcal{M}'$, then $X_\lambda(k_v) \neq \emptyset$.

(b) If $\tilde{\lambda}$ intersects $\tilde{M}$ for $M \in \mathcal{M}'$ at some place w of k_M (w unramified over v, of degree one over v), then $X_\lambda(k_v) \neq \emptyset$ if and only if w splits in K_M.

Proof. — Let $\mathcal{X}_\lambda/\mathfrak{o}_v$ be the $\mathfrak{o}_v$-scheme which is the inverse image of $\mathcal{X}/\mathbf{P}^1_{\mathfrak{o}_{S_0}}$ under $\tilde{\lambda}$.

(a) By the definition of S_0, the reduction mod v of the fibre $\mathcal{X}_\lambda/\mathfrak{o}_v$ contains a geometrically integral component of multiplicity one. As noted in Remark 1.3.3, this implies $X_\lambda(k_v) \neq \emptyset$.

(b) Since $\mathcal{X}_\lambda/\mathfrak{o}_v$ is proper we have $X_\lambda(k_v) = \mathcal{X}_\lambda(\mathfrak{o}_v)$. Thus a point of $X_\lambda(k_v)$ gives a local section over $\mathrm{Spec}(\mathfrak{o}_v)$ which intersects just one component of the closed fibre: that component must have multiplicity one and be geometrically integral. Thus the unramified extension K_M/k_M splits over w. Conversely, if this is so, the two conics which constitute the closed fibre of $\mathcal{X}_\lambda/\mathfrak{o}_v$ at v are individually defined over $\mathbf{F}_w = \mathbf{F}_v$. We may then apply Remark 1.3.3. We may also argue directly: Any conic contains at least 3 points, thus we can always find a smooth rational point in such a fibre. By Hensel's lemma this point can be lifted to a point over k_v. □

Given $M \in \mathcal{M}$, we may write $r_M(t) = \mathrm{Norm}_{k_M/k}(t - a_M)$, with $a_M \in k_M$ the class of t in $k[t]/r_M(t)$. For $M \in \mathcal{M}'$, let

$$(K_M/k_M, t - a_M) \in \mathrm{Br}(k_M(t))$$

be the class of the standard quaternion algebra associated to the element $t - a_M$ in $k_M(t)$ and the quadratic extension $K_M(t)/k_M(t)$. For such an M let

$$\mathfrak{a}_M(t) = \mathrm{Cores}_{k_M/k}(K_M/k_M, t - a_M) \in \mathrm{Br}(k(\mathbf{P}^1_k)).$$

The following Lemma is the analogue of Lemma 2.1.2 of [CSS3]. The hypotheses (a),(b) and (c) for the points $\{P_v\}_{v \in S_1}$ are in force.

Lemma 2.1.2. — *Let $\pi : X \to \mathbf{P}^1_k$ and S_0 (with $S_1 \subset S_0$) be as above, let $X_U = \pi^{-1}(U)$, and let $\{P_v\} \in \prod_{v \in \Omega} X(k_v)$ be an adelic point of X such that $\sum_{v \in \Omega} \mathrm{inv}_v(\mathscr{A}(P_v)) = 0$ for all $\mathscr{A} \in \mathrm{Br}_{\mathrm{vert}}(X)$. There exists a finite set S of places of k, containing S_0, and there exist points $Q_v \in X_U(k_v)$ for $v \in S$, with $Q_v = P_v$ for $v \in S_0$ (hence for $v \in S_1$), such that*

$$\sum_{v \in S} \mathrm{inv}_v(\mathfrak{a}_M(\pi(Q_v))) = 0$$

for all $M \in \mathcal{M}'$.

Proof. — This is just a special case of Harari's 'formal lemma' ([Ha1], Cor. 2.6.1; [CTSwD], Theorem 3.2.1). □

We now have a certain finite set S of places containing S_0.

The analogue of Prop. 2.1.3 of [CSS3] is the following proposition (*the freedom allowed at the archimedean places in* (6) *below should also have been allowed in* [CSS3].)

Proposition 2.1.3. — *Let $\pi : X \to \mathbf{P}^1_k$, the set S and the points $Q_v \in X(k_v)$, $v \in S$, be as in the conclusion of the previous lemma. For each point $M \in \mathcal{M} \cup \infty$, let T_M be a finite set of places of k such that $T_M \cap S = \emptyset$, $T_M \cap T_N = \emptyset$ for $M \neq N$. Let $\lambda \in U(k) \subset \mathbf{P}^1(k)$ and $\tilde{\lambda}$ the associated point in $\mathbf{P}^1(\mathfrak{o}_S)$. Assume:*

(1) *For each $M \in \mathcal{M}'$ and each place $v \in T_M$, there is an associated place w of k_M (of degree one over k) which splits in the quadratic extension $K_M = k_M(\sqrt{m(M)})/k_M$ (i.e. $m(M)$ is a square in $(k_M)_w$);*

(2) *each place $v \in T_\infty$ splits in each extension $k(\sqrt{\mathrm{Norm}_{k_M/k}(m(M))})/k$ for each $M \in \mathcal{M}'$ (i.e. for each such M, $\mathrm{Norm}_{k_M/k}(m(M))$ is a square in k_v for each $v \in T_\infty$);*

(3) *$\tilde{\lambda}$ is transversal over $\mathfrak{o}_S$;*

(4) *for any $M \in \mathcal{M}$, the (transversal) intersection of $\tilde{\lambda}$ and $\tilde{M}$, when viewed on $\tilde{M}$, consists of the places $w \in T_M$ and just one place w_M of k_M, of degree one over a place v_M of k; moreover, all v_M's are distinct from one another and none of them belongs to $\cup_{M \in \mathcal{M}} T_M$;*

(5) *the transversal intersection of $\tilde{\lambda}$ and the section at infinity of $\mathbf{P}^1_{\mathfrak{o}_S}$ consists exactly of the places $v \in T_\infty$ (hence for any such v, we have $v(\lambda) = -1$);*

(6) *λ is very close to $\lambda_v = \pi(Q_v) \in U(k_v)$ for nonarchimedean places $v \in S$ and lies in $\mathfrak{U}_v$ for $v \in S_1$.*

Let $T' = S \cup_{M \in \mathcal{M}'} T_M$ and $T'' = S \cup_{M \in \mathcal{M}''} T_M$, let

$$T'(\lambda) = T' \cup (\bigcup_{M \in \mathcal{M}'} \{v_M\}) \quad \text{and} \quad T''(\lambda) = T'' \cup (\bigcup_{M \in \mathcal{M}''} \{v_M\}).$$

Let $T = T' \cup T''$ and $T(\lambda) = T'(\lambda) \cup T''(\lambda)$.

Then one has:

(i) *$X_\lambda(k_v) \neq \emptyset$ for all places v of k;*

(ii.a) *the evaluation map $\mathrm{ev}_\lambda : \mathfrak{H}(\mathfrak{o}_{T' \cup T_\infty}[\mathcal{U}']) \to \mathfrak{H}(\mathfrak{o}_{T'(\lambda) \cup T_\infty})$ is an isomorphism, where $\mathcal{U}' = \mathcal{U}_{\mathfrak{o}_{T' \cup T_\infty}}$;*

(ii.b) *the evaluation map $\mathrm{ev}_\lambda : \mathfrak{H}(\mathfrak{o}_{T'' \cup T_\infty}[\mathcal{U}'']) \to \mathfrak{H}(\mathfrak{o}_{T''(\lambda) \cup T_\infty})$ is an isomorphism, where $\mathcal{U}'' = \mathcal{U}_{\mathfrak{o}_{T'' \cup T_\infty}}$;*

(iii.a) *the image of the point $P'(\lambda)$ of order 2 on E'_λ under the Kummer map $E'_\lambda(k) \to k^*/k^{*2}$ and the class of $m(\lambda)$ in k^*/k^{*2} lie in the subgroup $I''^{T''(\lambda)} = \mathfrak{H}(\mathfrak{o}_{T''(\lambda)})$ and are independent;*

(iii.b) *the image of the point $P''(\lambda)$ of order 2 on E''_λ under the Kummer map $E''_\lambda(k) \to k^*/k^{*2}$ lies in the subgroup $I'^{T'(\lambda)} = \mathfrak{H}(\mathfrak{o}_{T''(\lambda)})$ and is not trivial.*

Proof. — For (i), one uses Lemma 1.3.2 (or 2.1.1) together with Lemma 2.1.2, i.e. the hypothesis that there is no Brauer-Manin obstruction (this is what enables us to get the existence of a local solution at the 'Schinzel primes' v_M). For details, see [CSS3], p. 599/600.

To prove (ii), one uses two analogues of diagram (2.1.1) of [CSS3], namely one over U' and one over U''. For the first one, the right hand side group is $\oplus_{M \in \mathscr{M}'} \mathbf{Z}/2$, for the second one, it is $\oplus_{M \in \mathscr{M}''} \mathbf{Z}/2$. These are commutative diagrams of exact sequences.

$$\begin{array}{ccccccccc}
0 & \to & \mathfrak{H}(\mathfrak{o}_{T'' \cup T_\infty}) & \to & \mathfrak{H}(\mathfrak{o}_{T'' \cup T_\infty}[\mathscr{U}']) & \to & \oplus_{M \in \mathscr{M}'} \mathbf{Z}/2 & \to & 0 \\
& & \| & & \downarrow \mathrm{ev}_\lambda & & \| & & \\
0 & \to & \mathfrak{H}(\mathfrak{o}_{T'' \cup T_\infty}) & \to & \mathfrak{H}(\mathfrak{o}_{T'(\lambda) \cup T_\infty}) & \to & \oplus_{M \in \mathscr{M}'} \mathbf{Z}/2 & \to & 0
\end{array}$$

and

$$\begin{array}{ccccccccc}
0 & \to & \mathfrak{H}(\mathfrak{o}_{T'' \cup T_\infty}) & \to & \mathfrak{H}(\mathfrak{o}_{T'' \cup T_\infty}[\mathscr{U}'']) & \to & \oplus_{M \in \mathscr{M}''} \mathbf{Z}/2 & \to & 0 \\
& & \| & & \downarrow \mathrm{ev}_\lambda & & \| & & \\
0 & \to & \mathfrak{H}(\mathfrak{o}_{T'' \cup T_\infty}) & \to & \mathfrak{H}(\mathfrak{o}_{T''(\lambda) \cup T_\infty}) & \to & \oplus_{M \in \mathscr{M}''} \mathbf{Z}/2 & \to & 0
\end{array}$$

In these two diagrams, the bottom right hand side arrows are given by valuation at the places v_M.

To prove the independence statement in (iii.a), one then uses Assumption (2) in Theorem A, which is that m and $c^2 - d$ are independent (over $\mathbf{Z}/2$) in $\mathfrak{S}''$, that is to say that the class of m is not in the image of $\mathbf{Z}/2.P'$ under the Kummer map $E'(k(t)) \to k(t)^*/k(t)^{*2}$ coming from the isogeny $E'' \to E'$. (Recall that E' is given by $y^2 = (x - c(t))(x^2 - d(t))$.)

Similarly in (iii.b) one uses the fact that the class of d is not trivial in $k(t)^*/k(t)^{*2}$. That is automatic, since d is assumed nonconstant and squarefree. That the images lie in $\mathfrak{H}(\mathfrak{o}_{T''(\lambda)})$, resp. in $\mathfrak{H}(\mathfrak{o}_{T'(\lambda)})$, i.e. that they do not involve T_∞, is a consequence of the fact that $c^2 - d$ and m, resp. d, are of even degree, and that at any finite place $v \notin S_0$ their coefficients are integral and their leading coefficient a unit; this implies that their evaluation at any point λ with $v(\lambda) < 0$ has even v-adic valuation. □

2.2. The groups $\mathfrak{S}'$, $\mathfrak{S}''$ and the maps δ' and δ''.

Lemma 2.2.1. — *The maps*

$$\delta' : \mathfrak{S}' \to \oplus_{M \in \mathscr{M}''} k_M^* / k_M^{*2}$$

and

$$\delta'' : \mathfrak{S}'' \to \oplus_{M \in \mathscr{M}'} k_M^* / k_M^{*2}$$

given by evaluation at the relevant points M are homomorphisms. The kernels of these maps are finite.

Proof. — The group $\mathfrak{S}' = k[U']^*/k[U']^{*2}$ is generated by the classes of the r_M for $M \in \mathscr{M}'$ and by k^*/k^{*2}. Since $c^2 - d$ is not constant, $\mathscr{M}''$ is not empty. Now for any closed point M, the map $k^*/k^{*2} \to k_M^*/k_M^{*2}$ has finite kernel. Hence δ' has a finite kernel. The argument for the kernel of δ'' is the same (here we use the fact that d is not constant). □

2.3. Independence of the choice of λ: the spaces.

— Given $\lambda \in k$ as in Proposition 2.1.3, let us consider the curves E'_λ and E''_λ. They have good reduction at all places outside $T(\lambda)$ (they indeed have good reduction at places in T_∞). In Subsection 1.1 we defined $\mathbf{I}'^{T(\lambda)} = \mathfrak{H}(\mathfrak{o}_{T(\lambda)})$ and in Subsection 1.4 we considered the subspace $\mathbf{I}'^{T(\lambda)} \subset \mathbf{I}'^{T(\lambda)}$, and we defined similarly $\mathbf{I}''^{T(\lambda)}$ in $\mathbf{I}''^{T(\lambda)}$ (since λ determines the curves E'_λ and E''_λ, I only use the subscript λ).

By Lemma 1.4.1, the Selmer group $S'_\lambda = (S_{\phi'})_\lambda$, resp. $S''_\lambda = (S_{\phi''})_\lambda$ (we here simplify the notation of Section 1.5), can be computed as a subgroup of $\mathbf{I}'^{T(\lambda)}$, resp. $\mathbf{I}''^{T(\lambda)}$.

The group $\mathbf{I}''^{T(\lambda)}$ is defined by local conditions, namely, it consists of elements in $\mathfrak{H}(\mathfrak{o}_{T(\lambda)})$ whose image under the map $\mathfrak{H}(\mathfrak{o}_{T(\lambda)}) \to \mathfrak{H}(k_v)$ belongs to $W''_v(\lambda) + K_v$, this for all $v \in T(\lambda)$. Similarly for the group $\mathbf{I}'^{T(\lambda)}$.

The spaces K_v do not depend on λ. We shall fix the approximation condition (6) in Proposition 2.1.3 so that for $v \in S$, with the notation of Section 1.3, the subspaces $W''_v(\lambda)$ and $W''_v(\lambda_v)$ of V''_v are identical. For a real place v, the independence of these subspaces follows from the fact that $c(c^2 - d)$ does not vanish on the interval $\mathfrak{U}_v$. Hence the conditions defining $\mathbf{I}''^{T(\lambda)}$ at the places of S do not depend on λ.

For such a λ, from Remark 1.3.3, for $v \in T'(\lambda) \setminus S$, one has

$$W'_v(\lambda)/(W'_v(\lambda) \cap K'_v) = \mathbf{Z}/2,$$
$$W''_v(\lambda)/(W''_v(\lambda) \cap K''_v) = 0,$$
$$W''_v(\lambda) + K''_v = K''_v = \mathfrak{H}(\mathfrak{o}_v);$$

for $v \in T''(\lambda) \setminus S$, one has
$$W_v''(\lambda)/(W_v''(\lambda) \cap K_v'') = \mathbf{Z}/2,$$
$$W_v'(\lambda)/(W_v'(\lambda) \cap K_v') = 0,$$
$$W_v'(\lambda) + K_v' = K_v' = \mathfrak{H}(\mathfrak{o}_v);$$

for $v \notin T(\lambda) = T'(\lambda) \cup T''(\lambda)$, one has
$$W_v'(\lambda)/(W_v'(\lambda) \cap K_v') = 0,$$
$$W_v'(\lambda) + K_v' = K_v' = \mathfrak{H}(\mathfrak{o}_v),$$
$$W_v''(\lambda)/(W_v''(\lambda) \cap K_v'') = 0,$$
$$W_v''(\lambda) + K_v'' = K_v'' = \mathfrak{H}(\mathfrak{o}_v).$$

For such a λ, the natural embeddings $\mathbf{I}'^{T'(\lambda)} \subset \mathbf{I}'^{T(\lambda)}$ and $\mathbf{I}''^{T'(\lambda)} \subset \mathbf{I}''^{T(\lambda)}$ are isomorphisms, and the natural projection maps $\mathbf{W}'_{T(\lambda)} \to \mathbf{W}'_{T'(\lambda)}$ and $\mathbf{W}''_{T(\lambda)} \to \mathbf{W}''_{T''(\lambda)}$ are isomorphisms.

In particular we have $S'_\lambda \subset \mathbf{I}'^{T'(\lambda)} \subset \mathfrak{H}(\mathfrak{o}_{T'(\lambda)})$ and $S''_\lambda \subset \mathbf{I}''^{T''(\lambda)} \subset \mathfrak{H}(\mathfrak{o}_{T''(\lambda)})$.

From Section 1.4 we recall the isomorphisms $\tau'_\lambda : \mathbf{I}'^{T'(\lambda)} \simeq \mathbf{W}_{T'(\lambda)}$ and $\tau''_\lambda : \mathbf{I}''^{T''(\lambda)} \simeq \mathbf{W}_{T''(\lambda)}$.

From the above description of the sets $W'_v(\lambda), W''_v(\lambda), K'_v, K''_v$, one deduces the following lemma.

Lemma 2.3.1. — *Let T', T'', T and λ be as in Proposition 2.1.3.*

(a) *There are natural inclusions $\mathfrak{H}(\mathfrak{o}_S) \subset \mathfrak{H}(\mathfrak{o}_{T'}) \subset \mathfrak{H}(\mathfrak{o}_{T'(\lambda)}) \subset \mathfrak{H}(\mathfrak{o}_{T(\lambda)})$.*

The intersections of the subgroup $\mathbf{I}'^{T(\lambda)} \subset \mathfrak{H}(\mathfrak{o}_{T(\lambda)})$ with these various subgroups are precisely $\mathbf{I}'^S \subset \mathbf{I}'^{T'} \subset \mathbf{I}'^{T'(\lambda)} \subset \mathbf{I}'^{T(\lambda)}$.

The group $N'_0 := \mathbf{I}'^S$ depends neither on λ nor on the primes in $T(\lambda)$. The group $\mathbf{I}'^{T'}$ only depends on $T' = S \cup_{M \in \mathcal{M}'} T_M$, it depends neither on λ nor on the places v_M.

(b) *We have the analogous statements for the natural inclusions*
$$\mathfrak{H}(\mathfrak{o}_S) \subset \mathfrak{H}(\mathfrak{o}_{T''}) \subset \mathfrak{H}(\mathfrak{o}_{T''(\lambda)}) \subset \mathfrak{H}(\mathfrak{o}_{T(\lambda)})$$

for the group $N''_0 := \mathbf{I}''^S$, and for the group $\mathbf{I}''^{T''}$.

The proof to follow will involve the construction of 'subdiagrams' of the diagrams appearing in the proof of Prop. 2.1.3. Basically, we want to look at the inverse images under ev_λ of $\mathbf{I}'^{T'(\lambda)} \subset \mathfrak{H}(\mathfrak{o}_{T'(\lambda) \cup T_\infty}) = I'^{T'(\lambda) \cup T_\infty}$ and of $\mathbf{I}''^{T''(\lambda)} \subset \mathfrak{H}(\mathfrak{o}_{T''(\lambda) \cup T_\infty}) = I''^{T''(\lambda) \cup T_\infty}$.

Before doing this, we shall fix T_∞.

Consider the finite subgroup $\mathfrak{S}''_S$ of $\mathfrak{S}''$ spanned by I''^S and the elements $r_M(t)$ for M running through $\mathcal{M}''$. This subgroup contains $m(t)$. For each $M \in \mathcal{M}'$, we have the finite subgroup $\delta'_M(\mathfrak{S}''_S) \subset k_M^*/k_M^{*2}$, which defines a multiquadratic extension F_M/k_M. Consider also the finite subgroup $\mathfrak{S}'_S$ of $\mathfrak{S}'$ spanned by I'^S and the elements $r_M(t)$ for M running through $\mathcal{M}'$. For each $M \in \mathcal{M}''$, we have the finite subgroup $\delta'_M(\mathfrak{S}'_S) \subset k_M^*/k_M^{*2}$, which defines a multiquadratic extension F_M/k_M.

Let F/k be the composite of all the extensions F_M/k. It is a consequence of Tchebotarev's theorem that there exist infinitely many prime principal ideals (π_v) in the ring of integers $\mathfrak{o}_k$ such that the (chosen) generator π_v is a square (as a matter of fact, as close as one wishes to 1) in each completion k_w for $w \in S$, and such that moreover the place v splits in each quadratic extension $k(\sqrt{\mathrm{Norm}_{k_M/k}(m(M))})/k$. Choose $v_\infty \notin S$ a finite place which satisfies these properties (required in Proposition 2.1.3), and which moreover is unramified in the extension F/k (note: v_∞ is not an archimedean place, the index ∞ refers to the point at infinity on $\mathbf{P}^1$). Let $\mu = \pi_{v_\infty} \in \mathfrak{o}_k$ be a generator of the corresponding principal ideal. For each $w \in S$, μ is a square in k_w. For the rest of Section 2, we fix the set $T_\infty = \{v_\infty\}$, and we let $\lambda_\infty = \lambda_{v_\infty} = 1/\mu \in k_{v_\infty}$.

Lemma 2.3.2. — *Let notation be as in Proposition 2.1.3, with $T_\infty = \{v_\infty\}$ and μ as above.*

(a) For each $M \in \mathcal{M}'$, there exists a uniquely defined $a'_M \in I'^S$ such that for any T' and λ as in Proposition 2.1.3, $a'_M \cdot \mu^{\deg(r_M)} \cdot r_M(\lambda) \in I^{T'(\lambda) \cup T_\infty}$ is in the subspace of $\mathbf{I}'^{T'(\lambda)} \subset I^{T'(\lambda) \cup T_\infty}$ defined by the additional conditions that the local projection at $v \in S$ belongs to K'_v.

(b) For each $M \in \mathcal{M}''$, there exists a uniquely defined $a''_M \in I''^S$ such that for any T'' and λ as in Proposition 2.1.3, $a''_M \cdot \mu^{\deg(r_M)} \cdot r_M(\lambda) \in I^{T''(\lambda) \cup T_\infty}$ is in the subspace of $\mathbf{I}''^{T''(\lambda)} \subset I^{T''(\lambda) \cup T_\infty}$ defined by the additional conditions that the local projection at $v \in S$ belongs to K''_v.

Proof. — For any $a'_M \in I'^S$, the valuation of $a'_M \cdot \mu^{\deg(r_M)} \cdot r_M(\lambda)$ at a place $v \notin T'(\lambda)$ is even (for $v = v_\infty$ and r_M of odd degree, this follows from the transversality conditions (3) and (5) of Prop. 2.1.3). Thus any such $a'_M \cdot \mu^{\deg(r_M)} \cdot r_M(\lambda)$ belongs to $I'^{T'(\lambda)}$. For $M_1 \in \mathcal{M}'$ with $M_1 \neq M$, and $v \in T_{M_1} \cup \{v_{M_1}\}$, the local condition $a'_M \cdot \mu^{\deg(r_M)} \cdot r_M(\lambda) \in W'_v(\lambda) + K'_v$ is satisfied, since at any such place v the valuation $v(a'_M \cdot \mu^{\deg(r_M)} \cdot r_M(\lambda))$ is even and for $v \notin S$, K'_v consists of the classes in k_v^*/k_v^{*2} with even valuation. For any $v \in T_M \cup \{v_M\}$, the condition $a'_M \cdot \mu^{\deg(r_M)} \cdot r_M(\lambda) \in W'_v(\lambda) + K'_v$ is also satisfied. Indeed at such a place we have $W'_v(\lambda) = k_v^*/k_v^{*2}$ (Remark 1.3.3). Since λ is v-adically very close to λ_v for

$v \in S$ (condition (6) in Prop. 2.1.3), the class of $r_M(\lambda) \in k_v^*/k_v^{*2}$ does not depend on λ. Since μ is a square in each k_v for $v \in S$, the image of $\mu^{\deg(r_M)}.r_M(\lambda)$ in V'_S is independent of λ, T and T_∞. We have $I'^S \oplus K'_S = V'_S$. There thus exists a uniquely defined $a'_M \in I'^S$ such that $a'_M.\mu^{\deg(r_M)}.r_M(\lambda) \in I'^{T'(\lambda) \cup T_\infty}$ is in the subspace of $\mathbf{I}'^{T'(\lambda)} \subset I'^{T'(\lambda) \cup T_\infty}$ defined by the additional conditions that the local projection at each $v \in S$ belongs to K'_v. This completes the proof of (a). *Mutatis mutandis*, this also proves (b). □

For each $M \in \mathcal{M}'$, resp. $M \in \mathcal{M}''$, let a'_M, resp. a''_M, be the elements of I^S determined in the previous lemma.

We now define:

$$\mathbf{a}'_M(t) = (a'_M.\mu^{\deg(r_M)}.r_M(t)) \in \mathfrak{H}(\mathfrak{o}_{S \cup T_\infty}[\mathcal{U}']) \subset \mathfrak{S}'$$

for $M \in \mathcal{M}'$ and similarly

$$\mathbf{a}''_M(t) = (a''_M.\mu^{\deg(r_M)}.r_M(t)) \in \mathfrak{H}(\mathfrak{o}_{S \cup T_\infty}[\mathcal{U}'']) \subset \mathfrak{S}''$$

for $M \in \mathcal{M}''$.

Notation 2.3.3. — Let $A' \subset \mathfrak{H}(\mathfrak{o}_{S \cup T_\infty}[\mathcal{U}']) \subset \mathfrak{H}(k[U']) = \mathfrak{S}'$ be the subspace spanned by the $\mathbf{a}'_M(t)$ for $M \in \mathcal{M}'$. The subgroup A' and $\mathbf{I}'^{T'} = \mathbf{I}'^{T'(\lambda)} \cap \mathfrak{H}(\mathfrak{o}_{T'})$ of $\mathfrak{H}(\mathfrak{o}_{T' \cup T_\infty}[\mathcal{U}'])$ clearly have trivial intersection. Let $N' = A' \oplus \mathbf{I}'^{T'} \subset \mathfrak{H}(\mathfrak{o}_{T' \cup T_\infty}[\mathcal{U}'])$ be their direct sum. Let A'' and N'' be defined similarly.

The definition of N' depends on the choice of the family of subspaces K'_v (fixed once and for all) as well as on the points (Lemma 2.1.2) $Q_v \in X(k_v)$, $v \in S$ (points which determine the spaces $W_v(\lambda) = W_v(Q_v)$ for $v \in S$), as well as on T' and T_∞. But once these choices have been made, Lemma 2.3.1 shows that N' does not depend on the particular point λ chosen as in Proposition 2.1.3. More precisely, N' does not depend on the v_M's ($M \in \mathcal{M}'$).

If we fix T_∞ and μ as above, but still let the T_M's and v_M's for $M \in \mathcal{M}'$ vary, there is a subspace of $\mathfrak{H}(k[U'])$ which is contained in any N', namely the subspace $A' \oplus N'_0 \subset \mathfrak{H}(\mathfrak{o}_{S \cup T_\infty}[\mathcal{U}'])$, where $N'_0 = I'^S$ is as above and may be defined as $N'_0 = N' \cap \mathfrak{H}(\mathfrak{o}_S) \subset \mathfrak{H}(\mathfrak{o}_{T' \cup T_\infty}[\mathcal{U}'])$. Both N'_0 and A' are independent of T, and they clearly do not intersect.

The same considerations apply to N'', N''_0, A''.

As we observed in the proof of part (ii) of Proposition 2.1.3, the evaluation maps

$$\mathrm{ev}_\lambda : \mathfrak{H}(\mathfrak{o}_{T'\cup T_\infty}[\mathscr{U}']) \to \mathfrak{H}(\mathfrak{o}_{T'(\lambda)\cup T_\infty})$$

and

$$\mathrm{ev}_\lambda : \mathfrak{H}(\mathfrak{o}_{T''\cup T_\infty}[\mathscr{U}'']) \to \mathfrak{H}(\mathfrak{o}_{T''(\lambda)\cup T_\infty})$$

are isomorphisms as soon as properties (3) and (4) in that Proposition hold.

We have the exact analogue of Prop. 2.3.4 of [**CSS3**], to which we refer for the proof.

Proposition 2.3.4. — *Let T', T'', T_∞ and λ be as in Proposition 2.1.3, and let N' and N'' be as above.*

(a') *The evaluation map ev_λ from $\mathfrak{H}(\mathfrak{o}_{T'\cup T_\infty}[\mathscr{U}'])$ to $\mathfrak{H}(\mathfrak{o}_{T'(\lambda)\cup T_\infty})$ induces an isomorphism between N' and $\mathbf{I}'^{T'(\lambda)}$.*

(a'') *The evaluation map ev_λ from $\mathfrak{H}(\mathfrak{o}_{T''\cup T_\infty}[\mathscr{U}''])$ to $\mathfrak{H}(\mathfrak{o}_{T''(\lambda)\cup T_\infty})$ induces an isomorphism between N'' and $\mathbf{I}''^{T''(\lambda)}$.*

(b') *Denote by φ'_λ the composition of this isomorphism with the isomorphism $\mathbf{I}'^{T'(\lambda)} \xrightarrow{\simeq} \mathbf{W}'_{T'(\lambda)}$ in Section 1.4. There is a direct sum decomposition:*

$$N' = N'_0 \oplus A' \oplus \varphi_\lambda^{-1}(\oplus_{v\in T'\setminus S}(W'_v(\lambda)/W'_v(\lambda) \cap K'_v)).$$

(b'') *We have the analogous statement for φ''_λ.*

The class of $d(t) \in \mathfrak{H}(\mathfrak{o}_S[\mathscr{U}'])$ may clearly be written as

$$d(t) = \varepsilon . \prod_{M\in\mathscr{M}'} \mathbf{a}'_M(t)$$

with $\varepsilon \in I'^{S\cup T_\infty}$. Since the degree of d is even, one actually has $\varepsilon \in I'^S$. Since $d(\lambda) \in k^*/k^{*2}$ belongs to the image W'_λ of $E''_\lambda(k) \to k^*/k^{*2}$, ε belongs to $\mathbf{I}'^S$. From the above proposition we conclude that the class of d belongs to N', more precisely that d belongs to $N'_0 \oplus A'$.

The same argument shows that $(c^2 - d)(t) \in \mathfrak{H}(\mathfrak{o}_S[\mathscr{U}''])$ belongs to the group $N''_0 \oplus A' \subset N''$.

The same argument also applies to $m(t)$ viewed as an element in $\mathfrak{H}(\mathfrak{o}_S[\mathscr{U}''])$. Indeed, the degree of m is even, m is the product of some of the $r_M(t)$ for $M \in \mathscr{M}''$ by an element in $\mathfrak{o}_S^*$, and the curve $X_\lambda = X(m)_\lambda$ has points in each k_v, which implies that $m(\lambda)$ belongs to $W''_v(\lambda)$ for each place v, i.e. $m(\lambda) \in \mathbf{I}''^{T''(\lambda)}$. By the above proposition, the class of $m(t)$ belongs to N''. Now $m(t)$ divided by the appropriate product of elements $\mathbf{a}''_M(t)$ is an element of I''^S, and it belongs to N'', thus it lies in N''_0.

2.4. Independence of the choice of λ: the pairing.

At this point, for any λ as in Proposition 2.1.3, we have a pairing

$$\mathbf{e}: N' \times N'' \xrightarrow{(\mathrm{ev}'_\lambda, \mathrm{ev}''_\lambda)} \mathbf{I}'^{T'}(\lambda) \times \mathbf{I}''^{T''}(\lambda) \xrightarrow{(\mathrm{id}, \tau^{-1}_{T''(\lambda)})} \mathbf{I}'^{T'}(\lambda) \times \mathbf{W}''_{T''(\lambda)} \xrightarrow{\mathbf{e}_{T(\lambda)}} \mathbf{Z}/2$$

which may also be read

$$N' \times N'' \xrightarrow{(\mathrm{ev}'_\lambda, \mathrm{ev}''_\lambda)} \mathbf{I}'^{T'}(\lambda) \times \mathbf{I}''^{T''}(\lambda) \longrightarrow \mathbf{W}'_{T'(\lambda)} \times \mathbf{W}''_{T''(\lambda)} \xrightarrow{\mathbf{e}_{T(\lambda)}} \mathbf{Z}/2.$$

(Recall that we have $\mathbf{I}'^{T'(\lambda)} = \mathbf{I}'^{T(\lambda)}$, $\mathbf{I}''^{T''(\lambda)} = \mathbf{I}''^{T(\lambda)}$, $\mathbf{W}'_{T'(\lambda)} = \mathbf{W}'_{T(\lambda)}$ and $\mathbf{W}''_{T''(\lambda)} = \mathbf{W}''_{T(\lambda)}$.)

Proposition 2.4.1. — *Let T', T'', $T_\infty = \{v_\infty\}$ and λ be as in Proposition 2.1.3, and let N', N'' be as above. Then provided λ is close enough to $\lambda_\infty = 1/\mu$ at v_∞, the restriction of the pairing $\mathbf{e}$ to a pairing $(N'_0 \oplus A') \times (N''_0 \oplus A'')$ depends neither on T', T'' nor on λ.*

Proof. — It is identical to the one given in [**CSS3**], Section 2.4. The only differences are:

(1) We here have pairings $k_v^*/k_v^{*2} \times k_v^*/k_v^{*2} \to \mathbf{Z}/2$ rather than pairings $(k_v^*/k_v^{*2})^2 \times (k_v^*/k_v^{*2})^2 \to \mathbf{Z}/2$.

(2) We have to pair $\mathbf{a}'_P$ and $\mathbf{a}''_Q$ for $P \in \mathcal{M}'$ and $Q \in \mathcal{M}''$, hence in particular $P \neq Q$: so the only computation one has to reproduce is the one given on pages 610, 611 and the first lines of page 612 of [**CSS3**]. $\square$

2.5. Completion of the proof of Theorem A.

We have reached Section 2.5 of [**CSS3**] (page 612); I now transcribe the end of Section 2 of [**BSwD**].

For each $r \in N'$, if $r \neq 1, d$, there exists $M \in \mathcal{M}''$ such that $r(M) \neq 1 \in k_M^*/k_M^{*2}$. If r is of even degree, this is hypothesis (3.a) of the theorem. If r is of odd degree, one remarks that r is the product of an element of I'^S by an odd power of μ and a product of $r_M(t)$'s for $M \in \mathcal{M}'$. For any $M \in \mathcal{M}''$, the class $\delta'_M(r)$ is then nontrivial, since μ has been so chosen that it does not belong to the image under δ'_M of the group spanned by I'^S and the r_M's ($M \in \mathcal{M}'$).

For such an M there exist infinitely many primes v^M of k_M which are of degree one over k (we denote by v the induced place on k) and which remain inert in the quadratic extension $k_M(\sqrt{r(M)})/k_M$. We shall apply this to each element different from $1, d$ in the left kernel of the restriction of $\mathbf{e}$ to the group $(N'_0 \oplus A') \times (N''_0 \oplus A'')$. In this way, we produce distinct primes of the above type (one for each element of $N'_0 \oplus A'$ under consideration). Let v^M be one such prime, let v be the underlying prime of k and let $\lambda_v \in \mathfrak{o}_v$ be transversal to the corresponding $\tilde{M}$ at v^M. To such a λ_v are associated curves $E'_{\lambda_v}, E''_{\lambda_v}$ and groups $W'_v(\lambda_v), W''_v(\lambda_v)$.

Assertion: there exists $m''_v \in W''_v \subset k^*_v/k^{*2}_v$ which has a non-zero cup-product with $r(\lambda_v) \in k^*_v/k^{*2}_v$. This follows from the computations in Remark 1.3.3 (Lemma 4 in [**BSwD**]): for λ_v as above, one has $v((c^2-d)(\lambda_v)) = 1$, $v(d(\lambda_v)) = 0$ hence $v(r(\lambda_v)) = 0$ (all this modulo 2); the cup-product is simply given by $(r(\lambda_v), m''_v)$, it is non-zero if $v(m''_v) = 1$ (possible since $W''_v = k^*_v/k^{*2}_v$) and $r(\lambda_v) \in \mathfrak{o}^*_v$ is not a square, which it is not since v^M is inert in the quadratic extension $k_M(\sqrt{r(M)})/k_M$.

For each $r \in N'' \subset \mathfrak{H}(k(t))$, by hypothesis (3.b) of the theorem if the degree of r is even, or by the analogue of the above argument if the degree of r is odd, if r does not belong to the group spanned by $c^2 - d$ and m, there exists an $M \in \mathcal{M}'$ such that $r(M)$ does not belong to $(1, m(M)) \subset k^*_M/k^{*2}_M$. For this $M \in \mathcal{M}'$ there exist infinitely many places v^M of k_M of degree one over k (we denote by v the induced place on k), which are split in the quadratic extension $k_M(\sqrt{m(M)})/k_M$ but remain inert in the quadratic extension $k_M(\sqrt{r(M)})/k_M$.

(Note: at this point we do want the places v^M to split in $k_M(\sqrt{m(M)})/k_M$, see Proposition 2.1.3 (1): this is to ensure that $X_\lambda(k_v) \neq \emptyset$ at such places.)

We apply this to each element of $N''_0 \oplus A''$ which does not belong to the subgroup $\{c^2 - d, m\}$ and which is in the right kernel of the restriction of $\mathbf{e}$ to $(N'_0 \oplus A') \times (N''_0 \oplus A'')$. In this way, we produce distinct primes of the above type (one for each element of $N''_0 \oplus A''$ under consideration). Let v^M be one such prime, let v be the underlying prime of k and let $\lambda_v \in \mathfrak{o}_v$ be transversal to the corresponding $\tilde{M}$ at v^M. To such a λ_v are associated curves $E'_{\lambda_v}, E''_{\lambda_v}$ and groups W'_v, W''_v.

Assertion: there exists $m'_v \in W'_v \subset k^*_v/k^{*2}_v$ which has a nonzero cup-product with $r(\lambda_v) \in k^*_v/k^{*2}_v$. This follows from Remark 1.3.3: we here have $v(d(\lambda_v)) = 1$ hence $v((c^2-d)(\lambda_v)) = 0$ and $v(r(\lambda_v)) = 0$ (all this modulo 2), the cup-product is simply given by $(m'_v, r(\lambda_v))$. This is nonzero if $v(m'_v) = 1$ (which occurs, by Remark 1.3.3) and $r(\lambda_v) \in \mathfrak{o}^*_v$ is not a square, which it is not since v^M remains inert in the quadratic extension $k_M(\sqrt{r(M)})/k_M$.

Let us now consider the restriction of the pairing $\mathbf{e}$ to $(N'_0 \oplus A') \times (N''_0 \oplus A'')$. Recall that this restriction does not depend on the (not yet made) choice of T', T'', λ.

The classes of d, resp. $c^2 - d, m$ lie in $N'_0 \oplus A'$, resp $N''_0 \oplus A''$, as was pointed out after Prop. 2.3.4.

We let $N'_0 \oplus A' = F'_0 \oplus F'_1 \oplus F'_2$, where $F'_0 = \mathbf{Z}/2.d$ and $F'_0 \oplus F'_1$ is the left kernel of the pairing $(N'_0 \oplus A') \times (N''_0 \oplus A'') \to \mathbf{Z}/2$.

We let $N''_0 \oplus A'' = F''_0 \oplus F''_1 \oplus F''_2$, where $F''_0 = \mathbf{Z}/2.(c^2-d) \oplus \mathbf{Z}/2.m$ and $F''_0 \oplus F''_1$ is the right kernel of the pairing $(N'_0 \oplus A') \times (N''_0 \oplus A'') \to \mathbf{Z}/2$.

Note that the pairing then induces a pairing $F_2' \times F_2''$ which has trivial kernel on both sides.

We now proceed as on pages 615-616 of [**CSS3**].

To each nonzero element in $r \in F_1'$ we may by the above procedure associate an $M \in \mathcal{M}''$, a place v_M in k_M, of degree one over k, a transversal λ_v and an element m_v'' such that $(r(\lambda_v), m_v'') = 1 \in \mathbf{Z}/2$. Let S_2^0 be the set of places of k thus produced. The bilinear pairing $F_1' \times (\oplus_{v \in S_2^0} \mathbf{Z}/2) \to \mathbf{Z}/2$ sending the pair (r, c_v) to the cup-product $(r(\lambda_v), c_v) \in \mathbf{Z}/2$ is thus nondegenerate on the left hand side, one may therefore extract from S_2^0 a subset S_2 of order exactly the rank of the $\mathbf{Z}/2$-vector space F_1', such that the induced bilinear pairing $F_1' \times (\oplus_{v \in S_2} \mathbf{Z}/2) \to \mathbf{Z}/2$ (sending the pair $(r, 1_v)$ to $r(\lambda_v) \in \mathfrak{o}_v^*/\mathfrak{o}_v^{*2} = \mathbf{Z}/2$) is nondegenerate on both sides. We now fix such a set S_2 and the corresponding sets T_M for $M \in \mathcal{M}''$. We also fix the λ_v's as above.

We similarly proceed with F_1''. We thus get sets T_M's for suitable M in $\mathcal{M}'$, consisting of places v^M of degree one over the corresponding place of k (which we may assume distinct from the places appearing in $S \cup S_2$) and associated λ_v's. This defines a set S_3 of places of k such that the bilinear pairing $(\oplus_{v \in S_3} \mathbf{Z}/2) \times F_1'' \to \mathbf{Z}/2$ (sending the pair $(1_v, r)$ to $r(\lambda_v) \in \mathfrak{o}_v^*/\mathfrak{o}_v^{*2} = \mathbf{Z}/2$) is nondegenerate on both sides. We now fix such a set S_3 and the corresponding sets T_M for $M \in \mathcal{M}'$. We also fix the λ_v's as above.

If we now choose a λ as in Schinzel's hypothesis (in the (H_1) version, due to Serre, see [**CTSwD**], p. 71), close enough to the chosen λ_v's for v finite in $S \cup T_\infty \cup S_2 \cup S_3$, and positive at the real places of k (at the real places, only the analogous condition should have been imposed on pages 598 and 617 of [**CSS3**]), then the associated pairing $\mathbf{e}_\lambda$, whose kernels by Prop. 1.4.3 determine the Selmer groups S_λ' and S_λ'', breaks up as

$$(F_0' \oplus F_1' \oplus F_2' \oplus F_3') \times (F_0'' \oplus F_1'' \oplus F_2'' \oplus F_3'') \to \mathbf{Z}/2,$$

and we have the following properties.

F_0' is in the left kernel and F_0'' in the right kernel.

F_1' is orthogonal to $F_0'' \oplus F_1'' \oplus F_2''$ and F_1'' is orthogonal to $F_0' \oplus F_1' \oplus F_2'$.

The pairing $F_2' \times F_2'' \to \mathbf{Z}/2$ is nondegenerate.

The pairing $F_1' \times F_3'' \to \mathbf{Z}/2$ is nondegenerate.

The pairing $F_3' \times F_1'' \to \mathbf{Z}/2$ is nondegenerate.

This implies that the left kernel of the total pairing is $F_0' = \mathbf{Z}/2$, spanned by the image $d(\lambda) \in k^*/k^{*2}$ of $P''(\lambda) \in E_\lambda''(k)$ under the Kummer map, and that the right kernel of the total pairing is $F_0'' = \mathbf{Z}/2 \oplus \mathbf{Z}/2$, spanned by $m(\lambda)$ and the class $(c^2 - d)(\lambda) \in k^*/k^{*2}$ of the image of $P'(\lambda) \in E_\lambda'(k)$ under the Kummer map.

It then only remains to apply Proposition 1.5.2. $\square$

Remark The whole discussion with T_∞ and μ could be avoided if one made a stronger assumption in Theorem A, namely if one had hypothesis (3.a) for the map δ' on the whole group $\mathfrak{S}'$ and similarly for the hypothesis (3.b) for the map δ'' on the whole group $\mathfrak{S}''$.

3. Proof of Theorem B

The proof is very close to the proof of Theorem A. Only the points which differ will be described.

On the family $X/\mathbf{P}_k^1$ of Theorem B one first realizes a reduction analogous to the one described in 2.1.

We here have several sets of closed points $M \in \mathbf{A}_k^1$. The set $\mathscr{M}$ is defined by the vanishing of the separable polynomial

$$\Delta = d_{01}(d_{23}^2 + 4d_{24}d_{34})(d_{04}^2 - d_{02}d_{03})(d_{14}^2 - d_{12}d_{13}).$$

The set $\mathscr{M}_1'$ is defined by the vanishing of d_{01}. The set $\mathscr{M}_2'$ is defined by the vanishing of $(d_{23}^2 + 4d_{24}d_{34})$. Their union is the set $\mathscr{M}'$. The set $\mathscr{M}''$ is defined by the vanishing of the product $(d_{04}^2 - d_{02}d_{03})(d_{14}^2 - d_{12}d_{13})$.

To any closed point $M \in \mathscr{M}_1'$, with residue field k_M, we attach two quadratic extensions. The first one, K_M/k_M is obtained by adding the square root of the value at M of $(d_{04}^2 - d_{02}d_{03})$, or of $(d_{14}^2 - d_{12}d_{13})$ (at a point $M \in \mathscr{M}'$, the product of these two values is a square in k_M). The second extension, L_M/k_M, is the one obtained by adding the square root of the value of $-c$ in k_M (the value of c is given in the Introduction and repeated in Section 1.2).

To a closed point $M \in \mathscr{M}''$, we associate a quadratic extension K_M/k_M. If $(d_{04}^2 - d_{02}d_{03})$ vanishes at M, then K_M is obtained by adding the square root of the value of $-d_{01}d_{02}$. If $(d_{14}^2 - d_{12}d_{13})$ vanishes at M, then K_M is obtained by adding the square root of the value of $d_{01}d_{12}$.

We do not attach any quadratic extension to K_M if M belongs to $\mathscr{M}_2'$.

One defines a set S_0 as in Section 2.1 (containing any preassigned finite set of places), with an associated nice model $\mathscr{X}/\mathbf{P}_{\mathfrak{o}_{S_0}}^1$ over the ring $\mathfrak{o}_{S_0}$ of S_0-integers.

Lemma 3.1. — *Let $v \notin S_0$ and let $\lambda \in \mathbf{A}^1(k_v)$ be a transversal point. Let $\tilde{\lambda} \subset \mathbf{P}^1(\mathfrak{o}_v)$ be its closure.*

(a) If $\tilde{\lambda}$ does not intersect any of the $\tilde{M}$ for $M \in \mathscr{M}_1' \cup \mathscr{M}''$, then $X_\lambda(k_v) \neq \emptyset$.

(b) If $\tilde{\lambda}$ intersects $\tilde{M}$ for $M \in \mathscr{M}_1' \cup \mathscr{M}''$ at some place w of k_M (w unramified over v, of degree one over v), then $X_\lambda(k_v) \neq \emptyset$ if and only if w splits in K_M.

Note that the extension L_M/k_M does not play a rôle at this stage.

HASSE PRINCIPLE FOR PENCILS OF CURVES OF GENUS ONE 157

Given $M \in \mathcal{M}$, we write $r_M(t) = \text{Norm}_{k_M/k}(t - a_M)$, with $a_M \in k_M$. For $M \in \mathcal{M}'_1 \cup \mathcal{M}''$, let

$$(K_M/k_M, t - a_M) \in \text{Br}(k_M(t))$$

be the class of the standard cyclic algebra associated to the element $t - a_M$ in $k_M(t)$ and the cyclic extension $K_M(t)/k_M(t)$. Let

$$\mathfrak{a}_M(t) = \text{Cores}_{k_M/k}(K_M/k_M, t - a_M) \in \text{Br}(k(\mathbf{P}^1_k)).$$

For $M \in \mathcal{M}'_1$, let similarly

$$\mathfrak{b}_M(t) = \text{Cores}_{k_M/k}(L_M/k_M, t - a_M) \in \text{Br}(k(\mathbf{P}^1_k)).$$

Lemma 3.2. — *Let $\pi : X \to \mathbf{P}^1_k$ and S_0 be as above, let $X_U = \pi^{-1}(U)$, and let $\{P_v\} \in \prod_{v \in \Omega} X(k_v)$ be an adelic point of X such that $\sum_{v \in \Omega} \text{inv}_v(\mathscr{A}(P_v)) = 0$ for all $\mathscr{A} \in \text{Br}_{\text{vert}}(X)$. There exists a finite set S of places of k, containing S_0, and there exist points $Q_v \in X_U(k_v)$ for $v \in S$, with $Q_v = P_v$ for $v \in S_0$, such that*

$$\sum_{v \in S} \text{inv}_v(\mathfrak{a}_M(\pi(Q_v))) = 0$$

for all $M \in \mathcal{M}'_1 \cup \mathcal{M}''$ and such that

$$\sum_{v \in S} \text{inv}_v(\mathfrak{b}_M(\pi(Q_v))) = 0$$

for all $M \in \mathcal{M}'_1$.

Proposition 3.3. — *Let $\pi : X \to \mathbf{P}^1_k$, the set S of places and the $Q_v \in X(k_v)$, $v \in S$, be as in the conclusion of the previous lemma. For each point $M \in \mathcal{M} \cup \infty$, let T_M be a finite set of places of k such that $T_M \cap S = \emptyset$, $T_M \cap T_N = \emptyset$ for $M \neq N$. Let $\lambda \in U(k) \subset \mathbf{P}^1(k)$ and $\tilde{\lambda}$ the associated point in $\mathbf{P}^1(\mathfrak{o}_S)$. Assume:*

(1) *For each $M \in \mathcal{M}'_1$ and each place $v \in T_M$, there is an associated place w of k_M (of degree one over k) which splits in the quadratic extension K_M/k_M and in the quadratic extension L_M/k_M; for each $M \in \mathcal{M}''$ and each place $v \in T_M$, there is an associated place w of k_M (of degree one over k) which splits in the quadratic extension K_M/k_M;*

(2) *Write $K_M = k_M(\sqrt{\alpha_M})$ and $L_M = k_M(\sqrt{\beta_M})$; each place $v \in T_\infty$ splits in each extension $k(\sqrt{\text{Norm}_{k_M/k}(\alpha_M)})/k$ and $k(\sqrt{\text{Norm}_{k_M/k}(\beta_M)})/k$;*

(3) *$\tilde{\lambda}$ is transversal over $\mathfrak{o}_S$;*

(4) *for any $M \in \mathcal{M}$, the (transversal) intersection of $\tilde{\lambda}$ and $\tilde{M}$, when viewed on $\tilde{M}$, consists of the places $w \in T_M$ and just one place w_M of k_M, of degree one over a place v_M of k;*

(5) *the transversal intersection of $\tilde{\lambda}$ and the section at infinity of $\mathbf{P}^1_{o_S}$ consists exactly of the places $v \in T_\infty$ (hence for any such v, we have $v(\lambda) = -1$);*

(6) λ *is very close to* $\lambda_v = \pi(Q_v) \in U(k_v)$ *for nonarchimedean places* $v \in S$, *and lies in* $\mathfrak{U}_v$ *for* $v \in S_1$.

Let $T' = S \cup_{M \in \mathcal{M}'} T_M$ and $T'' = S \cup_{M \in \mathcal{M}''} T_M$, let

$$T'(\lambda) = T' \cup (\bigcup_{M \in \mathcal{M}'} \{v_M\}) \quad \text{and} \quad T''(\lambda) = T'' \cup (\bigcup_{M \in \mathcal{M}''} \{v_M\}).$$

Let $T = T' \cup T''$ and $T(\lambda) = T'(\lambda) \cup T''(\lambda)$.

Then one has:

(i) $X_\lambda(k_v) \neq \emptyset$ *for all places v of k;*

(ii.a) *the evaluation map* $\mathrm{ev}_\lambda : \mathfrak{H}(\mathfrak{o}_{T' \cup T_\infty}[\mathcal{U}']) \to \mathfrak{H}(\mathfrak{o}_{T'(\lambda) \cup T_\infty})$ *is an isomorphism, where* $\mathcal{U}' = \mathcal{U}_{\mathfrak{o}_{T' \cup T_\infty}}$;

(ii.b) *the evaluation map* $\mathrm{ev}_\lambda : \mathfrak{H}(\mathfrak{o}_{T'' \cup T_\infty}[\mathcal{U}'']) \to \mathfrak{H}(\mathfrak{o}_{T''(\lambda) \cup T_\infty})$ *is an isomorphism, where* $\mathcal{U}'' = \mathcal{U}_{\mathfrak{o}_{T'' \cup T_\infty}}$;

(iii.a) *the classes of* $(d_{04}^2 - d_{02}d_{03})(\lambda)$ *and* $(d_{14}^2 - d_{12}d_{13})(\lambda)$ *lie in the subgroup* $I^{\prime\prime T''(\lambda)} = \mathfrak{H}(\mathfrak{o}_{T''(\lambda)}) \subset k^*/k^{*2}$ *and are independent;*

(iii.b) *the class of* $(d_{23}^2 + 4d_{24}d_{34})(\lambda)$ *lies in the subgroup* $I^{\prime T'(\lambda)} = \mathfrak{H}(\mathfrak{o}_{T'(\lambda)})$ *of* k^*/k^{*2} *and is not trivial.*

Just as in Section 2, one uses two analogues of diagram (2.1.1) of [**CSS3**], namely one over U' and one over U'' (*same open sets* as in Section 2). For the first one, the right hand side group is $\oplus_{M \in \mathcal{M}'} \mathbf{Z}/2$, for the second one, it is $\oplus_{M \in \mathcal{M}''} \mathbf{Z}/2$. These are commutative diagrams of exact sequences:

$$\begin{array}{ccccccccc} 0 & \to & \mathfrak{H}(\mathfrak{o}_{T' \cup T_\infty}) & \to & \mathfrak{H}(\mathfrak{o}_{T' \cup T_\infty}[\mathcal{U}']) & \to & \oplus_{M \in \mathcal{M}'} \mathbf{Z}/2 & \to & 0 \\ & & \| & & \downarrow \mathrm{ev}_\lambda & & \| & & \\ 0 & \to & \mathfrak{H}(\mathfrak{o}_{T' \cup T_\infty}) & \to & \mathfrak{H}(\mathfrak{o}_{T'(\lambda) \cup T_\infty}) & \to & \oplus_{M \in \mathcal{M}'} \mathbf{Z}/2 & \to & 0 \end{array}$$

and

$$\begin{array}{ccccccccc} 0 & \to & \mathfrak{H}(\mathfrak{o}_{T'' \cup T_\infty}) & \to & \mathfrak{H}(\mathfrak{o}_{T'' \cup T_\infty}[\mathcal{U}'']) & \to & \oplus_{M \in \mathcal{M}''} \mathbf{Z}/2 & \to & 0 \\ & & \| & & \downarrow \mathrm{ev}_\lambda & & \| & & \\ 0 & \to & \mathfrak{H}(\mathfrak{o}_{T'' \cup T_\infty}) & \to & \mathfrak{H}(\mathfrak{o}_{T''(\lambda) \cup T_\infty}) & \to & \oplus_{M \in \mathcal{M}''} \mathbf{Z}/2 & \to & 0. \end{array}$$

In these two diagrams, the bottom right hand side arrows are given by valuation at the places v_M.

Statements (iii.a) and (iii.b) can be reformulated exactly as in Proposition 2.1.3, in terms of the curves E'_λ and E''_λ.

One now looks for a $\lambda \in \mathbf{A}^1(k)$ such that 3.3 holds (this will be provided by Schinzel's hypothesis) and such that the 2-torsion subgroup of the Tate-Shafarevich group of E'_λ is trivial. Propositions 1.5.2 and 3.3 then enable one to conclude.

We are nearly in the same situation as we were in Section 2.3, except that

$$d = d_{01}^2 \cdot (d_{23}^2 + 4 d_{24} d_{34}) \in k[t]$$

contains a square factor.

One first has to check the "independence of the spaces" on the choice of λ.

A key point here is to check that for each $M \in \mathscr{M}$ and each $v \in T(\lambda) \setminus S$, the spaces W'_v and W''_v are exactly the same as they were in Section 2.

Suppose $M \in \mathscr{M}'_1$, i.e. $d_{01}(M) = 0$. For each $v \in T_M \cup \{v_M\}$, we have $v(d(M)) = 2$, and the extension L_M/k_M is split at v (for $v = v_M$, the last statement follows from the reciprocity law), i.e. the class of $-c(M)$ is a square at the completion of k_M at v. It then follows from Remark 1.3.3 (iv) that $W''_v = 1$ and $W'_v = k^*_{M,v}/k^{*2}_{M,v}$, hence, since $K'_v = \mathfrak{o}^*_v/\mathfrak{o}^{*2}_v = K''_v$ for $v \notin S$, $W''_v/(K''_v + W''_v) = 0$ and $W'_v/(W'_v + K'_v) = \mathbf{Z}/2$.

Suppose $M \in \mathscr{M}'_2$, i.e. $(d_{23}^2 + 4 d_{24} d_{34})(M) = 0$. For each $v \in T_M \cup \{v_M\}$, we have $v(d(M)) = 1$. Remark 1.3.3 (ii) shows that we are exactly in the same situation as above.

Suppose $M \in \mathscr{M}''$. For each $v \in T_M \cup \{v_M\}$, we have $v(d(M)) = 0$ and $v((c^2 - d)(M)) = v(\Delta''(M)) = 1$. By Remark 1.3.3 (iii), we have $W'_v = 1$, $W''_v = k^*_{M,v}/k^{*2}_{M,v}$, hence finally $W'_v/(W'_v + K'_v) = 0$ and $W''_v/(K''_v + W''_v) = \mathbf{Z}/2$.

Now the situation is essentially identical to the one considered at the beginning of Section 2.3, and the rest of the proof of Theorem B is just the same as that of Theorem A, hence shall not be reproduced. □

Acknowledgements

This paper is a close variation on the paper [BSwD] of Bender and Swinnerton-Dyer, some parts of which have been simply reproduced.

Work for the paper was started on a hike in the Gers. Most of the writing was done in January 2000, while the author was staying at the Tata Institute of Fundamental Research (Mumbai, India), under the auspices of the Centre franco-indien pour la promotion de la recherche avancée (CEFIPRA/IFCPAR, Project 1601-2).

The author conveys his thanks to the referee for numerous and insightful remarks.

References

[BSwD] A. O. Bender and Sir Peter Swinnerton-Dyer, *Solubility of certain pencils of curves of genus 1, and of the intersection of two quadrics in* $\mathbf{P}^4$, Preprint, 1999. To appear in Proc. of the London Math. Soc.

[CT] J.-L. Colliot-Thélène, *The Hasse principle in a pencil of algebraic varieties*, in *Number Theory*, Proceedings of a conference held at Tiruchirapalli, India, January 1996, K. Murty and M. Waldschmidt ed., Contemp. math. **210** (1998), 19–39.

[CTSa] J.-L. Colliot-Thélène et J.-J. Sansuc, *Fibrés quadratiques et composantes connexes réelles*, Math. Ann. **244** (1979), 105–134.

[CTSaSwD] J.-L. Colliot-Thélène, J.-J. Sansuc and Sir Peter Swinnerton-Dyer, *Intersections of two quadrics and Châtelet surfaces, I*, J. reine angew. Math. (Crelle) **373** (1987), 37–107; II, J. reine angew. Math. (Crelle) **374** (1987), 72–168.

[CSS1] J.-L. Colliot-Thélène, A. N. Skorobogatov and Sir Peter Swinnerton-Dyer, *Double fibres and double covers: paucity of rational points*, Acta Arithmetica **LXXIX** (1997), 113–135.

[CSS2] J.-L. Colliot-Thélène, A. N. Skorobogatov and Sir Peter Swinnerton-Dyer, *Rational points and zero-cycles on fibred varieties: Schinzel's hypothesis and Salberger's device*, J. reine angew. Math. (Crelle) **495** (1998), 1–28.

[CSS3] J.-L. Colliot-Thélène, A. N. Skorobogatov and Sir Peter Swinnerton-Dyer, *Hasse principle for pencils of curves of genus one whose Jacobians have rational 2-division points*, Invent. Math. **134** (1998), 579–650.

[CTSwD] J.-L. Colliot-Thélène and Sir Peter Swinnerton-Dyer, *Hasse principle and weak approximation for pencils of Severi-Brauer and similar varieties*, J. reine angew. Math. (Crelle) **453** (1994), 49–112.

[Ha1] D. Harari, *Méthode des fibrations et obstruction de Manin*, Duke Math. J. **75** (1994), 221–260.

[Ha2] D. Harari, *Flèches de spécialisations en cohomologie étale et applications arithmétiques*, Bull. Soc. math. France **125** (1997), 143–166.

[Mi] J. S. Milne, *Arithmetic Duality Theorems* (Academic Press, 1986).

[Sk] A. N. Skorobogatov, *Descent on fibrations over the projective line*, Amer. J. Math. **118** (1996), 905–923.

[SwD1] Sir Peter Swinnerton-Dyer, *Rational points on certain intersections of two quadrics*, in *Abelian Varieties*, ed. Barth, Hulek and Lange, 273–292 (Walter de Gruyter, Berlin 1995).

[SwD2]　　　Sir Peter Swinnerton-Dyer, *Some applications of Schinzel's hypothesis to Diophantine equations*, Number theory in progress, Vol. 1 (Zakopane-Kościelisko, 1997), 503–530 (Walter de Gruyter, Berlin, 1999).

[SwD3]　　　Sir Peter Swinnerton-Dyer, *Arithmetic of diagonal quartic surfaces, II*, Proc. London Math. Soc. (3) **80** (2000), 513–544.

[SwD4]　　　Sir Peter Swinnerton-Dyer, *The solubility of diagonal cubic surfaces*, à paraître dans les Annales scientifiques de l'École Normale Supérieure.

ENRIQUES SURFACES WITH A DENSE SET OF RATIONAL POINTS,
APPENDIX TO THE PAPER BY J.-L. COLLIOT-THÉLÈNE

Alexei Skorobogatov

   Department of Mathematics, Imperial College, 180 Queen's Gate, London SW7 2BZ, U.K.

   The aim of this note is to deduce from the results of Bender and Swinnerton-Dyer [**BS**], as refined by Colliot-Thélène [**CT**], that rational points are Zariski dense on certain Enriques surfaces defined over a number field k, conditionally on the Schinzel Hypothesis (H) and the finiteness of Tate–Shafarevich groups of elliptic curves over k. It was shown by Bogomolov and Tschinkel [**BT**] that for any Enriques surface Y defined over a number field k there exists a finite extension K/k such that K-points are Zariski dense on Y ("potential density" of rational points). We intend to show that the results of [**BS**] and [**CT**] can be used to construct explicit families of Enriques surfaces over any number field k with the property that already k-points are Zariski dense. Although the general construction is conditional on the above mentioned conjectures, once an equation is written it is often possible to give a direct (and unconditional) proof of the Zariski density of rational points. We check this for the explicit example (4) below, using an idea from [**BT**]. Let us add that it is not known whether or not there exists an Enriques or $K3$-surface X over a number field k such that $X(k)$ is not empty but not Zariski dense in X.

Proposition 1. — *Let k be a field of characteristic zero. Let $c, d \in k[t]$ be polynomials, $deg(c) = 2$, $deg(d) = 4$, $deg(c^2 - d) = 4$, such that $td(c^2 - d)$ is*

separable. Then there exists a regular minimal model Y of the affine surface
$$y_1^2 = t(x - c(t)), \ y_2^2 = t(x^2 - d(t)) \tag{1}$$
which is an Enriques surface (hence Y is unique). Any K3 double covering of Y is given by $t = mu^2$, for some $m \in k^*$.

Proof. — A standard argument (cf. [CSS97], Section 4) shows that for any discrete valuation $v : k(Y)^* \to \mathbf{Z}$, which is trivial on k, the valuation $v(t)$ is even. Indeed, if $v(t) > 0$, then $v(d) = v(c^2 - d) = 0$ by our assumption. If $v(x) \neq 0$, then from the second equation (1) we see that $v(t)$ is also even. The case $v(x) = 0$, $v(x - c) > 0$, $v(x^2 - d) > 0$ leads to $v(c^2 - d) > 0$, which is a contradiction. Hence either $v(x - c) = 0$ or $v(x^2 - d) = 0$, implying that $v(t)$ is even. Now if $v(t) = -l < 0$, then $v(c) = -2l$, $v(d) = -4l$, $v(c^2 - d) = -4l$. It is clear from the second equation (1) that if $v(x) \neq -2l$, then $v(t)$ is even. Suppose $v(x) = -2l$. In the case when $v(x^2 - d) > -4l$ and $v(x - c) > -2l$, we have $v(c^2 - d) > -4l$, which is a contradiction. Hence at least one of $x^2 - d$ and $x - c$ has even valuation, implying that $v(t)$ is even in all cases.

It follows that the divisor (t) on any regular model of (1), in particular, on Y, is a double, that is, $(t) = 2D$ for some $D \in Div(Y)$. Hence the double covering $X(m) \to Y$ given by $t = mu^2$, $m \in k^*$, is unramified. Let $y_1 = uw_1$, $y_2 = uw_2$, then $X(m)$ is given by the affine equations
$$w_1^2 = m(x - c(mu^2)), \ w_2^2 = m(x^2 - d(mu^2)) \tag{2}$$
The affine surface (2) is isomorphic to
$$w_2^2 = m^{-1}w_1^4 + 2c(mu^2)w_1^2 + m(c(mu^2)^2 - d(mu^2)) \tag{3}$$
Note that 0 is not a root of $d(t)$ and $c(t)^2 - d(t)$, hence $d(mu^2)$ and $c(mu^2)^2 - d(mu^2)$ are separable. Then it is easy to check that the affine surface (3) is regular. The projection to the coordinate u gives rise to a proper morphism $\pi : X(m) \to \mathbf{P}_k^1$. It is an easy local calculation (cf. [CT], Lemma 1.3.2, as we are exactly in the situation of that paper) that the conditions imposed on c and d ensure that

1) each fibre of π at a root of $c(mu^2)^2 - d(mu^2) = 0$ is a rational curve with one node (fibre of type I_1);

2) each fibre at a root of $d(mu^2) = 0$ is the union of two rational curves tranversally intersecting each other in two points (fibre of type I_2);

3) all other fibres of π, including the fibre at infinity, are smooth.

We shall write $X = X(m)$, $\overline{X} = X \times_k \overline{k}$, where $\overline{k}$ is an algebraic closure of k. The topological Euler characteristic $e(\overline{X})$ equals $8 \times 2 + 8 = 24$ (see, e.g. [BPV], Prop. 11.4, p. 97). The canonical class K_X can be written as

$\pi^*(nP)$, where P is a k-point on $\mathbf{P}_k^1$, $n = \chi(\mathcal{O}_X) - 2$, and $\chi(\mathcal{O}_X)$ is the Euler characteristic of the structure sheaf on X ([**BPV**], Cor. 12.3, p. 162). In any case $(K_X)^2 = 0$, and hence the formula $(K_X)^2 + e(\overline{X}) = 12\chi(\mathcal{O}_X)$ implies that $\chi(\mathcal{O}_X) = 2$. Now the above formula for the canonical class gives $K_X = 0$. This implies $H^2(X, \mathcal{O}_X) = 1$, hence $H^1(X, \mathcal{O}_X) = 0$. This proves that $X = X(m)$ is a $K3$-surface. It is one of the equivalent definitions of Enriques surfaces that these are the surfaces with a $K3$ unramified double covering. In particular, Y is an Enriques surface. $\square$

Proposition 2. — *Suppose that $m \in k^* \setminus k^{*2}$, and that $d(mu^2)$ and $c(mu^2)^2 - d(mu^2)$ are irreducible polynomials, such that the fields $K_1 = k[u]/(d(mu^2))$ and $K_2 = k[u]/(c(mu^2)^2 - d(mu^2))$ do not contain quadratic extensions of k. Then the elliptic $K3$-surface $X(m)$ satisfies the assumptions (1), (2), (3.a) and (3.b) of Theorem A of [**CT**]. If $m(c(mr^2)^2 - d(mr^2)) \in k^{*2}$ for some $r \in k$, then $X(m)$ has a k-point.*

Proof. — Assumptions (1) and (2) clearly hold since m is a non-square constant. Assumption (3.a) says that the natural map $k^*/k^{*2} \to K_1^*/K_1^{*2}$ is injective. Non-trivial elements of its kernel correspond to quadratic extensions of k contained in K_1, and we assumed that there are none. For the same reason $k^*/k^{*2} \to K_2^*/K_2^{*2}$ is injective. Assumption (3.b) which asserts the injectivity of
$$k^*/\langle k^{*2}, m\rangle \to K_2^*/\langle K_2^{*2}, m\rangle,$$
is then obviously satisfied. If $r, s \in k$ are such that $m(c(mr^2)^2 - d(mr^2)) = s^2$, then $w_1 = 0$, $w_2 = s$, $u = r$, $x = c(mr^2)$ is a solution of (2). Since (2) defines a smooth affine surface, this point gives rise to a k-point on $X(m)$. $\square$

Lemma 3. — *Let $f(t)$ be an irreducible polynomial of even degree n with Galois group $G \subset S_n$. Then $k_f = k[t]/(f(t))$ contains a quadratic extension of k if and only if $G \cap S_{n-1}$ is contained in a subgroup of G of index 2.*

Proof. — Call K the splitting field of $f(t)$, then we have $\mathrm{Gal}(K/k) = G$. Let $H = G \cap S_{n-1}$, then the field k_f is recovered as the field of invariants $k_f \simeq K^H$. If $k \subset L \subset k_f \subset K$, where $[L : k] = 2$, then the corresponding sequence of subgroups of G reads as $G \supset F \supset H \supset 1$, for some index 2 subgroup $F \subset G$. $\square$

Corollary 4. — *Assume the Schinzel Hypothesis (H) and the finiteness of the Tate–Shafarevich groups of elliptic curves over $\mathbf{Q}$. Suppose that $c(t), d(t) \in \mathbf{Q}[t]$ satisfy the following conditions:*

(a) $\deg(c(t)) = 2$, $\deg(d(t)) = 4$, $\deg(c(t)^2 - d(t)) = 4$, and $td(t)(c(t)^2 - d(t))$ is separable;

(b) $d(t)$ and $c(t)^2 - d(t)$ are irreducible with Galois group isomorphic to S_4;

(c) $d(-u^2)$ and $c(-u^2)^2 - d(-u^2)$ are irreducible with Galois group isomorphic to the semi-direct product of $(\mathbf{Z}/2)^4$ and S_4, where S_4 acts on $(\mathbf{Z}/2)^4$ by permutations of factors;

(d) $d(0) - c(0)^2 \in \mathbf{Q}^{*2}$.

Then $\mathbf{Q}$-rational points are Zariski dense on the Enriques surface Y defined by (1).

Proof. — Set $m = -1$. In the notation of the previous proof G is the semi-direct product of $(\mathbf{Z}/2)^4$ and S_4, where S_4 acts on $(\mathbf{Z}/2)^4$ by permutations of factors. Hence H is the semi-direct product of $(\mathbf{Z}/2)^3$ and S_3. It is clear that H is not contained in a subgroup of G of index 2: such a subgroup is normal, but the conjugates of H generate G. Proposition 2 now shows that all the conditions of Theorem A of [CT] are satisfied for $X(-1)$. Our condition (d) implies that $X(-1)$ has a $\mathbf{Q}$-point with $u = 0$. Hence Theorem A of [CT] implies that $\mathbf{Q}$-points are Zariski dense on $X(-1)$, and hence also on Y. □

Conditions (a), (b) and (c) are satisfied for "generic" coefficients of $d(t)$ and $c(t)$. If one takes
$$d(t) = t^4 + t^2 + t + 5, \quad c(t) = 2(t^2 + 1),$$
then (a) and (d) hold. On the other hand, the Galois groups of $d(t)$ and $c(t)^2 - d(t)$ are of order 24, hence isomorphic to S_4. That of $d(-u^2)$ and $c(-u^2)^2 - d(-u^2)$ are of order $384 = 24 \times 16$ (computed using MAGMA package). The Galois group of any polynomial $f(u^2)$, where $deg(f) = 4$, is contained in the semi-direct product of $(\mathbf{Z}/2)^4$ and S_4. Hence the Galois groups of $d(-u^2)$ and $c(-u^2)^2 - d(-u^2)$ are equal to this semi-direct product. Thus (b) and (c) also hold. Condition (d) is obviously satisfied. We conclude that the Enriques surface given by

(4) $$y^2 = tx^4 + 4t^2(t^2 + 1)x^2 + t^3(3t^4 + 7t^2 - t - 1)$$

should have a Zariski dense set of $\mathbf{Q}$-rational points. In fact, this can be rigorously proved.

Proposition 5. — *The surface (4) has a Zariski dense set of rational points.*

Proof. — It will be easier to work with the $K3$-covering of (4) given by $t = -u^2$. This surface $X = X(-1)$ is given by

(5) $$w_2^2 = -w_1^4 + 4(u^4 + 1)w_1^2 - (3u^8 + 7u^4 + u^2 - 1)$$

The surface (5) contains the curve C of genus one given by $w_1 = u^2$. Its equation is

(6) $$w_2^2 = -3u^4 - u^2 + 1$$

Let us prove that $C(\mathbf{Q})$ is infinite. This curve contains the points $P = (1/2, 3/4)$ and $Q = (0, 1)$. Let us show that $P-Q$ has infinite order in the Jacobian of C. It is easier to work with the isogenous curve $C_1 : z^2 = t(-3t^2 - t + 1)$, where the unramified covering of degree 2 is $f : C \to C_1$ is $t = u^2$, $z = uw_2$ (f will become an isogeny after an appropriate choice of base points). The standard procedure yields the coordinate change $t = -3^{-3}(x+3)$, $z = 3^{-4}y$, and the equation of C_1 takes the standard form $y^2 = x^3 + Ax + B$ with $A = -270$, $B = -27 \times 29$. Then $M = f(P)$ has coordinates $(x_M, y_M) = (33, 6 \times 27)$, and $f(Q)$ is a point of order 2 (when the origin of the group law on C_1 is the point at infinity). Since $y_M \neq 0$ and y_M^2 (which is even) does not divide $4A^3 + 27B^2$ (which is odd), we conclude that $M = f(P)$ has infinite order in C_1. Thus $f(P) - f(Q)$ has infinite order in the Jacobian of C_1, and hence $P - Q$ has infinite order in the Jacobian of C.

Consider the double covering $\varphi : C \to \mathbf{P}_\mathbf{Q}^1$ given by projection to the coordinate u; this is just the restriction of $\pi : X \to \mathbf{P}_\mathbf{Q}^1$ to C. Let $\tau : C \to C$ be the corresponding involution, $\tau(u, w_2) = (u, -w_2)$. We note that $C(\mathbf{R})$ is connected, and since $C(\mathbf{Q})$ is infinite, $C(\mathbf{Q})$ is dense in $C(\mathbf{R})$ in the real topology. Let R be one of two real points of C with coordinate $w_2 = 0$ (ramification points of φ). There is a sequence $P_n \in C(\mathbf{Q})$ converging to R in the real topology. We observe that the fibre of π through R is a smooth curve of genus one, as $-3u^4 - u^2 + 1$ is coprime with the irreducible polynomials $d(-u^2)$ and $c(-u^2)^2 - d(-u^2)$. Then $\tau(P_n) - P_n$ has infinite order in the Jacobian of the fibre of π through P_n, provided $n > N$ for some positive N. Indeed, by Mazur's theorem the torsion of elliptic curves over $\mathbf{Q}$ is bounded. On the other hand, $\tau(P_n) - P_n$ tends to zero, hence cannot be torsion for $n > N$. (The key idea that when a ramification point lies on a smooth fibre, rational points are Zariski dense, is taken from [**BT**].)

We have found infinitely many $\mathbf{Q}$-fibres of $\pi : X \to \mathbf{P}_\mathbf{Q}^1$, each with infinitely many $\mathbf{Q}$-rational points. This implies the density of $\mathbf{Q}$-rational points of X in the Zariski topology. $\square$

Remarks 6. — 1. We used Mazur's theorem to simplify the argument, without any doubt one can find a proof not based on this theorem.

2. It should be possible to construct similar examples for the Enriques surfaces given by

$$y_1^2 = f(t)(x^2 - c(t)), \quad y_2^2 = f(t)(x^2 - d(t))$$

where f, c, d, $c - d$ are all of degree 2 such that $fcd(c-d)$ is a separable polynomial (cf. [**CSS97**], Example 4.1.2 and Remark 5.3.1). In this case the conditional result pointing to the Zariski density of rational points is proved in [CSS98].

3. The morphism $Y \to \mathbf{P}_k^1$ has two double fibres. It is proved in ([**CSS97**], Cor. 2.4) that if a pencil of curves of genus one defined over a number field k has at least 5 (geometric) double fibres, then all k-points are contained in finitely many k-fibres, and hence are not Zariski dense.

I would like to thank Sir Peter Swinnerton-Dyer for showing me the curve (6), and Jean-Louis Colliot-Thélène for useful discussions.

References

[BPV] W. Barth, C. Peters, A. Van de Ven, *Compact complex surfaces*, Springer-Verlag, 1984.

[BS] A.O. Bender and Sir Peter Swinnerton-Dyer, *Solubility of certain pencils of curves of genus 1, and of the intersection of two quadrics in $\mathbf{P}^4$*, Proc. London Math. Soc., to appear.

[BT] F.A. Bogomolov and Yu. Tschinkel, *Density of rational points on Enriques surfaces*, Math. Res. Lett. **5** (1998), 623–628.

[CT] J.-L. Colliot-Thélène, *Hasse principle for pencils of curves of genus one whose Jacobians have a rational 2-division point*, this volume.

[CSS97] J.-L. Colliot-Thélène, A.N. Skorobogatov and Sir Peter Swinnerton-Dyer, *Double fibres and double covers: paucity of rational points*, Acta Arithm. **79** (1997), 113–135.

[CSS98] J.-L. Colliot-Thélène, A.N. Skorobogatov and Sir Peter Swinnerton-Dyer, *Hasse principle for pencils of curves of genus one whose Jacobians have rational 2-division points*, Invent. Math. **134** (1998), 579–650.

Rational points on algebraic varieties
(E. PEYRE, Y. TSCHINKEL, ed.), p. 169–197

DENSITY OF INTEGRAL POINTS ON ALGEBRAIC VARIETIES

Brendan Hassett

Math Department–MS 136, Rice University, 6100 S. Main St., Houston TX 77005-1892 • *E-mail* : hassett@math.rice.edu

Yuri Tschinkel

Department of Mathematics, Princeton University, Washington Rd., Princeton, NJ 08544-1000, U.S.A. • *E-mail* : ytschink@math.princeton.edu

Abstract. — We study the Zariski density of integral points on quasi-projective algebraic varieties.

Introduction

Let K be a number field, S a finite set of valuations of K, including the archimedean valuations, and $\mathcal{O}_S$ the ring of S-integers. Let X be an algebraic variety defined over K and D a divisor on X. We will use $\mathcal{X}$ and $\mathcal{D}$ to denote models over $\mathrm{Spec}(\mathcal{O}_S)$.

We will say that integral points on (X, D) (see Section 1 for a precise definition) are potentially dense if they are Zariski dense on some model $(\mathcal{X}, \mathcal{D})$, after a finite extension of the ground field and after enlarging S. A central problem in arithmetic geometry is to find conditions insuring potential density (or nondensity) of integral points. This question motivates many interesting and concrete problems in classical number theory, transcendence theory and algebraic geometry, some of which will be presented below.

If we think about general reasons for the density of points - the first idea would be to look for the presence of a large automorphism group. There are many beautiful examples both for rational and integral points, like K3 surfaces given by a bihomogeneous $(2, 2, 2)$ form in $\mathbb{P}^1 \times \mathbb{P}^1 \times \mathbb{P}^1$ or the classical Markov equation $x^2 + y^2 + z^2 = 3xyz$. However, large automorphism groups are "sporadic" - they are hard to find and usually, they are not well behaved in families. There is one notable exception - namely automorphisms of algebraic groups, like tori and abelian varieties.

Thus it is not a surprise that the main geometric reason for the abundance of rational points on varieties treated in the recent papers [11], [3], [12] is the presence of elliptic or, more generally, abelian fibrations with *multisections* having a dense set of rational points and subject to some *nondegeneracy* conditions. Most of the effort goes into ensuring these conditions.

In this paper we focus on cases when D is nonempty. We give a systematic treatment of known approaches to potential density and present several new ideas for proofs. The analogs of elliptic fibrations in log geometry are conic bundles with a bisection removed. We develop the necessary techniques to translate the presence of such structures to statements about density of integral points and give a number of applications.

The paper is organized as follows: in Section 1 we introduce the main definitions and notations. Section 2 is geometrical - we introduce the relevant concepts from the log minimal model program and formulate several geometric problems inspired by questions about integral points. In Section 3, we recall the fibration method and nondegeneracy properties of multisections. We consider approximation methods in Section 4. Section 5 is devoted to the study of integral points on conic bundles with sections and bisections removed. In the final section, we survey the known results concerning potential density for integral point on log K3 surfaces.

Acknowledgements. The first author was partially supported by an NSF Postdoctoral Research Fellowship, NSF Continuing Grant 0070537, and the Institute of Mathematical Sciences of the Chinese University of Hong Kong. The second author was partially supported by the NSA. We benefitted from conversations with D. Abramovich, Y. André, F. Bogomolov, A. Chambert-Loir, J.-L. Colliot-Thélène, J. Kollár, D. McKinnon, and B. Mazur. We are grateful to P. Vojta for comments that improved the paper, especially Proposition 2.12, and to D.W. Masser for information on specialization of nondegenerate sections. Our approach in Section 5 is inspired by the work of F. Beukers (see [1] and [2]).

1. Generalities

1.1. Integral points. — Let $\pi : \mathscr{U} \to \mathrm{Spec}(\mathscr{O}_S)$ be a flat scheme over $\mathscr{O}_S$ with generic fiber U. An integral point on $\mathscr{U}$ is a section of π; the set of such points is denoted $\mathscr{U}(\mathscr{O}_S)$.

In the sequel, $\mathscr{U}$ will be the complement to a reduced effective Weil divisor $\mathscr{D}$ in a normal proper scheme $\mathscr{X}$, both generally flat over $\mathrm{Spec}(\mathscr{O}_S)$. Hence an S-integral point P of $(\mathscr{X}, \mathscr{D})$ is a section $s_P : \mathrm{Spec}(\mathscr{O}_S) \to \mathscr{X}$ of π, which does not intersect $\mathscr{D}$, that is, for each prime ideal $\mathfrak{p} \in \mathrm{Spec}(\mathscr{O}_S)$ we have $s_P(\mathfrak{p}) \notin \mathscr{D}_\mathfrak{p}$. We denote by X (resp. D) the corresponding generic fiber. We generally assume that X is a variety (i.e., a geometrically integral scheme); frequently X is smooth and D is normal crossings. Potential density of integral points on $(\mathscr{X}, \mathscr{D})$ does not depend on the choice of S or on the choices of models over $\mathrm{Spec}(\mathscr{O}_S)$, so we will not always specify them. Hopefully, this will not create any confusion.

If D is empty then every K-rational point of X is an S-integral point for $(\mathscr{X}, \mathscr{D})$ (on some model). Every K-rational point of X, not contained in D is S-integral on $(\mathscr{X}, \mathscr{D})$ for S large enough. Clearly, for any $\mathscr{X}$ and $\mathscr{D}$ there exists a finite extension K'/K and a finite set S' of prime ideals in $\mathscr{O}_{K'}$ such that there is an S'-integral point on $(\mathscr{X}', \mathscr{D}')$ (where $\mathscr{X}'$ is the base change of $\mathscr{X}$ to $\mathrm{Spec}(\mathscr{O}'_S)$).

The definition of integral points can be generalized as follows: let $\mathscr{Z}$ be any closed subscheme of $\mathscr{X}$. An S-integral point for $(\mathscr{X}, \mathscr{Z})$ is an $\mathscr{O}_S$-valued point of $\mathscr{X} \setminus \mathscr{Z}$.

1.2. Vojta's conjecture. — A *pair* consists of a proper normal variety X and a reduced effective Weil divisor $D \subset X$. A *morphism of pairs* $\varphi : (X_1, D_1) \to (X_2, D_2)$ is a morphism $\varphi : X_1 \to X_2$ such that (the support of) $\varphi^{-1}(D_2)$ is a subset of D_1. In particular, φ restricts to a morphism $X_1 \setminus D_1 \to X_2 \setminus D_2$. A morphism of pairs is *dominant* if $\varphi : X_1 \to X_2$ is dominant. If (X_1, D_1) dominates (X_2, D_2) then integral points are dense on (X_2, D_2) when they are dense on (X_1, D_1) (after choosing appropriate integral models). A morphism of pairs is *proper* if $\varphi : X_1 \to X_2$ is proper and the restriction $X_1 \setminus D_1 \to X_2 \setminus D_2$ is also proper; equivalently, we may assume that $\varphi : X_1 \to X_2$ is proper and D_1 is a subset of $\varphi^{-1}(D_2)$. A *resolution* of the pair (X, D) is a proper morphism of pairs $\rho : (\tilde{X}, \tilde{D}) \to (X, D)$ such that $\rho : \tilde{X} \to X$ is birational, $\tilde{X}$ is smooth, and $\tilde{D}$ is normal crossings.

Let X be a normal proper variety of dimension d. Recall that a Cartier divisor $D \subset X$ is *big* if $h^0(\mathscr{O}_X(nD)) > Cn^d$ for some $C > 0$ and all n sufficiently large and divisible.

Definition 1.1. — *A pair (X, D) is of log general type if it admits a resolution $\rho : (\tilde{X}, \tilde{D}) \to (X, D)$ with $\omega_{\tilde{X}}(\tilde{D})$ big.*

Let us remark that the definition does not depend on the resolution.

Conjecture 1.2. — *(Vojta, [30]) Let (X, D) be a pair of log general type. Then integral points on (X, D) are not potentially dense.*

This conjecture is known for semiabelian varieties and their subvarieties ([9], [31], [16]). Vojta's conjecture implies that a pair with dense integral points cannot dominate a pair of log general type.

We are interested in geometric conditions which would insure potential density of integral points. The most naive statement would be the direct converse to Vojta's conjecture. However this can't be true even when $D = \emptyset$. Indeed, varieties which are not of general type may dominate varieties of general type, or more generally, admit finite étale covers which dominate varieties of general type (see the examples in [7]). In the next section we will analyze other types of covers with the same arithmetic property.

2. Geometry

2.1. Morphisms of pairs. —

Definition 2.1. — *We will say that a class of dominant morphisms of pairs $\varphi : (X_1, D_1) \to (X_2, D_2)$ is arithmetically continuous if the density of integral points on (X_2, D_2) implies potential density of integral points on (X_1, D_1).*

For example, assume that $D = \emptyset$. Then any projective bundle in the Zariski topology $\mathbb{P} \to X$ is arithmetically continuous. In the following sections we present other examples of arithmetically continuous morphisms of pairs.

Definition 2.2. — *A pseudo-étale cover of pairs $\varphi : (X_1, D_1) \to (X_2, D_2)$ is a proper dominant morphism of pairs such that*
  a) $\varphi : X_1 \to X_2$ is generically finite, and
  b) the map from the normalization X_2^{norm} of X_2 (in the function field of X_1) onto X_2 is étale away from D_2.

Remark 2.3. — *For every pair (X, D) there exists a birational pseudo-étale morphism $\varphi : (\tilde{X}, \tilde{D}) \to (X, D)$ such that $\tilde{X}$ is smooth and $\tilde{D}$ is normal crossings.*

The following theorem is a formal generalization of the well-known theorem of Chevalley-Weil. It shows that potential density is stable under pseudo-étale covers of pairs.

Theorem 2.4. — *Let $\varphi : (X_1, D_1) \to (X_2, D_2)$ be a pseudo-étale cover of pairs. Then φ is arithmetically continuous.*

Remark 2.5. — An elliptic fibration $E \to X$, isotrivial on $X \setminus D$, is arithmetically continuous. Indeed, it splits after a pseudo-étale morphism of pairs and we can apply Theorem 2.4.

The following example is an integral analog of the example of Skorobogatov, Colliot-Thélène and Swinnerton-Dyer ([7]) of a variety which does not dominate a variety of general type but admits an étale cover which does.

Example 2.6. — Consider $\mathbb{P}^1 \times \mathbb{P}^1$ with coordinates $(x_1, y_1), (x_2, y_2)$ and involutions
$$j_1(x_1, y_1) = (-x_1, y_1) \quad j_2(x_2, y_2) = (y_2, x_2)$$
on the factors. Let j be the induced involution on the product; it has fixed points
$$\begin{array}{ll} x_1 = 0 & x_2 = y_2 \\ x_1 = 0 & x_2 = -y_2 \\ y_1 = 0 & x_2 = y_2 \\ y_1 = 0 & x_2 = -y_2 \end{array}.$$
The first projection induces a map of quotients
$$(\mathbb{P}^1 \times \mathbb{P}^1)/\langle j \rangle \to \mathbb{P}^1/\langle j_1 \rangle.$$
We use X to denote the source; the target is just $\mathrm{Proj}(\mathbb{C}[x_1^2, y_1^2]) \simeq \mathbb{P}^1$. Hence we obtain a fibration $f : X \to \mathbb{P}^1$. Note that f has two nonreduced fibers, corresponding to $x_1 = 0$ and $y_1 = 0$ respectively. Let D be the image in X of
$$(x_1 = 0) \cup (y_1 = 0) \cup (x_2 = m_1 y_2) \cup (x_2 = m_2 y_2)$$
where m_1 and m_2 are distinct, $m_1 m_2 \neq 1$, and $m_1, m_2 \neq 0, 1$. Since D intersects the general fiber of f in just two points, (X, D) is not of log general type.

We can represent X as a degenerate quartic Del Pezzo surface with four A1 singularities (see figure 1). If we fix invariants
$$a = x_1^2 x_2 y_2, \ b = x_1^2(x_2^2 + y_2^2), \ c = x_1 y_1(x_2^2 - y_2^2), \ d = y_1^2(x_2^2 + y_2^2), \ e = y_1^2 x_2 y_2$$
then X is given as a complete intersection of two quadrics:
$$ad = be, \qquad c^2 = bd - 4ae.$$

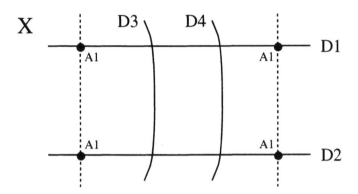

FIGURE 1. The log surface (X, D)

The components of D satisfy the equations

$$\begin{align}
D_1 &= \{a = b = c = 0\} \\
D_2 &= \{c = d = e = 0\} \\
D_3 &= \{(1 + m_1^2)a - m_1 b = (1 + m_1^2)e - m_1 d = 0\} \\
D_4 &= \{(1 + m_2^2)a - m_2 b = (1 + m_2^2)e - m_2 d = 0\}.
\end{align}$$

Our assumptions guarantee that D_3 and D_4 are distinct.

We claim that (X, D) does not admit a dominant map onto a variety of log general type and that there exists a pseudo-étale cover of (X, D) which does. Indeed, the preimage of $X \setminus D$ in $\mathbb{P}^1 \times \mathbb{P}^1$ is

$$(\mathbb{A}^1 \setminus 0) \times (\mathbb{P}^1 \setminus \{m_1, m_2, 1/m_1, 1/m_2\}),$$

which dominates a curve of log general type, namely, $\mathbb{P}^1$ minus four points. However, (X, D) itself cannot dominate a curve of log general type. Any such curve must be rational, with at least three points removed; however, the boundary D contains at most two mutually disjoint irreducible components.

The following was put forward as a possible converse to Vojta's conjecture.

Problem 2.7 (Strong converse to Vojta's conjecture)
Assume that the pair (X_2, D_2) does not admit a pseudo-étale cover

$$(X_1, D_1) \to (X_2, D_2)$$

such that (X_1, D_1) dominates a pair of log general type. Are integral points for (X_2, D_2) potentially dense?

2.2. Projective bundles in the étale topology.

— We would like to produce further classes of dominant arithmetically continuous morphisms

$$(X_1, D_1) \to (X_2, D_2).$$

Theorem 2.8. — *Let $\varphi : (X_1, D_1) \to (X_2, D_2)$ be a projective morphism of pairs such that φ is a projective bundle (in the étale topology) over $X_2 \setminus D_2$. We also assume that $\varphi^{-1}(D_2) = D_1$. Then φ is arithmetically continuous.*

Proof. We are very grateful to Prof. Colliot-Thélène for suggesting this proof.

Choose models $(\mathscr{X}_i, \mathscr{D}_i)$ ($i = 1, 2$) over some ring of integers $\mathscr{O}_S$, so that the morphism φ is well-defined and a projective bundle. (We enlarge S as necessary.)

We recall basic properties of the Brauer group $\mathrm{Br}(\mathscr{O}_S)$. Let v denote a place for the quotient field K and K_v the corresponding completion. Classfield theory gives the following exact sequences

$$0 \to \mathrm{Br}(K) \to \oplus_v \mathrm{Br}(K_v) \to \mathbb{Q}/\mathbb{Z} \to 0$$
$$0 \to \mathrm{Br}(\mathscr{O}_S) \to \mathrm{Br}(K) \to \oplus_{v \notin S} \mathrm{Br}(K_v).$$

The Brauer groups of the local fields corresponding to nonarchimedean valuations are isomorphic to $\mathbb{Q}/\mathbb{Z}$. Given a finite extension of K_w/K_v of degree n, the induced map on Brauer groups is multiplication by n.

Let p denote an S-integral point of (X_2, D_2). The fiber $\varphi^{-1}(p)$ is a Brauer-Severi variety over $\mathscr{O}_S$. If $r - 1$ denotes the relative dimension of φ then the corresponding element $\beta(p) \in \mathrm{Br}(\mathscr{O}_S)$ has order dividing r. Integral points in $\varphi^{-1}(p)$ are dense if rational points are dense, which is the case when $\beta(p) = 0$.

Our exact sequences imply that $\beta(p)$ yields elements of $\mathrm{Br}(K_v)$ which are zero unless $v \in S$, and are annihilated by r otherwise. It suffices to find an extension K'/K inducing a cyclic extension of K_v of order divisible by r for all $v \in S$. Indeed, such an extension necessarily kills $\beta(p)$ for *each* point p defined over $\mathscr{O}_S$. If $\mathscr{O}_{S'}$ is the integral closure of $\mathscr{O}_S$ in K' then $\varphi^{-1}(p)$ has dense S'-integral points. □

Remark 2.9. — Let X be a smooth simply connected projective variety which does not dominate a variety of general type. It may admit a projective bundle (in the étale topology) $\varphi : \mathbb{P} \to X$, for example if X is a K3 surface. However, $\mathbb{P}$ cannot dominate a variety of general type. Indeed, given such a dominant morphism $\pi : \mathbb{P} \to Y$, the fibers of φ are mapped to points by π. In particular, π necessarily factors through φ. (We are grateful to J. Kollár for emphasizing this point.)

Problem 2.10 (Geometric counterexamples to Problem 2.7)
Are there pairs which do not admit pseudo-étale covers dominating pairs of log general type but which do admit arithmetically continuous covers dominating pairs of log general type?

2.3. Punctured varieties. — In Section 2.1 we have seen that potential density of integral points is preserved under pseudo-étale covers. It is not an easy task, in general, to check whether or not some given variety (like an elliptic surface) admits a (pseudo-) étale cover dominating a variety of general type. What happens if we modify the variety (or pair) without changing the fundamental group?

Problem 2.11 (Geometric puncturing problem). — *Let X be a projective variety with canonical singularities and Z a subvariety of codimension ≥ 2. Assume that no (pseudo-) étale cover of $(X, \emptyset)$ dominates a variety of general type. Then (X, Z) admits no pseudo-étale covers dominating a pair of log general type. A weaker version would be to assume that X and Z are smooth.*

By definition, a pseudo-étale cover of (X, Z) is a pseudo-étale cover of a pair (X', D'), where X' is proper over X and $X' \setminus D' \simeq X \setminus Z$.

Proposition 2.12. — *Assume X and Z are as in Problem 2.11, and that X is smooth. Then*
  a) No pseudo-étale covers of (X, Z) dominate a curve of log general type.
  b) No pseudo-étale covers of (X, Z) dominate a variety of log general type of the same dimension.

Proof. Suppose we have a pseudo-étale cover $\rho : (X_1, D_1) \to (X, Z)$ and a dominant morphism $\varphi : (X_1, D_1) \to (X_2, D_2)$ to a variety of log general type. By Remark 2.3, we may take the X_i smooth and the D_i normal crossings. Since D_1 is exceptional with respect to ρ, Iitaka's Covering Theorem ([13] Theorem 10.5) yields an equality of Kodaira dimensions
$$\kappa(K_X) = \kappa(K_{X_1} + D_1).$$

Assume first that X_2 is a curve. We claim it has genus zero or one. Let X^{norm} be the normalization of X in the function field of X_1. The induced morphism $g : X^{\text{norm}} \to X$ is finite, surjective, and branched only over Z, a codimension ≥ 2 subset of X. Since X is smooth, it follows that g is étale (see SGA II X §3.4 [10]). If X_2 has genus ≥ 2 then $\varphi : X_1 \to X_2$ is constant along the fibers of $X_1 \to X^{\text{norm}}$, and thus descends to a map $X^{\text{norm}} \to X_2$. This would contradict our assumption that no étale cover of X dominates a variety of general type.

Choose a point $p \in D_2$ and consider the divisor $F = \varphi^{-1}(p)$. Note that $2F$ moves because $2p$ moves on X_2. However, $2F$ is supported in D_1, which lies in the exceptional locus for ρ, and we obtain a contradiction.

Now assume φ is generically finite. We apply the Logarithmic Ramification Formula to φ (see [13] Theorem 11.5)

$$K_{X_1} + D_1 = \varphi^*(K_{X_2} + D_2) + R$$

where R is the (effective) logarithmic ramification divisor. Applying the Covering Theorem again, we find that $\kappa(K_{X_1} + D_1 - R) = \kappa(K_{X_2} + D_2) = \dim(X)$. It follows that $K_{X_1} + D_1$ is also big, which contradicts the assumption that X is not of general type. □

Problem 2.13 (Arithmetic puncturing problem)

Let X be a projective variety with canonical singularities and Z a subvariety of codimension ≥ 2. Assume that rational points on X are potentially dense. Are integral points on (X, Z) potentially dense?

For simplicity, one might first assume that X and Z are smooth.

Remark 2.14. — Assume that Problem 2.13 has a positive solution. Then potential density of rational points holds for all K3 surfaces.

Indeed, if Y is a K3 surface of degree $2n$ then potential density of rational points holds for the symmetric product $X = Y^{(n)}$ (see [12]). Denote by Z the large diagonal in X and by Δ the large diagonal in Y^n (the ordinary product). Assume that integral points on (X, Z) are potentially dense. Then, by Theorem 2.4 integral points on (Y^n, Δ) are potentially dense. This implies potential density for rational points on Y.

3. The fibration method and nondegenerate multisections

This section is included as motivation. Let B be an algebraic variety, defined over a number field K and $\pi : G \to B$ be a group scheme over B. We will be mostly interested in the case when the generic fiber is an abelian variety or a split torus $\mathbb{G}_m^n$. Let s be a section of π. Shrinking the base we may assume that all fibers of G are smooth. We will say that s is *nondegenerate* if $\cup_n s^n$ is Zariski dense in G.

Problem 3.1 (Specialization). — *Assume that $G \to B$ has a nondegenerate section s. Describe the set of $b \in B(K)$ such that $s(b)$ is nondegenerate in the fiber G_b.*

For simple abelian varieties over a field a point of infinite order is nondegenerate. If $E \to B$ is a Jacobian elliptic fibration with a section s of infinite order then this section is automatically nondegenerate, and $s(b)$ is nondegenerate if it is nontorsion. By a result of Néron (see [26] 11.1), the set of $b \in B(K)$ such that $s(b)$ is not of infinite order is *thin*; this holds true for abelian fibrations of arbitrary dimension.

For abelian fibrations $A \to B$ with higher-dimensional fibers, one must also understand how rings of endomorphisms specialize. The set of $b \in B(K)$ for which the restriction

$$\mathrm{End}(A) \to \mathrm{End}(A(b))$$

fails to be surjective is also thin; this is a result of Noot [21] Corollary 1.5. In particular, a nondegenerate section of a family of generically simple abelian varieties specializes to a nondegenerate point outside a thin set of fibers.

More generally, given an arbitrary abelian fibration $A \to B$ and a nondegenerate section s, the set of $b \in B(K)$ such that $s(b)$ is degenerate is thin in B. (We are grateful to Prof. Masser for pointing out the proof.) After replacing A by an isogenous abelian variety and taking a finite extension of the function field $K(B)$, we obtain a family $A' \to B'$ with $A' \simeq A_1^{r_1} \times \ldots \times A_m^{r_m}$, where the A_j are (geometrically) simple and mutually non-isogenous. By the Theorems of Néron and Noot, the $A_j(b')$ are simple and mutually non-isogenous away from some thin subset of B'. A section s' of $A' \to B'$ is nondegenerate iff its projection onto each factor $A_j^{r_j}$ is nondegenerate; for b' not contained in our thin subset, $s'(b')$ is nondegenerate iff its projection onto each $A_j^{r_j}(b')$ is nondegenerate. Hence we are reduced to proving the claim for each $A_j^{r_j}$. Since A_j is simple, a section s_j of $A_j^{r_j}$ is nondegenerate iff its projections $s_{j,1}, \ldots, s_{j,r_j}$ are linearly independent over $\mathrm{End}(A_j)$. Away from a thin subset of B', the same statement holds for the specializations to b'. However, Néron's theorem implies that $s_{j,1}(b'), \ldots, s_{j,r_j}(b')$ are linearly independent away from a thin subset.

Remark 3.2. — There are more precise versions of Néron's Theorem due to Demyanenko, Manin and Silverman (see [28], for example). Masser has proposed another notion of what it means for a subset of $B(K)$ to be small, known as 'sparcity'. For instance, the endomorphism ring of a family of abelian varieties changes only on a 'sparse' set of rational points of the base (see [18]). For an analogue to Néron's Theorem, see [17].

Similar results hold for algebraic tori and are proved using a version of Néron's Theorem for $\mathbb{G}_m^n$-fibrations (see [26] pp. 154). A sharper result (for one-dimensional bases B) can be obtained from the following recent theorem:

Theorem 3.3. — ([4]) *Let C be an absolutely irreducible curve defined over a number field K and $x_1, ..., x_r$ rational functions in $K(C)$, multiplicatively independent modulo constants. Then the set of algebraic points $p \in C(\overline{\mathbb{Q}})$ such that $x_1(p), ..., x_r(p)$ are multiplicatively dependent has bounded height.*

The main idea of the papers [11], [3], [12] can be summarized as follows. We work over a number field K and we assume that all geometric data are defined over K. Let $\pi : E \to B$ be a Jacobian elliptic fibration over a one dimensional base B. This means that we have a family of curves of genus one and a global zero section so that every fiber is in fact an elliptic curve. Suppose that we have another section s which is of infinite order in the Mordell-Weil group of $E(K(B))$. The specialization results mentioned above show that for a Zariski dense set of $b \in B(K)$ the restriction $s(b)$ is of infinite order in the corresponding fiber E_b. If K-rational points on B are Zariski dense then rational points on E are Zariski dense as well.

Let us consider a situation when E does not have any sections but instead has a multisection M. By definition, a multisection (resp. rational multisection) M is irreducible and the induced map $M \to B$ is finite flat (resp. generically finite) of degree $\deg(M)$. The base-changed family $E \times_B M \to M$ has the identity section Id (i.e., the image of the diagonal under $M \times_B M \to E \times_B M$) and a (rational) section

$$\tau_M := \deg(M)\text{Id} - \text{Tr}(M \times_B M)$$

where $\text{Tr}(M \times_B M)$ is obtained (over the generic point) by summing all the points of $M \times_B M$. By definition, M is nondegenerate if τ_M is nondegenerate.

When we are concerned only with rational points, we will ignore the distinction between multisections and rational multisections, as every rational multisection is a multisection over an open subset of the base. However, this distinction is crucial when integral points are considered.

If M is nondegenerate and if rational points on M are Zariski dense then rational points on E are Zariski dense (see [3]).

Example 3.4. — ([11]) Let X be a quartic surface in $\mathbb{P}^3$ containing a line L. Consider planes $\mathbb{P}^2$ passing through this line. The residual curve has degree 3. Thus we obtain an elliptic fibration on X together with the trisection L. If L is ramified in a smooth fiber of this fibration then the multisection is nondegenerate and rational points are Zariski dense.

This argument generalizes to abelian fibrations $\pi : A \to B$. However, we do not know of any simple geometric conditions insuring nondegeneracy of a (multi)section in this case. We do know that for any abelian variety A over K there exists a finite extension K'/K with a nondegenerate point in $A(K')$ (see [12]). This allows us to produce nondegenerate sections over function fields.

Proposition 3.5. — *Let Y be a Fano threefold of type W_2, that is a double cover of $\mathbb{P}^3$ ramified in a smooth surface of degree 6. Then rational points on the symmetric square $Y^{(2)}$ are potentially dense.*

Proof. Observe that the symmetric square $Y^{(2)}$ is birational to an abelian surface fibration over the Grassmannian of lines in $\mathbb{P}^3$. This fibration is visualized as follows: consider two generic points in Y. Their images in $\mathbb{P}^3$ determine a line, which intersects the ramification locus in 6 points and lifts to a (hyperelliptic) genus two curve on Y. On $Y^{(2)}$ we have an abelian surface fibration corresponding to the degree two component of the relative Picard scheme. Now we need to produce a nondegenerate multisection. Pick two generic points b_1 and b_2 on the branch surface. The preimages in Y of the corresponding tangent planes are K3 surfaces Σ_1 and Σ_2, of degree two with ordinary double points at the points of tangency. The surfaces Σ_1 and Σ_2 therefore have potentially dense rational points (this was proved in [3]), as does $\Sigma_1 \times \Sigma_2$. This is our multisection; we claim it is nondegenerate for generic b_1 and b_2. Indeed, it suffices to show that given a (generic) point in $Y^{(2)}$, there exist b_1 and b_2 so that $\Sigma_1 \times \Sigma_2$ contains the point. Observe that through a (generic) point of $\mathbb{P}^3$, there pass many tangent planes to the branch surface. □

Remark 3.6. — Combining the above Proposition with the strong form of Problem 2.13 we obtain potential density of rational points on Fano threefolds of type W_2 - the last family of smooth Fano threefolds for which potential density is not known.

Here is a formulation of the fibration method useful for the analysis of integral points:

Proposition 3.7. — *Let B be a scheme over a number field K, $G \to B$ a flat group scheme, $T \to B$ an étale torsor for G, and $M \subset T$ a nondegenerate multisection over B. If M has potentially dense integral points then T has potentially dense integral points.*

Proof. Without loss of generality, we may assume that B is geometrically connected and smooth. The base-changed family $T \times_B M$ dominates T, so it suffices to prove density for $T \times_B M$. Note that since M is finite and flat over B, τ_M is a well-defined *section* over all of M (i.e., it is not just a rational section). Hence we may reduce to the case of a group scheme $G \to B$ with a nondegenerate section τ.

We may choose models $\mathscr{G}$ and $\mathscr{B}$ over $\operatorname{Spec}(\mathscr{O}_S)$ so that $\mathscr{G} \to \mathscr{B}$ is a group scheme with section τ. We may also assume that S-integral points of τ are Zariski dense. The set of multiples τ^n of τ, each a section of $\mathscr{G} \to \mathscr{B}$, is dense in $\mathscr{G}$ by the nondegeneracy assumption. Since each has dense S-integral points, it follows that S-integral points are Zariski dense. □

A similar argument proves the following

Proposition 3.8. — *Let $\varphi : X \to \mathbb{P}^1$ be a K3 surface with elliptic fibration. Let M be a multisection over its image $\varphi(M)$, nondegenerate and contained in the smooth locus of φ. Let $F_1, \ldots, F_n$ be fibers of φ and D a divisor supported in these fibers and disjoint from M. If M has potentially dense integral points then (X, D) has potentially dense integral points.*

Proof. We emphasize that X is automatically minimal and the fibers of φ are reduced (see [**3**]). Our assumptions imply that M is finite and flat over $\varphi(M)$.

After base-changing to M, we obtain a Jacobian elliptic fibration $X' := X \times_{\mathbb{P}^1} M$ with identity and a nondegenerate section τ_M. Let $G \subset X'$ be the open subset equal to the connected component of the identity. Since $D' := D \times_{\mathbb{P}^1} M$ is disjoint from the identity, it is disjoint from G. Hence it suffices to show that G has potentially dense integral points.

We assumed that M is contained in the smooth locus of φ, so τ_M is contained in the grouplike part of X', and some multiple of τ_M is contained in G. Repeating the argument for Proposition 3.7 gives the result. □

4. Approximation techniques

In this section we prove potential density of integral points for certain pairs (X, D) using congruence conditions to control intersections with the boundary. Several of these examples are included as support for the statement of Problem 2.13.

Proposition 4.1. — *Let $G = \prod_j^N G_j$ where G_j are algebraic tori $\mathbb{G}_m$ or geometrically simple abelian varieties. Let Z be a subvariety in G of codimension $> \mu = \max_j(\dim(G_j))$ and let $U = G \setminus Z$ be the complement. Then integral points on U are potentially dense.*

Proof. We are grateful to Prof. McKinnon for inspiring the following argument.

The proof proceeds by induction on the number of components N. The base case $N = 1$ follows from the fact that rational points on tori and abelian varieties are potentially dense, so we proceed with the inductive step. Consider the projections $\pi' : G \to G' = \prod_{j \neq N} G_j$ and $\pi_N : G \to G_N$. By assumption, generic fibers of π' are geometrically disjoint from Z.

Choose a ring of integers $\mathscr{O}_S$ and models $\mathcal{G}_j$ over $\mathrm{Spec}(\mathscr{O}_S)$. We assume that each $\mathcal{G}_j$ is smooth over $\mathrm{Spec}(\mathscr{O}_S)$ and that $\mathcal{G}_N$ has a nondegenerate S-integral point q (see [12], for example, for a proof of the existence of such points on abelian varieties).

Let $\mathcal{T}$ be any subscheme of $\mathcal{G}_N$ supported over a finite subset of $\mathrm{Spec}(\mathscr{O}_S)$ such that $\mathcal{G}_N$ has an S-integral point p_N disjoint from $\mathcal{T}$. We claim that such integral points are Zariski dense. Indeed, for some $m > 0$ we have

$$mq \equiv 0 \pmod{\mathfrak{p}}$$

for each $\mathfrak{p} \in \mathrm{Spec}(\mathscr{O}_S)$ over which $\mathcal{T}$ has support. Hence we may take the translations of p_N by multiples of mq.

After extending $\mathscr{O}_S$, we may assume U has at least one integral point $p = (p', p_N)$ so that $\pi'^{-1}(p')$ and $\pi_N^{-1}(p_N)$ intersect Z in the expected dimensions. In particular, $\pi'^{-1}(p')$ is disjoint from Z. By the inductive hypothesis, we may extend $\mathscr{O}_S$ so that

$$(\pi_N^{-1}(p_N) \simeq \mathcal{G}', \pi_N^{-1}(p_N) \cap \mathcal{Z})$$

has dense integral points. In particular, almost all such integral points are not contained in $\pi'(\mathcal{Z})$, a closed proper subscheme of $\mathcal{G}'$. Let r be such a point, so that $F_r = \pi'^{-1}(r) \simeq \mathcal{G}_N$ intersects $\mathcal{Z}$ in a subscheme $\mathcal{T}$ supported over a finite number of primes. Since $(r, p_N) \in F_r$ is disjoint from $\mathcal{T}$, the previous claim implies that the integral points of F_r disjoint from $\mathcal{T}$ are Zariski dense. As r varies, we obtain a Zariski dense set of integral points on $\mathcal{G} \setminus \mathcal{Z}$. □

Corollary 4.2. — *Let X be a toric variety and $Z \subset X$ a subvariety of codimension ≥ 2, defined over a number field. Then integral points on (X, Z) are potentially dense.*

Another special case of the Arithmetic puncturing problem 2.13 is the following:

Problem 4.3. — *Are integral points on punctured simple abelian varieties of dimension $n > 1$ potentially dense?*

Example 4.4. — Potential density of integral points holds for simple abelian varieties punctured in the origin, provided that their ring of endomorphisms contains units of infinite order.

5. Conic bundles and integral points

Let K be a number field, S a finite set of places for K (including all the infinite places), $\mathcal{O}_S$ the corresponding ring of S-integers, and $\eta \in \mathrm{Spec}(\mathcal{O}_S)$ the generic point. For each place v of K, let K_v be the corresponding complete field and $\mathfrak{o}_v$ the discrete valuation ring (if v is nonarchimedean). As before, we use calligraphic letters (e.g., $\mathscr{X}$) for schemes (usually flat) over $\mathcal{O}_S$ and roman letters (e.g., X) for the fiber over η.

5.1. Results on linear algebraic groups.

— Consider a linear algebraic group G/K. Choose a model $\mathscr{G}$ for G over $\mathcal{O}_S$, i.e., a flat group scheme of finite type $\mathscr{G}/\mathcal{O}_S$ restricting to G at the generic point. This may be obtained by fixing a representation $G \hookrightarrow \mathrm{GL}_n(K)$ (see also [32] §10-11). The S-rank of G (denoted $\mathrm{rank}(G, \mathcal{O}_S)$) is defined as the rank of the abelian group of sections of $\mathscr{G}(\mathcal{O}_S)$ over $\mathcal{O}_S$. This does not depend on the choice of a model. Indeed, consider two models $\mathscr{G}_1$ and $\mathscr{G}_2$ with a birational map $b : \mathscr{G}_1 \dashrightarrow \mathscr{G}_2$; of course, b is an isomorphism over the generic point and the proper transform of the identity section I_1 is the identity. There is a subscheme $Z \subset \mathrm{Spec}(\mathcal{O}_S)$ with finite support such that the indeterminacy of b is in the preimage of Z. It follows that the sections of $\mathscr{G}_1$ congruent to I_1 modulo Z have proper transforms which are sections of $\mathscr{G}_2$. Such sections form a finite-index subgroup of $\mathscr{G}_1(\mathcal{O}_S)$.

Let $\mathbb{G}_m$ be the multiplicative group over $\mathbb{Z}$, i.e., $\mathrm{Spec}(\mathbb{Z}[x,y]/\langle xy - 1\rangle)$; it can be defined over an arbitrary scheme by extension of scalars. There is a natural projection
$$\mathbb{G}_m(\mathbb{Z}) \to \mathrm{Spec}(\mathbb{Z}[x]) = \mathbb{A}^1_\mathbb{Z} \subset \mathbb{P}^1_\mathbb{Z}$$
so that $\mathbb{P}^1_\mathbb{Z} \setminus \mathbb{G}_m(\mathbb{Z}) = \{0, \infty\}$. A *form* of $\mathbb{G}_m$ over K is a group scheme G/K for which there exists a finite field extension K'/K and an isomorphism $G \times_K K' \simeq \mathbb{G}_m(K')$. These are classified as follows (see [23] for a complete account). Any group automorphism
$$\alpha : \mathbb{G}_m(K') \to \mathbb{G}_m(K')$$
is either inversion or the identity, depending on whether it exchanges 0 and ∞. The corresponding automorphism group is smooth, so we may work in the étale topology (see [19] Theorem 3.9). In particular,
$$K - \text{forms of } \mathbb{G}_m \simeq H^1_{\text{ét}}(\mathrm{Spec}(K), \mathbb{Z}/2\mathbb{Z}).$$

Each such form admits a natural open imbedding into a projective curve $G \hookrightarrow X$, generalizing the imbedding of $\mathbb{G}_m$ into $\mathbb{P}^1$. The complement $D = X \setminus G$ consists of two points. The Galois action on D is given by the cocycle in $H^1_{\text{ét}}(\mathrm{Spec}(K), \mathbb{Z}/2\mathbb{Z})$ classifying G.

There is a general formula for the rank of a torus T due to T. Ono and J.M. Shyr (see [22], Theorem 6 and [27]). Let $\hat{T}$ (resp. $\hat{T}_v$) the group of characters defined over K (resp. K_v), and $\rho(T)$ (resp. $\rho(T,v)$) the number of independent elements of $\hat{T}$ (resp. $\hat{T}_v$). The formula takes the form

$$\operatorname{rank}(T, \mathscr{O}_S) = \sum_{v \in S} \rho(T, v) - \rho(T).$$

For forms of $\mathbb{G}_m$ this is particularly simple. For split forms

$$\operatorname{rank}(\mathbb{G}_m, \mathscr{O}_S) = \#\{\text{places } v \in S\} - 1.$$

Now let G/K be a nonsplit form, corresponding to the quadratic extension K'/K, and S' the places of K' lying over the places of S. Then we have

$$\operatorname{rank}(G, \mathscr{O}_S) = \#\{\text{places } v \in S \text{ completely splitting in } S'\}.$$

5.2. Group actions and integral points. — Throughout this subsection, $\mathscr{X}$ is a normal, geometrically connected scheme and $\mathscr{X} \to \operatorname{Spec}(\mathscr{O}_S)$ a flat projective morphism. Let $\mathscr{D} \subset \mathscr{X}$ be an effective reduced Cartier divisor. Contrary to our previous conventions, we do not assume that $\mathscr{D}$ is flat over $\mathscr{O}_S$. Assume that a linear algebraic group G acts on X so that $X \setminus D$ is a G-torsor.

Proposition 5.1. — *There exists a model $\mathscr{G}$ for G such that $\mathscr{G}$ acts on $\mathscr{X}$ and stabilizes $\mathscr{D}$.*

Proof. Choose an imbedding $\mathscr{X} \hookrightarrow \mathbb{P}^n_{\mathscr{O}_S}$ and a compatible linearization $G \hookrightarrow \operatorname{GL}_{n+1}(K)$ (see [20], Ch. 1, Cor. 1.6 and Prop. 1.7). Let $\mathscr{G}' \hookrightarrow \operatorname{GL}_{n+1}(\mathscr{O}_S)$ be the resulting integral model of G, so that $\mathscr{G}'$ stabilizes the ideal of $\mathscr{X}$ and therefore acts on it. Furthermore, $\mathscr{G}'$ evidently stabilizes the irreducible components of $\mathscr{D}$ dominating $\mathscr{O}_S$. The fibral components of $\mathscr{D}$ are supported over a finite subset of $\operatorname{Spec}(\mathscr{O}_S)$. We take $\mathscr{G} \subset \mathscr{G}'$ to be the subgroup acting trivially over this subset; it has the desired properties. □

Proposition 5.2. — *Assume $(\mathscr{X}, \mathscr{D})$ has an S-integral point and that G has positive $\mathscr{O}_S$-rank. Then $(\mathscr{X}, \mathscr{D})$ has an infinite number of S-integral points.*

Proof. Consider the action of $\mathscr{G}(\mathscr{O}_S)$ on the integral point σ (which has trivial stabilizer). The orbit consists of S-integral points of $(\mathscr{X}, \mathscr{D})$, an infinite collection because $\mathscr{G}$ has positive rank. □

Now assume that X is a smooth rational curve. A rational section (resp. bisection) $\mathscr{D} \subset \mathscr{X}$ is a reduced effective Cartier divisor such that the generic fiber D is reduced of degree one (resp. two). Note that the open curve $X \setminus D$ is geometrically isomorphic to $\mathbb{P}^1 - \{\infty\}$ (resp. $\mathbb{P}^1 - \{0, \infty\}$), and thus is a torsor for some K-form G of $\mathbb{G}_a$ (resp. $\mathbb{G}_m$). This form is easily computed. Of course, $\mathbb{G}_a$ has no nontrivial forms. In the $\mathbb{G}_m$ case, we can regard D_η as an element of $H^1_{\acute{e}t}(\mathrm{Spec}(K), \mathbb{Z}/2\mathbb{Z})$, which gives the descent data for G.

The following result is essentially due to Beukers (see [1], Theorem 2.3):

Proposition 5.3. — *Let $(\mathscr{X}, \mathscr{D}) \to \mathrm{Spec}(\mathscr{O}_S)$ be a rational curve with rational bisection and G the corresponding form of $\mathbb{G}_m$ (as described above). Assume that $(\mathscr{X}, \mathscr{D})$ has an S-integral point and $\mathrm{rank}(G, \mathscr{O}_S) > 0$. Then S-integral points of $(\mathscr{X}, \mathscr{D})$ are Zariski dense.*

Proof. This follows from Proposition 5.2. Given an S-integral point σ of $(\mathscr{X}, \mathscr{D})$, the orbit $\mathscr{G}(\mathscr{O}_S)\sigma$ is infinite and thus Zariski dense. □

Combining with the formula for the rank, we obtain the following:

Corollary 5.4. — *Let $(\mathscr{X}, \mathscr{D}) \to \mathrm{Spec}(\mathscr{O}_S)$ be a rational curve with rational bisection such that $(\mathscr{X}, \mathscr{D})$ has an S-integral point. Assume that either*

a) D is reducible over $\mathrm{Spec}(K)$ and $|S| > 1$; or

b) D is irreducible over $\mathrm{Spec}(K)$ and at least one place in S splits completely in $K(D)$.

Then S-integral points of $(\mathscr{X}, \mathscr{D})$ are Zariski dense.

When D is a rational section we obtain a similar result (also essentially due to Beukers [1], Theorem 2.1):

Proposition 5.5. — *Let $(\mathscr{X}, \mathscr{D}) \to \mathrm{Spec}(\mathscr{O}_S)$ be a rational curve with rational section such that $(\mathscr{X}, \mathscr{D})$ has an S-integral point. Then S-integral points of $(\mathscr{X}, \mathscr{D})$ are Zariski dense.*

5.3. v-adic geometry. — For each place $v \in S$, consider the projective space $\mathbb{P}^1(K_v)$ as a manifold with respect to the topology induced by the v-adic absolute value on K_v. For simplicity, this will be called the v-adic topology; we will use the same term for the induced subspace topology on $\mathbb{P}^1(K)$. Given an étale morphism of curves $f : U \to \mathbb{P}^1$ defined over K_v, we will say that $f(U(K_v))$ is a *basic étale open subset*. These are open in the v-adic topology, either by the open mapping theorem (in the archimedean case) or by Hensel's lemma (in the nonarchimedean case).

Let
$$\chi_f(B) := \#\{z \in \mathscr{O}_{\{v\}} : |z|_v \le B \text{ and } z \in f(U(K_v))\}$$
where B is a positive integer and
$$\mathscr{O}_{\{v\}} := \{z \in K : |z|_w \le 1 \text{ for each } w \ne v\}.$$

We would like to estimate the quantity
$$\mu_f := \liminf_{B \to \infty} \chi_f(B)/\chi_{\mathrm{Id}}(B)$$
i.e., the fraction of the integers contained in the image of the v-adic points of U.

Proposition 5.6. — *Let $f : U \to \mathbb{P}^1$ be an étale morphism defined over K_v and $f_1 : C \to \mathbb{P}^1$ a finite morphism of smooth curves extending f. If there exists a point $q \in f_1^{-1}(\infty) \cap C(K_v)$ at which f_1 is unramified then $\mu_f = 1$.*

Proof. This follows from the fact that $f(U(K_v))$ is open if f is étale along U. □

As an illustrative example, we take $K = \mathbb{Q}$ and $K_v = \mathbb{R}$, so that $\mathcal{O}_{\{v\}} = \mathbb{Z}$. The set $f(U(\mathbb{R}))$ is a finite union of open intervals (r, s) with $r, s \in \mathbb{R} \cup \{\infty\}$, where the (finite) endpoints are branch points. We observe that

$$\mu_f = \begin{cases} 0 & \text{if } f(U(\mathbb{R})) \text{ is bounded;} \\ 1/2 & \text{if } \overline{f(U(\mathbb{R}))} \text{ contains a one-sided neighborhood of } \infty; \\ 1 & \text{if } \overline{f(U(\mathbb{R}))} \text{ contains a two-sided neighborhood of } \infty. \end{cases}$$

We can read off easily which alternative occurs in terms of the local behavior at infinity. Let $f_1 : C \to \mathbb{P}^1$ be a finite morphism of smooth curves extending f. If $f_1^{-1}(\infty)$ has no real points then $\mu_f = 0$. If $f_1^{-1}(\infty)$ has unramified (resp. ramified) real points then $\mu_f = 1$ (resp. $\mu_f > 0$.)

We specialize to the case of double covers:

Proposition 5.7. — *Let $U \to \mathbb{P}^1$ be an étale morphism defined over K_v and $f_1 : C \to \mathbb{P}^1$ a finite morphism of smooth curves extending f. Assume that f_1 has degree two and ramifies at $q \in f_1^{-1}(\infty)$. Then $\mu_f > 0$.*

Proof. Of course, q is necessarily defined over K_v. The archimedean case follows from the previous example, so we restrict to the nonarchimedean case. Assume f_1 is given by
$$y^2 = c_n z^n + c_{n-1} z^{n-1} + \ldots + c_0,$$
where z is a coordinate for the affine line in $\mathbb{P}^1(K_v)$, $c_n \neq 0$, and the $c_i \in \mathfrak{o}_v$. Substituting $z = 1/t$ and $y = x/t^{\lceil n/2 \rceil}$, we obtain the equation at infinity
$$\begin{cases} x^2 = c_n + c_{n-1} t + \ldots + c_0 t^n & \text{for } n \text{ even} \\ x^2 = c_n t + c_{n-1} t^2 + \ldots + c_0 t^n & \text{for } n \text{ odd} \end{cases}.$$

If n is even then $f_1^{-1}(\infty)$ consists of two non-ramified points, so we may assume n odd. Then $f_1^{-1}(\infty)$ consists of one ramification point q, necessarily defined over K_v.

Write $c_n = u_0 \pi^\alpha$ and $z = u_1 \pi^{-\beta}$, where u_0 and u_1 are units and π is a uniformizer in $\mathfrak{o}_v$. (We may assume that some power π^r is contained in $\mathcal{O}_K$.) Our equation takes the form

(1) $$y^2 \pi^{n\beta - \alpha} = u_0 u_1^n + c_{n-1} u_1^{n-1} \pi^{\beta - \alpha} + \ldots + c_0 u_1 \pi^{n\beta - \alpha}.$$

We review a property of the v-adic numbers (proved in [25], Ch. XIV §4). Consider the multiplicative group

$$U^{(m)} := \{u \in \mathfrak{o}_v : u \equiv 1 \pmod{\pi^m}\}.$$

Then for m sufficiently large we have $U^{(m)} \subset K_v^2$. In particular, to determine whether a unit u is a square, it suffices to consider its representative mod π^m.

Consequently, if β is sufficiently large and has the same parity as α, then we can solve Equation 1 for $y \in K_v$ precisely when $u_0 u_1$ is a square. For example, choose any $M \in \mathcal{O}_K$ so that $M \equiv u_0 \pi^{(r-1)\beta} \pmod{\pi^{r\beta}}$ and set $z = M/\pi^{r\beta} \in \mathcal{O}_{\{v\}}$. Hence, of the $z \in \mathcal{O}_{\{v\}}$ with $|z|_v \leq B$ (with $B \gg 0$), the fraction satisfying our conditions is bounded from below. It follows that $\mu_f > 0$. □

Now let $f : U \to \mathbb{P}^1$ be an étale morphism of curves defined over K. Consider the function

$$\omega_{f,S}(B) := \#\{z \in \mathcal{O}_S : |z|_v \leq B \text{ for each } v \in S \text{ and } z \in f(U(K))\}$$

and the quantity

$$\limsup_{B \to \infty} \omega_{f,\{v\}}(B)/\chi_f(B).$$

We expect that this is zero provided that f does not admit a rational section. We shall prove this is the case when f has degree two.

A key ingredient of our argument is a version of Hilbert's Irreducibility Theorem:

Proposition 5.8. — *Let $f : U \to \mathbb{P}^1$ be an étale morphism of curves, defined over K and admitting no rational section. Then we have*

$$\limsup_{B \to \infty} \omega_{f,\{v\}}(B)/\chi_{\mathrm{Id}}(B) = 0.$$

Proof. We refer the reader to Serre's discussion of Hilbert's irreducibility theorem ([26], §9.6, 9.7). Essentially the same argument applies in our situation. □

Combining Propositions 5.6, 5.7, and 5.8, we obtain:

Corollary 5.9. — *Let $f : C \to \mathbb{P}^1$ be a finite morphism of smooth curves defined over K. Assume that f admits no rational section and that $f^{-1}(\infty)$ contains a K_v-rational point. We also assume that f has degree two. Then we have*
$$\limsup_{B \to \infty} \omega_{f,\{v\}}(B)/\chi_f(B) = 0.$$
In particular, the set $\{z \in \mathcal{O}_{\{v\}} : z \in f(C(K_v)) \setminus f(C(K))\}$ is infinite.

5.4. A density theorem for surfaces.

Geometric assumptions: Let $\mathcal{X}$ and $\mathcal{B}$ be flat and projective over $\mathrm{Spec}(\mathcal{O}_S)$ and $\varphi : \mathcal{X} \to \mathcal{B}$ be a morphism. Let $\mathcal{L} \subset \mathcal{X}$ be a closed irreducible subscheme, $\mathcal{D} \subset \mathcal{X}$ a reduced effective Cartier divisor, and $\mathfrak{q} := \mathcal{D} \cap \mathcal{L}$. We assume the generic fibers satisfy the following: X is a geometrically connected surface, B a smooth curve, $\varphi : X \to B$ a flat morphism such that the generic fiber is a rational curve with bisection D. We also assume $L \simeq \mathbb{P}^1_K$, $\varphi|L$ is finite, and L meets D at a single point q, at which D is nonsingular. Write $\mathcal{X}'$ for $\mathcal{X} \times_{\mathcal{B}} \mathcal{L}$, $\mathcal{D}'$ for $\mathcal{D} \times_{\mathcal{B}} \mathcal{L}$, $\mathcal{L}'$ for the image of the diagonal in $\mathcal{X} \times_{\mathcal{B}} \mathcal{L}$ (now a section for $\varphi' : \mathcal{X}' \to \mathcal{L}$), and $\mathfrak{q}'$ for $\mathcal{L}' \cap \mathcal{D}'$. Finally, if $\mathcal{C}'$ denotes the normalization of the union of the irreducible components of $\mathcal{D}'$ dominating $\mathcal{L}$, we assume that $\mathcal{C}' \to \mathcal{L}$ has no rational section over K (i.e., that $\mathcal{C}'$ is irreducible over K).

Arithmetic assumptions: We assume that $(\mathcal{L}, \mathfrak{q})$ has an S-integral point. Furthermore, we assume that for some $v \in S$, C' has a K_v-rational point lying over $\varphi'(q')$.

Remark 5.10. — The second assumption is valid if any of the following are satisfied:

— $D \to B$ is unramified at q.
— $D \to B$ is finite (but perhaps ramified) at q and $L \to B$ has ramification at q of odd order.
— $D \to B$ is finite (but ramified) at q and $L \to B$ has ramification at q of order two. Choose local uniformizers $t, x,$ and y so that we have local analytic equations $t + ax^2 = 0$ and $t + by^2 = 0$ (with $a, b \in K$) for $D \to B$ and $L \to B$. We assume that ab is a square in K_v.

Note that in the last case, D' and C' have local analytic equations $ax^2 - by^2 = 0$ and $x/y = \pm\sqrt{b/a}$ respectively.

Theorem 5.11. — *Under the geometric and arithmetic assumptions made above, S-integral points of $(\mathcal{X}, \mathcal{D})$ are Zariski dense.*

Proof. It suffices to prove that S-integral points of $(\mathscr{X}', \mathscr{D}')$ are Zariski dense. These map to S-integral points $(\mathscr{X}, \mathscr{D})$.

Consider first S-integral points of $(\mathscr{L}', \mathfrak{q}')$. These are dense by Proposition 5.5, and contain a finite index subgroup of $\mathbb{G}_a(\mathscr{O}_S) \subset \mathbb{P}^1_K$. Corollary 5.9 and our geometric assumptions imply that infinitely many of these points lie in $\varphi'(C'(K_v)) \setminus \varphi'(C'(K))$.

Choose a generic S-integral point p of $(\mathscr{L}', \mathfrak{q}')$ as described above. Let $\mathscr{X}'_p = \varphi'^{-1}(p)$, $\mathscr{D}'_p = \mathscr{X}'_p \cap \mathscr{D}'$, and $\mathscr{L}'_p = \mathscr{X}'_p \cap \mathscr{L}'$, so that $(\mathscr{X}'_p, \mathscr{D}'_p)$ is a rational curve with rational bisection and integral point $\mathscr{L}'_p$. Combining the results of the previous paragraph with Proposition 5.3, we obtain that S-integral points of $(\mathscr{X}'_p, \mathscr{D}'_p)$ are Zariski dense. As we vary p, we obtain a Zariski dense collection of integral points for $(\mathscr{X}', \mathscr{D}')$. □

5.5. Cubic surfaces containing a line.

Let $\mathscr{X}_1$ be a cubic surface in $\mathbb{P}^3_{\mathscr{O}_S}$, $\mathscr{D}_1 \subset \mathscr{X}_1$ a hyperplane section, and $\mathscr{L}_1 \subset \mathscr{X}_1$ a line not contained in $\mathscr{D}_1$, all assumed to be flat over $\mathrm{Spec}(\mathscr{O}_S)$. Write $\mathfrak{q}_1 := \mathscr{D}_1 \cap \mathscr{L}_1$, a rational section over $\mathrm{Spec}(\mathscr{O}_S)$. Let $\mathbb{P}^3_{\mathscr{O}_S} \dashrightarrow \mathscr{B}$ be the projection associated with $\mathscr{L}_1$, $\mathscr{X} = \mathrm{Bl}_{\mathscr{L}_1} \mathscr{X}_1$, and $\varphi : \mathscr{X} \to \mathscr{B}$ the induced projection (of course, $\mathscr{B} = \mathbb{P}^1_{\mathscr{O}_S}$ if $\mathscr{O}_S$ is a UFD). Let $\mathscr{L} \subset \mathscr{X}$ be the proper transform of $\mathscr{L}_1$, $\mathscr{D} \subset \mathscr{X}$ the total transform of $\mathscr{D}_1$, and $\mathfrak{q} = \mathscr{L} \cap \mathscr{D}$. We shall apply Theorem 5.11 to obtain density results for S-integral points of $(\mathscr{X}_1, \mathscr{D}_1)$.

We will need to assume the following geometric conditions:

GA1 D_1 is reduced everywhere and nonsingular at q_1;

GA2 X_1 has only rational double points as singularities, with at most one singularity along L_1;

GA3 D_1 is not the union of a line and a conic containing q_1 (defined over K).

The first assumption and the fact that D_1 is Cartier imply that X_1 is nonsingular at q_1. We analyze the projection from the line L_1 using the first two assumptions. This induces a morphism

$$\varphi : X \to \mathbb{P}^1.$$

Of course, $X = X_1$ if and only if L_1 is Cartier in X_1, which is the case exactly when X_1 is smooth along L_1. We use L and D to denote the proper transforms of L_1 and D_1. Our three assumptions imply that D equals the total transform of D_1 and has a unique irreducible component C dominating $\mathbb{P}^1$. We also have that the generic fiber of φ is nonsingular, intersects D in two points, and intersects L in two points (if X_1 is smooth along L_1) or in one point (if X_1 has a singularity along L_1). In particular, L is a bisection (resp. section) of φ if X_1 is nonsingular (resp. singular) along L_1.

We emphasize that S-integral points of $(\mathscr{X}, \mathscr{D})$ map to S-integral points of $(\mathscr{X}_1, \mathscr{D}_1)$, and all the Geometric Assumptions of Theorem 5.11 are satisfied except for the last one. The last assumption is verified if any of the following hold:

GA4a The branch loci of $C \to \mathbb{P}^1$ and $L \to \mathbb{P}^1$ do not coincide.
GA4b The curve C has genus one.
GA4c X_1 has a singularity along L_1.

Clearly, either the second or the third condition implies the first.

We turn next to the Arithmetic Assumptions.

AA1 $(\mathscr{L}_1, \mathfrak{q}_1)$ has an S-integral point.

Note that S-integral points of $(\mathscr{L}_1, \mathfrak{q}_1)$ not lying in the singular locus of $\mathscr{X}_1 \to \text{Spec}(\mathscr{O}_S)$ lift naturally to S-integral points of $(\mathscr{L}, \mathfrak{q})$.

Our next task is to translate the conditions of Remark 5.10 to our situation. They are satisfied in any of the following contexts:

AA2a D_1 is irreducible over K and q_1 is not a flex of D_1;
AA2b X_1 has a singularity along L_1;
AA2c D_1 is irreducible over K and q_1 is a flex of D_1. Let H be the hyperplane section containing L_1 and the flex line. We assume that $H \cap X_1 = L_1 \cup M$, where M is a smooth conic.
AA2d D_1 is irreducible over K but q_1 is a flex so that the hyperplane H containing L_1 and the flex line F intersects X_1 in three coincident lines, i.e., $H \cap X_1 = L_1 \cup M_1 \cup M_2$. Choose local coordinates x and y for H so that $L_1 = \{x = 0\}, F = \{y = 0\}$, and $M_1 \cup M_2 = \{ax^2 + cxy + by^2 = 0\}$. Then we assume that ab is a square in K_v.
AA2e D_1 consists of a line and a conic C_1 irreducible over K, intersecting in two distinct points, each defined over K_v.

In the first case, the map $D \to B$ is unramified at q. Note that in the second case L is a section for φ. In the third case, our assumption implies that $L \to B$ is unramified at q. In the last case, we observe that the points of L lying over $\varphi(q)$ are defined over K_v, hence C' has a K_v-rational point over $\varphi'(q')$.

It remains to show that AA2d allows us to apply case 3 of Remark 5.10. We fix projective coordinates on $\mathbb{P}^3$ compatibly with the coordinates already chosen on H: $y = 0$ is the linear equation for the hyperplane containing D_1, $z = 0$ the equation for H, $x = z = 0$ the equations for L_1, and $x = z = y = 0$ the equations for q_1. Under our assumptions, the equations for D_1 and X_1 take

the form
$$g := zw^2 + ax^3 + c_1wxz + c_2wz^2 + c_4x^2z + c_5xz^2 + c_6z^3 = 0$$
$$f := g + cx^2y + bxy^2 + yz\ell(w,x,y,z) = 0$$
where ℓ is linear in the variables. The conic bundle structure $\varphi : X \to B$ is obtained by making the substitution $z = tx$
$$g' = tw^2 + x(wc_1t + wc_2t^2) + x^2(a + c_4t + c_5t^2 + c_6t^3) = 0$$
$$f' = g' + cxy + by^2 + ty\ell(w,x,y,tx).$$
We analyze the local behavior of $D \to B$ at q using x as a coordinate for D. First dehomogenize
$$g'' = t + x(c_1t + c_2t^2) + x^2(a + c_4t + c_5t^2 + c_6t^3) = 0$$
and then take a suitable analytic change of coordinate on D to obtain $t+aX^2 = 0$. To analyze $L \to B$, we set $x = 0$ and use y as a coordinate
$$f'' = t + by^2 + ty\ell(1,0,y,0) = 0.$$
After a suitable analytic change of coordinate on L, we obtain $t + bY^2 = 0$.

Remark 5.12. — We further analyze condition AA2d when $K_v = \mathbb{R}$. Then ab is a square if and only if $ab \geq 0$. This is necessarily the case if $c^2 - 4ab < 0$, i.e., if the lines M_1 and M_2 are defined over an imaginary quadratic extension.

We summarize our discussion in the following theorem:

Theorem 5.13. — *Let $\mathscr{X}_1$ be a cubic surface, $\mathscr{D}_1 \subset \mathscr{X}_1$ a hyperplane section, and $\mathscr{L}_1 \subset \mathscr{X}_1$ a line not contained in $\mathscr{D}_1$, all assumed to be flat over $\mathrm{Spec}(\mathscr{O}_S)$. Write $\mathfrak{q}_1 := \mathscr{D}_1 \cap \mathscr{L}_1$. Assume the following:*
(1) *GA1, GA2, GA3, and AA1;*
(2) *at least one of the assumptions GA4a, GA4b, or GA4c;*
(3) *at least one of the assumptions AA2a, AA2b, AA2c, AA2d, or AA2e.*
Then S-integral points of $(\mathscr{X}_1, \mathscr{D}_1)$ are Zariski dense.

We recover the following result (essentially Theorem 2 of Beukers [2]):

Corollary 5.14. — *Let $\mathscr{X}_1$ be a cubic surface, $\mathscr{D}_1 \subset \mathscr{X}_1$ a hyperplane section, and $\mathscr{L}_1 \subset \mathscr{X}_1$ a line not contained in $\mathscr{D}_1$, all assumed to be flat over $\mathrm{Spec}(\mathbb{Z})$. Write $\mathfrak{q}_1 := \mathscr{D}_1 \cap \mathscr{L}_1$. Assume that*
(1) *X_1 and D_1 are smooth;*
(2) *there exists an $\mathbb{Z}$-integral point of $(\mathscr{L}_1, \mathfrak{q}_1)$;*

(3) if q is a flex of D_1, we assume that the hyperplane containing L_1 and the flex line intersects X_1 in a smooth conic and L_1.

Then $\mathbb{Z}$-integral points of $(\mathscr{X}_1, \mathscr{D}_1)$ are Zariski dense.

We also recover a weak version of Theorem 1 of [2]. (This theorem is asserted to be true but the proof is not quite complete; the problem occurs in the argument for the second part of Lemma 2.)

Corollary 5.15. — *Retain all the hypotheses of Corollary 5.14, except that we allow the existence of a hyperplane H intersecting X_1 in three lines L_1, M_1, and M_2 and containing a flex line F for D_1 at q. Let p be a place for $\mathbb{Z}$ (either infinite or finite). Choose local coordinates x and y for H so that $L_1 = \{x = 0\}, F = \{y = 0\}$, and $M_1 \cup M_2 = \{ax^2 + cxy + by^2 = 0\}$, and assume that ab is a square in $\mathbb{Q}_p$. Then $\mathbb{Z}[1/p]$-integral points of $(\mathscr{X}_1, \mathscr{D}_1)$ are Zariski dense (where $\mathbb{Z}[1/\infty] = \mathbb{Z}$ and $\mathbb{Q}_\infty = \mathbb{R}$.)*

Of course, there are infinitely many primes p such that ab is a square in $\mathbb{Q}_p$. When $p = \infty$, by Remark 5.12 it suffices to verify that M_1 and M_2 are defined over an imaginary quadratic extension.

We also obtain results in cases where the boundary is reducible:

Corollary 5.16. — *Let $\mathscr{X}_1$ be a cubic surface, $\mathscr{D}_1 \subset \mathscr{X}_1$ a hyperplane section, and $\mathscr{L}_1 \subset \mathscr{X}_1$ a line not contained in $\mathscr{D}_1$, all assumed to be flat over $\mathrm{Spec}(\mathbb{Z})$. Write $\mathfrak{q}_1 := \mathscr{D}_1 \cap \mathscr{L}_1$. Assume that*

(1) X_1 is smooth;
(2) there exists an S-integral point of $(\mathscr{L}_1, \mathfrak{q}_1)$;
(3) $D_1 = E \cup C$, where E is a line intersecting L_1 and C is a conic irreducible over K;
(4) C intersects E in two points, defined over K_v where v is some place in S;
(5) there exists at most one conic in X_1 tangent to both L_1 and C.

Then S-integral points of $(\mathscr{X}, \mathscr{D})$ are Zariski dense.

Note that the assumption on the conics tangent to L_1 and C is used to verify GA4a.

5.6. Other applications. — Theorem 5.11 can be applied in many situations.

Theorem 5.17. — *Let $\mathscr{X} = \mathbb{P}^1_{\mathscr{O}_S} \times \mathbb{P}^1_{\mathscr{O}_S}$, $\mathscr{D} \subset \mathscr{X}$ a divisor of type $(2, 2)$, and $\mathscr{L} \subset \mathscr{X}$ a ruling of $\mathscr{X}$, all flat over $\mathscr{O}_S$. Assume that*

(1) D is nonsingular;

(2) L is tangent to D at q;
(3) S-integral points of $(\mathscr{L}, q)$ are Zariski dense.

Then S-integral points of $(\mathscr{X}, \mathscr{D})$ are Zariski dense.

Proof. Let φ be the projection for which $\mathscr{L}$ is a section. Since $\mathscr{C} = \mathscr{D}$ in this case, the second arithmetic assumption of Theorem 5.11 is easily satisfied. □

We also obtain the following potential density result:

Theorem 5.18. — *Let X be a smooth Del Pezzo surface of degree K_X^2 and index one (i.e., $\mathbb{Z}K_X$ is saturated in $\mathrm{Pic}(X)$). Let $D \subset X$ be a smooth anticanonical divisor. If X and D are defined over a number field K then integral points of (X, D) are potentially dense.*

Proof. Applying the classification theory of surfaces (and enlarging the base field), we may represent X as a conic bundle $\varphi : X \to \mathbb{P}^1$. First express X as a blow-up of $\mathbb{P}^2$ in $9 - K_X^2$ points. The pencil of lines in $\mathbb{P}^2$ containing one of the points w gives the conic bundle structure. Let L denote a (-1)-curve not contained in a fiber of φ, e.g., the exceptional curve lying over w. By adjunction, D is a bisection for φ and intersects L in one point q. Finally, the irreducible components of the normalization of $D \times_{\mathbb{P}^1} L$ all have positive genus, because D has positive genus. In particular, X, D, L, and q satisfy all the geometric assumptions of Theorem 5.11. On the other hand, its arithmetic assumptions may always be satisfied after judicious extensions of $\mathscr{O}_S$. It follows that integral points of (X, D) are potentially dense. □

6. Potential density for log K3 surfaces

We consider the following general situation:

Problem 6.1 (Integral points of log K3 surfaces)
Let X be a surface and D a reduced effective Weil divisor such that (X, D) has log terminal singularities and $K_X + D$ is trivial. Are integral points on (X, D) potentially dense?

Problem 6.1 has been studied when $D = \emptyset$ (see, for example, [3]). In this case density holds if X has infinite automorphisms or an elliptic fibration.

The case $X = \mathbb{P}^2$ and D a plane cubic has also attracted significant attention. Silverman [29] proved potential density in the case where D is singular and raised the general case as an open question. Beukers [1] established this by considering the cubic surface X_1 obtained as the triple cover of X totally branched over D.

Implicit in [2] is a proof of potential density when X_1 is a smooth cubic surface and D_1 is a smooth hyperplane section. Note that this also follows from Theorem 5.13 (cf. also Corollaries 5.14 and 5.15.) After suitable extensions of K and additions to S, there exists a line $L \subset X$ defined over K and the relevant arithmetic assumptions are satisfied. Similarly, the case of $X = \mathbb{P}^1 \times \mathbb{P}^1$ and D a smooth divisor of type $(2,2)$ follows from Theorem 5.17. Finally, Theorem 5.18 gives potential density when X is an index-one Del Pezzo surface and D is a smooth anticanonical divisor.

We summarize our results as follows:

Theorem 6.2. — *Let X be a smooth Del Pezzo surface and D a smooth anticanonical divisor. Then integral points for (X, D) are potentially dense.*

We close this section with a list of open special cases of Problem 6.1.

(1) Let X be a Del Pezzo surface and D a singular anticanonical cycle. Show that integral points for (X, D) potentially dense.

(2) Let X be a Hirzebruch surface and D an anticanonical cycle. Find a smooth rational curve L, intersecting D in exactly one point p, so that the induced map $\varphi : L \to \mathbb{P}^1$ is finite surjective.

6.1. Appendix: some geometric remarks. — The reader will observe that the methods employed to prove density for integral points on conic bundles (with bisection removed) are not quite analogous to the methods used for elliptic fibrations. The discrepancy can be seen in a number of ways. First, given a multisection M for a conic bundle (with bisection removed), we can pull-back the conic bundle to the multisection. The resulting fibration has two *rational* sections, Id and τ_M (see section 3). However, *a priori* one cannot control how τ_M intersects the boundary divisor (clearly, this is irrelevant if the boundary is empty). A second explanation may be found in the lack of a good theory of (finite type) Néron models for algebraic tori (see chapter 10 of [5]).

We should remark that in some special cases these difficulties can be overcome, so that integral points may be obtained by geometric methods completely analogous to those used for rational points. Consider the cubic surface

$$x^3 + y^3 + z^3 = 1$$

with distinguished hyperplane at infinity. This surface contains a line with equations $x+y = z-1 = 0$. Euler showed that the resulting conic bundle admits a multisection $(x_0, y_0, z_0) = (9t^4, 3t - 9t^4, 1 - 9t^3)$, which may be reparametrized as $(x_1, y_1, z_1) = (9t^4, -3t - 9t^4, 1 + 9t^3)$. Lehmer [15] showed that this is the

first in a sequence of multisections, given recursively by
$$(x_{n+1}, y_{n+1}, z_{n+1}) = 2(216t^6 - 1)(x_n, y_n, z_n) - (x_{n-1}, y_{n-1}, z_{n-1})$$
$$+ (-108t^4, -108t^4, 216t^6 + 4)$$
This is related to the fact that the norm group scheme
$$u^2 - 3(108t^6 - 1)v^2 = 1,$$
admits a section of infinite order $(u, v) = (216t^6 - 1, 12t^3)$. Remarkably, this group acts *regularly* on the conic bundle, i.e., the coordinate transformations are integral polynomials in t. In general, one would only expect a rational action, defined over the generic point of the t-line.

References

[1] F. Beukers, *Ternary forms equations*, J. Number Theory **54** (1995), no. 1, 113–133.

[2] F. Beukers, *Integral points on cubic surfaces*, Number theory (Ottawa, ON, 1996), 25–33, CRM Proc. Lecture Notes, **19**, AMS, Providence, RI, (1999).

[3] F. Bogomolov and Yu. Tschinkel, *Density of rational points on elliptic K3 surfaces*, Asian Journ. of Math. **4** (2000), no. 2, 351–368.

[4] E. Bombieri, D. Masser and U. Zannier, *Intersecting a Curve with Algebraic Subgroups of Multiplicative Groups*, Internat. Math. Res. Notices no. **20** (1999), 1119–1140.

[5] S. Bosch, W. Lütkebohmert and M. Raynaud, *Néron models*, Ergebnisse der Mathematik und ihrer Grenzgebiete (3), 21, Springer-Verlag, Berlin, 1990.

[6] J.-L. Colliot-Thélène, A. N. Skorobogatov and P. Swinnerton-Dyer, *Rational points and zero-cycles on fibred varieties: Schinzel's hypothesis and Salberger's device*, J. Reine Angew. Math. **495** (1998), 1–28.

[7] J.-L. Colliot-Thélène, A. N. Skorobogatov and P. Swinnerton-Dyer, *Double fibres and double covers: paucity of rational points*, Acta Arith. **79** (1997), no. 2, 113–135.

[8] J.-L. Colliot-Thélène and P. Swinnerton-Dyer, *Hasse principle and weak approximation for pencils of Severi-Brauer and similar varieties*, J. Reine Angew. Math. **453** (1994), 49–112.

[9] G. Faltings, *Diophantine approximation on abelian varieties*, Ann. of Math. (2) **133** (1991), no. 3, 549–576.

[10] A. Grothendieck, *Cohomologie locale des faisceaux cohérents et théorèmes de Lefschetz locaux et globaux* (SGA 2), Advanced Studies in Pure Mathematics, Vol. 2. North-Holland Publishing Co., Amsterdam; Masson & Cie, Éditeur, Paris, 1968.

[11] J. Harris and Yu. Tschinkel, *Rational points on quartics*, Duke Math. Journ. **104** (2000), no. 3, 477–500.

[12] B. Hassett and Yu. Tschinkel, *Abelian fibrations and rational points on symmetric products*, Intern. Journ. of Math. **11** (2000), no. 9, 1163–1176.

[13] S. Iitaka, *Algebraic geometry: An introduction to birational geometry of algebraic varieties*, Springer-Verlag, New York-Berlin, 1982.

[14] Y. Kawamata, K. Matsuda and K. Matsuki, *Introduction to the minimal model problem*, Algebraic geometry, Sendai, 1985, 283–360, Adv. Stud. Pure Math., 10, North-Holland, Amsterdam-New York, 1987.

[15] D. H. Lehmer, *On the diophantine equation $x^3 + y^3 + z^3 = 1$*, J. Lond. Math. Soc. **31** (1956), 275–280.

[16] M. McQuillan, *Division points on semi-abelian varieties*, Invent. Math. **120** (1995), no. 1, 143–159.

[17] D. W. Masser, *Specializations of finitely generated subgroups of abelian varieties*, Trans. Amer. Math. Soc. **311** (1989), no. 1, 413–424.

[18] D. W. Masser, *Specializations of endomorphism rings of abelian varieties*, Bull. Soc. Math. France **124** (1996), no. 3, 457–476.

[19] J.S. Milne, *Étale cohomology*, Princeton University Press, Princeton New Jersey, 1980.

[20] D. Mumford, J. Fogarty and F. Kirwan, *Geometric invariant theory*, Third edition. Ergebnisse der Mathematik und ihrer Grenzgebiete (2), 34, Springer-Verlag, Berlin, 1994.

[21] R. Noot, *Abelian varieties—Galois representation and properties of ordinary reduction*, Compositio Math. **97** (1995), no. 1, 161–171.

[22] T. Ono, *On some arithmetic properties of linear algebraic groups*, Ann. of Math. (2) **70** (1959), 266–290.

[23] T. Ono, *Arithmetic of algebraic tori*, Ann. of Math. (2) **74** (1961), 101–139.

[24] J. P. Serre, *Algebraic groups and class fields*, Graduate Texts in Mathematics, 117, Springer-Verlag, New York-Berlin, 1988.

[25] J. P. Serre, *Local fields*, Graduate Texts in Mathematics, 67, Springer-Verlag, New York-Berlin, 1979.

[26] J. P. Serre, *Lectures on the Mordell-Weil theorem*, Aspects of Mathematics, E15. Friedr. Vieweg & Sohn, Braunschweig, 1989.

[27] J.M. Shyr, *A generalization of Dirichlet's unit theorem*, J. Number Theory **9** (1977), no. 2, 213–217.

[28] J. Silverman, *Heights and the specialization map for families of abelian varieties*, J. reine und angew. Math. **342** (1983), 197–211.

[29] J. Silverman, *Integral points on curves and surfaces*, Number theory (Ulm, 1987), 202–241, LNM, 1380, Springer, New York-Berlin, 1989.

[30] P. Vojta, *Diophantine approximations and value distribution theory*, LNM, 1239, Springer-Verlag, Berlin, 1987.

[31] P. Vojta, *Integral points on subvarieties of semiabelian varieties. II*, Amer. J. Math. **121** (1999), no. 2, 283–313.

[32] V. E. Voskresenskii, *Algebraic groups and their birational invariants*, Translations of Mathematical Monographs, 179. American Mathematical Society, Providence, RI, 1998.

Rational points on algebraic varieties
(E. PEYRE, Y. TSCHINKEL, ed.), p. 199–219

COMPOSITION OF POINTS AND THE MORDELL–WEIL PROBLEM FOR CUBIC SURFACES

Dimitri Kanevsky
  T.J. Watson Research Center, P.O. Box 218, 23-116A, Yorktown Heights, New York 10598, US • *E-mail* : kanevsky@us.ibm.com

Yuri Manin
  Max–Planck–Institut für Mathematik, Bonn, Germany
  E-mail : manin@mpim-bonn.mpg.de

Abstract. — Let V be a plane smooth cubic curve over a finitely generated field k. The Mordell–Weil theorem for V states that there is a finite subset $P \subset V(k)$ such that the whole $V(k)$ can be obtained from P by drawing secants and tangents through pairs of previously constructed points and consecutively adding their new intersection points with V. Equivalently, the group of birational transformations of V generated by reflections with respect to k-points is finitely generated. In this paper, elaborating an idea from [M3], we establish a Mordell–Weil type finite generation result for some birationally trivial cubic surfaces W. To the contrary, we prove that the birational automorphism group generated by reflections cannot be finitely generated if $W(k)$ is infinite.

1. Introduction

1.1. Composition of points. — Let V be a cubic hypersurface without multiple components over a field k in $\mathbf{P}^d$, $d \geq 2$. Three points $x, y, z \in V(k)$ (possibly coinciding) are called *collinear* if either $x + y + z$ is the intersection cycle of V with a line in $\mathbf{P}^d$ (with correct multiplicities), or x, y, z lie on a k–line belonging to V. If x, y, z are collinear, we write $x = y \circ z$. Thus $\circ$ is a (partial and multivalued) composition law on $V(k)$. We will also consider its restriction on subsets of $V(k)$, e.g. that of smooth points.

If $x \in V(k)$ is smooth, and does not lie on a hyperplane component of V, the birational map $t_x : V \to V$, $y \mapsto x \circ y$, is well defined. It is called reflection with respect to x. Denote by Bir V the full group of birational automorphisms of V.

The following two results summarize the properties of $\{t_x\}$ for curves and surfaces respectively. The first one is classical, and the second is proved in [M1], Chapter V.

Theorem 1.1.1. — *Let V be a smooth cubic curve. Then:*

(a) Bir V is a semidirect product of a finite group and the subgroup consisting of products of an even number of reflections $\{t_x \mid x \in V(k)\}$.

(b) We have identically

$$t_x^2 = (t_x t_y t_z)^2 = 1 \tag{1}$$

for all $x, y, z \in V(k)$.

If in addition k is finitely generated over a prime field, then:

(c) Bir V is finitely generated.

(d) All points of $V(k)$ can be obtained from a finite subset of them by drawing secants and tangents and adding the intersection points.

Theorem 1.1.2. — *Let V be a minimal smooth cubic surface over a perfect non-closed field k. Then:*

(a) Bir V is a semi-direct product of the group of projective automorphisms and the subgroup generated by

$$\{t_x \mid x \in V(k)\} \text{ and } \{s_{u,v} \mid u, v \in V(K); [K:k]=2;\ u,v \text{ are conjugate over } k\}$$

where

$$s_{u,v} := t_u t_{u \circ v} t_v,$$

and u, v do not lie on lines of V.

(b) We have identically

$$t_x^2 = (t_x t_{x \circ y} t_y)^2 = (s_{u,v})^2 = 1, \quad s t_x s^{-1} = t_{s(x)}, \tag{2}$$

for all pairs u, v not lying on lines in V, and projective automorphisms s.

(c) The relations (2) form a presentation of Bir V.

We remind that V is called *minimal* if one cannot blow down some lines of V by a birational morphism defined over k. The opposite class consists of *split* surfaces upon which all lines are k-rational.

1.2. Main results of the paper. — Although Theorems 1.1.1 and 1.1.2 look very similar, there is an important difference between finiteness properties in one– and two–dimensional cases.

Basically, (1) means only that $x+y := e \circ (x \circ y)$ is an Abelian group law with identity e: see [M1], Theorem I.2.1. The statements c) and d) of the Theorem 1.1.1 additionally assert that this group is finitely generated. Therefore, (1) generally is not a complete system of relations between $\{t_x\}$.

On the contrary, (2) is complete, and in §2 we will see that this prevents Bir V from being finitely generated if $V(k)$ is infinite. This answers one of the questions raised in [M3].

Therefore, any reasonable analog of the Mordell–Weil problem must address the problem of finite generation for $(V(k), \circ)$ or of quotients of $V(k)$ with respect to various equivalence relations compatible with $\circ$. This is the subject of §§3–5.

As in [M1], Chapter II, we can start with the universal equivalence relation U. By definition, this is the finest equivalence relation compatible with collinearity and such that $\circ$ induces a well defined operation on $V(k)/U$ also denoted $\circ$. Then one of the Mordell–Weil type questions asks about finite generation (= finiteness) of the CH–quasigroup $(V(k)/U, \circ)$ (see [M1], Chapter I.).

In §3 and §4 we give a description of U refining earlier results of [M1]. Consider the set of intersections of V with tangent planes at points of $V(k)$ and add to it all images of these curves with respect to the group generated by all $t_x, x \in V(k)$. Then one class of U consists of points that can be pairwise joined by a chain of curves belonging to this set of curves. This is the content of Theorem 3.3 below. We then discuss various versions of finite generation of $(V(k), \circ)$. One essential choice is whether to allow to apply $\circ$ only to the different previously constructed points (for minimal surfaces, the result will then be uniquely defined). Another option giving more flexibility is to allow expressions $x \circ x$ and treat them as multivalued, thus adding at one step all the intersection points of V with a tangent plane at x. Finally, in §4 we extend the group–theoretic description of U given in [M1], II.13.10.

The results of §3 and §4 are essentially algebraic and do not add any new cases of finite generation of $(V(k), \circ)$ to the short list of locally compact local fields already treated in [M1]. (In fact, [M1] proves the finiteness of $V(k)/U$ over such fields by establishing that $V(k)$ is covered by a finite number of sets of the form $(x \circ x) \circ (y \circ y)$).

In §5 we study modified composition laws of points introduced in [M3]. The idea behind this development is to reinterpret the classical theorem on the structure of abstract projective planes as a finiteness result.

Namely, let k be a finitely generated field. Start with a finite subset $S \subset \mathbf{P}^2(k)$ and add to it pairwise intersections of all lines passing through two points of S thus getting a new finite set S'. Apply the same procedure to S', and so on. If S is large enough, in the limit we will get the whole $\mathbf{P}^2(k)$. This easily follows from the fact that if we start with S consisting of ≥ 4 points in general position, in the limit we will get an abstract projective plane satisfying the Desargues axiom and therefore coinciding with $\mathbf{P}^2(k')$ for $k' \subset k$ up to a projective coordinate change.

A trick, first introduced in [M3], allowed us to translate this remark into a finiteness theorem for $V(k)$ assuming the existence of a birational morphism $p: V \to \mathbf{P}^2$ defined over k. However, this required dealing with modified composition laws: roughly speaking, instead of looking at the collinearity relation induced by that in $\mathbf{P}^3$, we now have to use the collinearity relations determined by the morphism p.

In this paper we make some steps towards eliminating this complication. Although the final result falls short of what we would like to prove, we feel that the connection and analogies with the theory of abstract projective planes deserve further study.

Acknowledgement. The first named author would like to thank V. Berkovich and J-L. Colliot-Thélène for useful discussions. The work was partially supported by the Humboldt Foundation during the author's stay at the Max-Planck-Institut für Mathematik.

2. Cardinality of generators of subgroups in a reflection group

Notation 2.1. — We shall call an *abstract cubic* a set S with a ternary relation $L \subset S \times S \times S$, satisfying the following axioms:

(a) L is invariant with respect to permutations of factors.

(b) If $(x, y, z), (x, y, z') \in L$ and $x \neq y$, then $z = z'$.

The *reflection group* G_S of an abstract cubic S is generated by symbols $t_x, x \in S$, subject to the following relations:

$t_x^2 = 1$ for all $x \in S$;

$(t_x t_y t_z)^2 = 1$ for all $(x, y, z) \in L$.

The following result is proved in [K1].

Theorem 2.2. — *(a). Any element of finite order in G_S is conjugate to either t_x or to $t_x t_y t_z$ for appropriate $x, y, z \in S$. Let S be given effectively and $L \subset S \times S \times S$ be decidable. Then: (b) The word problem in G_S is decidable. (c) The conjugacy problem in G_S is decidable.*

The proof is based on a direct description of G_S as a limit of amalgamated sums. In [**K2**] it is shown that S can be sometimes reconstructed from G_S. Moreover, under some additional assumptions it is proved that Aut G_S is generated by G_S and permutations of S preserving L.

A different interesting description of G_S and another proof of the Theorem 2.2 is given in [**P**].

For the purposes of our paper we need the following description of G_S that is a special case of the general structure theorem 1.4 in [**K1**].

Structural Theorem 2.3. — *Let $x \in S$ be an arbitrary point and $S' := \{S \setminus x\}$. Then G_S is canonically isomorphic to $G_{S'} *_\Pi K$ (the free product of $G_{S'}$ and K with the amalgamated subgroup Π). The groups in this product can be described as follows.*

(a) $G_{S'}$ is the reflection group of the cubic S' with the ternary relation induced by L on S'.

(b) The amalgamated subgroup Π is a free group generated by free generators $a_{u,v} = t_u t_v$ for all distinct pairs $u, v \in S'$ such that $(u, v, x) \in L$ and $u < v$ (for some fixed ordering of S).

(c) $K \xrightarrow{\sim} Z_2 * Z_2 * \cdots * Z_2$. Generators of the subgroups Z_2 in this free product are t_x and $t_x a_{u,v}$.

(d) Π is of index 2 in K. The quotient group K/Π is generated by the class t_x.

This structural result leads to the following auxiliary statement, which we need to prove our main results in this section.

Definition–Lemma 2.4. — *(a) In the situation of Theorem 2.3 a family $W = \langle R_1, t_x, R_2, t_x, \ldots, t_x, R_n \rangle$, where $R_i \in G_{S'}$, is called a reduced t_x-partition of $g = R_1 t_x R_2 t_x \ldots R_n$ if $R_i \notin \Pi$ for $1 < i < n$.*

(b) Let W be a reduced t_x-partition of $g \in G_S$. Let us define $\mathrm{ord}_x(g)$ as the number of t_x in W. This number depends on g and x and is the same for different reduced t_x-partitions of g.

(c) Let $g \in G_S$ be such that $\mathrm{ord}_x(g) = 0$. Then $g \in G_{S'}$.

(d) $\mathrm{ord}_x(g_1 g_2) \equiv (\mathrm{ord}_x(g_1) + \mathrm{ord}_x(g_2)) \mod 2$.

(e) $\mathrm{ord}_x(aga^{-1}) \equiv \mathrm{ord}_x(g) \mod 2$ for any $a, g \in G_S$.

(f) Let $g \in G_S$. We put $\delta(g) := \{x \in S \mid \mathrm{ord}_x(g) \neq 0\}$. The set $\delta(g)$ is finite.

(g) $\delta(g_1 g_2) \subset \delta(g_1) \cup \delta(g_2)$.

(i) Let $\langle h_1, h_2, \ldots \rangle$ be a family generating a subgroup H. Then $\cup_{h \in H} \delta(h) = \cup_i \delta(h_i)$.

Now we can formulate the main theorem of this section.

Theorem 2.5. — *Consider a subgroup $H = \langle g_1, g_2, \ldots g_n, \ldots \rangle \subset G_S$ generated by an infinite family of elements such that $\delta = \cup_i \delta(g_i)$ is infinite. Then H is not finitely generated.*

Proof. — Assume that H is finitely generated by $h_1, \ldots, h_k$. Then the set $\delta' = \cup_{i=1,\ldots k} \delta(h_i)$ is finite. Therefore there exists some g_r and $x \in S$ such that the following holds: $ord_x(g_r) \neq 0$ and $x \notin \delta'$. Hence $ord_x(h_i) = 0$ for all $i = 1, \ldots, k$. By 2.4(i), $H \subset G_{S'}$ if H is generated by $h_1, \ldots, h_k$. Since $ord_x(g_r) \neq 0$, $g_r \notin H$. This contradiction proves the theorem. □

The following extension of Theorem 2.5 can be applied to various subgroups of Bir V.

Corollary 2.6. — *Let G_S be the reflection group of an abstract cubic S and let W be the group of permutations of S, preserving its ternary relation L. Let $G \cong W * G_S$ be the semi-direct product of W and G_S, such that $wt_x w^{-1} = t_{w(x)}$ for any $w \in W$ and $x \in S$. Let a subgroup $H \subset G$ be generated by a finite subgroup $W' \subset W$ and an infinite family of elements $g_i \in G_S$ such that $\delta = \cup_{i=1,2,\ldots} \delta(g_i)$ is infinite. Then H is not finitely generated.*

Proof. — Let us assume that H is generated by a finite number of elements $h_1, \ldots, h_k \in G_S$ and a finite number of elements from W'. Let $\delta = \cup_i \delta(g_i)$ and let $\delta' = \cup_{w \in W'} w\delta$ be obtained by applications of all $w \in W'$ to δ. Since δ and W' are finite, δ' is finite. Therefore there exists a generator $g_i \in H$ such that $\delta(g_i) \not\subset \delta'$. Therefore, $g_i \notin G_{\delta'}$, i.e. it cannot be obtained as a product of elements from $h_1, \ldots, h_k$ and $w \in W'$. This contradiction proves the corollary. □

Example 2.7. — In the situation of Theorem 1.1.2 assume that $S = V(k)$ is infinite. Then the following subgroups of Bir V cannot be finitely generated:

(a) Bir V, $B(V) := \langle t_x \,|\, x \in V(k) \rangle$ and

$G := \langle t_x, s_{u,v} \,|\, x \in V(k),\, u, v \in V(K);\, [K:k] = 2;\, u, v$ are conjugate over $k \rangle$.

(b) The commutant of any of subgroups described in (a).

(c) Let $B_0(V)$ denote the normal subgroup of $B(V)$ generated by elements of the form $t_x t_y t_z t_{x'} t_y t_{z'}$, where (x, y, z) and (x', y, z') run through triples of collinear points of $V(k)$. This subgroup was introduced in [**M1**] (II.13.9; beware of a misprint there: the second y carries a superfluous prime). It is closely related to the universal equivalence on $V(k)$ (see the section 4 below).

(d) Let $B_1(V)$ denote the normal subgroup of $B(V)$ generated by elements of the form $t_x t_y t_z$, where (x, y, z) run through all possible triples of collinear points of $V(k)$. This subgroup was introduced in [M2] because it is closely related to some admissible equivalence relations on $V(k)$

We will now show that Theorem 2.3 implies the statement for the case (b). Other cases will be discussed later and stronger statements will be proved.

Proof of (b). — The commutant of $B(V)$ contains elements
$$t_x t_y t_x^{-1} t_y^{-1} = t_x t_y t_x t_y = a_{x,y}^2.$$
Let us consider an infinite family of elements $a_{x_i y_i}^2$, $x_i, y_i \in S$. The statement for (b) will follow if we show that $\delta(a_{x,y}^2)$ contains x, since it will follow that $\cup_i \delta(a_{x_i y_i})$ is infinite. For this it is enough to show that $t_x t_y t_x t_y$ has the following reduced t_x-partition: (t_x, t_y, t_x, t_y). Indeed, in the notation of 2.4 one has to check that $t_y \notin \Pi$. But this fact follows immediately from 2.3 if one notes that Π is a free group (hence it contains no nontrivial elements of finite order) and $t_y^2 = 1$. This implies that $ord_x(a_{x,y}^2) > 0$, i.e. $x \in \delta(a_{x,y}^2)$. □

Our next theorem provides a lower bound for the number of generators in the normal closure.

Theorem 2.8. — *For any $g \in G_S$ let $\tilde{\delta}(g) = \{x \in S \,|\, ord_x(g) \not\equiv 0 \mod 2\}$. Let H be the normal closure in G_S generated by a family of elements that contains a subfamily of elements $h = (h_1, \ldots, h_i, \ldots)$ such that the following condition holds. (J): For any i there exist $x_i \in \tilde{\delta}(h_i)$ such that $x_i \notin \tilde{\delta}(h_j)$ if $i \neq j$. Then H cannot be the normal closure in G_S of less than $\mathrm{card}\, h$ generators.*

This theorem immediately implies

Corollary 2.9. — *In the situation of Theorem 2.8, H cannot be the normal closure of a finite number of elements if there is an infinite subsystem h satisfying (J).*

Proof of Theorem 2.8. — Define a map of G_S into the vector space $\mathbf{F}_2^S$ as follows:
$$\psi : G_S \to V, \quad \psi(g) = (\ldots, ord_x(g) \mod 2, \ldots).$$
It follows from 2.4(d) that ψ is a group homomorphism so that it maps conjugacy classes in G_S into one element. The theorem will follow if one shows that the image $\psi(H)$ cannot be generated by less than $\mathrm{card}\, h$ vectors. But this follows immediately from the condition (J) in the theorem that guarantees that each image $\psi(h_i)$ has a non–zero x_i–component while all other vectors $\psi(h_j)$ have a zero x_i–component. □

Corollary 2.10. — *None of the subgroups that are defined in (a),(c) and (d) in 2.7 can be obtained as the normal closure of a finite number of generators.*

Proof. — (a) follows from the fact that $\tilde{\delta}(t_x) = x$. Since G_S contains the infinite number of t_x, G_S cannot be obtained as the normal closure of a finite number of elements.

(c) will follow similarly to (a) if we show that $\tilde{\delta}(t_x t_y t_z t_{x'} t_y t_{z'})$ contains x for $x \neq y, z, x', z'$. This follows from the fact that the t_x-partition of $t_x t_y t_z t_{x'} t_y t_{z'}$ is $(t_x, t_y t_z t_{x'} t_y t_{z'})$ (where $t_y t_z t_{x'} t_y t_{z'} \in G_{S'}$). Since $V(k)$ has infinitely many collinear triples (x, y, z), such that $x \neq y \neq z \neq x$, one can find infinitely many generators in $B_0(V)$ satisfying the condition (**J**).

The case (d) can be treated similarly. $\square$

3. Structure of universal equivalence

3.1. Setup. — Let P be an abstract cubic with the collinearity relation $L \subset P \times P \times P$, such that for any $x, y \in P$, there exists $z \in P$ with $(x, y, z) \in P$.

An equivalence relation R on P is called *admissible* if the relation L/R induced on P/R has the following property: for any $X, Y \in P/R$, there exists a unique Z with $(X, Y, Z) \in L/R$. An admissible equivalence relation is called *universal* if it is finer that any other admissible relation.

In [**M1**] it was proved that the universal relation exists (and of course, is unique) by a simple argument: just take the intersection of all admissible relations. Here we will clarify its structure by representing it as a limit of a sequence of explicitly constructed equivalence relations of which every next one is less fine than the previous one.

3.2. Approximations. — For every $i \geq 0$, we will describe inductively a symmetric and reflexive binary relation $\sim_i$ on P and its transitive closure $\approx_i$. By definition, $\sim_0$ and $\approx_0$ are simply identical relations $x = x'$.

Definition 3.2.1. — *If $\sim_i$ and $\approx_i$ are already defined, we put $x \sim_{i+1} x'$ iff $x = x'$ or there exist $u, v, u', v' \in P$ such that $u \approx_i u'$, $v \approx_i v'$, and $(u, v, x) \in L$, $(u', v', x') \in L$.*

Furthermore, we put $x \approx_{i+1} x'$ iff there is a sequence of points $x = y_0, y_1, \ldots, y_r = x'$ such that $y_a \sim_{i+1} y_{a+1}$ for all $a < r$.

Let us consider the case $i = 1$. By definition, $x \sim_1 x'$ iff there exist $u, v \in P$ such that $(u, v, x), (u, v, x') \in L$. Let P be the set of k–points of a cubic surface V and L the usual collinearity relation. Assume for simplicity that V does not contain lines defined over k. Then $x \sim_1 x'$ means that $x = x'$ or x and x'

lie on the intersection of V with the tangent plane at some k–point u (with u deleted if the double tangent lines to u in this plane are not defined over k). So one equivalence class for $\approx_1$ consists of one point or of a maximal connected union of such quasiprojective curves, two of them being connected if they have an intersection point defined over k. The case of general cubic surface allows a similar description, but points of k–lines in V must be added as subsets of equivalence classes.

Theorem 3.3. — (a) If $x \approx_i x'$ then $x \approx_{i+1} x'$.
(b) Denote by $\approx$ the equivalence relation

$$x \approx x' \iff \exists i,\ x \approx_i x'.$$

Then it is admissible and universal.

Proof. — (a) It suffices to prove that if $x \neq x'$, $x \sim_i x'$ then $x \approx_{i+1} x'$. For $i = 0$ this is clear. Assume that we have proved that $u \sim_{i-1} u'$ implies $u \approx_i u'$.

If $x \sim_i x'$ then by definition $(u, v, x) \in L$ and $(u', v', x') \in L$ for some $u \approx_{i-1} u'$ and $v \approx_{i-1} v'$. From the inductive assumption it follows that $u \approx_i u'$ and $v \approx_i v'$. By definition, then $x \approx_{i+1} x'$.

(b) Let us first prove that $\approx$ is admissible, in other words, if $(u, v, x) \in L$, $(u', v', x') \in L$, and $u \approx u'$, $v \approx v'$, then $x \approx x'$. In fact, for some i we have $u \approx_i u'$, $v \approx_i v'$, so that $x \approx_{i+1} x'$ and $x \approx x'$.

Now denote temporarily the universal equivalence relation by $\approx_U$. The previous argument shows that $x \approx_U x' \Rightarrow x \approx x'$. It remains to prove that $x \approx_i x' \Rightarrow x \approx_U x'$. We argue by induction. Again, it suffices to check that $x \sim_i x' \Rightarrow x \approx_U x'$ assuming $x \neq x'$. We can then find $(u, v, x) \in L$, $(u', v', x') \in L$ such that $u \approx_{i-1} u'$, $v \approx_{i-1} v'$. Therefore $u \approx_U u'$, $v \approx_U v'$, and finally $x \approx_U x'$. □

3.4. Types of finite generation. — Let us say, as in [M3], that P is ○–generated by $(x_\alpha \,|\, \alpha \in A)$ if for any $y \in P$ there is a non–associative commutative word in x_α's such that, informally, y is one of the values of this word. This means that when we calculate this word in the order determined by the brackets, every time that we have to calculate some $u \circ v$, we may replace it by any x such that $(u, v, x) \in L$.

Claim 3.4.1. — If P is ○–generated by $(x_\alpha | \alpha \in A)$, then the CH–quasigroup P/U is generated by the classes X_α of x_α.

We consider the following different types of ○–generation.

3.4.2. Values of nonassociative words. — Let W be a non-associative commutative word in finite number of variables X_i, P as in 3.1, and x_i a family of elements of P with the same set of indices. We define different rules of computing values of W on (x_i) in the order determined by the brackets inductively as follows for $i = 0, 1, \ldots \infty$. We set $x \approx_\infty y$ if $x \approx y$ (i.e. $x \approx_j y$ for some j).

Rule A_i. If the word $W = X$ has length 1, then a value of W at any point $x \in P$ is any $y \in P$ such that $x \approx_i y$. In particular, A_0 means that the value of W at x coincides with x. The rule A_1 means that the set of values of W consists of those y for which which there are points u_j, y_j, $j = 0, \ldots, r$, $y_0 = x, y_r = y$ such that the following holds: $(u_j, u_j, y_{j-1}) \in L, (u_j, u_j, y_j) \in L$ for $j = 1, \ldots, r$.

If the word $W = X \circ Y$ has length 2, its set $P(x, y)$ of values of W at $x, y \in P^2$ is defined as follows.

$$P(x, y) = \{z \in P \mid z \approx_i z', (x, y, z') \in L\}.$$

If the word W has length more than two, it is a product of two non empty words $W = W_1 \circ W_2$. Let $P(W_i)$ be a set of values of W_i that is defined inductively. Then the set of values $P(W)$ is defined as $\cup P(x, y)$ for all $(x, y) \in P(W_1) \times P(W_2)$.

We say that P is $\circ_{A_i}$ generated by $P' = (x_\alpha \mid \alpha \in A)$ if it is generated by application of the rule A_i to points in P'.

The inverse statement of 3.4.1 is valid for $\circ_{A_\infty}$ by trivial reasons.

Claim 3.4.3. — *If the CH-quasigroup $P/U \approx$ is $\circ$-generated by classes of $(x_\alpha \mid \alpha \in A)$, then P is $\circ_{A_\infty}$ generated by x_α.*

3.4.4. Questions. — Let us define the *generation index* $i(P)$ of P as the smallest i such that P is $\circ_{A_i}$-generated by a finite number of points in P. Let $P = V(k)$ for some cubic surface.

(1) For which fields k and for which classes of cubic surfaces $i(P)$ is finite? In particular, is $i(P) = 0$ for V defined over a number field (the original Mordell-Weil problem)?

(2) If the CH-quasigroup P/U is finite, is the index $i(P)$ finite?

It would be worthwhile to study (2) for an abstract cubic P that has an additional property: every three points of it generate an Abelian group like points on a plane cubic curve.

4. A group–theoretic description of universal equivalence

In [M1], II.13.10 a group–theoretic description of universal equivalence was given for a cubic surface that is defined over an infinite field and has a point of general type. In this section we extend this description of universal equivalence. We relate the sequence of explicitly constructed equivalence relations from §3 to a filtration by subgroups in the reflection group associated with a minimal cubic surface.

Let $B(V)$ and $B_0(V)$ be the groups described in the examples 2.7. Here the field k over which the cubic surface V is defined can be finite and therefore we do not assume that $V(k)$ is infinite.

Define $x \sim y \mod U$ if $t_x t_y \in B_0(V)$. It is clear that U is an equivalence relation on $V(k)$. The proof of the following theorem differs from the proof of the corresponding theorem 13.10 in [M1] in the following respects. It uses the explicit description of the universal admissible equivalence from the section 3 and the structural description of the reflection group of $S = V(k)$.

Theorem 4.1. — *U is the universal admissible equivalence relation.*

Proof. — We will check in turn that each of the equivalence relations is finer than the other one.

Assume first that z' and z are universally equivalent. We want to show that $z' \sim z \mod U$.

According to Theorem 3.3, $z' \approx_i z$ for some i. Since U is an equivalence relation, it is sufficient to treat the case $z' \sim_i z$. The following Lemmma does the job.

Lemma 4.2. — *Denote by $B^i(V), i = 0, 1, \ldots,$ the normal closure of the family $\{t_x t_{x'} \mid x \sim_i x'\}$ in $B(V)$. Let $x \sim_i x'$, $y \sim_i y'$, $(x, y, z) \in L$ and $(x', y', z') \in L$. Then the following holds:*

$$t_z t_{z'} \in t_z t_{z'} B^i(V) = t_x t_y t_z t_{x'} t_{y'} t_{z'} B^i(V) \subset B^{i+1} \subset B_0(V)$$

Proof of Lemma 4.2. — Using relations $t_x^2 = 1$ and $t_x t_y t_z = t_z t_y t_x$ we get $b = t_z t_y t_x t_{x'} t_{y'} t_{z'} = t_z t_{z'} b'$ where $b' = t_{z'} t_y t_x t_{x'} t_{y'} t_{z'}$. Next, b' is conjugate to $b'' = t_y t_x t_{x'} t_{y'}$. And, finally, b'' is a product of $t_y t_{y'} \in B^i(V)$ and $t_y t_x t_{x'} t_y$ which is conjugate to $t_x t_{x'} \in B^i(V)$. This proves the equality $t_z t_{z'} B^i(V) = t_x t_y t_z t_{x'} t_{y'} t_{z'} B^i(V)$. It remains to show the inclusion $B^{i+1}(V) \subset B_0(V)$. We will prove this inductively.

$B^1(V)$ is generated by $t_z t_{z'}$ such that z' and z lie on the intersection of V with a tangent plane at some k-point u. In this case $t_{z'} t_z = t_{z'} t_u t_u t_z t_u t_u \in B_0(V)$.

Assume that we already proved that $B^i(V) \subset B_0(V)$ and let us prove that $t_z t_{z'} \in B_0(V)$. Let $z'' \in V(k)$ be such that $(x', y, z'') \in L$. Then $t_z t_{z'} = t_z t_{z''} t_{z''} t_{z'}$ and the following inclusions hold:

$$t_z t_{z''} \in t_z t_y t_x t_{x'} t_y t_{z''} B^i(V) \subset B_0(V) B^i(V) \subset B_0(V),$$

$$t_{z''}t_{z'} \in t_{z''}t_{x'}t_y t_{z'}t_{x'}t_{y'} B^i(V) \subset B_0(V)B^i(V) \subset B_0(V).$$

Since $t_z t_{z'} \in B^{i+1}(V)$, this proves the inductive statement, establishes the Lemma and the first part of the Theorem. □

We turn now to the second part. Let A be any admissible equivalence relation. We shall show that $x \sim y \mod U$ implies $x \sim y \mod A$. Let X, Y, Z be the A-classes of x, y, z. Then $Z = X \circ Y$ in the sense of the composition law induced by collinearity relation on $S = V(k)$. Denote by $E = V(k)/A$ the set of classes with the induced structure of the symmetric quasigroup. Let $t_X : E \to E$ be the map $t_X(Y) = X \circ Y$. The map $t_x \mapsto t_X$ extends to an epimorphism of groups $\varphi : B(V) \to T(E)$. We will show that its kernel contains $B_0(V)$. Therefore if $t_x t_y \in B_0(V)$ then $\varphi(t_x t_y) = t_X t_Y = 1$. This implies that $t_X = t_Y$ and that $X = Y$. To prove this property of φ we need to extend the Theorem 13.1 (ii),(iii) in [**M1**] to our case. Recall that the Theorem 13.1 uses assumptions for cubic hypersurfaces that implies the fact that every equivalence class is dense in the Zariski topology. This is not true any more in general in our case.

Lemma 4.3. — (a) $\varphi : B(V) \to T(E)$ *is well defined and is an epimorphism of groups.*
(b) *In $T(E)$ the following equality holds:* $t_X t_Y t_Z = t_{Y \circ Y}$.

Proof. — (a) Our proof is based on the representation of elements in $B_0(V)$ as "minimal" words in the group K^S, the free product of groups Z_2 generated by symbols T_x, one for each point x with the relations $T_x^2 = 1$ (cf. [**K1**], 2.6 and §6). In order to construct the homomorphism $B(V) \to T(E)$, we first define the action of $B(V)$ on E. Denote by $T_{x_1} T_{x_2} \ldots T_{x_n}$ a minimal representation in K^S of some $s \in B(V)$. Choose $Y \in E$ and put $s(Y) = X_1 \circ (X_2 \circ \ldots (X_n \circ Y) \ldots)$ where X_i are classes of x_i in E.

One can show that this definition does not depend on the choice of a minimal representation of s in K^S. This can be done inductively on the length of minimal words in K^S. All minimal words of length one representing the same element in $B(V)$ coincide. Let us assume that the statement is proved for minimal words of the length $i - 1$. Consider now two different minimal words $w = T_1 \ldots T_i$, $w' = T_1' \ldots T_i'$ of the length i representing $s \in B(V)$. (Minimal words representing the same element have the same length). If $T_i = T_i'$ then the action of w (resp. w') on E can be factored through the actions of T_i and $w_1 = T_1 \ldots T_{i-1}$ (resp. $w_1' = T_1' \ldots T_{i-1}'$). Since w_1 and w_1' represent the same element in $B(V)$ and have the length $i - 1$, the statement follows by the inductive assumption.

Otherwise, if $T_i \neq T_i'$, consider a T_i-partition of w' (it is defined in the same way as t_x-partition above): $(R_1, T_i, \ldots R_{k-1}, T_i, R_k)$. From [K1] it follows that $R_k = T_{u_1} T_{v_1} T_{u_2} T_{v_2} \ldots T_{u_r} T_{v_r}$ and $(u_j, v_j, u) \in L$ for all $j = 1, \ldots r$ and $T_u = T_i$. Moreover, if we replace $T_i R_k$ in w' with $R_k' T_i$ where $R_k' = T_{v_1} T_{u_1} T_{v_2} T_{u_2} \ldots T_{v_r} T_{u_r}$, then we get a new word w'' that is already a minimal representation of s. Since w'' and w both end with the same element $T_i = T_u$, they act in the same way on $T(E)$. In order to prove that w' and w'' also act identically on $T(E)$ it is enough to check that $T_u R_k$ and $R_k' T_u$ act in the same way on $T(E)$. This can be shown using the fact that $t_{u_j} t_{v_j} t_u = t_u t_{v_j} t_{u_j}$.

To complete (a) we need to show that for any two elements $s_1, s_2 \in B(V)$ and $Z \in E$ we have $s_1(s_2(Z)) = (s_1 s_2)(Z)$. We will prove this statement by induction on the sum of lengths of minimal representation of s_1 and s_2. The statement is obvious if s_1 has length 0. Assume now that s_1 has a minimal representation $w_1 = T_{x_1} \ldots T_{x_i}$, $i \geq 1$, and s_2 has a minimal representation $w_2 = T_{y_1} \ldots T_{y_k}$. If $w = w_1 w_2$ is the minimal representation of $s = s_1 s_2$ than the action of s on E is defined via the action of w by the rule $X_1 \circ (\ldots X_i \circ (Y_1 \circ \ldots (Y_k \circ Z) \ldots)$ where X_i (resp. Y_j) are the classes of x_i (resp. y_j) and $Z \in E$. Therefore $s_1(s_2(Z)) = (s_1 s_2)(Z)$. Assume now that $w_1 w_2$ is not minimal.

Consider first the case when there exists such minimal representation of w_1, w_2 that $T_{x_i} = T_{y_1}$ (i.e. the last element in w_1 coincides with the first element in w_2). Let $s_1' \in B(V)$ be represented by $w_1 = T_{x_1} \ldots T_{x_{i-1}}$ and $s_2' \in B(V)$ be represented by $w_2' = T_{y_1} \ldots T_{y_{k-1}}$. Then $s_1'(s_2'(Z)) = s_1(s_2(Z))$ and one can apply the inductive statement to s_1' and s_2'.

Otherwise, let us assume that the word $w_1 w_2$ has the following T_x-partition;

$$R_1 T_x R_2 \ldots T_x R_l R_{l+1} T_x R_{l+2} T_x \ldots T_x R_m$$

where $R_1 T_x R_2 \ldots T_x R_l$ (resp. $R_{l+1} T_x R_{l+2} T_x \ldots T_x R_m$) is a minimal partition of w_1 (resp. w_2). Since $w_1 w_2$ is not minimal, T_x can be chosen in such a way that $R_l R_{l+1} = T_{u_1} T_{v_1} T_{u_2} T_{v_2} \ldots T_{u_r} T_{v_r}$, where $(u_s, v_s, x) \in L$ for $s = 1, \ldots, r$. As in the case of minimal words above one can replace $T_x R_l R_{l+1}$ in $w_1 w_2$ with

$$T_{v_1} T_{u_1} T_{v_2} T_{u_2} \ldots T_{v_r} T_{u_r} T_x$$

and obtain a new word w' that has the same action on E that $w_1 w_2$. Since w' has two subsequent elements T_x, we can split it into a product of w_1' that ends with T_x and w_2' that starts with T_x. This case was already considered in this proof.

(b) follows from properties of the group law on plane cubic curves. This proves the Lemma 4.2. □

To finish the proof of Theorem 4.1, we use the following identity:

$$\varphi(t_x t_y t_z t_{x'} t_y t_{z'}) = t_X t_Y t_{X \circ Y} t_{X'} t_Y t_{X' \circ Y'} = t_{Y \circ Y}^2 = 1.$$

Here $X, Y, \ldots$ are the classes of $x, y, \ldots$ mod A. As a consequence, $B_0(V) \subset Ker\,\varphi$, proving the theorem. □

Corollary 4.4. — *Let V be a minimal cubic surface over a finite field with q elements. Then $B(V)/B_0(V) = Z_2$, except when all points of $V(k)$ are Eckardt points. In the later case we have either $q = 2$, card $V(k) = 3$, or $q = 4$, card $V(k) = 9$.*

Proof. — This follows from the description of the universal equivalence for V over finite fileds in [Sw–D]. □

4.5. Remarks. — (a) As it follows from the proof of Theorem 4.1, it can be extended to an abstract cubic for which every three points generate an abelian group, in the same sense as for a plane cubic curve. We believe that this theorem can be proved also for an abstract cubic using only a structural description of G_S without this additional assumption. We plan to address this problem elsewhere.

(b) Groups G_S were studied in [**P**] using different methods. [**P**] asked whether *the dependency problem $DP(n)$ is decidable for reflection groups of an abstract cubic* for $n \geq 3$ or $n = \infty$. $DP(n)$ can be formulated as follows.

We will say that g_0 is *dependent* on $(g_1, \ldots, g_k)$ if there is a family $(g_{i_1}, \ldots, g_{i_p})$ and elements $u_1, \ldots, u_p$ of G such that

$$g_0(u_1 g_{i_1} u_1^{-1}) \ldots (u_p g_{i_p} u_p^{-1}) = 1.$$

If n is a positive number or infinity then the *dependence problem $DP(n)$* asks for an algorithm to decide for any sequence $(g_0, \ldots, g_k)$, $0 \leqslant k < n$, of elements of G whether or not g_0 is dependent on $(g_1, \ldots, g_k)$. The problems $D(1), D(2)$ are usually called the *word problem* and the *conjugacy problem*.

A special case of the dependence problem for $t_x t_y \in B_0(V)$ can be related to the decidability of universal equivalence. Namely, if $DP(\infty)$ is decidable for $g_i = t_{x_i} t_{y_i} t_{z_i} t_{x_i'} t_{y_i} t_{z_i'}$ and $g_0 = t_x t_y$ than one can efficiently define whether x, y are universally equivalent.

Since the decidability of the universal equivalence seems to be a very difficult problem in general, one can infer about the difficulty of the $DP(\infty)$ for $B_0(V)$.

Question. Let an abstract cubic S be decidable. Is $DP(n)$ decidable for arbitrary $t_x t_y$ and generators of the subgroup $B_0(V)$ described in 2.7(c)?

(c) Another construction of a filtration of the group of birational automorphism of V reflecting the structure of admissible equivalences is given in [**M2**].

One can apply the method from [M2] to the classes of universal equivalence. One can show that there exist classes of universal equivalence that are abstract cubics. One can consider universal equivalence on the set of points of such a class (considered as the abstract cubic). Applying this construction iteratively one can get a set of abstract cubics that corresponds to a filtration of subgroups in reflection groups. As in [M2] one can ask whether this sequence of subgroups stabilizes and what is its intersection.

5. Birationally trivial cubic surfaces: a finiteness theorem

5.1. Modified composition. — Let V be a smooth cubic surface, and $x, y \in V(k)$. Let $C \subset V$ be a curve on V passing through x, y, and $p : C \to \mathbf{P}^2$ an embedding of C into a projective plane such that $p(C)$ is again a cubic, and $p(x) \circ p(y)$ is defined in $p(C)$. We assume that C and p are defined over k. In this situation, following [M3], we will put

$$x \circ_{(C,p)} y := p^{-1}(p(x) \circ p(y)).$$

Example 1. Choose $C = $ a plane section of V containing x, y. If p is the embedding of C into the secant plane, then $x \circ_{(C,p)} y = x \circ y$ in the standard notation. Notice that the result does not depend on C if $x \neq y$. If $x = y$, then the choice of C determines a choice of one or two tangent lines to V at x so that the multivaluedness of $\circ$ is taken care of by the introduction of this new parameter.

Example 2. Assume now that V admits a birational morphism $p : V \to \mathbf{P}^2$ defined over k (e.g., V is split). We will choose and fix p once for all. Then any plane section C of V not containing one of the blown down lines as a component is embedded by p into $\mathbf{P}^2$ as a cubic curve. Therefore we can apply to (C, p) the previous construction. Notice that this time $x \circ_{(C,p)} y$ depends on C even if $x \neq y$.

Theorem 5.2. — *Assume that k is a finitely generated field. In the situation of Example 2, the complement to the blown down lines in $V(k)$ is finitely generated with respect to operations $\circ_{(C,p)}$ with the additional restriction:*

(C) the operation $x \circ_{(C,p)} y$ is applied only to the different previously constructed points.

Proof. — This theorem was stated and proved in [M3] without the additional condition (C). It uses the following auxiliary construction. Choose a k-rational line $l \subset \mathbf{P}^2$. Then $\Gamma := p^{-1}(l)$ is a twisted rational cubic in V. The family of all such cubics reflects properties of that of lines: a) any two different points a, b of $V(k)$ belong to a unique $\Gamma(a, b)$; b) any two different Γ's either have one common k–point, or intersect a common blown down line. The proof of this theorem is based on generation of points by adding intersections of lines l passing through pairs of previously constructed points in a projective plane. This induces generation of points on V that are intersections of $p^{-1}(l)$. Analysis of this proof in [**M3**] shows that it considers only *different* points in pairs of previously constructed points hereby providing the statement of the theorem with the condition (**C**). □

If one drops the condition (**C**) one can prove the stronger statement.

Theorem 5.3. — *Let V be a smooth cubic surface over an arbitrary field k. Assume that V admits a birational morphism $p : V \to \mathbf{P}^2$. Then the complement P to all blown down lines in $V(k)$ is generated by any single point from P (in the sense of the composition $\circ_{(C,p)}$).*

Proof. — Let us choose a point $x \in P$. The theorem will follow if we prove that the set of points $x \circ_{(C,p)} x$ contains P (here C runs through all k–rational plane sections of V passing through x). Let us show that for any other point y in P there exists such C that $y = x \circ_{(C,p)} x$. Indeed, following arguments of [**M3**], for $y \in P$ there exists a twisted cubic curve $G(x,y) := p^{-1}(l)$ where l is the line through $p(x), p(y)$ in $\mathbf{P}^2$. Let l_1 be the tangent line to $G(x,y)$ at x. Let a plane through points x, y and l_1 cut a curve C on V. Then l_1 is a tangent line to C at x, i.e. $G(x,y)$ is tangent to C at x. Hence l in $\mathbf{P}^2$ is tangent to $p(C)$ at $p(x)$. Since this line l passes through $p(y)$, on $p(C)$ we have $p(y) = p(x) \circ p(x)$. This gives $y \in x \circ_{(C,p)} x$ proving the statement. □

One can apply this theorem to the proof of the triviality of the 3–component of the universal equivalence on $P = V(k)$. 3–component of the universal equivalence can be defined as the finest admissible equivalence U_3 for which the following condition holds:

For any class $X \in P/U_3, X \circ X = X$.

Similarly one can define the 2–component of the universal equivalence as the finest admissible equivalence for which the following condition holds:

For any class $X \in P/U_2, X \circ X = O$ for some fixed class $O \in P$.

It follows from [**M1**] that $U = U_3 \cap U_2$, where U denotes the universal equivalence.

Corollary 5.4. — *Let V be a smooth cubic surface over an arbitrary field k. Assume that V admits a birational morphism $p : V \to P^2$. Then U_3 is trivial on $V(k)$.*

The corollary can be deduced from the following two lemmas.

Lemma 5.5. — *Let C be a smooth plane cubic curve defined over a field k such that $C(k)$ is non-empty. Let p be another plane embedding of C over k. Then*
$$x \circ_{(C,p)} y := p^{-1}(p(x) \circ p(y)) = t^{-1}((t(x) \circ t(y))$$
where $t \in \operatorname{Bir} C$ is some birational automorphism of C over k which can be represented as a product of reflections of C defined over k.

Proof. — The statement easily follows from the following fact: p can be decomposed into a product of reflections of C over k and a projective isomorphism of C and $p(C)$. Indeed, let us choose a point $0 \in C(k)$. Isomorphism classes of invertible sheaves of degree 3 are parametrized by the jacobian of C of degree 3, say, T, and T is a principal homogeneous space over C. This means that $C(k)$ acts transitively on $T(k)$, i.e. any two sheaves L_1, L_2 differ by a translation by a point $a \in C(k)$. Any translation is a product of two reflections, whereas a projective isomorphism preserves collinearity. □

Lemma 5.6. — *In the same notation, for any two points $x, y \in C(k)$ the following holds:*
$$t^{-1}(t(x) \circ t(y)) \sim x \circ y \mod U_3.$$

Proof. — Let $t = t_{x_1} \ldots t_{x_n}$ where $x_i \in C(k)$. It is enough to check the statement for $n = 1$ since the general statement can be obtained by induction. Let $t = t_z$. We have: $t^{-1}(t(x) \circ t(y)) = t_z(t_z(x) \circ t_z(y)) = z \circ ((z \circ x) \circ (z \circ y)) = z \circ ((z \circ z) \circ (x \circ y)) \sim z \circ (z \circ (x \circ y)) \mod U_3 \sim x \circ y \mod U_3$. Here we used $z \circ z \sim z \mod U_3$. □

We can now deduce the Corollary 5.4

Fix some $x \in P$, where P is the complement to all blown down lines in $V(k)$. By the Theorem 5.3, any point $z \in P$ can be represented as $x \circ_{(C,p)} x$. Let $z = x \circ_{C,p} x$ for some C. If C is singular then all points on $C(k)$ are equivalent mod U_3 (this is a general property of any singular plane cubic curve that does not have a line as a component). Otherwise, by lemmas 5.5 and 5.6
$$z = x \circ_{(C,p)} x \sim x \circ x \mod U_3 \sim x \mod U_3.$$

5.7. Elimination of $\circ_{(C,p)}$. — The use of the modified operation $\circ_{(C,p)}$ is somewhat annoying, and we would like to replace it by the standard composition $\circ$. For example, in the setup of the Theorem 5.2 for any three points x, y, z on a plane smooth section $C \subset V$ the following equality holds:
$$(x \circ_{(C,p)} y) \circ_{(C,p)} z = (x \circ y) \circ z.$$

This naturally leads to the question whether one can obtain the traditional Mordell–Weil statement for the composition ∘ using our finiteness results for $\circ_{(C,p)}$ and some tricks like the formula above.

The remaining part of the paper is dedicated to the description of our, not altogether successful, attempts to eliminate $\circ_{(C,p)}$. We reformulate the finiteness theorem above in terms that do not use explicitly compositions $\circ_{(C,p)}$ and a morphism p of a cubic surface into a projective plane. We only use the standard operation ∘ and implicitly use some intersections of planes with lines that belong to this cubic surface.

Before we can state a new statement we need to define a new kind of operation on a cubic surface that involves lines belonging to this cubic surface.

Definition 5.7.1. — Let V be a smooth cubic surface over an arbitrary field k. Let $\Lambda = \{l_1, l_2, m\}$ be three (not necessary k-rational) lines belonging to V and such that the following properties hold:

(**A**) l_1 and l_2 are skew lines (i.e. they do not have a common point) and m intersects l_1 and l_2.

Given a triple of lines Λ satisfying (**A**) and an arbitrary plane T not containing lines in V, let us define a new composition of points u, and w on $T \cap V$ as follows:

(**B**) $u \circ_{(T,\Lambda)} w = (x \circ y) \circ [z \circ (u \circ w)]$,
where $x = l_1 \cap T$, $y = l_2 \cap T$ and $z = m \cap T$.

Of course, the point $u \circ_{(T,\Lambda)} w$ is not necessarily k–rational even u, w, and T are k–rational. But there is a special case when the composition $\circ_{(T,\Lambda)}$ produces rational points (over k) when u, w, and T are defined over k (whereas lines in Λ are not necessarily defined over k). This case is described in the following statement that reformulates the Theorem 5.2 in terms of the composition $\circ_{(T,\Lambda)}$.

Theorem 5.7.2. — *Let V be a smooth cubic surface. Assume that V admits a birational morphism to a projective plane defined over k. Assume that k is finitely generated field. Then there exists a triplet of lines on V satisfying the property* (**A**) *such that the following statement holds: the complement to the blown down lines in $V(k)$ is finitely generated with respect to operations $\circ_{(T,\Lambda)}$ with the additional restriction:*

(**D**) *the operation $x \circ_{(T,\Lambda)} y$ is applied only to different previously constructed points. (Here Λ is fixed and T runs through some set of k-rational planes).*

Similarly, one can reformulate Theorem 5.3 in terms of new operations.

Theorem 5.7.3. — *Let V be a smooth cubic surface over an arbitrary field k. Assume that V admits a birational morphism to a projective plane defined* over k. Then the complement P to all blown down lines in $V(k)$ is generated by any single point from P in the sense of compositions $\circ_{(T,\Lambda)}$ for some fixed triple of lines Λ in V.

Below we will show how to replace operations $\circ_{(C,p)}$ by operations $\circ_{(T,\Lambda)}$.

Lemma 5.7.4. — *Let V be a smooth cubic surface defined over a field k and $\bar{k}$ be an algebraic closure of k. Let $p : V \to \mathbf{P}^2$ be a birational morphism over $\bar{k}$. Then there exists a triplet of lines Λ satisfying the property (**A**) such that for any plane section C of V not containing one of the blown down lines as a component and for any two points $u, w \in V(\bar{k})$ lying on C the following holds: $u \circ_{(C,p)} w = u \circ_{(T,\Lambda)} w$ where T is a plane that cuts the curve C on V.*

Corollary 5.7.5. — *Assume that the birational morphism p in Lemma 5.7.4 is defined over k. Then a triplet Λ can be chosen in such a way that the point $u \circ_{(T,\Lambda)} w$ is k-rational if u, w and the plane T are k-rational.*

The proof of Lemma 5.7.4 is a consequence of the following claims which might be of independent interest.

Claim 5.7.6. — *In the conditions of Lemma 5.7.4, let x, y, u, w be points on C. Then the following equality holds:*

$$u \circ_{(C,p)} w = (x \circ y) \circ [(x \circ_{(C,p)} y) \circ (u \circ w)].$$

In other words, if we know how to compute $z = x \circ_{(C,p)} y$ at least for some two points x, y in C then operation $\circ_{(C,p)}$ for all other points in C can be computed in terms of $\circ$ only.

Claim 5.7.7. — *In the conditions of Lemma 5.7.4, let $\Lambda = \{l_1, l_2, m\}$ be a triplet of lines satisfying (**A**) and such that $p(m)$ is a line on the plane $\mathbf{P}^2$, and l_1, l_2 are blown down lines. Let $x = l_1 \cap T$, $y = l_2 \cap T$ and $z = m \cap T$, where the plane T cuts a curve C on V. Then $z = x \circ_{(C,p)} y$.*

In other words, one can easily compute an operation $\circ_{(C,p)}$ for intersection of lines l_1 and l_2 with a plane T. The result of this composition is an intersection of a third line m with T !

To show that the Lemma 5.7.4 follows from these claims, it is sufficient to note the following. By Claim 5.7.6, the operation $u \circ_{(C,p)} w$ can be replaced by $(x \circ y) \circ [(x \circ_{(C,p)} y) \circ (u \circ w)]$ where x, y are any points on C. There exists a triplet of lines Λ on V satisfying (**A**), such that $p(m)$ is a line on the plane $\mathbf{P}^2$, and l_1, l_2 are the blown down lines. By the Claim 5.7.7, x, y can be chosen as intersections of lines l_1, l_2 with a plane T that cuts C on V and in this case $x \circ_{(C,p)} y = m \cap T$.

Now we prove our Claims.

Proof of the Claim 5.7.6. — *Step 1:* Since C and p are fixed, one can simplify our notation by putting $x * y =: x \circ_{(C,p)} y$. In this step we show that for any points x, y, u, w on C the following equality holds:

(3) $$u * w = (x * y)(x \circ y)^{-1}(u \circ w),$$

where the expressions in brackets are multiplied by using an Abelian structure on C: $xy = a \circ (x \circ y)$ for some point a in $C(k)$.

First, we consider the case when C is smooth. In this case by the Lemma 5.5 p in the formula $p(p^{-1}(u) \circ p^{-1}(w))$ can be replaced by a product of reflections of C. Let us check (3) for the case when p can be replaced by one reflection t_b:

$$u * w = p(p^{-1}(u) \circ p^{-1}(w)) = b \circ ((b \circ u) \circ (b \circ w)) = b \circ ((b \circ b) \circ (u \circ w)).$$

The general case can be obtained by iterating this argument.

Using the identity $u \circ w = (a \circ a) u^{-1} w^{-1}$ we get:

$$u * w = b \circ ((b \circ b) \circ (u \circ w)) = b^{-1}(b \circ b)(u \circ w)$$

Similarly we have for other two points: $x * y = b^{-1}(b \circ b)(x \circ y)$. Replacing $b^{-1}(b \circ b)$ with $(x * y)(x \circ y)^{-1}$ in $b \circ ((b \circ b) \circ (u \circ w))$ gives (3).

Step 2: Replacing the Abelian multiplication operation in (3) by $a \circ (\ldots)$ we can rewrite (3) as:

$$u * w = a \circ (r \circ (u \circ w)),$$

where $r = a \circ \{(x * y) \circ [(a \circ a) \circ (x \circ y)]\}$. Since the point a is arbitrary, one can choose $a = x \circ y$. This gives $r = x * y$ and immediately implies the formula in the Claim 5.7.6.

In order to complete the proof of the Claim we need to consider the case when C is a singular plane cubic curve that does not contain a line. This can be done by appealing to an obvious limiting construction in the case of topological field k, or to a similar argument using the Zariski topology in general. □

Proof of the Claim 5.7.7. — Since l_1, l_2 are blown down lines and $p(m)$ is a line in $\mathbf{P}^2$, the points $p(x), p(y), p(z)$ are intersections of the line $p(m)$ with the curve $p(C)$ in $\mathbf{P}^2$. This means that on $p(C)$ we have $p(x) \circ p(y) = p(z)$. This is equivalent to the equality $z = x \circ_{(C,p)} y$ in the Claim. □

References

[K1] D. S. Kanevski, *Structure of groups, related to cubic surfaces*, Mat. Sb. **103**:2, (1977), 292–308 (in Russian); English. transl. in Mat. USSR Sbornik, Vol. **32**:2 (1977), 252–264.

[K2]   D. S. Kanevsky, *On cubic planes and groups connected with cubic surfaces*, J. Algebra **80:2** (1983), 559–565.

[M1]   Yu. I. Manin, *Cubic Forms: Algebra, Geometry, Arithmetic*, North Holland, 1974 and 1986.

[M2]   Yu. I. Manin, *On some groups related to cubic surfaces*, In: Algebraic Geometry. Tata Press, Bombay, 1968, 255–263.

[M3]   Yu. I. Manin, *Mordell-Weil problem for cubic surfaces*, in: Advances in the Mathematical Sciences—CRM's 25 Years (L. Vinet, ed.) CRM Proc. and Lecture Notes, vol. 11, Amer. Math. Soc., Providence, RI, 1997, 313–318.

[P]   S. J. Pride, *Involutary presentations, with applications to Coxeter groups, NEC-Groups, and groups of Kanevsky*, J. of Algebra **120** (1989), 200–223.

[Sw–D] H. P. F. Swinnerton–Dyer, *Universal equivalence for cubic surfaces over finite and local fields*, Symp. Math., Bologna **24** (1981), 111–143.

Rational points on algebraic varieties
(E. PEYRE, Y. TSCHINKEL, ed.), p. 221–274

TORSEURS UNIVERSELS
ET MÉTHODE DU CERCLE

Emmanuel Peyre

Institut Fourier, UFR de Mathématiques, UMR 5582, Université de Grenoble I et CNRS, BP 74, 38402 Saint-Martin d'Hères CEDEX, France
Url : http://www-fourier.ujf-grenoble.fr/~peyre
E-mail : Emmanuel.Peyre@ujf-grenoble.fr

Résumé. — Ce texte décrit les premières étapes d'une généralisation de la méthode du cercle au cas d'une hypersurface lisse dans une variété presque de Fano.

En effet, sous certaines conditions, il est possible d'exprimer dans ce cas les deux membres d'une version raffinée de la conjecture de Manin sur le comportement asymptotique du nombre de points de hauteur bornée de l'hypersurface en termes des torseurs universels de la variété ambiante qui jouent, dans ce cadre, le rôle de l'espace affine.

Introduction

L'objet de ce texte est le comportement asymptotique du nombre de points de hauteur bornée sur des variétés dont le faisceau anticanonique vérifie certaines conditions de positivité.

De nombreux progrès ont été réalisés dans la compréhension de ce comportement asymptotique. Une interprétation géométrique de la puissance et de la puissance du logarithme qui interviennent a été proposée dans les articles de Franke, Manin et Tschinkel [**FMT**] et de Batyrev et Manin [**BM**]. Des descriptions adéliques de la constante ont été proposées lorsque la hauteur est

---

Classification mathématique par sujets (2000). — primaire 14G05.

associée au faisceau anticanonique dans [**Pe1**] puis, dans un cadre plus général, par Batyrev et Tschinkel dans [**BT4**].

Plusieurs stratégies ont été développées pour attaquer ces conjectures.

Une première famille de méthodes est basée sur des techniques d'analyse harmonique fine qui s'appliquent notamment lorsque la variété est équipée d'une action non triviale d'un groupe algébrique. Parmi les cas traités par ce type de méthodes, on peut citer celui des variétés de drapeaux généralisées étudiées dans [**FMT**] et [**Pe1**] à l'aide des travaux de Langlands sur les séries d'Eisenstein, le cas des variétés toriques considéré par Batyrev et Tschinkel dans [**BT1**], [**BT2**] et [**BT3**] et celui des fibrations en variétés toriques au-dessus de variétés de drapeaux généralisées par Strauch et Tschinkel [**ST**], ainsi que diverses compactifications de l'espace affine dues à Chambert-Loir et Tschinkel [**CLT1**], [**CLT2**].

Parallèlement des techniques de descente ont été mises au point dans ce cadre. Elles apparaissent de manière implicite dans le cas des intersections complètes lisses dans $\mathbf{P}^n$ et dans l'étude de quelques variétés toriques (cf. [**Pe1**] et [**Ro**]). Salberger les a rendues explicites dans [**Sa**], redémontrant ainsi en partie les résultats de Batyrev et Tschinkel sur les variétés toriques. La méthode introduite par Salberger fut ensuite exploitée par de la Bretèche qui put, à l'aide d'outils d'analyse complexe, améliorer les estimations pour les variétés toriques [**Bre**].

Une autre famille de méthodes, issue de la méthode du cercle, qui a depuis longtemps prouvée son efficacité pour les intersections complètes dans l'espace projectif, a été tout récemment utilisée par Robbiani pour l'étude d'un cas sortant de ce cadre, à savoir celui d'une hypersurface dans $\mathbf{P}^m \times \mathbf{P}^m$ définie par l'annulation d'une section de $\mathscr{O}(1,1)$. Bien que la variété considérée par Robbiani soit une variété de drapeaux pour laquelle la conjecture de Manin avait été démontrée, le fait qu'il ait étendu la méthode du cercle à ce cas laisse espérer que celle-ci puisse également s'appliquer à des cas où le rang du groupe de Picard n'est pas égal à un.

Le but de ce texte est d'étendre à un cadre plus général quelques étapes de la méthode du cercle en exploitant un principe de descente présenté dans [**Pe2**]. Il reste toutefois un important et difficile travail à faire concernant le cœur même de la méthode du cercle, à savoir la majoration de sommes d'exponentielle.

Le paragraphe 1 de ce texte rappelle la description conjecturale du comportement asymptotique du nombre de points de hauteur bornée. Le troisième a pour objet le passage aux torseurs universels au niveau desquels le problème se décrit naturellement comme passage d'une somme à une intégrale. Dans le quatrième nous décrivons comment, dans le cas d'une hypersurface vérifiant certaines conditions, on peut passer du torseur universel de la variété ambiante à celui de la sous-variété à l'aide de formules inspirées de la formule d'inversion de Fourier.

1. Une version raffinée d'une conjecture de Manin

1.1. Variétés presque de Fano. — Nous utiliserons dans ce texte les notations suivantes :

Notations 1.1.1. — Si $\mathscr{X}$ est un schéma sur un anneau commutatif A et B une A-algèbre commutative, $\mathscr{X}(B)$ désigne l'ensemble $\mathrm{Hom}_{\mathrm{Spec}\,A}(\mathrm{Spec}\,B, \mathscr{X})$ et $\mathscr{X}_B$ le produit de schémas $\mathscr{X} \times_{\mathrm{Spec}\,A} \mathrm{Spec}\,B$. Si C est un monoïde, alors $A[C]$ désigne la A-algèbre associée. Si X est une variété lisse sur un corps E, son groupe de Picard est noté $\mathrm{Pic}\,X$, son groupe de Neron-Severi $\mathrm{NS}(X)$ et son faisceau canonique ω_X. On désigne par $C_{\mathrm{eff}}(X)$ le cône des classes de diviseurs effectifs dans $\mathrm{NS}(X) \otimes \mathbf{R}$. On note $\overline{E}$ une clôture algébrique de E et E^s sa clôture séparable dans $\overline{E}$. On pose alors $\overline{X} = X_{\overline{E}}$ et $X^s = X_{E^s}$.

Le dual d'un module M est noté $M^\vee$.

Définition 1.1.2. — Une variété V sur un corps k de caractéristique nulle sera dite presque de Fano si elle est projective, lisse et géométriquement intègre et si elle vérifie les conditions suivantes :
  (i) les groupes de cohomologie $H^i(V, \mathcal{O}_V)$ sont nuls pour $i = 1$ ou 2,
  (ii) le groupe de Néron-Severi géométrique, qui sous l'hypothèse (i) coïncide avec $\mathrm{Pic}\,\overline{V}$, est sans torsion,
  (iii) la classe $[\omega_V^{-1}]$ de ω_V^{-1} dans $\mathrm{NS}(V) \otimes \mathbf{R}$ appartient à l'intérieur du cône des diviseurs effectifs.

Exemple 1.1.3. — Si V est une variété de Fano, alors V est presque de Fano. En effet, par le théorème de Kodaira, la condition (i) est vérifiée, la condition (ii) résulte de [**Pe1**, lemme 1.2.1] et (iii) découle du fait que, par définition, ω_V^{-1} est ample.

Exemple 1.1.4. — Si V est une variété torique projective et lisse, alors par [**Da**, corollary 7.4], les groupes $H^i(V, \mathcal{O}_V)$ sont nuls pour $i > 0$, et par [**Oda**, lemma 2.3] tout fibré en droites a une base de sections équivariantes sous l'action du tore et donc le cône des diviseurs effectifs dans $\mathrm{Pic}\,\overline{V} \otimes \mathbf{R}$ est engendrée par les $[D]$ où D décrit l'ensemble des sous-variétés irréductibles invariantes de codimension 1 dans $\overline{V}$. La classe $[\omega_V^{-1}]$ étant la somme de ces $[D]$ par [**Oda**, page 70, example], il est à l'intérieur du cône et la condition (iii) est vérifiée. La variété V est donc presque de Fano.

Proposition 1.1.5. — *Soit X une compactification équivariante projective et lisse d'un tore T sur $\mathbf{C}$, $L_1, \ldots, L_m$ des faisceaux inversibles amples sur X et $s_1, \ldots, s_m$ des sections non nulles de ces faisceaux. On note $X_T^{(1)}$ l'ensemble des sous-variétés irréductibles invariantes de codimension un de X. On suppose que $\dim X \geqslant m + 3$, que*

$$\left[\sum_{D \in X_T^{(1)}} D - \sum_{i=1}^m L_i\right] \in \overset{\circ}{\overline{C_{\mathrm{eff}}(X)}},$$

que les hypersurfaces définies par les s_i se coupent transversalement, que leurs intersections successives sont connexes et qu'elles coupent proprement les diviseurs D de $X_T^{(1)}$.

Alors la sous-variété V définie par l'annulation des s_i est presque de Fano. En outre, la restriction induit un isomorphisme

$$\mathrm{Pic}\, X \xrightarrow{\sim} \mathrm{Pic}\, V$$

qui envoie $C_{\mathrm{eff}}(X)$ dans $C_{\mathrm{eff}}(V)$ et la classe de $\sum_{D \in X_T^{(1)}} D - \sum_{i=1}^m L_i$ sur celle de ω_V^{-1}.

Démonstration. — Nous allons démontrer par récurrence sur n que V vérifie les assertions de la proposition et que si $L = \sum_{j=1}^m \varepsilon_i L_j$ avec $\varepsilon_j \in \mathbf{Z}$ et $\varepsilon_j \leqslant 0$ pour $1 \leqslant j \leqslant m$, alors le groupe $H^i(V, L)$ est nul si $0 < i < \dim V$.

Si $m = 0$, l'énoncé de la proposition résulte de l'exemple précédent, l'assertion de nullité pour $\mathcal{O}_V$ résulte de [**Da**, corollary 7.4] et celle pour les sommes de fibrés L_i de [**Da**, theorem 7.5.2] et du théorème de dualité de Serre (cf. [**Ha**, corollary III.7.7]).

Supposons le résultat démontré pour $m-1$ et soit V' la sous-variété de X définie par l'annulation de $s_1, \ldots s_{m-1}$. La variété V' vérifie alors les assertions ci-dessus.

La variété V est alors définie dans V' comme lieu des zéros de s_m. Par l'hypothèse de transversalité, V est lisse et étant connexe, elle est intègre. Par définition elle est projective. Par ailleurs, on a une suite exacte de faisceaux de Zariski sur V'

(1.1) $$0 \to L_m^{-1} \otimes \mathcal{O}_{V'} \to \mathcal{O}_{V'} \to \mathcal{O}_V \to 0.$$

D'où une suite exacte longue de cohomologie (cf. [**Ha**, lemma III.2.10])

$$H^i(V', L_m^{-1}) \to H^i(V', \mathcal{O}_{V'}) \to H^i(V, \mathcal{O}_V) \to H^{i+1}(V', L_m^{-1}).$$

On obtient donc que $H^i(V, \mathcal{O}_V)$ est nul pour $0 < i < \dim V = \dim V' - 1$. Comme $\dim V \geqslant 3$, cela entraîne l'assertion (i) de la définition. De même,

on obtient l'annulation des groupes de cohomologie de l'hypothèse de récurrence. Par le théorème de Lefschetz classique [**Bo**, corollary, page 212] on a un isomorphisme :
$$H^2(V(\mathbf{C}), \mathbf{Z}) \xrightarrow{\sim} H^2(V'(\mathbf{C}), \mathbf{Z}).$$
En utilisant la suite exacte de faisceaux analytiques
$$0 \to \mathbf{Z} \to \mathcal{O}_V \xrightarrow{\exp} \mathcal{O}_V{}^\times \to 0$$
et des théorèmes de comparaison entre géométrie algébrique et géométrie algébrique, on obtient un diagramme commutatif

$$\begin{array}{ccccc} 0 \to & H^1(V', \mathcal{O}_{V'}^\times) & \xrightarrow{\sim} & H^2(V'(\mathbf{C}), \mathbf{Z}) & \to 0 \\ & \downarrow & & \downarrow \wr & \\ 0 \to & H^1(V, \mathcal{O}_V{}^\times) & \xrightarrow{\sim} & H^2(V(\mathbf{C}), \mathbf{Z}) & \to 0 \end{array}$$

et on obtient que la restriction de Pic V' à Pic V est un isomorphisme.

Le cône des classes de diviseurs effectifs de X étant engendré par les classes des diviseurs D de $X_T^{(1)}$, l'assertion sur les cônes effectifs résulte de l'hypothèse sur la propreté des intersections avec ces diviseurs.

Enfin $[\omega_{V'}^{-1}] = \sum_{D \in X_T^{(1)}} [D] - \sum_{i=1}^{m-1} [L_i]$ et l'assertion correspondante pour V résulte de [**Ha**, proposition II.8.20]. □

Remarque 1.1.6. — A priori le cône des diviseurs effectifs de V pourrait être plus grand que celui de X. Toutefois, si X est de la forme $\prod_{i=1}^t \mathbf{P}_\mathbf{C}^{n_i}$ et si $m < \inf_{1 \leqslant i \leqslant t} n_i$, alors il y a égalité entre les cônes de diviseurs effectifs. En effet la formule de Künneth implique que
$$\forall L \in \operatorname{Pic} X, \quad H^i(X, L) = 0 \quad \text{si} \quad 0 < i < \inf_{1 \leqslant i \leqslant t} n_i.$$
On obtient alors par récurrence sur m que
$$\forall L \in \operatorname{Pic} V, \quad H^i(V, L) = 0 \quad \text{si} \quad 0 < i < \inf_{1 \leqslant i \leqslant t} n_i - m.$$
et la suite exacte (1.1) tensorisée par L fournit une suite exacte
$$H^0(V', L) \to H^0(V, L) \to H^1(V', L \otimes L_m^{-1})$$
ce qui implique que les deux cônes coïncident.

1.2. Hauteurs d'Arakelov. — La donnée naturelle pour construire des fonctions de comptage sur l'ensemble des points rationnels de variétés propres est une hauteur d'Arakelov dont nous allons rappeler la définition.

Notations 1.2.1. — Dans la suite, k désigne un corps de nombres, $\mathcal{O}_k$ son anneau des entiers, d son discriminant, M_k l'ensemble de ses places, M_f celui de ses places finies et M_∞ celui de ses places archimédiennes. Pour toute place v de k, on note k_v le complété correspondant et $|.|_v$ la norme sur k_v normalisée par

$$\forall v | p, \quad \forall x \in k_v, \quad |x|_v = \left| N_{k_v/\mathbf{Q}_p}(x) \right|_p.$$

Si v est une place finie, $\mathcal{O}_v$ est l'anneau des entiers de k_v et $\mathbf{F}_v$ le corps résiduel.

Définition 1.2.2. — Soit V une variété projective lisse et géométriquement intègre sur k, L un faisceau inversible sur V. Si v est une place de k, une *métrique v-adique* sur L est une application associant à un point x de $V(k_v)$ une norme $\|.\|_v$ sur $L(x) = L_x \otimes_{\mathcal{O}_{V,x}} k_v$ de sorte que pour toute section s de L définie sur un ouvert W de V l'application

$$x \mapsto \|\mathrm{s}(x)\|_v$$

soit continue pour la topologie v-adique.

Si v est un place finie de k, $\mathscr{V}$ un modèle projectif et lisse de V sur $\mathcal{O}_v$ et $\mathscr{L}$ un modèle de L, alors on peut lui associer une métrique v-adique sur L de la manière suivante : tout point x de $V(k_v)$ définit un point $\tilde{x}$ de $\mathscr{V}(\mathcal{O}_v)$ et $\tilde{x}^*(\mathscr{L})$ fournit une $\mathcal{O}_v$-structure sur $L(x)$ dont on peut choisir un générateur y_0 ; la norme d'un élément y de L est alors donnée par la formule

$$\|y\|_v = \left| \frac{y}{y_0} \right|_v.$$

Une *métrique adélique* sur L est une famille de métriques $(\|.\|_v)_{v \in M_k}$ telle qu'il existe un ensemble fini de places finies S, un modèle projectif et lisse $\mathscr{V}$ de V sur l'anneau $\mathcal{O}_S$ des S-entiers et un modèle $\mathscr{L}$ de L sur cet anneau tel que pour tout v de $M_f - S$, $\|.\|_v$ soit la métrique définie par $\mathscr{L} \otimes_{\mathcal{O}_{\mathscr{V}}} \mathcal{O}_v$.

Nous appellerons *hauteur d'Arakelov* sur V la donné d'une paire

$$\mathbf{h} = (L, (\|.\|_v)_{v \in M_k})$$

où L est un faisceau inversible sur V et $(\|.\|_v)_{v \in M_k}$ une métrique adélique sur ce fibré.

Pour toute hauteur $\mathbf{h}$ sur V et tout point rationnel x de V, la *hauteur de x relativement à* $\mathbf{h}$ est définie par

$$\forall y \in L(x), \quad \mathbf{h}(x) = \prod_{v \in M_k} \|y\|_v^{-1}.$$

Remarque 1.2.3. — La formule du produit assure que le produit ci-dessus est indépendant de y.

Rappelons quelques exemples de hauteurs (cf. également [**Sa**, exemples 1.7]).

Exemple 1.2.4. — Si $\mathbf{h}_i = (L_i, (\|.\|_v^i)_{v \in M_k})$ pour $i = 1$ ou 2 sont deux hauteurs d'Arakelov, leur produit tensoriel $\mathbf{h}_1 \otimes \mathbf{h}_2$ est $(L_1 \otimes L_2, (\|.\|_v)_{v \in M_k})$ où

$$\forall v \in M_k, \ \forall x \in V(k_v), \ \forall y \in L_1(x), \ \forall z \in L_2(x), \ \|y \otimes z\|_v = \|y\|_v^1 \|z\|_v^2.$$

On en déduit immédiatement l'égalité

$$\forall x \in V(k), \quad \mathbf{h}_1 \otimes \mathbf{h}_2(x) = \mathbf{h}_1(x)\mathbf{h}_2(x).$$

Exemple 1.2.5. — Donnons-nous une hauteur $\mathbf{h} = (L, (\|.\|_v)_{v \in M_k})$ sur V et une famille de fonctions $\boldsymbol{f} = (f_v)_{v \in M_k}$ strictement positives sur $V(k_v)$ telle que pour presque toute place v de k la fonction f_v soit constante et égale à 1, alors

$$\boldsymbol{f}.\mathbf{h} = (L, (f_v\|.\|_v)_{v \in M_k})$$

est une hauteur sur V. Réciproquement, si $\mathbf{h}' = (L, (\|.\|_v')_{v \in M_k})$ est une autre hauteur sur V relative au même faisceau, alors pour toute place v de k le quotient $\|.\|_v'/\|.\|_v$ définit une fonction f_v sur $V(k_v)$ qui, pour presque toute place, est constante et égale à 1. On a bien sûr $\mathbf{h}' = \boldsymbol{f}.\mathbf{h}$.

Exemple 1.2.6. — Si $\phi : V \to W$ est un morphisme de variétés projectives lisses et géométriquement intègres et $\mathbf{h} = (L, (\|.\|_v)_{v \in M_k})$ une hauteur sur W alors $\phi^*(\mathbf{h})$ est la hauteur $(\phi^*L, (\|\phi(.)\|_v)_{v \in M_k})$ où l'on note également ϕ l'application induite $\phi^*L(x) \to L(\phi(x))$ pour tout x de V.

En particulier si L est un faisceau inversible très ample, il définit un morphisme

$$\phi : V \to \mathbf{P}(\Gamma(V, L)^\vee)$$

de sorte que $L = \phi^*(\mathcal{O}(1))$ et tout système de métriques sur $\mathcal{O}(1)$ induit une hauteur sur V.

Exemple 1.2.7. — Si K/k est une extension de corps de nombres, V une variété projective lisse et géométriquement intègre sur k, L un faisceau inversible sur V et $\mathbf{h}_K = (L \otimes K, (\|.\|_v)_{v \in M_K})$ une hauteur sur V_K, alors la hauteur induite $\mathbf{h} = (L, (\|.\|'_v)_{v \in M_k})$ est définie par

$$\forall \mathfrak{p} \in M_k, \quad \forall x \in V(k_\mathfrak{p}), \quad \forall y \in L(x), \quad \|y\|'_\mathfrak{p} = \left(\prod_{\mathfrak{P}|\mathfrak{p}} \|y_\mathfrak{P}\|_\mathfrak{P}\right)^{[K:k]^{-1}}.$$

Cela permet également d'associer à tout hauteur $\mathbf{h}$ relative à un faisceau inversible L sur V_K une hauteur $N_{K/k}\mathbf{h}$ relative au faisceau $N_{K/k}L$ sur V.

Exemple 1.2.8. — Soit $\mathscr{V}$ un schéma plat projectif et régulier sur $\mathcal{O}_k$ et $(\mathscr{L}, h)$ un fibré en droites hermitien sur $\mathscr{V}$ (cf. [**BGS**, §2.1.2]), $\mathscr{L}$ désigne donc un fibré inversible sur $\mathscr{V}$ et h une forme hermitienne C^∞ sur le fibré en droites holomorphe $L_\mathbf{C}$ sur $\sqcup_{\sigma:k \to \mathbf{C}} \mathscr{V}_\sigma(\mathbf{C})$ invariante sous l'action de la conjugaison. Cette forme correspond donc à des formes hermitiennes C^∞ sur $\mathscr{V}_\sigma$ que l'on notera h_σ. On a en outre que $h_{\overline{\sigma}}$ est la conjuguée de h_σ.

Pour toute place finie v de k, $\mathscr{L}$ induit comme ci-dessus une métrique $\|.\|_v$ sur $L = \mathscr{L} \otimes k$ et pour toute place archimédienne v de k, on a un plongement σ de k dans $\mathbf{C}$ et la forme hermitienne h_σ définit une métrique v-adique $\|.\|_v$ sur L. Par définition, $\mathbf{h} = (L, (\|.\|_v)_{v \in M_k})$ est une hauteur sur V, et la hauteur d'un point rationnel est donné par la formule

$$\forall x \in V(k), \quad \mathbf{h}(x) = \exp(\widehat{\deg}(\hat{c}_1(\mathscr{L})|\overline{x}))$$

où $\overline{x}$ est l'adhérence de x dans $\mathscr{V}$, $\hat{c}_1(\mathscr{L})$ le caractère de Chern arithmétique de $\mathscr{L}$ (cf. [**BGS**, page 932]), $(.|.)$ l'accouplement

$$\widehat{\mathrm{CH}}^*(\mathscr{V}) \times Z_*(\mathscr{V}) \to \widehat{\mathrm{CH}}^*(\mathrm{Spec}\,\mathcal{O}_k)_\mathbf{Q}$$

défini par Bost, Gillet et Soulé (cf. [**BGS**, §2.3]) et $\widehat{\deg}$ l'application degré sur le groupe de Chow arithmétique $\widehat{\mathrm{CH}}^*(\mathrm{Spec}\,\mathcal{O}_k)$.

En effet par [**BGS**, §3.1.2.1 et (2.1.15)],

$$\widehat{\deg}(\hat{c}_1(\mathscr{L})|\overline{x}) = \widehat{\deg}(\tilde{x}^*(\mathscr{L}))$$
$$= \log(\#(\tilde{x}^*\mathscr{L}/\mathcal{O}_k y)) - \sum_{\sigma: k \to \mathbf{C}} \log h_\sigma(y,y)^{1/2}$$

où $\tilde{x} : \mathrm{Spec}\,\mathcal{O}_k \to \mathscr{V}$ est définie par x et y un élément de $\tilde{x}^*(\mathscr{L}) \subset L(x)$. En suivant les définition on obtient

$$\widehat{\deg}(\hat{c}_1(\mathscr{L})|\overline{x}) = -\sum_{v \in M_k} \log \|y\|_v.$$

Définition 1.2.9. — On note $\mathscr{H}(V)$ l'ensemble des classes d'isomorphismes de hauteurs d'Arakelov quotienté par la relation d'équivalence définie par

$$(L, (\|.\|_v)_{v \in M_k}) \sim (L, (\lambda_v\|.\|_v)_{v \in M_k})$$

pour toute famille de réels $(\lambda_v)_{v \in M_k} \in \bigoplus_{v \in M_k} \mathbf{R}_{>0}$ telle que $\prod_{v \in M_k} \lambda_v = 1$.

L'ensemble $\mathscr{H}(V)$ est un groupe pour le produit tensoriel des hauteurs, il est muni d'une structure de $\mathbf{R}_{>0}$-ensemble donnée par

$$\lambda.(L, (\|.\|_v)_{v \in M_k}) = (L, (\lambda_v \|.\|_v)_{v \in M_k})$$

si $(\lambda_v)_{v \in M_k} \in \bigoplus_{v \in M_k} \mathbf{R}_{>0}$ vérifie $\prod_{v \in M_k} \lambda_v = \lambda$. On dispose d'un morphisme d'oubli $o : \mathscr{H}(V) \to \operatorname{Pic} V$. Si $\phi : V \to W$ est un morphimes de variétés projectives, lisses et géométriquement intègres sur k, alors ϕ^* définit un morphisme $\mathscr{H}(W) \to \mathscr{H}(V)$ qui s'insère dans un diagramme commutatif :

$$\begin{array}{ccc} \mathscr{H}(W) & \longrightarrow & \mathscr{H}(V) \\ \downarrow & & \downarrow \\ \operatorname{Pic}(W) & \longrightarrow & \operatorname{Pic}(V). \end{array}$$

Enfin si K/k est une extension de corps de nombres on dispose d'un morphisme de norme

$$N_{K/k} : \mathscr{H}(V_K) \to \mathscr{H}(V).$$

Remarque 1.2.10. — Si x est un point rationnel et $\mathbf{h}$ une hauteur, $\mathbf{h}(x)$ ne dépend que de la classe de $\mathbf{h}$ dans $\mathscr{H}(V)$. On notera ev_x le morphisme $\mathscr{H}(V) \to \mathbf{R}_{>0}$ obtenu.

Exemple 1.2.11. — Si $V = \operatorname{Spec} k$, alors une hauteur d'Arakelov est la donnée d'un espace vectoriel L de dimension un sur k et d'une famille de normes $(\|.\|_v)_{v \in M_k}$ sur L telle qu'il existe une $\mathcal{O}_k$-structure de $\mathscr{L}$ de L de sorte que pour tout place finie v de k en-dehors d'un ensemble fini S, on ait

$$\forall y \in \mathscr{L}, \quad \|y\|_v = (\#(\mathcal{O}_v y / \mathscr{L} \otimes \mathcal{O}_v))^{-1}.$$

Cette description explicite montre que le morphisme $\operatorname{ev}_{\operatorname{Spec} k}$ est un isomorphisme.

Notons qu'en outre on a pour toute variété V projective lisse et géométriquement intègre sur k et tout point x de $V(k)$ un diagramme commutatif.

$$\begin{array}{ccc} \mathscr{H}(V) & \xrightarrow{\operatorname{ev}_x} & \mathbf{R}_{>0} \\ \downarrow x^* & & \| \\ \mathscr{H}(\operatorname{Spec} k) & \xrightarrow{\sim} & \mathbf{R}_{>0}. \end{array}$$

Définition 1.2.12. — On appelle système de hauteurs une section de l'application composée

$$\mathscr{H}(V) \xrightarrow{o} \operatorname{Pic} V \to \operatorname{NS}(V).$$

Un système de hauteurs **H** sur V induit un accouplement
$$\mathbf{H} : \mathrm{NS}(V) \otimes \mathbf{C} \times V(k) \to \mathbf{C}$$
qui est l'exponentielle d'une fonction linéaire en la première variable et telle que
$$\forall L \in \mathrm{NS}(V), \quad \forall x \in V(k), \quad \mathbf{H}(L, x) = \mathbf{H}(L)(x).$$

Comme l'ont souligné Batyrev et Manin [**BM**], l'existence de sous-variétés accumulatrices susceptibles d'occulter certains phénomènes globaux dans le comportement asymptotique du nombre de points de hauteur bornée amène à se restreindre à un ouvert non vide assez petit de la variété. On utilisera donc la définition qui suit.

Définition 1.2.13. — Soit V une variété projective, lisse et géométriquement intègre sur k et W un sous-espace localement fermé de V. Alors pour toute hauteur **h** sur V et tout nombre réel H strictement positif
$$n_{W,\mathbf{h}}(H) = \#\{x \in W(k) \mid \mathbf{h}(x) \leqslant H\}.$$

Si **H** est un système de hauteurs sur V alors la fonction zêta associée est définie par
$$\forall s \in \mathrm{NS}(V) \otimes_{\mathbf{Z}} \mathbf{C}, \quad \zeta_{\mathbf{H}}(s) = \sum_{x \in W(k)} \mathbf{H}(s, x)^{-1}.$$

Remarque 1.2.14. — Si $[o(\mathbf{h})]$ appartient à l'intérieur de $C_{\mathrm{eff}}(V)$, alors il existe un ouvert U de V tel que $n_{U,\mathbf{h}}(H)$ soit fini pour tout H.

1.3. Mesure de Tamagawa. — Dans la suite V désigne une variété presque de Fano sur k. Dans ce cas toute métrique adélique sur le fibré anticanonique ω_V^{-1} définit une mesure de Tamagawa qui permet de donner une interprétation conjecturale du terme principal du nombre de points de hauteur bornée.

Notations 1.3.1. — Si X est un variété sur k, $X(\boldsymbol{A}_k)$ désigne l'espace adélique qui lui est associé. (cf. [**We**, §1]).

Pour toute place v de k, la mesure de Haar $\mathrm{d}x_v$ sur k_v est normalisée de la manière suivante :

- Si v est finie, alors $\int_{\mathcal{O}_v} \mathrm{d}x_v = 1$,
- si $k_v \xrightarrow{\sim} \mathbf{R}$, alors $\mathrm{d}x_v$ est la mesure de Lebesgue usuelle,
- si $k_v \xrightarrow{\sim} \mathbf{C}$, alors $\mathrm{d}x_v = i\mathrm{d}z \wedge \mathrm{d}\bar{z}$.

Soit $\mathbf{h} = (\omega_V^{-1}, (\|.\|_v)_{v \in M_k})$ une hauteur sur une variété presque de Fano V. En toute place v de k on lui associe la mesure borélienne $\omega_{\mathbf{h},v}$ sur $V(k_v)$ définie par la relation (cf. [**We**], [**Pe1**, §2.2.1])

$$\omega_{\mathbf{h},v} = \left\| \frac{\partial}{\partial x_1} \wedge \cdots \wedge \frac{\partial}{\partial x_n} \right\|_v \mathrm{d}x_{1,v} \ldots \mathrm{d}x_{n,v}$$

où $x_1, \ldots, x_n$ désignent des coordonnées locales analytiques au voisinage d'un point x de $V(k_v)$ et $\frac{\partial}{\partial x_1} \wedge \cdots \wedge \frac{\partial}{\partial x_n}$ est vu comme section locale de ω_V^{-1}.

D'après [**Pe1**, lemme 2.1.1], on peut se donner un ensemble fini S de places finies et un modèle projectif et lisse $\mathscr{V}$ de V sur $\mathcal{O}_S$ dont les fibres sont géométriquement intègres et tel que pour toute place finie $\mathfrak{p}$ en-dehors de S, le groupe de Picard géométrique $\operatorname{Pic} \mathscr{V}_{\overline{\mathbf{F}_\mathfrak{p}}}$ soit isomorphe à $\operatorname{Pic} \overline{V}$ de façon compatible aux actions des groupes de Galois et la partie l-primaire du groupe de Brauer $\operatorname{Br}(\overline{V})$ soit finie pour tout nombre premier l n'appartenant pas à $\mathfrak{p}$.

Pour tout $\mathfrak{p}$ de $M_k - S$, le terme local de la fonction L associée à $\operatorname{Pic} \overline{V}$ est défini par

$$L_\mathfrak{p}(s, \operatorname{Pic} \overline{V}) = \frac{1}{\operatorname{Det}(1 - (\#\mathbf{F}_\mathfrak{p})^{-s} \operatorname{Fr}_\mathfrak{p} \mid \operatorname{Pic} \mathscr{V}_{\overline{\mathbf{F}_\mathfrak{p}}} \otimes \mathbf{Q})}$$

où $\operatorname{Fr}_\mathfrak{p}$ est le Frobenius en $\mathfrak{p}$. La fonction L globale est définie par le produit eulérien

$$L_S(s, \operatorname{Pic} \overline{V}) = \prod_{\mathfrak{p} \in M_f - S} L_\mathfrak{p}(s, \operatorname{Pic} \overline{V})$$

qui par [**Pe1**, lemme 2.2.5] converge absolument pour $\operatorname{Re} s > 1$ et s'étend en une fonction méromorphe sur $\mathbf{C}$ avec un pôle d'ordre $t = \operatorname{rg} \operatorname{Pic} V$ en 1.

Les facteurs de convergence $(\lambda_v)_{v \in M_k}$ pour la mesure de Tamagawa sont définis par

$$\lambda_v = \begin{cases} L_v(1, \operatorname{Pic} \overline{V}) \text{ si } v \in M_f - S \\ 1 \text{ sinon.} \end{cases}$$

Les conjectures de Weil montrées par Deligne impliquent la convergence de la mesure adélique $\prod_{v \in M_k} \lambda_v^{-1} \omega_{\mathbf{h},v}$ (cf. [**Pe1**, proposition 2.2.2]).

Définition 1.3.2. — Avec les notation qui précèdent, la *mesure de Tamagawa* associée à $\mathbf{h}$ est définie par

$$\omega_\mathbf{h} = \lim_{s \to 1} (s-1)^t L_S(s, \operatorname{Pic} \overline{V}) \frac{1}{\sqrt{d}^{\dim V}} \prod_{v \in M_k} \lambda_v^{-1} \omega_{\mathbf{h},v}.$$

Remarque 1.3.3. — Par construction elle est indépendante du choix de S et ne dépend que de l'image de $\mathbf{h}$ dans $\mathscr{H}(V)$.

Exemple 1.3.4. — Si $\boldsymbol{f} = (f_v)_{v \in M_k}$ est une famille de fonction comme dans l'exemple 1.2.5, alors
$$\omega_{\boldsymbol{f}.\mathbf{h}} = \Big(\prod_{v \in M_k} f_v \Big) \omega_{\mathbf{h}}.$$

Notation 1.3.5. — On pose $\tau_{\mathbf{h}}(V) = \omega_{\mathbf{h}}\big(\overline{V(k)}\big)$ où $\overline{V(k)}$ désigne l'adhérence des points rationnels de V dans $V(\boldsymbol{A}_k)$.

1.4. Énoncé d'une question. — Pour énoncer notre question qui est une version raffinée d'une conjecture de Manin [**BM**, conjecture C′], nous utiliserons la notion d'accumulation qui suit :

Définition 1.4.1. — Soit $\mathbf{h}$ une hauteur d'Arakelov sur V telle que $[o(\mathbf{h})]$ appartienne à l'intérieur du cône effectif. Un fermé irréductible strict F de V est dit modérément accumulateur pour $\mathbf{h}$ si et seulement si pour tout ouvert non vide W de F, il existe un ouvert non vide U de V tel que
$$\varlimsup_{H \to +\infty} \frac{n_{W,\mathbf{h}}(H)}{n_{U,\mathbf{h}}(H)} > 0.$$
Nous renvoyons à [**BT4**] et [**Pe2**, §2.4] pour des exemples de telles sous-variétés.

Notation 1.4.2. — Si V est une variété presque de Fano, on considère l'hyperplan affine $\mathscr{P}$ de $\mathrm{NS}(V)^{\vee} \otimes \mathbf{R}$ d'équation $\langle y, \omega_V^{-1} \rangle = 1$. Cet hyperplan est muni d'une mesure canonique θ définie par ω_V^{-1} (cf. [**Pe1**, page 120]). On note $C_{\mathrm{eff}}(V)^{\vee}$ le cône dual de $C_{\mathrm{eff}}(V)$ défini par
$$C_{\mathrm{eff}}(V)^{\vee} = \{ y \in \mathrm{NS}(V)^{\vee} \otimes \mathbf{R} \mid \forall x \in C_{\mathrm{eff}}(V), \langle x, y \rangle > 0 \}$$
et on pose
$$\alpha(V) = \theta(C_{\mathrm{eff}}(V)^{\vee} \cap \mathscr{P}).$$
On note également
$$\beta(V) = \#H^1(k, \mathrm{Pic}\,\overline{V}).$$

Remarque 1.4.3. — On pourra noter que la constante $\alpha(V)$ définie par Batyrev et Tschinkel [**BT1**] est obtenue en multipliant par $(t-1)!$ celle considérée ici.

Question 1.4.4. — *Soit V une variété presque de Fano sur k et $\mathbf{h}$ une hauteur sur V définie par une métrique adélique sur ω_V^{-1}. On suppose que $V(k)$ est dense pour la topologie de Zariski et que le complémentaire U dans V des* *sous-variétés modérément accumulatrices est un ouvert de Zariski non vide de V. À quelle condition a-t-on l'équivalence*

(1.2) $$n_{U,\mathbf{h}}(H) \sim \alpha(V)\beta(V)\tau_{\mathbf{h}}(V)H(\log H)^{t-1}$$

lorsque H tend vers l'infini ?

Remarques 1.4.5. — (i) L'introduction du facteur $\beta(V)$ est due à Batyrev et Tschinkel [**BT1**].

(ii) L'équivalence (1.2) est compatible avec le produit de variétés [**FMT**, §1.2, proposition], [**Pe1**, corollaire 4.3].

(iii) Elle est vérifiée dans les cas suivants :

- Si V est une intersection complète lisse dans $\mathbf{P}_{\mathbf{Q}}^N$ définie par m équations homogènes de degré $d \geqslant 2$ si

$$N > 2^{d-1}m(m+1)(d-1)$$

[**Bi**], [**Pe1**, proposition 5.5.3],

- Si V est une variété de drapeaux généralisée [**FMT**], [**Pe1**, théorèmes 6.1.1 et 6.2.2],
- Si V est une variété torique lisse [**Pe1**, §8-11], [**BT1**], [**BT3**], [**Sa**],
- pour certains fibrés en variétés toriques au-dessus de variétés de drapeaux généralisées [**ST**].

(iv) Comme me l'a signalé Tschinkel, la question 2.6.1 dans [**Pe2**] est mal posée. En général on peut seulement espérer que la fonction

$$s \mapsto \zeta_{\mathbf{H}}(s\omega_V^{-1})/\chi_{C_{\text{eff}}(V)}((s-1)\omega_V^{-1})$$

s'étende en une fonction holomorphe au voisinage de 1 et prenne la valeur $\beta(V)\tau_{\mathbf{H}}(V)$ en ce point.

2. Passage au torseur universel

L'objectif de ce paragraphe est de relever au torseur universel chaque coté de (1.2). C'est l'objet des propositions 2.4.2 et 2.5.2.

2.1. Structures sur les torseurs universels. — Nous allons commencer par rappeler la définition des torseurs universels qui est due à Colliot-Thélène et Sansuc [**CTS1**] [**CTS3**].

Définition 2.1.1. — Soient G un groupe algébrique linéaire sur un corps E et Y une variété sur E. Un G-*torseur au-dessus de* Y est la donnée d'un

morphisme fidèlement plat $\pi : X \to Y$ au-dessus de E et d'une action $\mu :$
$X \times G \to X$ de G sur X au-dessus de Y telle que l'application

$$(g, x) \mapsto (gx, x)$$

définisse un isomorphisme de variétés de $G \times_E X$ sur $X \times_Y X$.

Par [**Mi**, théorème III.3.9 et corollaire III.4.7], si G est lisse et abélien, les classes d'isomorphismes de G-torseurs au-dessus de Y sont classifiées par le groupe de cohomologie étale $H^1_{\text{ét}}(Y, G)$ et par [**CTS3**, (2.0.2) et proposition 2.2.8], si T est un tore sur E, c'est-à-dire une E-forme de $\mathbf{G}_m^n$ et si X est une variété propre, lisse et géométriquement intègre ayant un point rationnel sur E, alors on dispose d'une suite exacte naturelle

$$0 \to H^1(E, T) \to H^1_{\text{ét}}(X, T) \xrightarrow{\rho} \text{Hom}_{\text{Gal}(E^s/E)}(X^*(T), \text{Pic}\, X_{E^s}) \to 0$$

où $X^*(T)$ désigne le groupe des caractères de T^s et où pour tout torseur $\mathcal{T}$ et tout caractère ξ de T, $\rho(\mathcal{T})(\xi)$ est la classe du $\mathbf{G}_m$-torseur $\xi_*(\mathcal{T})$ dans $\text{Pic}\, X_{E^s}$ qui est isomorphe à $H^1_{\text{ét}}(X_{E^s}, \mathbf{G}_m)$.

Soit X une variété propre, lisse et géométriquement intègre sur un corps E. On suppose que le groupe de Picard géométrique $\text{Pic}\, X^s$ est de type fini et sans torsion. On note alors T_{NS} le tore dont le groupe de caractères est le $\text{Gal}(E^s/E)$-module $\text{Pic}\, X^s$. Un *torseur universel* pour X est un T_{NS}-torseur $\mathcal{T}$ au-dessus de X dont l'invariant $\rho(\mathcal{T})$ coïncide avec $\text{Id}_{\text{Pic}(X^s)}$.

Remarque 2.1.2. — Nous renvoyons à [**CTS3**, §2.5, §2.6] et [**Pe2**, §3.3] pour des exemples de torseurs universels. Rappelons seulement qu'il résulte de [**CTS1**, proposition 6] et de [**Sa**, §8] qu'un torseur universel au-dessus d'une compactification équivariante lisse d'un tore T est un ouvert d'un espace affine.

Si Y est une intersection complète lisse dans une variété presque de Fano X ayant un point rationnel et si la restriction de $\text{Pic}\, X^s$ à $\text{Pic}\, Y^s$ est un isomorphisme, alors on a un diagramme commutatif

$$\begin{array}{ccccccccc} 0 & \to & H^1(E, T_{\text{NS}}) & \to & H^1_{\text{ét}}(X, T_{\text{NS}}) & \to & \text{End}_{\text{Gal}(E^s/E)}(\text{Pic}\, X^s) & \to & 0 \\ & & \| & & \downarrow j^* & & \downarrow \wr & & \\ 0 & \to & H^1(E, T_{\text{NS}}) & \to & H^1_{\text{ét}}(Y, T_{\text{NS}}) & \to & \text{End}_{\text{Gal}(E^s/E)}(\text{Pic}\, Y^s) & \to & 0 \end{array}$$

où j désigne le plongement de Y dans X. Il en résulte que les torseurs universels au-dessus de Y sont obtenus en prenant l'image inverse de Y dans les torseurs universels au-dessus de X. On dispose donc de diagrammes commutatifs de la

forme :
$$\begin{array}{ccc} \mathcal{T}_Y & \longrightarrow & \mathcal{T}_X \\ \downarrow & & \downarrow \\ Y & \longrightarrow & X \end{array}$$
où l'application du haut est une immersion fermée T_{NS}-équivariante. Si, en outre, X est une compactification équivariante lisse d'un tore, alors $\mathcal{T}_X$ se plonge comme ouvert dans un espace affine $\mathbf{A}_E^N$ et l'action de T_{NS} s'étend à cet espace affine.

A chaque torseur universel au-dessus d'une variété presque de Fano sont associées deux structures canoniques, à savoir un espace d'adèles et une mesure sur cet espace. Ces structures ont été définies dans [**Pe2**, §4.2 et 4.4] mais nous allons maintenant en redonner une construction intrinsèque.

Notation 2.1.3. — Si L appartient à $C_{\mathrm{eff}}(V)$, on pose
$$\delta(L) = \inf\{\langle x, L\rangle,\, x \in C_{\mathrm{eff}}(\overline{V})^{\vee} \cap \mathrm{Pic}\,\overline{V}^{\vee} - \{0\}\}$$
et on note $\delta(V) = \delta(\omega_V^{-1})$.

Hypothèses 2.1.4. — Dans la suite V désigne une variété presque de Fano sur k dont le cône des diviseurs effectifs $C_{\mathrm{eff}}(\overline{V})$ est un cône polyédral rationnel de $\mathrm{Pic}\,\overline{V} \otimes \mathbf{R}$. On suppose en outre que $\delta(V) > 1$.
On note U une ouvert non vide de V.

Remarque 2.1.5. — La condition (iii) dans la définition 1.1.2 assure que pour toute variété presque de Fano $\delta(V) > 0$ et donc $\delta(V) \geqslant 1$.

Exemple 2.1.6. — Si V est une intersection complète lisse dans $\mathbf{P}^N$ définie par m équations $f_1, \ldots, f_m$ de degrés respectifs $d_1, \ldots, d_m$, alors
$$\omega_V^{-1} = \mathcal{O}_V\left(N + 1 - \sum_{i=1}^m d_i\right)$$
et la condition s'écrit $\delta(V) - 1 = N - \sum_{i=1}^m d_i > 0$, qui est exactement l'hypothèse faite dans [**Pe1**, page 131]. La raison pour laquelle cette condition apparaît dans [**Pe1**] est exactement la même qu'ici : elle assure la convergence de sommations liées à la formule d'inversion de Möbius.

Exemple 2.1.7. — Si V est une compactification équivariante lisse d'un tore T sur k et $\overline{V}_T^{(1)}$ désigne l'ensemble des sous-variétés irréductibles invariantes de codimension un de $\overline{V}$, on a une suite exacte canonique

(2.1) $$0 \to X^*(T) \xrightarrow{j} \bigoplus_{D \in \overline{V}_T^{(1)}} \mathbf{Z}D \xrightarrow{\pi} \operatorname{Pic}\overline{V} \to 0$$

où $X^*(T)$ désigne le groupe des $\overline{k}$-caractères de T ; le cône $C_{\text{eff}}(\overline{V})$ est engendré par les $\pi(D)$ pour $D \in \overline{V}_T^{(1)}$ et

$$\omega_V^{-1} = \sum_{D \in \overline{V}_T^{(1)}} \pi(D).$$

Supposons qu'il existe λ de $C_{\text{eff}}(\overline{V})^\vee \cap \operatorname{Pic}\overline{V}^\vee - \{0\}$ vérifiant $\langle \lambda, \omega_V^{-1} \rangle = 1$. On a alors

$$\left\langle \lambda, \sum_{D \in \overline{V}_T^{(1)}} \pi(D) \right\rangle = 1 \quad \text{et} \quad \forall D \in \overline{V}_T^{(1)}, \langle \lambda, \pi(D) \rangle \geqslant 0$$

et donc il existe $D_0 \in \overline{V}_D^{(1)}$ tel que

$$\langle \lambda, \pi(D) \rangle = \begin{cases} 1 & \text{si } D = D_0, \\ 0 & \text{sinon.} \end{cases}$$

Si on considère la suite exacte duale de (2.1),

$$0 \to \operatorname{Pic}\overline{V}^\vee \xrightarrow{\pi^\vee} \bigoplus_{P \in \overline{V}_T^{(1)}} \mathbf{Z}D^\vee \xrightarrow{j^\vee} X^*(T)^\vee \to 0,$$

on obtient que $D_0^\vee = \pi^\vee(\lambda)$ et donc $D_0^\vee \in \operatorname{Ker} j^\vee$. Mais il résulte de [**Da**, §6] que, par définition de j, l'application $j^\vee$ est non nulle en D_0, ce qui est contradictoire. Par conséquent, les variétés toriques projectives et lisses vérifient les conditions ci-dessus.

Exemple 2.1.8. — Si V est la surface obtenue en éclatant quatre points en position générale sur $\mathbf{P}_k^2$, alors

$$\operatorname{Pic}\overline{V} = \mathbf{Z}\Lambda \oplus \bigoplus_{i=1}^{4} \mathbf{Z}E_i$$

où on note Λ le relevé strict d'une droite de $\mathbf{P}_k^2$ et E_i les diviseurs obtenus par éclatement. Le cône effectif est engendré par les diviseurs $F_{i,5} = E_i$ pour

$1 \leqslant i \leqslant 4$ et $F_{k,l} = \Lambda - E_i - E_j$ pour $\{i,j,k,l\} = \{1,2,3,4\}$ et le faisceau canonique est donné par
$$\omega_V^{-1} = 3\Lambda - \sum_{i=1}^{4} E_i = 2F_{1,2} + F_{3,4} + F_{3,5} + F_{4,5}.$$
Comme le groupe des automorphismes de $\overline{V}$ agit transitivement sur les diviseurs $F_{i,j}$, on obtient que pour tout i,j avec $1 \leqslant i < j \leqslant 5$, $\omega_V^{-1} - 2F_{i,j}$ appartient au cône effectif. Par conséquent cette surface vérifie également la condition précédente.

Notation 2.1.9. — On note $\mathbf{A}_{-C_{\mathrm{eff}}(\overline{V}),k}$ le schéma affine
$$\mathrm{Spec}(\overline{k}[-C_{\mathrm{eff}}(\overline{V}) \cap X^*(T_{\mathrm{NS}})]^{\mathscr{G}})$$
où $\mathscr{G}$ désigne le groupe de Galois absolu de k. Pour tout torseur universel $\mathcal{T}$ au-dessus de V, on note $\widehat{\mathcal{T}}_{C_{\mathrm{eff}}(\overline{V})}$ le produit contracté
$$\mathcal{T} \times^{T_{\mathrm{NS}}} \mathbf{A}_{-C_{\mathrm{eff}}(\overline{V}),k}.$$
On dispose d'une immersion ouverte $\mathcal{T} \to \widehat{\mathcal{T}}_{C_{\mathrm{eff}}(\overline{V})}$, l'action de T_{NS} s'étend à $\widehat{\mathcal{T}}_{C_{\mathrm{eff}}(\overline{V})}$ et on a une fibration $\widehat{\mathcal{T}}_{C_{\mathrm{eff}}(\overline{V})} \to V$ en variétés toriques affines géométriquement isomorphes à la variété $\mathbf{A}_{-C_{\mathrm{eff}}(\overline{V}),k}$.

On appelle espace adélique associé à $\mathcal{T}$ et $C_{\mathrm{eff}}(\overline{V})$ l'intersection
$$\mathcal{T}_{C_{\mathrm{eff}}(\overline{V})}(\mathbf{A}_k) = \Big(\prod_{v \in M_k} \mathcal{T}(k_v) \Big) \cap \widehat{\mathcal{T}}_{C_{\mathrm{eff}}(\overline{V})}(\mathbf{A}_k)$$
qui peut être explicitement décrit comme produit restreint des $\mathcal{T}(k_v)$ (cf. également [**Pe2**, §4.2]).

Nous allons maintenant démontrer la trivialité de $\omega_{\mathcal{T}}$.

Lemme 2.1.10. — *Si Y est une variété lisse sur un corps algébriquement clos E et si $\pi : X \to Y$ est un T-torseur où T est un tore, alors il existe un isomorphisme*
$$\omega_X \xrightarrow{\sim} \pi^*(\omega_Y).$$
En outre cet isomorphisme est canonique au signe près.

Remarque 2.1.11. — Ce lemme est en fait un généralisation facile de l'existence d'une forme volume naturelle, bien définie, au signe près, sur le tore T. En fait, on pourrait aussi le voir comme une conséquence de la description du fibré cotangent relatif $\Omega^1_{X/Y}$ pour les torseurs sous un groupe algébrique lisse (cf. [**Sa**, proposition 3.8]).

Démonstration. — Soit $(\xi_i)_{1 \leqslant i \leqslant t}$ une base de $X^*(T)$. Comme E est algébriquement clos, cette base induit un isomorphisme de T sur $\mathbf{G}_m^t$. Les classes d'isomorphismes de T-torseurs sur Y sont classifiés par

$$H^1_{\text{ét}}(Y,T) \xrightarrow{\sim} H^1_{\text{ét}}(Y,\mathbf{G}_m)^t \xrightarrow{\sim} H^1_{\text{Zar}}(Y,\mathbf{G}_m)^t.$$

Par conséquent π est localement triviale pour la topologie de Zariski. Soit $U = \operatorname{Spec} A$ un ouvert affine de Y sur lequel π se trivialise, c'est-à-dire sur lequel il existe une section $s : U \to X$ de π, l'isomorphisme correspondant $\phi : \pi^{-1}(U) \xrightarrow{\sim} U \times T$ étant caractérisé par

$$\phi \circ s(y) = (y, e).$$

Soit $X_i = \xi_i \circ \operatorname{pr}_2 \circ \phi : \pi^{-1}(U) \to \mathbf{G}_m$. On a donc que X_i appartient au groupe $\Gamma(\pi^{-1}(U), \mathcal{O}_Y)^\times$. Alors, par [**Ha**, remarque II.8.9.2 et exemple II.8.12.1] la famille $(\frac{dX_i}{X_i})_{1 \leqslant i \leqslant t}$ est une base de $\Omega^1_{X/Y}$ en tant que $\mathcal{O}_X$ module et donc $\wedge_{i=1}^t \frac{dX_i}{X_i}$ fournit une trivialisation de $\det(\Omega^1_{X/Y})$ sur $\pi^{-1}(U)$. D'autre part on a une suite exacte de fibrés vectoriels (cf. [**Ha**, proposition II.8.11])

$$0 \to \pi^* \Omega^1_{Y/E} \xrightarrow{j} \Omega^1_{X/E} \to \Omega^1_{X/Y} \to 0,$$

où l'injectivité résulte de l'hypothèse de lissité. Par conséquent on a un isomorphisme canonique

$$\omega_{X/E} \xrightarrow{\sim} \pi^*(\omega_{Y/E}) \otimes \det(\Omega^1_{X/Y}).$$

Donc $\wedge_{i=1}^t \frac{dX_i}{X_i}$ fournit un isomorphisme

(2.2) $$\omega_{X/E}|_{\pi^{-1}(U)} \xrightarrow{\sim} \pi^*(\omega_{Y/E})|_{\pi^{-1}(U)}.$$

Si s' est une autre section trivialisante de π et $X'_i : \pi^{-1}(U) \to \mathbf{G}_m$ les fonctions correspondantes, on a alors $X'_i = a_i X_i$ où a_i est définie par

$$a_i = X'_i \circ s \in \Gamma(U, \mathcal{O}_U)^\times = A^\times.$$

On obtient que

$$\frac{dX'_i}{X'_i} = \frac{d(a_i X_i)}{a_i X_i} = \frac{(da_i) X_i + a_i (dX_i)}{a_i X_i} = \frac{dX_i}{X_i}$$

puisque $da_i = 0$ dans $\Omega^1_{B/A}$ où $B = \Gamma(\pi^{-1}(U), \mathcal{O}_X)$. Donc l'isomorphisme (2.2) est indépendant de la section choisie et, par recollement, définit un isomorphisme

$$\omega_{X/E} \xrightarrow{\sim} \pi^*(\omega_{Y/E})$$

Si $(\xi'_i)_{1 \leqslant i \leqslant t}$ est une autre base de $X^*(T)$, alors on note $M \in GL_n(\mathbf{Z})$ la matrice de changement de base. La section $\wedge_{i=1}^t \frac{dX_i}{X_i}$ sera remplacée par la section

$\det(M)\wedge_{i=1}^{t}\frac{dX_i}{X_i}$ ce qui montre qu'au signe près l'isomorphisme est indépendant de la base choisie. □

Lemme 2.1.12. — *Avec les notations ci-dessus, le fibré canonique $\omega_{\mathcal{T}}$ est trivial.*

Démonstration. — On a une suite exacte
$$0 \to H^1(k, \overline{k}[\mathcal{T}]^\times) \to \mathrm{Pic}(\mathcal{T}) \to \mathrm{Pic}(\overline{\mathcal{T}}).$$
Mais il découle de [**CTS3**, proposition 2.1.1] que
$$\Gamma(\mathcal{T}, \mathcal{O}_{\mathcal{T}}^\times) = \Gamma(V, \mathcal{O}_V^\times) = k^\times.$$
Et, par le théorème d'Hilbert 90, $\mathrm{Pic}(\mathcal{T})$ s'injecte dans $\mathrm{Pic}(\overline{\mathcal{T}})$ et il suffit de montrer le résultat sur $\overline{k}$. Mais par le lemme précédent, on a un isomorphisme
$$\omega_{\overline{\mathcal{T}}} \xrightarrow{\sim} \pi^*(\omega_{\overline{V}}).$$
En appliquant à nouveau [**CTS3**, proposition 2.1.1], l'application π^* de $\mathrm{Pic}(\overline{V})$ à $\mathrm{Pic}(\overline{\mathcal{T}})$ est triviale et, par conséquent, $\omega_{\overline{\mathcal{T}}}$ est triviale. □

Notation 2.1.13. — Par conséquent, il existe une section $\breve{\omega}_{\mathcal{T}}$ de $\omega_{\mathcal{T}}$ partout non nulle et, comme $\Gamma(\mathcal{T}, \mathcal{O}_{\mathcal{T}}^\times) = k^\times$, cette section est unique à une constante multiplicative près. Par [**We**, §2] cette section $\breve{\omega}_{\mathcal{T}}$ définit pour toute place v de k une mesure $\boldsymbol{\omega}_{\mathcal{T},v}$ sur $\mathcal{T}(k_v)$.

Le résultat suivant est annoncé dans [**Pe2**, remarque 4.4.4].

Lemme 2.1.14. — *Avec les hypothèses ci-dessus, si $\mathcal{T}$ a un point rationnel, le produit des mesures $\boldsymbol{\omega}_{\mathcal{T},v}$ converge et redonne la mesure $\boldsymbol{\omega}_{\mathcal{T}}$ définie dans [**Pe2**, définition 4.4.3]*.

Démonstration. — Il suffit de montrer que l'on peut choisir la section $\breve{\omega}_{\mathcal{T}}$ de sorte que la mesure $\boldsymbol{\omega}_{\mathcal{T},v}$ coïncide avec celle définie dans [**Pe2**, notations 4.4.1]. Or, par définition, $\boldsymbol{\omega}_{\mathcal{T},v}$ est localement donnée par la formule
$$\boldsymbol{\omega}_{\mathcal{T},v} = \left| \left\langle \frac{\partial}{\partial x_1} \wedge \cdots \wedge \frac{\partial}{\partial x_N}, \breve{\omega}_{\mathcal{T}} \right\rangle \right|_v dx_{1,v} \ldots dx_{N,v}$$
où $x_1, \ldots, x_n$ désignent des coordonnés locales analytiques au voisinage d'un point x de $V(k_v)$.

D'un autre coté, la mesure $\boldsymbol{\omega}'_{\mathcal{T},v}$ définie par [**Pe2**, notations 4.4.1] est construite de la manière suivante : on note $\boldsymbol{\omega}_{T_{\mathrm{NS}},v}$ la mesure définie par la forme différentielle canonique $\breve{\omega}_{T_{\mathrm{NS}}}$ sur T_{NS} et on se donne un morphisme $\psi_{\omega_V^{-1}}$ de $\mathcal{T}$ dans ω_V^{-1} dont l'image ne rencontre pas la section nulle et qui est compatible avec le morphisme de tore de $T_{\mathrm{NS}} \to \mathbf{G}_m$ induit par l'injection $\mathbf{Z} \to \mathrm{Pic}\, V$

envoyant 1 sur la classe de ω_V^{-1}. Pour tout point x de $V(k_v)$, on considère sur la fibre $\mathcal{T}_x(k_v)$ la mesure $\omega_{\mathcal{T}_x,v}$ donnée par

$$\int_{\mathcal{T}_x(k_v)} f(r)\omega_{\mathcal{T}_x,v}(r) = \int_{T_{\mathrm{NS}}(k_v)} f(r.y)\left\|\psi_{\omega_V^{-1}}(r.y)\right\|_v^{-1} \omega_{T_{\mathrm{NS}},v}(r)$$

où y est un point arbitraire de $\mathcal{T}_x(k_v)$. La mesure $\omega'_{\mathcal{T},v}$ est alors définie par la relation

$$\int_{\mathcal{T}(k_v)} f(y)\omega'_{\mathcal{T},v}(y) = \int_{V(k_v)} \omega_{\mathrm{h},v}(x) \int_{\mathcal{T}_x(k_v)} f(y)\omega_{\mathcal{T}_x,v}(y).$$

Mais, par la démonstration du lemme 2.1.10, $\check{\omega}_{T_{\mathrm{NS}}}$ fournit une trivialisation $\tilde{\omega}_{T_{\mathrm{NS}}}$ du faisceau $\det(\Omega^1_{\mathcal{T}/V})$ et donc un isomorphisme $\omega_{\mathcal{T}} \xrightarrow{\sim} \pi^*\omega_V$. Par conséquent, $\psi^\vee_{\omega_V^{-1}} : \mathcal{T} \to \omega_{\mathcal{T}}$ fournit une section partout non nulle de $\omega_{\mathcal{T}}$, qu'on peut supposer égale à $\check{\omega}_{\mathcal{T}}$. D'autre part, on peut choisir des coordonnés locales $x_1, \ldots, x_n$ sur un ouvert W de $V(k_v)$ sur lequel $\mathcal{T}$ se trivialise et fixer cette trivialisation

$$\mathcal{T}(k_v)_{|W} \xrightarrow{\sim} W \times T_{\mathrm{NS}}(k_v).$$

Des coordonnées locales $x_{n+1}, \ldots, x_N$ sur $T_{\mathrm{NS}}(k_v)$ fournissent alors des coordonnées locales sur $\mathcal{T}(k_v)$. On a alors les relations

$$\begin{aligned}\omega'_{\mathcal{T},v} &= \left\|\frac{\partial}{\partial x_1} \wedge \cdots \wedge \frac{\partial}{\partial x_n}\right\|_v \|\psi_{\omega_V^{-1}}(x_1, \ldots, x_N)\|_v^{-1} \\ &\quad \times \left|\left\langle \frac{\partial}{\partial x_{n+1}} \wedge \cdots \wedge \frac{\partial}{\partial x_N}, \check{\omega}_{T_{\mathrm{NS}}}\right\rangle\right|_v \mathrm{d}x_{1,v} \ldots \mathrm{d}x_{N,v} \\ &= \left|\left\langle \frac{\partial}{\partial x_1} \wedge \cdots \wedge \frac{\partial}{\partial x_N}, \psi^\vee_{\omega_V^{-1}}(x_1, \ldots, x_N) \otimes \tilde{\omega}_{T_{\mathrm{NS}}}(x_1, \ldots, x_N)\right\rangle\right|_v \\ &\quad \times \mathrm{d}x_{1,v} \ldots \mathrm{d}x_{N,v} \\ &= \omega_{\mathcal{T},v}. \quad \square\end{aligned}$$

Définition 2.1.15. — La mesure

$$\omega_{\mathcal{T}} = \frac{1}{\sqrt{d}^{\dim \mathcal{T}}} \prod_{v \in M_k} \omega_{\mathcal{T},v}$$

est, par la formule du produit, indépendante du choix de $\check{\omega}_{\mathcal{T}}$. On l'appelle *mesure canonique* de $\mathcal{T}_{C_{\mathrm{eff}}(\overline{V})}(\boldsymbol{A}_k)$.

Exemple 2.1.16. — Si V est une intersection complète dans une variété X, définie par l'annulation de sections $s_1, \ldots, s_m$ de fibrés en droites $L_1, \ldots, L_m$

de sorte que la restriction donne un isomorphisme
$$\operatorname{Pic}(\overline{X}) \to \operatorname{Pic}(\overline{V})$$
et que X et V vérifient la convention 2.1.4 et si V a un point rationnel, alors par la remarque 2.1.2, un torseur universel $\mathcal{T}_V$ est l'image inverse de V dans un torseur universel $\pi: \mathcal{T}_X \to X$. Comme les faisceaux inversibles $\pi^*(L_i)$ sont triviaux pour $1 \leqslant i \leqslant m$, $\mathcal{T}_V$ est donc défini dans $\mathcal{T}_X$ par l'annulation de m fonctions $f_1, \ldots, f_m$ qui vérifient
$$\forall y \in \mathcal{T}_X(\overline{k}), \quad \forall t \in T_{\mathrm{NS}}(\overline{k}), \quad f_i(t.y) = [L_i](t) f_i(y),$$
où $[L_i] \in \operatorname{Pic} \overline{V} = X^*(T_{\mathrm{NS}})$. Si $\breve{\omega}_{\mathcal{T}_X}$ est une trivialisation de $\omega_{\mathcal{T}_X}$, on dispose alors d'une forme différentielle de Leray $\breve{\omega}_{\mathrm{L},\mathcal{T}_V}$ section de $\omega_{\mathcal{T}_V}$ et définie par la relation
$$\forall y \in \mathcal{T}_V(\overline{k}), \quad \breve{\omega}_{\mathrm{L},\mathcal{T}_V}(y) \wedge \boldsymbol{f}^* \left(\bigwedge_{i=1}^m \mathrm{d}x_i \right)(y) = \breve{\omega}_{\mathcal{T}_X}(y).$$
Cette forme différentielle est une section partout non nulle de $\omega_{\mathcal{T}_V}$, on peut donc poser $\breve{\omega}_{\mathcal{T}_V} = \breve{\omega}_{\mathrm{L},\mathcal{T}_V}$.

Si, en outre, X est une compactification projective lisse d'un tore T, alors $\mathcal{T}_X$ est un ouvert d'un espace affine $\mathbf{A}_k^N$ et on peut prendre
$$\breve{\omega}_{\mathcal{T}_X} = \mathrm{d}x_1 \wedge \cdots \wedge \mathrm{d}x_N.$$
La forme pour $\mathcal{T}_V$ est alors donnée localement, au signe près, par l'expression explicite
$$\breve{\omega}_{\mathrm{L},\mathcal{T}_V}(x) = \det \left(\frac{\partial f_i}{\partial x_{l_j}}(x) \right)^{-1}_{1 \leqslant i,j \leqslant m} \mathrm{d}x_0 \wedge \cdots \wedge \widehat{\mathrm{d}x_{l_1}} \wedge \cdots \wedge \widehat{\mathrm{d}x_{l_m}} \wedge \cdots \wedge \mathrm{d}x_N$$
pour $1 \leqslant l_1 < \cdots < l_m \leqslant N$.

2.2. Fonctions de comptage. — Nous souhaitons maintenant expliciter et démontrer la description en termes des torseurs universels de la formule asymptotique (1.2) telle qu'elle est annoncée dans [**Pe2**, §5.4].

Le passage aux torseurs universels nécessite la construction d'un domaine fondamental dans le produit $\prod_{v \in S} \mathcal{T}(k_v)$ sous l'action de $T_{\mathrm{NS}}(\mathcal{O}_S)$, qui permettra en fait de construire un domaine fondamental de $\mathcal{T}_{C_{\mathrm{eff}}(\overline{V})}(\boldsymbol{A}_k)$ sous l'action de $T_{\mathrm{NS}}(k)$. Nous allons rappeler la construction d'un tel domaine donnée dans [**Pe2**].

On peut rapprocher cette construction du lien entre systèmes de métriques et sections des applications quotients
$$\mathcal{T}(\boldsymbol{A}_k)/K_{T_{\mathrm{NS}}} \to V(\boldsymbol{A}_k),$$

où $K_{T_{\text{NS}}}$ est le sous-groupe maximal de $T_{\text{NS}}(\mathbf{A}_k)$, indiquée par Salberger [**Sa**, page 94].

Notations 2.2.1. — Pour tout tore T sur k, on note $X_*(T)$ le $\text{Gal}(\overline{k}/k)$-réseau dual de $X^*(T)$ et pour tout place v de k, $X_*(T)_v$ désigne le groupe $X_*(T)^{\text{Gal}(\overline{k_v}/k_v)}$. En outre $T(\mathcal{O}_v)$ désigne le sous-groupe compact maximal de $T(k_v)$ et on pose

$$K_T = \prod_{v \in M_k} T(\mathcal{O}_v) \quad \text{et} \quad W(T) = K_T \cap T(k).$$

Le groupe $W(T)$ est le groupe fini des éléments de torsion dans $T(k)$. On dispose d'une injection canonique

$$\log_v : T(k_v)/T(\mathcal{O}_v) \to X_*(T)_v \otimes \mathbf{R}.$$

Quitte à augmenter l'ensemble des mauvaises places S, on peut supposer qu'il contient les places archimédiennes et les places ramifiées dans une extension galoisienne fixée K/k qui déploie le tore T_{NS}. Par [**Ono1**, theorem 4] et [**Ono2**, §3], on peut en outre supposer que l'application naturelle

$$T_{\text{NS}}(k) \to \bigoplus_{v \in M_k - S} X_*(T_{\text{NS}})_v$$

est surjective et qu'on a une suite exacte

$$0 \to W(T_{\text{NS}}) \to T_{\text{NS}}(\mathcal{O}_S) \xrightarrow{\log_S} \prod_{v \in S} X_*(T_{\text{NS}})_v \otimes \mathbf{R}$$

où $\log_S$ est induite par les applications $\log_v$ pour $v \in S$. En outre l'image M de $\log_S$ est un réseau dans le noyau du morphisme

$$\prod_{v \in S} X^*(T_{\text{NS}})_v^{\vee} \otimes \mathbf{R} \to X^*(T_{\text{NS}})_k^{\vee} \otimes \mathbf{R}$$

où $X^*(T_{\text{NS}})_k = X^*(T_{\text{NS}})^{\text{Gal}(\overline{k}/k)}$. On fixe une base de M et on note Δ le domaine fondamental correspondant de M dans ce noyau et pr une projection du groupe de gauche sur ce noyau.

On se donne alors un système de hauteurs $\mathbf{H}_K$ sur K et on note $\mathbf{H}$ le système de hauteurs défini par le diagramme commutatif

$$\begin{array}{ccc} \text{NS}(V) & \xrightarrow{[K:k]\mathbf{H}} & \mathcal{H}(V) \\ \downarrow & & \uparrow N_{K/k} \\ \text{NS}(V_K) & \xrightarrow{\mathbf{H}_K} & \mathcal{H}(V_K). \end{array}$$

On suppose en outre que $\mathbf{h} = \mathbf{H}([\omega_V^{-1}])$ et que

(2.3) $\quad\quad \forall L \in C_{\text{eff}}(V) \cap \operatorname{Pic} V, \quad \forall x \in V(k), \quad \mathbf{H}(L,x) \geqslant 1.$

Soit $\mathcal{T}$ un torseur universel au-dessus de V ayant un point rationnel y_0. Si L est un fibré en droites sur V_K, $L^\times$ désigne le complémentaire de la section nulle dans L. Le morphisme $\mathbf{Z} \to \operatorname{Pic} V_K$ envoyant 1 sur la classe de L induit un morphisme $\phi_L : T_{\text{NS}\,K} \to \mathbf{G}_{m,K}$ et $\phi_{L*}(\mathcal{T})$ est isomorphe à $L^\times$. On note $\psi_L : \mathcal{T} \to L$ un morphisme partout non nul obtenu de cette manière. On fixe une place $\mathfrak{p}_0$ de k, et on suppose que la hauteur $(L, (\|.\|_\mathfrak{P})_{\mathfrak{P} \in M_K})$ représente $\mathbf{H}_K([L])$, on note alors

$$\forall \mathfrak{P} \in M_K, \ \forall y \in \mathcal{T}(K_\mathfrak{P}), \ \|y\|_\mathfrak{P}^L = \begin{cases} \dfrac{\|\psi_L(y)\|_\mathfrak{P}}{\|\psi_L(y_0)\|_\mathfrak{P}} & \text{si } \mathfrak{P} \nmid \mathfrak{p}_0 \\ \dfrac{\|\psi_L(y)\|_\mathfrak{P}}{\|\psi_L(y_0)\|_\mathfrak{P}} \mathbf{H}_K(L, \pi(y_0))^{-\frac{[K_\mathfrak{P}:k_{\mathfrak{p}_0}]}{[K:k]}} & \text{sinon.} \end{cases}$$

Les fonctions $\|.\|_\mathfrak{P}^L$ ne dépendent que de $\mathbf{H}_K([L])$, de y_0 et de $\mathfrak{p}_0$. Elles induisent des fonctions $\|.\|_v^L$ pour toute place v de k et tout L de $\operatorname{Pic} V_v$. On obtient des fonctions

$$\tilde{\mathbf{H}}_{\mathcal{T},v}^{\log} : \mathcal{T}(k_v) \to (\operatorname{Pic} V_v)^\vee \otimes \mathbf{R}$$

caractérisées par les relations

$$\forall y \in \mathcal{T}(k_v), \quad \forall L \in \operatorname{Pic} V_v, \quad \|y\|_v^L = q_v^{-\tilde{\mathbf{H}}_{\mathcal{T},v}^{\log}(y)(L)}$$

où $q_v = \#\mathbf{F}_v$ si $v \in M_f$, $q_v = e$ si k_v est isomorphe à $\mathbf{R}$ et $q_v = e^2$ sinon.

On considère alors

(2.4) $\quad\quad \Delta_{\mathbf{H}_K}(\mathcal{T}) = \left\{ y \in \prod_{v \in S} \mathcal{T}(k_v) \,\Big|\, \operatorname{pr}((\tilde{\mathbf{H}}_{\mathcal{T},v}^{\log}(y_v))_{v \in S}) \in \Delta \right\}$

qui, par [**Pe2**, proposition 4.3.1] est, sous réserve d'une augmentation de S, un domaine fondamental de $\prod_{v \in S} \mathcal{T}(k_v)$ sous $T_{\text{NS}}(\mathcal{O}_S)/W(T_{\text{NS}})$.

Nous pouvons maintenant définir les fonctions de comptage.

Notations 2.2.2. — Quitte à agrandir S, on peut fixer un modèle lisse $\widehat{\mathscr{T}}$ de $\widehat{\mathcal{T}}_{C_{\text{eff}}(\overline{V})}$. Pour toute place $\mathfrak{p}$ de k en-dehors de S, on note

$$\mathcal{T}_{C_{\text{eff}}(\overline{V})}(\mathcal{O}_\mathfrak{p}) = \widehat{\mathscr{T}}(\mathcal{O}_\mathfrak{p}) \cap \mathcal{T}(k_\mathfrak{p}).$$

et on considère

$$\mathcal{T}_{C_{\text{eff}}(\overline{V}), S}(\mathbf{A}_k) = \prod_{v \in S} \mathcal{T}(k_v) \times \prod_{\mathfrak{p} \notin S} \mathcal{T}_{C_{\text{eff}}(\overline{V})}(\mathcal{O}_\mathfrak{p}).$$

Pour tout élément $\mathfrak{b}$ de $\bigoplus_{\mathfrak{p} \in M_k - S} X_*(T_{\mathrm{NS}})_{\mathfrak{p}}$, on note $\mathfrak{b}_{\mathfrak{p}}$ la composante de $\mathfrak{b}$ dans $X_*(T_{\mathrm{NS}})_{\mathfrak{p}}$ et on pose

$$T_{\mathrm{NS}}(-C_{\mathrm{eff}}(\overline{V}), \mathfrak{b}_{\mathfrak{p}}) = \{t \in T_{\mathrm{NS}}(k_{\mathfrak{p}}) \mid \forall y \in C_{\mathrm{eff}}(\overline{V}) \cap \mathrm{Pic}\, V_{\mathfrak{p}},\ v_{\mathfrak{p}}(y(t)) \leqslant \langle y, \mathfrak{b}_{\mathfrak{p}} \rangle \}.$$

Notons que

$$T_{\mathrm{NS}}(-C_{\mathrm{eff}}(\overline{V}), 0). T_{\mathrm{NS}}(-C_{\mathrm{eff}}(\overline{V}), \mathfrak{b}_{\mathfrak{p}}) = T_{\mathrm{NS}}(-C_{\mathrm{eff}}(\overline{V}), \mathfrak{b}_{\mathfrak{p}})$$

et si $b_{\mathfrak{p}} \in T_{\mathrm{NS}}(k_{\mathfrak{p}})$ est tel que $\log_{\mathfrak{p}}(b_{\mathfrak{p}}) = \mathfrak{b}_{\mathfrak{p}}$, alors

$$T_{\mathrm{NS}}(-C_{\mathrm{eff}}(\overline{V}), \mathfrak{b}_{\mathfrak{p}}) = b_{\mathfrak{p}} T_{\mathrm{NS}}(-C_{\mathrm{eff}}(\overline{V}), 0)$$

et

$$T_{\mathrm{NS}}(-C_{\mathrm{eff}}(\overline{V}), \mathfrak{b}_{\mathfrak{p}}). \mathcal{T}_{C_{\mathrm{eff}}(\overline{V})}(\mathcal{O}_{\mathfrak{p}}) = b_{\mathfrak{p}}. \mathcal{T}_{C_{\mathrm{eff}}(\overline{V})}(\mathcal{O}_{\mathfrak{p}}).$$

En réalité les $T_{\mathrm{NS}}(-C_{\mathrm{eff}}(\overline{V}), \mathfrak{b}_{\mathfrak{p}})$ vont jouer le rôle d'idéaux dans notre cadre. On considère alors

$$\mathfrak{b}.\mathcal{T}_{C_{\mathrm{eff}}(\overline{V}), S}(\boldsymbol{A}_k) = \prod_{v \in S} \mathcal{T}(k_v) . \prod_{\mathfrak{p} \notin S} T_{\mathrm{NS}}(-C_{\mathrm{eff}}(\overline{V}), \mathfrak{b}_{\mathfrak{p}}) \mathcal{T}_{C_{\mathrm{eff}}(\overline{V})}(\mathcal{O}_{\mathfrak{p}}).$$

La fonction de comptage sur le torseur universel $\mathcal{T}$ associée au système de hauteurs $\mathbf{H}_K$, au nombre réel positif H et à l'élément $\mathfrak{b}$ de $\bigoplus_{v \in M_k - S} X_*(T_{\mathrm{NS}})_v$ est alors la fonction $\Phi_{\mathcal{T}}^{\mathbf{H}}(H, \mathfrak{b}, .)$ indicatrice de l'ensemble des $\boldsymbol{y} = (y_v)_{v \in M_k}$ de $\mathcal{T}_{C_{\mathrm{eff}}(\overline{V})}(\boldsymbol{A}_k)$ vérifiant les conditions qui suivent :

(2.5) $\qquad \forall v \in S, \quad \pi(y_v) \in U(k_v),$

(2.6) $\qquad (y_v)_{v \in S} \in \Delta_{\mathbf{H}_K}(\mathcal{T}),$

(2.7) $\qquad \forall L \in C_{\mathrm{eff}}(V), \quad \prod_{v \in S} \|y_v\|_v^L \leqslant 1,$

(2.8) $\qquad \prod_{v \in S} (\|y_v\|_v^{\omega_v^{-1}})^{-1} \leqslant H,$

(2.9) $\qquad \boldsymbol{y} \in \mathfrak{b}.\mathcal{T}_{C_{\mathrm{eff}}(\overline{V}), S}(\boldsymbol{A}_k).$

2.3. Fonctions de Möbius. — Nous aurons besoin dans le prochain paragraphe de fonctions de Möbius que nous allons maintenant définir et étudier.

Notations 2.3.1. — Soit M un $\mathbf{Z}$-module libre de type fini et $C \subset M \otimes \mathbf{R}$ un cône polyédral rationnel strictement convexe, c'est-à-dire de la forme

$$\sum_{i=0}^{N} \mathbf{R}_{\geqslant 0} m_i$$

avec $m_i \in M$ et tel que $C \cap -C = \{0\}$. Si R est un anneau commutatif, on note $R[[C]]$ (respectivement $R((C))$) l'ensemble des fonctions $M \to R$ dont le support est contenu dans C (respectivement dans un translaté de C). On dispose sur ces R-modules d'un produit (de convolution) défini par

$$\forall x \in M, \quad fg(x) = \sum_{y+z=x} f(y)g(z).$$

En effet si $\mathrm{Supp}(f) \subset m + C$ et $\mathrm{Supp}(g) \subset n + C$ alors le support de fg est contenu dans $m + n + C$. La fonction δ_0 indicatrice de $\{0\}$ est une unité pour ce produit. Si A est une partie de M, on note $\mathbf{1}_A$ sa fonction indicatrice.

Exemple 2.3.2. — Si C est un cône régulier c'est-à-dire de la forme

$$\sum_{i=0}^{m} \mathbf{R}_{\geqslant 0} m_i$$

où $(m_i)_{0 \leqslant i \leqslant m}$ peut être complété en une base de M, alors on a des isomorphismes évidents

$$\mathbf{Z}[[C]] = \mathbf{Z}[[T_1, \ldots, T_m]] \quad \text{et} \quad \mathbf{Z}((C)) = \mathbf{Z}[[T_1, \ldots, T_m]][T_1^{-1}, \ldots, T_m^{-1}]$$

où $T_1, \ldots, T_m$ sont des indéterminées.

Remarque 2.3.3. — Géométriquement, $\mathbf{Q}[[C]]$ peut être vu comme complété de l'anneau local à l'origine de la variété torique affine

$$\mathrm{Spec}\, \mathbf{Q}[C \cap M]$$

pour la topologie définie par l'idéal maximal, l'origine étant définie par l'annulation des fonctions correspondant aux éléments de $C \cap M - \{0\}$.

Notations 2.3.4. — On a un plongement canonique $R[M] \subset R((C))$ et on pose

$$R[C] = R[M] \cap R[[C]]$$

qui coïncide en fait avec l'algèbre du monoïde $C \cap M$. On note T une indéterminée. Si $m \in M$, T^m désigne l'élément correspondant de $R[M]$. Si $f \in R[[C]]$, on pose

$$\sum_{m \in M} f(m) T^m = f.$$

Si $\phi : M \to M'$ est un morphisme de $\mathbf{Z}$-modules libres de type fini envoyant C dans un cône polyédral rationnel strictement convexe C' de $M' \otimes \mathbf{R}$ et tel que $\mathrm{Ker}\, \phi \cap C = \{0\}$, alors on dispose d'un morphisme de R-modules

$$\phi_* : R((C)) \to R((C'))$$

envoyant $R[[C]]$ dans $R[[C']]$ défini par
$$\forall x \in M', \quad \phi_* f(x) = \sum_{\phi(y)=x} f(y).$$

Lemme 2.3.5. — *Avec les notation ci-dessus, ϕ_* est un morphisme d'anneaux.*

Démonstration. — Si $f, g \in R((C))$ et $x \in M'$, on a les relations
$$\phi_*(fg)(x) = \sum_{\phi(y)+\phi(z)=x} f(y)g(z)$$
$$= \sum_{y+z=x} \Big(\sum_{\phi(y')=y} f(y')\Big)\Big(\sum_{\phi(z')=z} g(z')\Big)$$
$$= (\phi_* f)(\phi_* g)(x). \quad \square$$

Exemple 2.3.6. — Si $\lambda \in M^\vee$ appartient à l'intérieur du cône $C^\vee$ défini par
$$C^\vee = \{x \in M^\vee \otimes \mathbf{R} \mid \forall y \in C, \langle x, y \rangle \geqslant 0\}$$
alors $\lambda : M \to \mathbf{Z}$ envoie C dans $\mathbf{R}_{\geqslant 0}$ et on dispose d'un morphisme
$$\lambda_* : R((C)) \to R((T)).$$

Lemme 2.3.7. — *Avec les notations qui précèdent, si R est intègre, alors $R((C))$ est un anneau intègre.*

Démonstration. — Soient f et g deux éléments non nuls de $R((C))$. On peut choisir λ de $M^\vee$ à l'intérieur de $C^\vee$, $x_0 \in \operatorname{Supp} f$ et $y_0 \in \operatorname{Supp} g$ de sorte que
$$\forall x \in \operatorname{Supp} f - \{x_0\}, \lambda(x) > \lambda(x_0) \quad \text{et} \quad \forall y \in \operatorname{Supp} g - \{y_0\}, \lambda(y) > \lambda(y_0).$$
On en déduit que $\lambda_*(f)$ et $\lambda_*(g)$ sont non nuls et le lemme découle de l'intégrité de $R((T))$. $\quad \square$

Lemme 2.3.8. — *Avec les notations ci-dessus, si R est intègre et si un élément f de $R[[C]]$ vérifie $f(0) \in R^\times$, alors f est inversible dans $R[[C]]$.*

Démonstration. — La fonction g est un inverse de f si et seulement si elle vérifie la relation
$$\forall y \in M, \quad \sum_{x \in C} g(y-x)f(x) = \delta_0(y).$$
Soit $C^\times = C - \{0\}$, alors cette équation s'écrit également
$$(2.10) \qquad \forall y \in M, \quad g(y) = f(0)^{-1}\Big(\delta_0(y) - \sum_{x \in C^\times} g(y-x)f(x)\Big).$$

Or pour tout m de $M^\vee$ à l'intérieur du cône $C^\vee$, on a
$$\forall x \in C^\times, \quad \langle x, m \rangle > 0.$$
Un récurrence sur $\langle x, m \rangle$ montre alors que (2.10) défini une fonction g dont le support est contenu dans C. $\square$

Notation 2.3.9. — On note $\boldsymbol{\mu}_C$ l'inverse de $\mathbf{1}_C$ dans $\mathbf{Z}[[C]]$.

Lemme 2.3.10. — *En conservant les notations qui précèdent, Il existe un élément P de $\mathbf{Z}[C]$ et une famille finie $(m_j)_{j \in J}$ d'éléments de M tels que*
$$\mathbf{1}_C = \frac{P}{\prod\limits_{j \in J} (1 - T^{m_j})}.$$

Démonstration. — Si C est un cône régulier de la forme $\sum_{i=0}^{m} \mathbf{R}_{\geq 0} m_i$, la fonction $\mathbf{1}_C$ peut s'écrire
$$\mathbf{1}_C = \sum_{n \in \mathbf{Z}_{\geq 0}^m} T^{\sum\limits_{i=1}^{m} n_i m_i} = \frac{1}{\prod\limits_{i=1}^{m}(1 - T^{m_i})}.$$

Dans le cas général (cf. par exemple [**Oda**, page 23]), on écrit C comme support d'un éventail régulier Σ, c'est-à-dire que Σ est un ensemble de cônes polyédraux rationnels strictement convexes de $M \otimes \mathbf{R}$ tel que
(i) si $\sigma \in \Sigma$ et σ' est un face de σ, alors $\sigma' \in \Sigma$,
(ii) si $\sigma, \sigma' \in \Sigma$ alors $\sigma \cap \sigma'$ est une face de σ et de σ',
(iii) $C = \cup_{\sigma \in \Sigma} \sigma$,
(iv) tout σ de Σ est régulier.
La fonction $\mathbf{1}_C$ s'écrit alors
$$\mathbf{1}_C = \sum_{\sigma \in \Sigma} \alpha_\sigma \mathbf{1}_\sigma$$
avec $\alpha_\sigma \in \mathbf{Z}$ et le résultat découle du cas précédent. $\square$

Proposition 2.3.11. — *Pour tout élément λ de $M^\vee$ à l'intérieur de $C^\vee$, il existe une constante R telle que*
$$\forall x \in C, \quad |\boldsymbol{\mu}_C(x)| < R^{\langle \lambda, x \rangle}.$$

Démonstration. — L'élément P de $\mathbf{Z}[C]$ du lemme 2.3.10 peut s'écrire
$$P = 1 + \sum_{m \in C^\times} \alpha_m T^m.$$

On pose $Q = 1 + \sum_{m \in C^\times} -|\alpha_m|T^m$. La relation (2.10) montre alors que les coefficients de P^{-1} vérifient
$$\forall x \in M, \quad |P^{-1}(x)| \leqslant Q^{-1}(x).$$
Mais par le lemme 2.3.10 la fonction de Möbius s'écrit
$$\mu_C = P^{-1} \prod_{j \in J}(1 - T^{m_i})$$
et donc
$$\forall x \in C, \quad |\mu_C(x)| \leqslant \left(Q^{-1} \prod_{j \in J}(1 + T^{m_i})\right)(x).$$
on en déduit l'inégalité
$$\forall x \in C, \quad |\mu_C(x)| \leqslant \lambda_*\left(Q^{-1} \prod_{j \in J}(1 + T^{m_j})\right)(\lambda(x)).$$
Soit R_0 l'inverse de la plus petite des valeurs absolues des racines de λ_*Q. Dans $\mathbf{C}((T))$ si λ_*Q s'écrit $\prod_{i=1}^d (1 - \alpha_i T)$ alors
$$\lambda_*\left(Q^{-1} \prod_{j \in J}(1 + T^{m_j})\right) = \prod_{j \in J}(1 + T^{\lambda(m_j)}) \times \prod_{i=1}^n \left(\sum_{n \geqslant 0} \alpha_i^n T^n\right).$$
Pour tout nombre réel $\varepsilon > 0$, on obtient que les coefficients de la série vérifient
$$\frac{\lambda_*\left(Q^{-1} \prod_{j \in J}(1 + T^{m_j})\right)(x)}{(R_0 + \varepsilon)^x} \underset{x \to +\infty}{\to} 0$$
en outre $\mu_C(0) = 1$ et le lemme est démontré. □

Remarque 2.3.12. — L'utilisation de fonction de Moebius dans des situations similaires apparaît dans [**Sc**], [**Pe1**] et [**Sa**, §11].

2.4. Montée du nombre de points.
— Notre but est maintenant d'exprimer le nombre $n_{U,\mathbf{h}}(H)$ en termes des torseurs universels.

Notations 2.4.1. — Une famille de représentants des classes d'isomorphisme de torseurs universels ayant un point rationnel au-dessus de V, qui est finie par [**CTS2**, proposition 2], est notée $(\mathcal{T}_i)_{i \in I}$.

Pour toute place v de $M_f - S$, on considère le cône
$$C_v = \{x \in X_*(T_{\mathrm{NS}})_v \mid \forall y \in C_{\mathrm{eff}}(V_v), \langle x, y \rangle \leqslant 0\}.$$

On pose $\mu_v = \mu_{C_v}$ et
$$\mu = \prod_{v \in M_k - S} \mu_v : \bigoplus_{v \in M_k - S} X_*(T_{\mathrm{NS}})_v \to \mathbf{R}.$$

Proposition 2.4.2. — *Avec les notations qui précèdent, quitte à augmenter S, pour tout nombre réel positif H, on a la relation :*
$$n_{U,\mathbf{H}}(H) = \frac{1}{\#W(T_{\mathrm{NS}})} \sum_{i \in I} \sum_{\mathfrak{b} \in \bigoplus_{v \in M_k - S} X_*(T_{\mathrm{NS}})_v} \mu(\mathfrak{b}) \sum_{y \in \mathcal{T}_i(k)} \Phi_{\mathcal{T}_i}^{\mathbf{H}}(H, \mathfrak{b}, y).$$

Remarques 2.4.3. — (i) Les sommations du terme de droite ne font intervenir qu'un nombre fini de termes non nuls.

(ii) Si on remplace ω_V^{-1} par un autre fibré en droites L à l'intérieur de $C_{\mathrm{eff}}(V)$ dans la condition (2.8) de la définition des fonctions de comptage, la démonstration reste valide et on obtient une expression de $n_{U,\mathbf{H}(L)}$ en termes des torseurs universels.

Démonstration. — Soit x un point rationnel de U et i l'unique élément de I pour lequel $\mathcal{T}_i$ a un point rationnel au-dessus de x. Il nous faut montrer que :

(2.11)
$$\frac{1}{\#W(T_{\mathrm{NS}})} \sum_{\mathfrak{b} \in \bigoplus_{v \in M_k - S} X_*(T_{\mathrm{NS}})_v} \mu(\mathfrak{b}) \sum_{y \in \mathcal{T}_i(k)} \Phi_{\mathcal{T}_i}^{\mathbf{H}}(H, \mathfrak{b}, y) = \begin{cases} 1 \text{ si } \mathbf{h}(x) \leqslant H, \\ 0 \text{ sinon.} \end{cases}$$

Mais, pour S assez gros, il résulte de [**Pe2**, proposition 4.2.2] que $\mathcal{T}_{i_{C_{\mathrm{eff}}(\overline{V})}}(\mathcal{O}_{\mathfrak{p}})$ peut être décrit de la manière suivante :
$$\mathcal{T}_{i_{C_{\mathrm{eff}}(\overline{V})}}(\mathcal{O}_{\mathfrak{p}}) = \{y \in \mathcal{T}_i(k_{\mathfrak{p}}) \mid \forall \mathfrak{P} \in \{\mathfrak{P} \in M_K \mid \mathfrak{P}|\mathfrak{p}\}, \forall L \in C_{\mathrm{eff}}(\overline{V}), \|y\|_{\mathfrak{P}}^L \geqslant 1\}.$$

Il en résulte que
$$\sum_{\mathfrak{b} \in X_*(T_{\mathrm{NS}})_{\mathfrak{p}}} \mu_{\mathfrak{p}}(\mathfrak{b}) \mathbf{1}_{T_{\mathrm{NS}}(-C_{\mathrm{eff}}(\overline{V}), \mathfrak{b}) \cdot \mathcal{T}_{i_{C_{\mathrm{eff}}(\overline{V})}}(\mathcal{O}_{\mathfrak{p}})}$$

est la fonction indicatrice de
$$\{y \in \mathcal{T}_i(k_{\mathfrak{p}}) \mid \forall \mathfrak{P}|\mathfrak{p}, \forall L \in C_{\mathrm{eff}}(\overline{V}), \|y\|_{\mathfrak{P}}^L = 1\}.$$

Le terme de gauche de (2.11) est donc $\#W(T_{\mathrm{NS}})^{-1}$ fois la somme des valeurs de la fonction caractéristique de l'ensemble des y de $\mathcal{T}_{i_x}(k)$ vérifiant les conditions suivantes :

(2.12) $$(y_v)_{v \in S} \in \Delta_{\mathbf{H}_K}(\mathfrak{T}_i),$$

(2.13) $$\forall L \in C_{\text{eff}}(V), \quad \prod_{v \in S} \|y_v\|_v^L \leqslant 1,$$

(2.14) $$\prod_{v \in S} (\|y_v\|_v^{\omega_v^{-1}})^{-1} \leqslant H,$$

(2.15) $$\forall \mathfrak{P} \in M_K - S_K, \quad \forall L \in C_{\text{eff}}(\overline{V}), \quad \|y_{\mathfrak{P}}\|_{\mathfrak{P}}^L = 1,$$

où S_K désigne l'ensemble des places de K au-dessus de S. Comme, par (2.3)

(2.16) $$\prod_{v \in M_k} \|y_v\|_v^L = \mathbf{H}(L, \pi(y))^{-1} \leqslant 1,$$

la condition (2.15) implique (2.13). Mais on a une suite exacte

$$0 \to T_{\text{NS}}(\mathcal{O}_S) \to T_{\text{NS}}(k) \to \prod_{v \in M_k - S} X_*(T_{\text{NS}})_v \to 0$$

et donc les conditions (2.12) et (2.15) définissent, d'après (2.4) un domaine fondamental pour l'ensemble $\mathfrak{T}_{C_{\text{eff}}(\overline{V})}(\boldsymbol{A}_k)$ sous $T_{\text{NS}}(k)/W(T_{\text{NS}})$. D'autre part la condition (2.14), compte tenu de (2.15) et de (2.16), peut être remplacée par $\mathbf{h}(\pi(y)) \leqslant H$. □

2.5. Montée de la constante. — Nous allons maintenant montrer l'analogue intégral du résultat précédent.

Hypothèses 2.5.1. — Dans la suite nous supposons également que les torseurs universels au-dessus de V vérifient le principe de Hasse et l'approximation faible.

Proposition 2.5.2. — *Avec les notations qui précèdent, sous les hypothèses 2.1.4 et 2.5.1 il existe S tel que pour tout nombre réel positif H on ait*

(2.17) $$\alpha(V)\beta(V)\tau_{\mathbf{h}}(V) \int_0^{\log H} u^{t-1} e^u du$$
$$= \frac{1}{\#W(T_{\text{NS}})} \sum_{i \in I} \sum_{\mathfrak{b} \in \bigoplus_{v \in M_k - S} X_*(T_{\text{NS}})_v} \mu(\mathfrak{b}) \int_{\mathfrak{T}_{i\,C_{\text{eff}}(\overline{V})}(\boldsymbol{A}_k)} \Phi_{\mathfrak{T}_i}^{\mathbf{H}}(H, \mathfrak{b}, y) \omega_{\mathfrak{T}_i}(y).$$

Remarques 2.5.3. — (i) Le terme principal de $\int_0^{\log H} u^{t-1} e^u du$ est égal à $H(\log H)^{t-1}$; c'est en fait le seul ayant une signification pour le comportement asymptotique.

(ii) L'intégrale converge par le lemme 2.1.14. Il résultera de la démonstration que la sommation sur $\mathfrak{b}$ converge absolument.

(iii) En rapprochant la proposition 2.4.2 de la proposition précédente, on constate que la question 1.4.4 se ramène à des majorations de la forme

$$\left| \sum_{y \in \mathcal{T}_i(k)} \Phi_{\mathcal{T}_i}^{\mathbf{H}}(H, \mathfrak{b}, y) - \int_{\mathcal{T}_{iC_{\text{eff}}(\overline{V})}(\mathbf{A}_k)} \Phi_{\mathcal{T}_i}^{\mathbf{H}}(H, \mathfrak{b}, y) \omega_{\mathcal{T}_i}(y) \right|$$

comme c'était le cas dans [**Pe2**] pour les fonctions zêta associées.

Notons que l'équivalence entre ces deux termes lorsque H tend vers l'infini est l'analogue, dans notre cadre, de la notion de variété strictement d'Hardy-Littlewood introduite par Borovoi et Rudnick dans [**BR**].

Démonstration. — Remarquons tout d'abord que si v est une place de k, $\mathcal{T}$ un torseur au-dessus de V, b un élément de $T_{\text{NS}}(k_v)$ et U un ouvert de $\mathcal{T}(k_v)$, alors par le lemme 2.1.14 et [**Pe2**, (4.4.1)]

$$\omega_{\mathcal{T},v}(bU) = \left| [\omega_V^{-1}](b) \right|_v^{-1} \omega_{\mathcal{T},v}(U).$$

On en déduit que

$$\int_{\mathcal{T}_{iC_{\text{eff}}(\overline{V})}(\mathbf{A}_k)} \Phi_{\mathcal{T}_i}^{\mathbf{H}}(H, \mathfrak{b}, y) \omega_{\mathcal{T}_i}(y)$$

$$= \frac{1}{\prod_{\mathfrak{p} \in M_k - S} \#\mathbf{F}_{\mathfrak{p}}^{\langle \omega_V^{-1}, -\mathfrak{b}_{\mathfrak{p}} \rangle}} \int_{\mathcal{T}_{iC_{\text{eff}}(\overline{V})}(\mathbf{A}_k)} \Phi_{\mathcal{T}_i}^{\mathbf{H}}(H, 0, y) \omega_{\mathcal{T}_i}(y).$$

Mais par l'hypothèse 2.1.4,

$$\forall x \in C_{\text{eff}}(\overline{V})^{\vee} \cap \text{Pic}\,\overline{V}^{\vee} - \{0\}, \quad \langle x, \omega_V^{-1} \rangle > 1$$

et, C_v étant l'opposé du cône dual de $C_{\text{eff}}(V_v)$, il en résulte que

(2.18) $\qquad \forall v \in M_f - S, \quad \forall \mathfrak{b} \in C_v - \{0\}, \quad \langle \mathfrak{b}, \omega_V \rangle \geq 2.$

Or il découle du lemme 2.3.7 et de la proposition 2.3.11 que, pour toute place v de k, $\boldsymbol{\mu}_{C_v}$ est supportée par C_v et il existe une constante R_v telle que

$$\forall \mathfrak{b} \in C_v, \quad \left| \boldsymbol{\mu}_{C_v}(\mathfrak{b}) \right| < R_v^{\langle \mathfrak{b}, \omega_v \rangle}.$$

En outre, l'ensemble décrit par les paires $(X_*(T_{\text{NS}})_v, C_v)$ étant fini, on peut choisir une constante R indépendante de v. Quitte à agrandir S, la série

$$\sum_{\mathfrak{b} \in \bigoplus X_*(T_{\text{NS}})_v} \boldsymbol{\mu}_v(\mathfrak{b}) (\#\mathbf{F}_v)^{\langle \mathfrak{b}, \omega_v^{-1} \rangle}$$

converge absolument et par (2.18), il existe une constante R' telle que

$$\left|1 - \sum_{\mathfrak{b}\in X_*(T_{\mathrm{NS}})_v} |\boldsymbol{\mu}_v(\mathfrak{b})|(\#\mathbf{F}_v)^{\langle \mathfrak{b},\omega_V^{-1}\rangle}\right| < \sum_{\mathfrak{b}\in C_v-\{0\}} (R\#\mathbf{F}_v^{-1})^{\langle \mathfrak{b},\omega_V\rangle} < R'\#\mathbf{F}_v^{-2}.$$

Par conséquent,

$$\sum_{\mathfrak{b}\in\bigoplus_{v\in M_k-S} X_*(T_{\mathrm{NS}})_v} \boldsymbol{\mu}(\mathfrak{b}) \prod_{v\in M_k-S} (\#\mathbf{F}_v)^{\langle \omega_V^{-1},\mathfrak{b}_v\rangle}$$

converge absolument et le terme de droite de (2.17) se met sous la forme

$$(2.19) \quad \frac{1}{\#W(T_{\mathrm{NS}})} \prod_{\mathfrak{p}\in M_k-S} \left(\sum_{\mathfrak{b}\in X_*(T_{\mathrm{NS}})_\mathfrak{p}} \frac{\boldsymbol{\mu}_\mathfrak{p}(\mathfrak{b})}{\#\mathbf{F}_\mathfrak{p}^{\langle \omega_V^{-1},-\mathfrak{b}\rangle}} \right) \frac{1}{\sqrt{d}^{\dim \mathcal{T}_i}}$$
$$\times \sum_{i\in I} \int_{\prod_{v\in S} \mathcal{T}_i(k_v)} \Phi_{\mathcal{T}_i,S}^{\mathbf{H}}(H,y) \prod_{v\in S} \omega_{\mathcal{T}_i,v}(y) \prod_{v\in M_k-S} \omega_{\mathcal{T}_i,v}(\mathcal{T}_{i_{C_{\mathrm{eff}}(\overline{V})}}(\mathcal{O}_v))$$

où $\Phi_{\mathcal{T}_i,S}^{\mathbf{H}}(H,y)$ désigne la fonction caractéristique de l'ensemble des éléments y de $\Delta_{\mathbf{H}_K}(\mathcal{T}_i)$ tels que

$$\begin{cases} \prod_{v\in S}(\|y_v\|_v^{\omega_V^{-1}})^{-1} \leqslant H, \\ \forall L \in C_{\mathrm{eff}}(V), \quad \prod_{v\in S}(\|y_v\|_v^{L}) \leqslant 1. \end{cases}$$

Mais la relation $\boldsymbol{\mu}_v \mathbf{1}_{C_v} = \delta_0$ implique que

$$(2.20) \quad \sum_{\mathfrak{b}\in X_*(T_{\mathrm{NS}})_v} \frac{\boldsymbol{\mu}_v(\mathfrak{b})}{\#\mathbf{F}_v^{\langle \omega_V^{-1},-\mathfrak{b}\rangle}} = \left(\sum_{\mathfrak{b}\in -C_v} \frac{1}{\#\mathbf{F}_v^{\langle \omega_V^{-1},\mathfrak{b}\rangle}}\right)^{-1}$$
$$= L_v(T_{\mathrm{NS}}, C_{\mathrm{eff}}(\overline{V}), \omega_V^{-1})^{-1}$$

où $L_v(T_{\mathrm{NS}}, C_{\mathrm{eff}}(\overline{V}), \omega_V^{-1})$ désigne le terme local de la fonction L de Draxl [**Dr**, proposition 4]. D'autre part, il résulte de la démonstration de [**Pe2**, théorème

5.3.1, page 293] qu'on a la relation

(2.21)
$$\prod_{v \in M_k - S} \omega_{\mathcal{T}_i,v}(\mathcal{T}_{iC_{\text{eff}}(\overline{V})}(\mathcal{O}_v)) = \left[\prod_{v \in M_k - S} L_v(T_{\text{NS}}, C_{\text{eff}}(\overline{V}), \omega_V^{-1}) \right]$$
$$\times \left[\prod_{v \in M_k - S} \omega_{T_{\text{NS}},v}(T_{\text{NS}}(\mathcal{O}_v)) L_v(1, \text{Pic}\,\overline{V}) \right]$$
$$\times \left[\prod_{v \in M_k - S} L_v(1, \text{Pic}\,\overline{V})^{-1} \omega_{\mathbf{h},v}(V(k_v)) \right].$$

Fixons un élément i de I et $(x_v)_{v \in S} \in \prod_{v \in S} V(k_v)$ un point appartenant à l'image de l'ensemble $\prod_{v \in S} \mathcal{T}_i(k_v)$. Il découle de [**Ono2**, pages 120-122] qu'on a une suite exacte

$$0 \to T^1_{\text{NS}}\left(\prod_{v \in S} k_v\right) \to \prod_{v \in S} T_{\text{NS}}(k_v) \to (\text{Pic}\,V)^{\vee} \otimes \mathbf{R} \to 0$$

qui définit le groupe $T^1_{\text{NS}}(\prod_{v \in S} k_v)$. Il existe donc un élément y_0 de $\prod_{v \in S} \mathcal{T}_i(k_v)$ tel qu'on ait $\prod_{v \in S} \|y_0\|_v^L = 1$ pour tout L de $\text{Pic}\,V$. En utilisant la démonstration du lemme 2.1.14 on obtient comme dans la démonstration de [**Pe2**, théorème 5.3.1] que

(2.22)
$$\int_{\prod_{v \in S} \mathcal{T}_i(k_v)} \Phi^{\mathbf{H}}_{\mathcal{T}_i,S}(H,y) \prod_{v \in S} \omega_{\mathcal{T}_i,v}(y)$$
$$= \#W(T_{\text{NS}}) \int_{\{y \in C_{\text{eff}}(V)^{\vee} | \langle y, \omega_V^{-1} \rangle \leqslant \log H\}} e^{\langle \omega_V^{-1}, y \rangle} dy$$
$$\times \omega_{T^1_{\text{NS}},S}\left(T^1_{\text{NS}}\left(\prod_{v \in S} k_v\right)/T_{\text{NS}}(\mathcal{O}_S)\right)\left(\prod_{v \in S} \omega_{\mathbf{h},v}\right)\left(\pi\left(\prod_{v \in S} \mathcal{T}_i(k_v)\right)\right)$$

où $\omega_{T^1_{\text{NS}},S}$ est la mesure induite par la forme canonique sur le groupe quotient $T^1_{\text{NS}}(\prod_{v \in S} k_v)/T_{\text{NS}}(\mathcal{O}_S)$. Mais on obtient directement l'égalité

(2.23)
$$\int_{\{y \in C_{\text{eff}}(V)^{\vee} | \langle y, \omega_V^{-1} \rangle \leqslant \log H\}} e^{\langle \omega_V^{-1}, y \rangle} dy = \alpha(V) \int_0^{\log H} u^{t-1} e^u du$$

et on déduit du théorème d'Ono [**Ono3**, main theorem] l'égalité suivante

(2.24)
$$\frac{\#H^1(k,\operatorname{Pic}\overline{V})}{\#\mathrm{III}^1(k,T_{\mathrm{NS}})} = \frac{1}{\sqrt{d}^t}\left[\lim_{s\to 1}(s-1)^t L_S(s,\operatorname{Pic}\overline{V})\right]^{-1}$$
$$\omega_{T^1_{\mathrm{NS}},S}\Big(T^1_{\mathrm{NS}}\big(\prod_{v\in S}k_v\big)/T_{\mathrm{NS}}(\mathcal{O}_S)\Big)$$
$$\prod_{v\in M_k - S}\omega_{T_{\mathrm{NS}},v}(T_{\mathrm{NS}}(\mathcal{O}_v))L_v(1,\operatorname{Pic}\overline{V})$$

où

$$\mathrm{III}^1(k,T_{\mathrm{NS}}) = \operatorname{Ker}\Big(H^1(k,T_{\mathrm{NS}}) \to \prod_{v\in M_k}H^1(k_v,T_{\mathrm{NS}})\Big).$$

En réunissant les formules (2.19) à (2.24), on obtient que le terme de droite de la proposition peut se réécrire

$$\alpha(V)\beta(V)\sum_{i\in I}\frac{\omega_{\mathbf{h}}\Big(\pi\big(\prod_{v\in M_k}\mathcal{T}_i(k_v)\big)\Big)}{\mathrm{III}^1(k,T_{\mathrm{NS}})}\int_0^{\log H}u^{t-1}e^u\mathrm{d}u\,.$$

Mais il résulte de l'hypothèse 2.5.1 sur les torseurs universels que tout point de l'adhérence $\overline{V(k)}$ appartient exactement à $\#\mathrm{III}^1(k,T_{\mathrm{NS}})$ ensembles de la forme $\pi(\prod_{v\in M_k}\mathcal{T}_i(k_v))$. La somme coïncide donc avec $\tau_{\mathbf{h}}(V)$. □

3. Intersections complètes

Dans ce paragraphe, on se place dans le cas où V est une intersection complète dans une variété X et on cherche à écrire la somme (resp. l'intégrale) obtenue pour $\mathcal{T}_V$ en termes de celle de $\mathcal{T}_X$. C'est l'objet de la proposition 3.1.4 et du théorème 3.7.2.

3.1. Encerclement du nombre de points.

Hypothèses 3.1.1. — Dans ce paragraphe, on suppose que V est une intersection complète dans une variété X presque de Fano sur k de sorte que V elle-même soit presque de Fano et qu'on ait un isomorphisme

$$\operatorname{Pic}\overline{X} \xrightarrow{\sim} \operatorname{Pic}\overline{V}$$

qui envoie le cône des diviseurs effectifs de $\overline{X}$ exactement sur celui de $\overline{V}$. On suppose en outre que

$$\forall L \in C_{\mathrm{eff}}(\overline{V})^{\vee} \cap (\operatorname{Pic}\overline{V}^{\vee} - \{0\}),\quad \langle L,\omega_V^{-1}\rangle > 1.$$

***Remarque* 3.1.2**. — Si V est une intersection complète dans une variété torique projective et lisse sur k vérifiant les hypothèses de la proposition 1.1.5 alors les conditions qui précèdent, à l'exception des deux dernières sont automatiquement vérifiées.

***Notations* 3.1.3**. — On fixe un torseur universel $\mathcal{T}_V$ au-dessus de V et on suppose V définie par l'annulation de m sections $s_1, \ldots, s_m$ de fibrés $[L_i]$ de Pic X. D'après l'exemple 2.1.16, on a un diagramme commutatif de la forme

$$\begin{array}{ccc} \mathcal{T}_V & \xrightarrow{\tilde{j}} & \mathcal{T}_X \\ \downarrow \pi_V & & \downarrow \pi_X \\ V & \xrightarrow{j} & X \end{array}$$

où $\tilde{j}$ est une immersion fermée de sorte que $\mathcal{T}_V$ soit défini par l'annulation de m sections $f_1, \ldots, f_m$ du faisceau structural $\mathcal{O}_{\mathcal{T}_X}$, vérifiant les relations

$$(3.1) \quad \forall y \in \mathcal{T}_X(\overline{k}), \quad \forall t \in T_{\mathrm{NS}}(\overline{k}), \quad f_i(t.y) = [L_i](t) f_i(y).$$

On note $\boldsymbol{f} : \mathcal{T}_X \to \boldsymbol{A}_k^m$ l'application induite. On peut noter que l'application $\boldsymbol{f}$ s'étend au schéma $\widehat{\mathcal{T}_X} = \widehat{\mathcal{T}_X}_{C_{\mathrm{eff}}(\overline{X})}$.

Soit χ le caractère de Tate $\boldsymbol{A}_k/k \to \boldsymbol{S}^1$. Il est défini par

$$\xi \mapsto e^{2i\pi \Lambda(\xi)}$$

où $\Lambda = \sum_{v \in M_k} \Lambda_v$ et $\Lambda_v : k_v \to \boldsymbol{R}/\boldsymbol{Z}$ est donnée par $\Lambda_v = \Lambda_p \circ \mathrm{Tr}_{k_v/\boldsymbol{Q}_p}$ où p est l'unique place de $\boldsymbol{Q}$ sous v, $\mathrm{Tr}_{k_v/\boldsymbol{Q}_p}$ l'application trace et Λ_p le caractère défini par :

- si $\boldsymbol{Q}_p = \boldsymbol{R}$, $\lambda_p(x) = [-x]$, la classe de $-x$ dans $\boldsymbol{R}/\boldsymbol{Z}$,
- sinon, Λ_p est la composée des applications naturelles

$$\boldsymbol{Q}_p \to \boldsymbol{Q}_p/\boldsymbol{Z}_p \to \boldsymbol{R}/\boldsymbol{Z}.$$

Par [**Ta**, theorem 4.1.4], l'application qui à un élément x de k associe $\xi \mapsto \chi(x\xi)$ définit un isomorphisme de k sur le groupe $\widehat{\boldsymbol{A}_k/k}$ des caractères de $\boldsymbol{A}_k/k$, et par [**Ta**, theorem 4.1.1], l'application qui à un élément η de $\boldsymbol{A}_k$ associe $\xi \mapsto \chi(\eta\xi)$ définit un isomorphisme de $\boldsymbol{A}_k$ sur son groupe des caractères $\widehat{\boldsymbol{A}_k}$. On obtient donc des dualités

$$\mathbf{e}(\langle .,. \rangle) : (\boldsymbol{A}_k/k)^m \times k^m \to \boldsymbol{S}^1 \text{ et } \mathbf{e}(\langle .,. \rangle) : \boldsymbol{A}_k^m \times \boldsymbol{A}_k^m \to \boldsymbol{S}^1.$$

On note également $\mathbf{e}_v$ le caractère défini par $\xi \mapsto e^{2i\pi\Lambda_v(\xi)}$. Par [**Ta**, §2.2], la mesure autoduale $\mathrm{d}\xi_v$ sur k_v pour $\mathbf{e}_v$ est donnée par

$$\mathrm{d}\xi_v = \begin{cases} \mathrm{d}x_v & \text{si } v \in M_\infty, \\ \#(\mathcal{O}_v/\mathfrak{d}_v)^{-1/2}\mathrm{d}x_v & \text{sinon,} \end{cases}$$

où, pour toute place finie v de k, $\mathfrak{d}_v$ désigne la différente absolue de k_v. On note $\mathrm{d}\boldsymbol{\xi}$ la mesure autoduale $\prod_{v \in M_k} \mathrm{d}\xi_v = \frac{1}{\sqrt{d}} \prod_{v \in M_k} \mathrm{d}x_v$ sur $\mathbf{A}_k$.

On fixe un système de hauteurs $\mathbf{H}_X$ sur X et on note $\mathbf{H}_V$ le système induit sur V, on note U_X un ouvert de X et $U_V = U_X \cap V$ que l'on supposera dense. On remplacera dans la suite U par U_X (respectivement U_V) dans la définition des fonctions $\Phi^{\mathbf{H}_X}_{\mathcal{T}_X}(H, \mathfrak{b}, .)$ (respectivement $\Phi^{\mathbf{H}_V}_{\mathcal{T}_V}(H, \mathfrak{b}, .)$) ainsi que ω_X^{-1} par ω_V^{-1} dans celle de $\Phi^{\mathbf{H}_X}_{\mathcal{T}_X}(H, \mathfrak{b}, .)$. On se donne également une fonction $\rho : \mathbf{A}_k^m \to \mathbf{R}$ continue, qui envoie 0 sur 1 et dont l'introduction sera justifiée partiellement au paragraphe 4.

Proposition 3.1.4. — *Avec les notations et conventions qui précèdent, pour tout $\mathfrak{b}$ du produit $\prod_{v \in M_k - S} X_*(T_{\mathrm{NS}})_v$ et tout $H \in \mathbf{R}_{>0}$, on a la relation*

$$\sum_{y \in \mathcal{T}_V(k)} \Phi^{\mathbf{H}_V}_{\mathcal{T}_V}(H, \mathfrak{b}, y) = \int_{\mathbf{A}_k^m/k^m} \sum_{y \in \mathcal{T}_X(k)} \Phi^{\mathbf{H}_X}_{\mathcal{T}_X}(H, \mathfrak{b}, y) \rho(\boldsymbol{f}(y)) \mathbf{e}(\langle \boldsymbol{\xi}, \boldsymbol{f}(y) \rangle) \mathrm{d}\boldsymbol{\xi}.$$

Démonstration. — Soit g la fonction de k^m dans $\mathbf{R}$ définie par

$$g(z) = \rho(z) \sum_{\{y \in \mathcal{T}_X(k) | \boldsymbol{f}(y) = z\}} \Phi^{\mathbf{H}_X}_{\mathcal{T}_X}(H, \mathfrak{b}, y).$$

Alors, comme $\{y \in \mathcal{T}_X(k) \mid \Phi^{\mathbf{H}_X}_{\mathcal{T}_X}(H, \mathfrak{b}, y) \neq 0\}$ est fini, $\boldsymbol{f}$ ne prend qu'un nombre fini de valeurs sur cet ensemble et donc g est à support fini. Par la formule d'inversion de Fourier, on a donc

$$g(0) = \int_{\mathbf{A}_k^m/k^m} \sum_{z \in k^m} g(z) \mathbf{e}(\langle \boldsymbol{\xi}, z \rangle) \mathbf{e}(-\langle \boldsymbol{\xi}, 0 \rangle) \mathrm{d}\boldsymbol{\xi}.$$

La proposition s'obtient alors à l'aide de la définition de g. □

3.2. Encerclement de la constante : introduction. — Notre objectif est maintenant d'obtenir un analogue intégral du résultat précédent, c'est-à-dire une formule de la forme

$$\int_{\mathcal{T}_{V_{C_{\text{eff}}(\overline{V})}(A_k)}} \Phi_{\mathcal{T}_V}^{\mathbf{H}_V}(H,\mathfrak{b},y)\boldsymbol{\omega}_{\mathcal{T}_V}(y) =$$

$$\int_{\mathbf{A}_k^m}\int_{\mathcal{T}_{X_{C_{\text{eff}}(\overline{X})}(A_k)}} \Phi_{\mathcal{T}_X}^{\mathbf{H}_X}(H,\mathfrak{b},y)\rho(\boldsymbol{f}(y))\mathbf{e}(\langle\boldsymbol{\xi},\boldsymbol{f}(y)\rangle)\boldsymbol{\omega}_{\mathcal{T}_X}(y)\mathrm{d}\boldsymbol{\xi}\,.$$

Nous allons nous inspirer pour cela du travail d'Igusa [**Ig2**, §IV.6]. Pour cela il nous faut tout d'abord construire l'analogue intégral de g et nous utiliserons la définition suivante :

Définition 3.2.1. — Soit z un élément de k_v^m ; soit $\mathcal{T}_X^z$ la sous-variété de $\mathcal{T}_{X_{k_v}}$ définie par le système d'équations

$$\boldsymbol{f}(y) = \boldsymbol{z}.$$

On considère alors la forme différentielle $\breve{\omega}_z$ définie par la relation :

$$\breve{\omega}_z(y) \wedge \boldsymbol{f}^*\Big(\bigwedge_{i=1}^m \mathrm{d}x_i\Big)(y) = \breve{\omega}_{\mathcal{T}_X}(y)$$

dans $\omega_{\mathcal{T}_X|\mathcal{T}_X^z}$.

Remarque 3.2.2. — Comme indiqué dans l'exemple 2.1.16, si $z = 0$ alors $\breve{\omega}_z$ coïncide avec la mesure $\breve{\omega}_{\mathcal{T}_V}$.

Exemple 3.2.3. — Si X est en outre une compactification équivariante projective lisse d'un tore T, alors $\mathcal{T}_X$ est un ouvert d'un espace affine $\mathbf{A}_k^N$ et la forme obtenue coïncide avec la forme de Leray usuelle.

Définition 3.2.4. — On note $\omega_{z,v}$ la mesure sur $\mathcal{T}_X^z(k_v)$ définie par la forme différentielle $\breve{\omega}_z$.

Remarque 3.2.5. — Si z est un élément de k^m et si $\mathcal{T}_X^z$ est une variété non-singulière sur k, ce qui résulte de nos hypothèses si $m = 1$, alors il résulte de [**We**, theorem 2.2.5] que pour presque toute place finie $\mathfrak{p}$ de k, on a la relation

$$\int_{\mathscr{T}_X^z(\mathcal{O}_\mathfrak{p})} \omega_{z,\mathfrak{p}} = \frac{\#\mathscr{T}_X^z(\mathbf{F}_\mathfrak{p})}{\#\mathbf{F}_\mathfrak{p}^{\dim \mathcal{T}_X^z}}$$

où $\mathscr{T}_X^z$ désigne un modèle de $\mathcal{T}_X^z$ sur $\mathcal{O}_S$.

Avant de passer à la formule adélique, il convient de considérer la formule locale correspondante qui s'écrit pour une fonction Φ convenable de $\mathcal{T}_X(k_v)$ dans $\mathbf{R}$:

$$(3.2) \quad \int_{\mathcal{T}_V(k_v)} \Phi(y) \omega_{\mathcal{T}_V,v}(y) = \int_{k_v^m} \int_{\mathcal{T}_X(k_v)} \Phi(y) e_v(\langle \boldsymbol{\xi}_v, \boldsymbol{f}(y) \rangle) \omega_{\mathcal{T}_X,v}(y) \mathrm{d} \boldsymbol{\xi}_v.$$

Pour cela nous étudions la fonction $g_v : k_v^m \to \mathbf{R}$ définie par la relation :

$$g_v(z) = \int_{\{y \in \mathcal{T}_X(k_v) | f(y) = z\}} \Phi(y) \omega_{z,v}(y).$$

Comme dans [**Ig2**, §IV.6], cette étude comprend deux parties indépendantes et de natures différentes. D'une part il faut montrer que g_v est continue en 0, ce qui se ramène à un problème de nature géométrique. D'autre part, il faut déterminer si la transformation de Fourier de g_v appartient à $L^1(k_v^m)$, qui est un problème analytique associé à des majorations de sommes d'exponentielles.

Finalement le passage du local à l'adélique nécessite la construction pour tout z de $\mathbf{A}_k^m$ d'une mesure $\boldsymbol{\omega}_z$ sur le produit restreint des $\mathcal{T}_X^{z_v}(k_v)$ relativement aux

$$\mathcal{T}_X^{z_v}(k_v) \cap \mathcal{T}_{X_{C_{\mathrm{eff}}(\overline{X})}}(\mathcal{O}_v).$$

La convergence de cette mesure passe également par une majoration de la transformée de Fourier de g_v.

Le prochain paragraphe est consacré à l'aspect géométrique de la question, à savoir l'étude de ce qu'Igusa nomme "données numériques".

3.3. Aspect géométrique. — Dans [**Ig2**, §IV.6], la partie géométrique est liée à des résolutions des singularités ; ici ce rôle est joué par des plongements équivariants de $\mathcal{T}_X$.

Pour cette étude nous ferons en outre l'hypothèse suivante :

Hypothèses 3.3.1. — La variété V est une hypersurface de X.

Remarque 3.3.2. — (i) On peut a priori se ramener à cette hypothèse par un argument d'itération.

(ii) Cette hypothèse entraîne que la variété $\mathcal{T}_X^z$ est lisse pour tout z de $\overline{k}$. Cela résulte de l'hypothèse sur V si $z = 0$ et de la formule (3.1) sinon.

Définition 3.3.3. — On dira que V *vérifie l'hypothèse* (G) s'il existe une désingularisation $\mathbf{A}_\Sigma$ de $\mathbf{A}_{C_{\mathrm{eff}}(\overline{X})}$ qui est équivariante sous l'action de T_{NS}.

Remarque 3.3.4. — L'existence d'une telle désingularisation a été annoncée par Brylinski [**Br**].

Proposition 3.3.5. — *Si V vérifie les hypothèses 2.1.4, 3.3.1 et (G) et si Φ est une fonction continue à support compact dans $\widehat{\mathcal{T}_X}(k_v)$, alors la fonction*

$$g_v : k_v \to \mathbf{R}$$
$$z \mapsto \int_{\{y \in \mathcal{T}_X(k_v) | f(y) = z\}} \Phi(y) \omega_{z,v}(y)$$

est continue.

Démonstration. — La continuité en-dehors de 0 est immédiate. Pour $z = 0$, il nous faut désingulariser $\widehat{\mathcal{T}_X}$. On considère donc une désingularisation équivariante $\mathbf{A}_\Sigma$ de $\mathbf{A}_{C_{\text{eff}}(\overline{X})}$ qui correspond à un éventail régulier Σ dont $C_{\text{eff}}(\overline{X})^\vee$ est le support (cf. [**Oda**, §1.5] et la démonstration du lemme 2.3.10). L'ensemble $\Sigma(1)$ des générateurs des cônes de dimension un dans Σ est en bijection avec l'ensemble des $\overline{k}$-hypersurfaces $\overline{T}_{\text{NS}}$-invariantes de $\mathbf{A}_\Sigma$. Quitte à éclater des sous-variétés toriques conjuguées de dimensions croissantes, on peut supposer que si deux de ces hypersurfaces sont conjuguées sous l'action du groupe de Galois, alors leur intersection est vide. On pose alors pour tout L de $\text{Pic}(X)$

$$\delta_\Sigma(L) = \inf_{\sigma \in \Sigma(1)} \langle \sigma, L \rangle$$

et $\delta_\Sigma(V) = \delta_\Sigma(\omega_V^{-1})$. Comme $\Sigma(1) \subset C_{\text{eff}}(\overline{V})^\vee \cap \text{Pic}\,\overline{V}^\vee - \{0\}$, on a

$$\delta_\Sigma(V) \geqslant \delta(V) > 1.$$

Le produit contracté

$$\mathcal{T}_{X,\Sigma} = \mathcal{T}_X \times^{T_{\text{NS}}} \mathbf{A}_\Sigma$$

définit alors une désingularisation de la variété $\widehat{\mathcal{T}_X}$ et on dispose d'un diagramme commutatif

$$\begin{array}{ccc} \mathcal{T}_X & \xrightarrow{\hat{\jmath}} & \mathcal{T}_{X,\Sigma} \\ & \searrow^j & \downarrow \pi \\ & & \widehat{\mathcal{T}_X} \end{array}$$

où j et $\hat{\jmath}$ sont des immersions ouvertes. Les composantes géométriques irréductibles du complémentaire de $\mathcal{T}_X$ dans $\mathcal{T}_{X,\Sigma}$ sont en bijection avec les éléments de $\Sigma(1)$. Soit $\mathscr{E}$ l'ensemble des composantes géométriques irréductibles du lieu des zéros de $f \circ \pi$. On a donc

$$\mathscr{E} = \left\{ \overline{\hat{\jmath}(\mathcal{T}_V)} \right\} \cup \{D_\sigma, \sigma \in \Sigma(1)\}.$$

Pour chaque E de $\mathscr{E}$, on définit comme dans [**Ig2**, §III.2.2], les données numériques (N_E, ν_E) le long de E dans $\mathcal{T}_{X,\Sigma}$ de la manière suivante : on se place sur une extension K_v de k_v sur laquelle E est défini, N_E désigne la multiplicité de

$f \circ \pi$ sur E et si $X_1, \dots, X_N$ sont des coordonnées v-adiques au voisinage d'un point x_0 de E, en-dehors des intersections $E \cap E'$ pour $E' \neq E$, de sorte que E ait pour équation $X_1 = 0$ au voisinage de x_0, alors la forme $\pi^* \breve\omega_{\mathcal{T}_X}$ s'écrit

$$(3.3) \qquad \eta X_1^{\nu_E - 1} \bigwedge_{i=1}^{N} \mathrm{d}X_i$$

au voisinage de x_0, où η désigne une fonction localement inversible au voisinage de x_0.

Supposons que Φ soit à valeurs positives ou nulles. Comme dans [**Ig1**, p. 53-54], lorsque z tend vers 0, la fonction g_v tend vers

$$\int_{\mathcal{T}_{X, \Sigma}(k_v)} \Phi \circ \pi(y) \boldsymbol{\omega}_{(0,v)}(y) \leqslant +\infty$$

où $\boldsymbol{\omega}_{(0,v)}$ est une mesure supportée par

$$\bigcup_{E \in \mathscr{E}^{\mathrm{Gal}(\overline{k}_v / k_v)}} E.$$

La restriction à chacun des diviseurs E est donnée par

$$\left| \frac{\eta Y_1^{\nu_E - 1}}{\frac{\partial f \circ \pi}{\partial Y_1}} \right|_{v \mid Y_1 = 0} \bigwedge_{i=2}^{N} \mathrm{d}Y_i.$$

Elle est donc finie si $N_E - 1 \leqslant \nu_E - 1$ et nulle si $\nu_E - N_E > 0$. Si $E = \overline{\hat{\jmath}(\mathcal{T}_V)}$, $\nu_E = N_E = 1$ et l'intégrale obtenue coïncide avec $g_v(0)$. Par conséquent, si $\nu_{D_\sigma} - N_{D_\sigma} > 0$ pour tout σ de $\Sigma(1)$, la fonction g_v est continue en 0.

Il reste à déterminer $\nu_\sigma = \nu_{D_\sigma}$ et $N_\sigma = N_{D_\sigma}$, ce qui peut être fait dans une fibre au-dessus de X. Fixons donc un élément σ de $\Sigma(1)$ et donnons-nous une base $\xi_1, \dots, \xi_r$ de $X^*(T_{\mathrm{NS}})$, de sorte que $\sigma(\xi_i) = \delta_{1,i}$. A une constante multiplicative inversible près la mesure sur cette fibre correspond à la forme différentielle

$$\prod_{i=1}^{n} \xi_i^{\langle \xi_i^\vee, \omega_X^{-1} \rangle} \frac{\mathrm{d}\xi_1}{\xi_1} \wedge \dots \wedge \frac{\mathrm{d}\xi_n}{\xi_n}$$

où $(\xi_1^\vee, \dots, \xi_r^\vee)$ est la base duale de $(\xi_1, \dots, \xi_r)$. Comme D_σ est donné par $\xi_1 = 0$ et $\sigma = \xi_1^\vee$, on obtient par (3.3) que

$$\nu_\sigma = \langle \sigma, \omega_X^{-1} \rangle.$$

Par ailleurs, par la relation (3.1), la restriction de $f \circ \pi$ à la fibre est, à une constante éventuellement nulle près, de la forme

$$\prod_{i=1}^{r} \xi_i^{\langle \xi_i^\vee, L \rangle}$$

où $L = L_1$. On obtient donc l'égalité

$$N_\sigma = \langle \sigma, L \rangle$$

donc

$$\nu_\sigma - N_\sigma = \langle \omega_X^{-1} - [L], \sigma \rangle = \langle \omega_V^{-1}, \sigma \rangle. \qquad \square$$

Proposition 3.3.6. — *Si V vérifie les hypothèses 2.1.4, 3.3.1 et* (G), *si*

$$\delta(\omega_X^{-1} - (k+1)L) > 0$$

et si Φ est une fonction C^∞ à support compact dans $\widehat{\mathcal{T}_X}(k_v)$ pour une place archimédienne v, alors la fonction g_v définie dans la proposition précédente est de classe C^k.

Démonstration. — On reprend les notations de la proposition précédente. Soit $x \in \mathcal{T}_{X,\Sigma}(k_v)$ tel que $f \circ \pi(x) = 0$. Soit

$$\mathscr{E}_x = \{ E \in \mathscr{E}^{\mathrm{Gal}(\overline{k}_v/k_v)} \mid x \in E(k_v) \}.$$

Notons $E_1, \ldots, E_r$ les éléments de $\mathscr{E}_x$ et (N_i, ν_i) les données numériques associées à E_i. On peut choisir des coordonnées v-adiques $Y_1, \ldots, Y_n$ sur $\mathcal{T}_{X,\Sigma}(k_v)$ au voisinage de x de sorte que le diviseur E_i soit donné par l'équation $Y_i = 0$ pour $i = 1, \ldots, r$. Localement $f \circ \pi$ s'écrit

$$f \circ \pi = \varepsilon \prod_{1 \leqslant i \leqslant r} Y_i^{N_i}$$

où ε est localement inversible et la forme $\pi^* \breve{\omega}_{\mathcal{T}_X}$

$$\pi^* \breve{\omega}_{\mathcal{T}_X} = \varepsilon' \prod_{1 \leqslant i \leqslant r} Y_i^{\nu_i - 1} dY_1 \wedge \cdots \wedge dY_n,$$

où ε' désigne une fonction localement inversible. Sur une boule ouverte W de centre x, la mesure $\pi^*(\omega_{z,v})$ pour $z \in k_v^\times$ peut être décrite comme la restriction à $f \circ \pi^{-1}(z)$ de la forme

$$\varepsilon'' Y_1^{\nu_1 - N_1} \prod_{i=2}^{r} Y_i^{\nu_i - N_i - 1} dY_2 \wedge \cdots \wedge dY_n$$

où
$$\varepsilon'' = (N_1\varepsilon + \frac{\partial \varepsilon}{\partial Y_1}Y_1)^{-1}\varepsilon'$$
est inversible au voisinage de x. Quitte a réordonner les E_i, on peut en outre supposer que
$$\nu_1 - N_1 = \inf_{1 \leqslant i \leqslant r}(\nu_i - N_i).$$
Soit donc C_0 un cône de k_v^r ne rencontrant pas les hyperplans de coordonnées et C l'ensemble $C_0 \times k_v^{n-r}$. On considère la fonction
$$\rho(z) = \int_{\{x \in W \cap C \mid f \circ \pi = z\}} |\varepsilon''|_v |Y_1|_v^{\nu_1 - N_1} \prod_{i=2}^{r} |Y_i|_v^{\nu_i - N_i - 1} dY_2 \wedge \cdots \wedge dY_n.$$
Si $r = 1$ et $E_1 = \overline{j(\mathcal{T}_V)}$, alors on obtient
$$\rho(z) = \int_{\{x \in W \mid f \circ \pi = z\}} |\varepsilon''|_v dY_2 \wedge \cdots \wedge dY_n$$
et on obtient que ρ est C^∞ en 0. Sinon en utilisant un changement de coordonnées homogènes en les r premières variables, on obtient que la dérivée k-ième de ρ admet une majoration de la forme
$$\rho^{(k)}(z) = O\left(|z|_v^{\frac{\sum_{i=1}^r \nu_i}{\sum_{i=1}^r N_i} - k - 1}\right)$$
au voisinage de 0. En conséquence, en décomposant un voisinage du fermé $f \circ \pi^{-1}(0)$ en domaines de la forme précédente, si $\delta(\omega_X^{-1} - (k+1)L) > 0$ et donc $\nu_i > (k+1)N_i$, on montre donc que ρ est C^k en 0. □

3.4. Aspect analytique. — Le but de ce paragraphe est de donner un exemple pour la condition analytique suivante :

Définition 3.4.1. — On se place dans les hypothèses du paragraphe précédent. On dira que la paire (X, V) *vérifie la condition* (A) si et seulement s'il existe $\sigma > 2$ tel que, quitte à agrandir S, pour toute place finie v de k et tout sous-ensemble W ouvert compact de $\widehat{\mathcal{T}_X}(k_v)$ on ait
$$\forall \xi \in k_v - \mathcal{O}_v, \quad \left| \int_{\mathcal{T}_X(k_v) \cap W} \mathbf{e}_v(\xi f(x)) \boldsymbol{\omega}_{\mathcal{T}_X}(x) \right| \leqslant C|\xi|_v^{-\sigma}$$
pour une constante C égale à 1 si $v \notin S$ et $W \subset \widehat{\mathcal{T}_X}(\mathcal{O}_v)$.

Remarque 3.4.2. — Il s'agit en fait d'une majoration de somme d'exponentielles.

Nous allons maintenant décrire un cas particulier où cette condition est impliquée par le cas classique des hypersurfaces projectives.

Proposition 3.4.3. — *Si, en outre, X est le produit de m espaces projectifs $\prod_{i=0}^{m} \mathbf{P}_{\mathbf{Q}}^{n_i}$ sur $\mathbf{Q}$, f est alors donné par un polynôme multihomogène dont le degré total est noté d. On suppose que la dimension de V est supérieure ou égale à 4 et que*
$$\inf_{0 \leqslant i \leqslant m} n_i \geqslant 2^d d.$$
Alors la condition (A) *est vérifiée pour* (X, V).

Démonstration. — On pose $N = \sum_{i=0}^{m} n_i + 1$. Par [**Pe2**, exemple 4.2.2], on sait que l'ensemble $\mathcal{T}_X(\mathbf{Q}_p) \cap \widehat{\mathcal{T}_X}(\mathbf{Z}_p)$ coïncide avec $\prod_{i=0}^{m}(\mathbf{Z}_p^{n_i+1} - \{0\})$. Mais la fonction indicatrice d'un ouvert compact de $\prod_{i=0}^{m} \mathbf{A}^N(\mathbf{Q}_p)$ s'écrit comme combinaison linéaire d'indicatrices d'images de $\mathbf{Z}_p^N$ par des homothéties-translations. On est donc ramené à la majoration classique
$$\left| \int_{\mathbf{Z}_p^N} \mathbf{e}(\xi f(x)) \mathrm{d}x \right|.$$
Mais les singularités de $f(x) = 0$ sont contenues dans la réunion
$$\bigcup_{i=0}^{m} \{0\} \times \prod_{j \neq i} \mathbf{A}_{\mathbf{Q}}^{n_i+1}.$$
La codimension du lieu singulier est donc minorée par le plus petit des entiers $n_i + 1$. La proposition résulte alors de [**Ig1**, lemma 6, page 63]. □

3.5. Transformation de Fourier locale. — Dans ce paragraphe nous étudions la validité de la formule locale (3.2) sous les conditions décrites dans les deux paragraphes qui précèdent.

Proposition 3.5.1. — *Si V vérifie l'hypothèse* (G), *si $\delta(\omega_X^{-1} - 3L) > 0$ et si la paire (X, V) satisfait la condition* (A), *alors pour toute place v de k et toute fonction complexe Φ sur $\widehat{\mathcal{T}_X}(k_v)$ continue, à support compact, C^∞ si v est archimédienne et localement constante si v est finie, on a la relation*
$$\int_{\mathcal{T}_V(k_v)} \Phi(y) \boldsymbol{\omega}_{\mathcal{T}_V,v}(y) = \int_{k_v} \int_{\mathcal{T}_X(k_v)} \Phi(y) \mathbf{e}_v(\xi_v f(y)) \boldsymbol{\omega}_{\mathcal{T}_X,v}(y) \mathrm{d}\xi_v.$$

Démonstration. — Il résulte de la proposition 3.3.5 que la fonction
$$g_v : k_v \to \mathbf{R}$$
$$z \mapsto \int_{\{y \in \mathcal{T}_X(k_v) \mid f(y) = z\}} \Phi(y) \boldsymbol{\omega}_{z,v}(y)$$

est continue. Elle est en outre nulle en dehors de l'image par f du support de Φ, et donc à support compact. A fortiori, g_v appartient à $L^1(k_v)$. La transformée de Fourier $\widehat{g_v}$ de g_v est bien définie et se met sous la forme

$$\widehat{g_v}(\xi) = \int_{k_v} g_v(z)\mathbf{e}_v(\xi z)\mathrm{d}z$$
$$= \int_{\mathcal{T}_X(k_v)} \mathbf{e}_v(\xi f(y))\Phi(y)\boldsymbol{\omega}_{\mathcal{T}_X}(y).$$

Mais par hypothèse si $v \in M_f$, Φ s'écrit comme combinaison linéaire de fonctions indicatrices d'ensembles ouverts compacts W_i de $\widehat{\mathcal{T}_X}$. Cette transformée de Fourier se récrit donc

$$\widehat{g_v}(\xi) = \sum_{i \in J} \lambda_i \int_{\mathcal{T}_X(k_v) \cap W_i} \mathbf{e}(\xi f(y))\boldsymbol{\omega}_{\mathcal{T}_X,v}(y)$$
$$\leqslant C|\xi|_v^{-\sigma}$$

avec $\sigma > 2$. Donc $\widehat{g_v}$ est une fonction L^1. Si v est réelle, l'hypothèse que le nombre $\delta(\omega_X^{-1} - 3L)$ soit positif assure que que g_v est C^2 à support compact et, par conséquent, $\hat{g}_v \in L^1(k_v)$.

Dans tous les cas on peut appliquer la formule d'inversion de Fourier et on obtient

$$g_v(0) = \int_{k_v} \widehat{g_v}(\xi)\mathrm{d}\xi$$

et par conséquent

$$\int_{\mathcal{T}_V(k_v)} \Phi(y)\boldsymbol{\omega}_{\mathcal{T}_V,v}(y) = \int_{k_v}\int_{\mathcal{T}_X(k_v)} \Phi(y)\mathbf{e}_v(\xi f(y))\boldsymbol{\omega}_{\mathcal{T}_X,v}(y)\mathrm{d}\xi. \quad \square$$

3.6. Mesures adéliques. — Pour passer de la transformation de Fourier sur k_v à celle sur les adèles, il nous faut démontrer la convergence de certaines mesures adéliques.

Notation 3.6.1. — Si z est un élément de $\boldsymbol{A}_k$, on note $\mathcal{T}_X^z(\boldsymbol{A}_k)$ le produit restreint de $\mathcal{T}_X^{z_v}(k_v)$ relativement aux intersections

$$\mathcal{T}_X^{z_v}(k_v) \cap \widehat{\mathscr{T}_X}(\mathcal{O}_v)$$

où $\widehat{\mathscr{T}_X}$ est un modèle de $\widehat{\mathcal{T}_X}$.

Proposition 3.6.2. — *Si V vérifie la condition* (G), *si $\delta(\omega_X^{-1} - 3L) > 0$ et si la paire (X, V) vérifie la propriété* (A), *alors pour tout z de $\boldsymbol{A}_k$, le produit des mesures*
$$\boldsymbol{\omega}_z = \prod_{v \in M_k} \boldsymbol{\omega}_{z_v, v}$$
converge sur $\mathcal{T}_X^z(\boldsymbol{A}_k)$ où $\boldsymbol{\omega}_{z_v, v}$ a été définie au paragraphe 3.2.

Cette proposition repose sur le lemme suivant, analogue du lemme 6.6 de [**Ig2**, page 165].

Lemme 3.6.3. — *Si V vérifie* (G), *si $\delta(\omega_X^{-1} - 3L) > 0$ et s'il existe $\sigma > 2$ tel que pour presque tout $v \in M_f - S$ on ait*
$$\left| \int_{\mathcal{T}_X(k_v) \cap \widehat{\mathcal{F}_X}(\mathcal{O}_v)} \mathbf{e}_v(\xi f(y)) \boldsymbol{\omega}_{\mathcal{T}_X, v}(y) \right| < |\xi|_v^{-\sigma},$$
alors pour presque toute place v de $M_f - S$ on a
$$\left| \int_{\mathcal{T}_X^{z_v}(k_v) \cap \widehat{\mathcal{F}_X}(\mathcal{O}_v)} \boldsymbol{\omega}_{\mathcal{T}_X, v}(y) - 1 \right| < C q_v^{-\frac{3}{2}} + C' q_v^{-(\sigma - 1)}$$
où C et C' sont indépendantes de v et $q_v = \#\mathbf{F}_v$.

Remarque 3.6.4. — Il résulte du lemme précédent que la proposition 3.6.2 est valide sous la condition analytique plus faible notée (AF) qui suit : il existe un ensemble S_1 de places et un nombre réel $\sigma > 2$, tel que pour tout $v \in M_f - S_1$ on ait
$$\left| \int_{\mathcal{T}_X(k_v) \cap \widehat{\mathcal{F}_X}(\mathcal{O}_v)} \mathbf{e}_v(\xi f(y)) \boldsymbol{\omega}_{\mathcal{T}_X, v}(y) \right| < |\xi|^{-\sigma}.$$

Démonstration du lemme 3.6.3. — On considère la fonction
$$\begin{aligned} g_v : k_v &\to \mathbf{R} \\ z &\mapsto \int_{\mathcal{T}_X^{z_v}(k_v) \cap \widehat{\mathcal{F}_X}(\mathcal{O}_v)} \boldsymbol{\omega}_{z_v, v}. \end{aligned}$$
Il résulte de la démonstration de la proposition 3.5.1 qu'on a les relations
$$\begin{aligned} g_v(z_v) &= \int_{k_v} \int_{\mathcal{T}_X(k_v) \cap \widehat{\mathcal{F}_X}(\mathcal{O}_v)} \mathbf{e}_v(\xi_v(f(y) - z_v)) \boldsymbol{\omega}_{\mathcal{T}_X, v}(y) d\xi_v \\ &= \int_{k_v} \widehat{g_v}(\xi_v) \mathbf{e}_v(-\xi_v z_v) d\xi_v. \end{aligned}$$

Mais, en-dehors d'un nombre fini de places, si $\xi \in \mathcal{O}_v$, on a

$$\hat{g}_v(\xi) = \int_{\mathcal{T}_X(k_v) \cap \widehat{\mathcal{F}_X}(\mathcal{O}_v)} \omega_{\mathcal{T}_X}$$
$$= L_v(T_{\mathrm{NS}}, C_{\mathrm{eff}}(\overline{V}), \omega_X^{-1}) L_v(1, \mathrm{Pic}\,\overline{V})^{-1} \boldsymbol{\omega}_{\mathbf{H}(\omega_X^{-1}),v}(X(k_v))$$

où la deuxième égalité résulte de [**Pe2**, démonstration du théorème 5.3.1] et de [**BT1**, p. 604].

Il résulte alors de l'hypothèse 2.1.4 et de [**Dr**, proposition 3] que

$$|L_v(T_{\mathrm{NS}}, C_{\mathrm{eff}}(\overline{V}), \omega_V^{-1}) - 1| \leqslant C_1 q_v^{-2}$$

et de [**Pe1**, page 117] que

$$\left| \frac{\boldsymbol{\omega}_{\mathbf{H}(\omega_X^{-1}),v}(X(k_v))}{L_v(1, \mathrm{Pic}\,\overline{V})} - 1 \right| < C_2 q_v^{-3/2}.$$

Par conséquent $\hat{g}_v(\xi)$ est constant pour $\xi \in \mathcal{O}_v$ et vérifie sur cet ensemble

$$|\hat{g}_v(\xi) - 1| < C q_v^{-3/2}.$$

On en déduit les inégalités

$$|g_v(z_v) - 1| \leqslant C q_v^{-3/2} + \int_{k_v - \mathcal{O}_v} |\xi_v|^{-\sigma} d\xi_v$$
$$\leqslant C q_v^{-3/2} + (1 - 2^{-(\sigma-1)})^{-1} q_v^{-(\sigma-1)}$$

où la dernière inégalité est donnée par [**Ig2**, page 164]. □

3.7. Transformation adélique et encerclement de la constante. —
Nous pouvons maintenant énoncer un des principaux résultats de ce texte.

Théorème 3.7.1. — *Soit X une variété presque de Fano vérifiant les hypothèses 2.1.4 ainsi que la condition* (G). *Soit V une hypersurface de X définie par l'annulation d'une section d'un fibré L vérifiant également ces conditions et telle qu'en outre la restriction induise un isomorphisme*

$$\mathrm{Pic}\,\overline{X} \to \mathrm{Pic}\,\overline{V}$$

envoyant le cône effectif de X sur celui de V. Soit $\mathcal{T}_V$ un torseur universel au-dessus de V ayant un point rationnel et s'inscrivant dans un diagramme

$$\begin{array}{ccc} \mathcal{T}_V & \longrightarrow & \mathcal{T}_X \\ \downarrow & & \downarrow \\ V & \stackrel{j}{\longrightarrow} & X. \end{array}$$

On suppose que $\delta(\omega_X^{-1} - 3L) > 0$ et que la paire (X, V) vérifie (A). On fixe en outre une fonction continue $\rho_S : \prod_{v \in S} k_v \to \mathbf{R}$ envoyant 0 sur 1 et telle que l'application

$$(y_v)_{v \in S \cap M_f} \mapsto \begin{pmatrix} \rho_\infty^x : \prod_{v \in S \cap M_\infty} k_v & \to & \mathbf{R} \\ (x_v)_{v \in S \cap M_\infty} & \to & \rho(\boldsymbol{y}, \boldsymbol{x}) \end{pmatrix}$$

soit localement constante à valeurs dans les fonctions C^∞. On note $\rho : \boldsymbol{A}_k \to \mathbf{R}$ la fonction induite. Alors pour toute fonction

$$\Phi = \Phi_\infty \cdot \prod_{v \in M_f} \Phi_v : \mathcal{T}_{X_{C_{\mathrm{eff}}}(\overline{X})}(\boldsymbol{A}_k) \to \mathbf{R}$$

où Φ_∞ est C^∞ à support compact dans $\prod_{v \in M_\infty} \widehat{\mathcal{T}_X}(k_v)$, et Φ_v localement constante pour v finie et coïncide avec la fonction caractéristique de l'intersection $\mathcal{T}_X(k_v) \cap \widehat{\mathcal{T}_X(\mathcal{O}_v)}$ pour presque toute place finie, on a la relation

(3.4)
$$\int_{\mathcal{T}_{V_{C_{\mathrm{eff}}}(\overline{V})}(\boldsymbol{A}_k)} \Phi(y) \boldsymbol{\omega}_{\mathcal{T}_V}(y) = \int_{\boldsymbol{A}_k} \int_{\mathcal{T}_{X_{C_{\mathrm{eff}}}(\overline{X})}(\boldsymbol{A}_k)} \Phi(y) \rho(f(y)) \mathrm{e}(\boldsymbol{\xi} f(y)) \boldsymbol{\omega}_{\mathcal{T}_X}(y) \mathrm{d}\boldsymbol{\xi}.$$

Démonstration. — On considère la fonction

$$\begin{array}{rcl} g : \boldsymbol{A}_k & \to & \mathbf{R} \\ \boldsymbol{z} & \mapsto & \rho(\boldsymbol{z}) \int_{\mathcal{T}_X^z(\boldsymbol{A}_k)} \Phi(y) \boldsymbol{\omega}_z(y). \end{array}$$

La fonction g est à support compact et donc a fortiori $g \in L^1(\boldsymbol{A}_k)$. Il résulte de la proposition 3.3.5 et de la démonstration de la convergence de $\boldsymbol{\omega}_z$ que la fonction g est continue.

Sa transformée de Fourier est donnée par la formule

$$\hat{g}(\boldsymbol{\xi}) = \int_{\mathcal{T}_{X_{C_{\mathrm{eff}}}(\overline{X})}(\boldsymbol{A}_k)} \rho(f(y)) \Phi(y) \mathrm{e}(\langle \boldsymbol{\xi}, f(y) \rangle) \boldsymbol{\omega}_{\mathcal{T}_X}(y)$$

$$= \prod_{v \in M_f - S} \int_{\mathcal{T}_X(k_v)} \Phi_v(y) \mathrm{e}_v(\xi_v f(y)) \boldsymbol{\omega}_{\mathcal{T}_{X,v}}(y)$$

$$\times \int_{\prod_{v \in S} \mathcal{T}_X(k_v)} \rho_S(f(y)) \Phi_S(y) \mathrm{e}_S(\xi_S f(y)) \prod_{v \in S} \boldsymbol{\omega}_{\mathcal{T}_{X,v}}(y)$$

où Φ_S et e_S sont définis par produit sur les places de S. Il en résulte que

$$|\hat{g}(\boldsymbol{\xi})| \leqslant C' \left(\prod_{|\xi_v|_v > 1} |\xi_v|_v^{-\sigma} \right) \left(\prod_{v \in M_f} (1 + C \# \mathbf{F}_v)^{-3/2} \right)$$

où $\sigma > 2$ et donc $\hat{g}$ est une fonction L^1. En appliquant le formule d'inversion de Fourier, on obtient (3.4) □

Théorème 3.7.2. — *On suppose que X, V et ρ vérifient les hypothèses du théorème précédent, que $\delta(\omega_X^{-1} - ([k:\mathbf{Q}]+2)L) > 0$ et que $\dim V > 2[k:\mathbf{Q}]$ et on considère les fonctions $\Phi_{\mathcal{T}_X}^{\mathbf{H}_X}(H,\mathfrak{b},.)$ définies au paragraphe 3.1. On suppose en outre que pour toute place archimédienne v de k les métriques choisies sur $X(k_v)$ sont C^∞ et la fonction définie sur $X(k_v)$ par*

$$\tilde{f}_v : y \mapsto \frac{|f(y)|_v}{\|y\|_v^L}$$

est telle qu'en tout point en lequel $\mathrm{d}\tilde{f}_v = 0$ la forme quadratique $\mathrm{d}^2 \tilde{f}_v$ est non dégénérée. Alors pour tout $\mathbf{H}$ de $\mathbf{R}_{>0}$ et tout $\mathfrak{b}$ de $\bigoplus_{v \in M_f - S} X_*(T_{\mathrm{NS}})_v$, on a

$$(3.5) \quad \int_{\mathcal{T}_{V_{C_{\mathrm{eff}}(\overline{V})}}(\mathbf{A}_k)} \Phi_{\mathcal{T}_V}^{\mathbf{H}_V}(H,\mathfrak{b},y)\omega_{\mathcal{T}_V}(y)$$
$$= \int_{\mathbf{A}_k} \int_{\mathcal{T}_{X_{C_{\mathrm{eff}}(\overline{X})}}(\mathbf{A}_k)} \Phi_{\mathcal{T}_X}^{\mathbf{H}_X}(H,\mathfrak{b},y)\rho(f(y))\mathbf{e}(\langle \boldsymbol{\xi}, f(y)\rangle)\omega_{\mathcal{T}_X}(y)\mathrm{d}\boldsymbol{\xi}.$$

Démonstration. — Rappelons que la fonction $\Phi_{\mathcal{T}_X}^{\mathbf{H}_X}(H,\mathfrak{b},.)$ est la fonction indicatrice de l'ensemble $\mathcal{E}$ défini par

(3.6) $\quad\quad\quad\quad \forall v \in S, \quad \pi_X(y_v) \in U_X(k_v),$

(3.7) $\quad\quad\quad\quad (y_v)_{v \in S} \in \Delta_{\mathbf{H}_K}(\mathcal{T}_X),$

(3.8) $\quad\quad\quad\quad \forall L \in C_{\mathrm{eff}}(V), \quad \prod_{v \in S} \|y_v\|_v^L \leqslant 1,$

(3.9) $\quad\quad\quad\quad \prod_{v \in S}(\|y_v\|_v^{\omega_v^{-1}})^{-1} \leqslant H,$

(3.10) $\quad\quad\quad\quad y \in \mathfrak{b}.\mathcal{T}_{X_{C_{\mathrm{eff}}(\overline{V})},S}(\mathbf{A}_k).$

Elles s'écrivent donc

$$\Phi_S \times \prod_{v \notin S} \Phi_v$$

où les fonctions Φ_v sont localement constantes, à support compact et coïncident pour presque toute place v avec la fonction caractéristique de $\mathcal{T}_X(k_v) \cap \widehat{\mathcal{T}_X}(\mathcal{O}_v)$. D'autre part, la fonction Φ_S est à support compact dans $\prod_{v \in S} \widehat{\mathcal{T}_X}(k_v)$. En effet il résulte des conditions (3.7), (3.8) et (3.9) qui précèdent que l'image par

$\prod_{v\in S} \tilde{\mathbf{H}}^{\log}_{\mathfrak{T}_X,v}$ du domaine $\mathscr{E}$ est bornée dans
$$\prod_{v\in S} X^*(T_{\mathrm{NS}})^\vee_v \otimes \mathbf{R}.$$

En outre la fonction
$$(x_v)_{v\in M_f\cap S} \mapsto \Phi^{\boldsymbol{x}}_\infty = ((y_v)_{v\in M_\infty} \mapsto \Phi_S(\boldsymbol{y},\boldsymbol{x}))$$
est localement constante. Compte tenu de la démonstration du théorème précédent, il suffit de montrer que pour $\boldsymbol{x} = (x_v)_{v\in M_f\cap S}$ fixé la fonction

$$g_\infty : \prod_{v\in M_\infty} k_v \to \mathbf{R}$$
$$\boldsymbol{z} = (z_v)_{v\in M_\infty} \mapsto \int_{\prod_{v\in M_\infty} \mathfrak{T}^{z_v}_X(k_v)} \Phi^{\boldsymbol{x}}_\infty(y) \rho^{\boldsymbol{x}}_\infty(f(y)) \omega_{\mathfrak{T}_X,\infty}(y)$$

est $C^{1+[k:\mathbf{Q}]}$ ce qui entraînera que sa transformée de Fourier admet une majoration de la forme $C\|\boldsymbol{\xi}\|^{-(1+[k:\mathbf{Q}])}$ et donc que la fonction $\hat{g}$ définie dans la démonstration précédente appartient à $L^1(\boldsymbol{A}_k)$. Comme $\delta(\omega_X^{-1} - ([k:\mathbf{Q}]+2)L) > 0$, la difficulté provient uniquement des discontinuités de la fonction $\Phi^{\boldsymbol{x}}_\infty$.

Mais il résulte des hypothèses faites sur les métriques réelles que l'ensemble $\mathscr{E}^{\boldsymbol{x}}_\infty$ défini par les conditions (3.7), (3.8) et (3.9) peut être décrit comme l'image inverse par l'application C^∞ dominante $\prod_{v\in M_\infty} \tilde{\mathbf{H}}^{\log}_{\mathfrak{T}_X,v}$ d'un polyèdre $P_{\boldsymbol{x}}$ de $\prod_{v\in M_\infty} X^*(T_{\mathrm{NS}})_v \otimes \mathbf{R}$. On considère l'application

$$\prod_{v\in M_\infty} \tilde{\mathbf{H}}^{\log}_{\mathfrak{T}_X,v} \times \prod_{v\in M_\infty} \tilde{f}_v : \prod_{v\in M_\infty} \mathfrak{T}_X(k_v) \to \prod_{v\in M_\infty} X^*(T_{\mathrm{NS}})_v \otimes \mathbf{R} \times \prod_{v\in M_\infty} \mathbf{R}.$$

Au voisinage d'un point en lequel les $\mathrm{d}\tilde{f}_v$ ne s'annulent pas, on peut à l'aide de paramétrisations C^∞ se ramener à l'intersection d'un domaine de la forme

$$P_{\boldsymbol{x}} \times \prod_{v\in M_\infty} \mathbf{R} \subset \prod_{v\in M_\infty} X^*(T_{\mathrm{NS}})_v \otimes \mathbf{R} \times \prod_{v\in M_\infty} \mathbf{R}$$

avec un espace affine dépendant de $(z_v)_{v\in M_\infty}$ donné par un système d'équations de la forme
$$z_v l_v(t) = y_v \quad \text{pour} \quad v \in M_\infty$$
où l_v est une forme linéaire sur $\prod_{v\in M_\infty} X^*(T_{\mathrm{NS}})_v \otimes \mathbf{R}$. Au voisinage d'un point en lequel $\mathrm{d}\tilde{f}_v$ s'annule, une réduction analogue amène à considérer en la place v une équation de la forme
$$\sum_{i=1}^{[k_v:\mathbf{R}]\dim X} \varepsilon_i y_{i,v}^2 = z_v l_v(t).$$

avec $\varepsilon_i \in \{-1,1\}$. Notons que ce cas ne peut pas se produire lorsque $z_v = 0$ puisqu'on a supposé V lisse et que l'adhérence du domaine choisi est contenue dans $\prod_{v \in M_\infty} \mathcal{T}_X(k_v)$. Dans tous les cas, l'assertion de dérivabilité résulte de l'hypothèse sur la dimension de V et de calculs de volumes élémentaires.

On obtient donc que $\hat{g}$ est L^1 et la fin de la démonstration est similaire à celle du théorème précédent. $\square$

***Corollaire* 3.7.3.** — *Sous les hypothèses 2.5.1 et celles du théorème précédent, on a les relations*

(3.11) $\#W(T_{\mathrm{NS}}) n_{U,\mathbf{H}}(H)$
$$= \sum_{\substack{i \in I \\ \mathfrak{b} \in \bigoplus_{v \in M_f - S} X_*(T_{\mathrm{NS}})_v}} \mu(\mathfrak{b}) \int_{A_k/k} \sum_{y \in \mathcal{T}_{i,X}(k)} \Phi_{\mathcal{T}_{i,X}}^{\mathbf{H}_X}(H, \mathfrak{b}, y) \rho(f(y)) \mathbf{e}(\langle \boldsymbol{\xi}, f(y) \rangle) \mathrm{d}\boldsymbol{\xi}$$

(3.12) $\#W(T_{\mathrm{NS}}) \alpha(V) \beta(V) \tau_{\mathbf{h}}(V) \int_0^{\log H} u^{t-1} e^u \mathrm{d}u$
$$= \sum_{\substack{i \in I \\ \mathfrak{b} \in \bigoplus_{v \in M_f - S} X_*(T_{\mathrm{NS}})_v}} \mu(\mathfrak{b}) \int_{A_k} \int_{\mathcal{T}_{i,X_{C_{\mathrm{eff}}(\overline{X})}}(A_k)} \Phi_{\mathcal{T}_{i,X}}^{\mathbf{H}_X}(H, \mathfrak{b}, y) \rho(f(y)) \mathbf{e}(\langle \boldsymbol{\xi}, f(y) \rangle) \omega_{\mathcal{T}_{i,X}}(y) \mathrm{d}\boldsymbol{\xi}.$$

***Corollaire* 3.7.4.** — *Si* $X = \prod_{i=1}^m \mathbf{P}_{\mathbf{Q}}^{n_i}$ *et* f *est donnée par un polynôme homogène de degré total* $d \geqslant 2$ *de sorte que*
$$\inf_{1 \leqslant i \leqslant m} n_i \geqslant 2^d d,$$
si V *est lisse de dimension supérieure ou égale à 4 et si les métriques sont choisies de manière à vérifier les conditions ci-dessus, alors on pose* $N = \sum_{i=1}^m n_i + 1$ *et on a les relations*

(3.13)
$$n_{U,\mathbf{H}}(H) = \frac{1}{2^m} \sum_{\substack{\mathfrak{b} \in \bigoplus_{v \in M_f - S} \mathbf{Z}^m}} \mu(\mathfrak{b}) \int_{A_{\mathbf{Q}}/\mathbf{Q}} \sum_{y \in \mathbf{Q}^N} \Phi_{A_{\mathbf{Q}}^N}^{\mathbf{H}_X}(H, \mathfrak{b}, y) \rho(f(y)) \mathbf{e}(\langle \boldsymbol{\xi}, f(y) \rangle) \mathrm{d}\boldsymbol{\xi}$$

$$(3.14) \quad \alpha(V)\beta(V)\tau_{\mathbf{h}}(V)\int_0^{\log H} u^{t-1}e^u du$$
$$= \frac{1}{2^m}\sum_{\mathbf{b}\in\bigoplus_{v\in M_f - S}\mathbf{Z}^m}\mu(\mathbf{b})\int_{A_\mathbf{Q}}\int_{A_\mathbf{Q}^N}\Phi_{A_\mathbf{Q}^N}^{\mathbf{H}_X}(H,\mathbf{b},y)\rho(f(y))\mathbf{e}(\langle\boldsymbol{\xi},f(y)\rangle)\mathrm{d}y\,\mathrm{d}\boldsymbol{\xi}.$$

Démonstration. — Par la remarque 1.1.6, le cône des diviseurs effectifs de V est donné par celui de X. Il résulte de le proposition 3.4.3 que la paire (X,V) vérifie la condition (A). La condition (G) est automatique dans ce cas. □

4. Conclusion

Les deux derniers corollaires du paragraphe précédent permettent de se réduire à des majorations de différences de la forme

$$(4.1) \quad \left|\int_{A_k/k}\sum_{y\in\mathcal{T}_X(k)}\Phi_{\mathcal{T}_X}^{\mathbf{H}_X}(H,\mathbf{b},y)\rho(f(y))\mathbf{e}(\langle\boldsymbol{\xi},f(y)\rangle)\mathrm{d}\boldsymbol{\xi}\right.$$
$$\left.-\int_{A_k}\int_{\mathcal{T}_X C_{\mathrm{eff}}(\overline{X})(A_k)}\Phi_{\mathcal{T}_X}^{\mathbf{H}_X}(H,\mathbf{b},y)\rho(f(y))\mathbf{e}(\langle\boldsymbol{\xi},f(y)\rangle)\omega_{\mathcal{T}_X}(y)\mathrm{d}\boldsymbol{\xi}\right|$$

En s'inspirant de la méthode du cercle, on est alors amené à décomposer l'espace quotient $\prod_{v\in M_f}\mathcal{O}_v\backslash\mathbf{A}_k/k$, qui est homéomorphe à un produit de cercles, en arcs majeurs et arcs mineurs puis à majorer sur les arcs mineurs chacun des deux termes obtenus ce qui passe par des majorations de sommes d'exponentielles comme celles qui sont au cœur de la méthode du cercle, et sur les arcs majeurs la différence entre ces termes. A ce propos il convient de noter que pour $\boldsymbol{\xi}=0$ et $\rho=1$, majorer la différence

$$(4.2) \quad \left|\sum_{y\in\mathcal{T}_X(k)}\Phi_{\mathcal{T}_X}^{\mathbf{H}_X}(H,\mathbf{b},y)\rho(f(y))\mathbf{e}(\langle\boldsymbol{\xi},f(y)\rangle)\right.$$
$$\left.-\int_{\mathcal{T}_X C_{\mathrm{eff}}(\overline{X})(A_k)}\Phi_{\mathcal{T}_X}^{\mathbf{H}_X}(H,\mathbf{b},y)\rho(f(y))\mathbf{e}(\langle\boldsymbol{\xi},f(y)\rangle)\omega_{\mathcal{T}_X}(y)\right|.$$

revient à comparer $n_{U,\mathbf{H}_X(\omega_V^{-1})}$ avec une formule intégrale. On ne peut donc espérer que cette différence ne soit négligeable que si U_X est fortement saturé pour ω_V^{-1} au sens de Batyrev et Tschinkel [**BT4**, definition S_2], c'est à dire si pour tout fermé strict W de U_X, on a :

$$\frac{n_{W,\mathbf{H}_X(\omega_V^{-1})}(H)}{n_{U_X,\mathbf{H}_X(\omega_V^{-1})}(H)} \to 0$$
$$H \to +\infty.$$

En particulier si V est une hypersurface de produits d'espaces projectifs, cela implique $L \in \mathbf{Q}\omega_X^{-1}$. Il est toutefois envisageable que ce problème puisse être évité par un choix judicieux de la fonction auxiliaire ρ.

Notons par contre que si $L \in \mathbf{Q}\omega_X^{-1}$, alors l'équivalence des deux termes de (4.2) pour $\xi = 0$ résulte de la conjecture pour X.

Références

[BM] V. V. Batyrev et Y. I. Manin, *Sur le nombre des points rationnels de hauteur bornée des variétés algébriques*, Math. Ann. **286** (1990), 27–43.

[BT1] V. V. Batyrev and Y. Tschinkel, *Rational points of bounded height on compactifications of anisotropic tori*, Internat. Math. Res. Notices **12** (1995), 591–635.

[BT2] _____, *Height zeta functions of toric varieties*, J. Math. Sci. **82** (1996), n° 1, 3220–3239.

[BT3] _____, *Manin's conjecture for toric varieties*, J. Algebraic Geom. **7** (1998), n° 1, 15–53.

[BT4] _____, *Tamagawa numbers of polarized algebraic varieties*, Nombre et répartition de points de hauteur bornée, Astérisque, vol. 251, SMF, Paris, 1998, pp. 299–340.

[Bi] B. J. Birch, *Forms in many variables*, Proc. Roy. Soc. London **265A** (1962), 245–263.

[BR] M. Borovoi and Z. Rudnick, *Hardy-Littlewood varieties and semi-simple groups*, Invent. math. **119** (1995), 37–66.

[BGS] J.-B. Bost, H. Gillet, and C. Soulé, *Heights of projective varieties and positive Green forms*, J. Amer. Math. Soc. **7** (1994), 903–1027.

[Bo] R. Bott, *On a theorem of Lefschetz*, Mich. Math. J. **6** (1959), 211–216.

[Bre] R. de la Bretèche, *Compter des points d'une variété torique*, J. Number Theory **87** (2001), n° 2, 315–331.

[Br] J.-L. Brylinski, *Décomposition simpliciale d'un réseau, invariante par un groupe fini d'automorphismes*, C. R. Acad. Sci. Paris Sér. I Math. **288** (1979), 137–139.

[CLT1] A. Chambert-Loir and Y. Tschinkel, *Points of bounded height on equivariant compactifications of vector groups, I*, Compositio Math. **124(1)** (2000), 65–93.

[CLT2] _____, *Points of bounded height on equivariant compactifications of vector groups, II*, J. Number Theory **85** (2000), no. 2, 172–188.

[CTS1] J.-L. Colliot-Thélène et J.-J. Sansuc, *Torseurs sous des groupes de type multiplicatif; applications à l'étude des points rationnels de certaines variétés algébriques*, C. R. Acad. Sci. Paris Sér. I Math. **282** (1976), 1113–1116.

[CTS2] _____, *La descente sur les variétés rationnelles*, Journées de géométrie algébrique d'Angers (1979) (A. Beauville, éd.), Sijthoff & Noordhoff, Alphen aan den Rijn, 1980, pp. 223–237.

[CTS3] _____, *La descente sur les variétés rationnelles, II*, Duke Math. J. **54** (1987), n° 2, 375–492.

[Da] V. I. Danilov, *The geometry of toric varieties*, Uspekhi. Mat. Nauk **33** (1978), n° 2, 85–134; English transl. in Russian Math. Surveys **33** (1978), n° 2, 97–154.

[Dr] P. K. J. Draxl, *L-Funktionen algebraischer Tori*, J. Number Theory **3** (1971), 444–467.

[FMT] J. Franke, Y. I. Manin, and Y. Tschinkel, *Rational points of bounded height on Fano varieties*, Invent. Math. **95** (1989), 421–435.

[Ha] R. Hartshorne, *Algebraic geometry*, Graduate Texts in Math., vol. 52, Springer-Verlag, Berlin, Heidelberg and New York, 1977.

[Ig1] J.-I. Igusa, *Criteria for the validity of a certain Poisson formula*, Algebraic number theory (Kyoto, 1976) (S. Iyanaga, ed.), Japan society for the promotion of science, Tokio, 1977, pp. 43–65.

[Ig2] _____, *Lectures on forms of higher degree*, Tata institute of fundamental research, Bombay and Springer-Verlag, Berlin, 1978.

[Mi] J. S. Milne, *Étale cohomology*, Princeton Math. Series, vol. 33, Princeton University Press, 1980.

[Oda] T. Oda, *Convex bodies and algebraic geometry*, Ergebnisse der Mathematik und ihrer Grenzgebiete, 3. Folge, vol. 15, Springer-Verlag, Berlin, Heidelberg and New York, 1988.

[Ono1] T. Ono, *On some arithmetic properties of linear algebraic groups*, Ann. of Math. (2) **70** (1959), n° 2, 266–290.

[Ono2] _____, *Arithmetic of algebraic tori*, Ann. of Math. (2) **74** (1961), n° 1, 101–139.

[Ono3] _____, *On the Tamagawa number of algebraic tori*, Ann. of Math. (2) **78** (1963), n° 1, 47–73.

[Pe1] E. Peyre, *Hauteurs et mesures de Tamagawa sur les variétés de Fano*, Duke Math. J. **79** (1995), n° 1, 101–218.

[Pe2] _____, *Terme principal de la fonction zêta des hauteurs et torseurs universels*, Nombre et répartition de points de hauteur bornée, Astérisque, vol. 251, SMF, Paris, 1998, pp. 259–298.

[Ro] M. Robbiani, *Rational points of bounded height on Del Pezzo surfaces of degree six*, Comment. Math. Helv. **70** (1995), 403–422.

[Sa] P. Salberger, *Tamagawa measures on universal torsors and points of bounded height on Fano varieties*, Nombre et répartition de points de hauteur bornée, Astérisque, vol. 251, SMF, Paris, 1998, pp. 91–258.

[Sc]   S. H. Schanuel, *Heights in number fields*, Bull. Soc. Math. France **107** (1979), 433–449.

[ST]   M. Strauch and Y. Tschinkel, *Height zeta functions of twisted products*, Math. Res. Lett. **4** (1997), 273–282.

[Ta]   J. T. Tate, *Fourier analysis in number fields and Hecke's zeta functions*, Algebraic number theory (J. W. S. Cassels and A. Fröhlich, eds.), Academic press, London, 1967, pp. 305–347.

[We]   A. Weil, *Adèles and algebraic groups*, Progress in Mathematics, vol. 23, Birkhaüser, Boston, Basel, Stuttgart, 1982.

Rational points on algebraic varieties
(E. PEYRE, Y. TSCHINKEL, ed.), p. 275–305
Progress in Mathematics, Vol. 199, © 2001 Birkhäuser Verlag Basel/Switzerland

TAMAGAWA NUMBERS OF DIAGONAL CUBIC SURFACES OF HIGHER RANK

Emmanuel Peyre

  Institut Fourier, UFR de Mathématiques, UMR 5582, Université de Grenoble I et CNRS, BP 74, 38402 Saint-Martin d'Hères CEDEX, France
  Url : http://www-fourier.ujf-grenoble.fr/~peyre
  E-mail : Emmanuel.Peyre@ujf-grenoble.fr

Yuri Tschinkel

  Department of Mathematics, Princeton University, Washington Rd., Princeton, NJ 08544-1000, U.S.A. • *E-mail* : ytschink@math.princeton.edu

Abstract. — We consider diagonal cubic surfaces defined by an equation of the form
$$ax^3 + by^3 + cz^3 + dt^3 = 0.$$
Numerically, one can find all rational points of height $\leqslant B$ for B in the range of up to 10^5, thanks to a program due to D. J. Bernstein. On the other hand, there are precise conjectures concerning the constants in the asymptotics of rational points of bounded height due to Manin, Batyrev and the authors. Changing the coefficients one can obtain cubic surfaces with rank of the Picard group varying between 1 and 4. We check that numerical data are compatible with the above conjectures. In a previous paper we considered cubic surfaces with Picard groups of rank one with or without Brauer-Manin obstruction to weak approximation. In this paper, we test the conjectures for diagonal cubic surfaces with Picard groups of higher rank.

Introduction

This paper is devoted to numerical tests of a refined version of a conjecture of Manin about the number of points of bounded height on Fano varieties (see [**BM**], [**FMT**], [**Pe**], or [**BT**] for a description of the conjectures). The choice of diagonal cubic surfaces to test these conjectures was motivated by the work of Heath-Brown [**H-B**] in which he treated the cases
$$X^3 + Y^3 + Z^3 + aT^3 = 0$$

for $a = 2$ or 3. The results he obtained were used as a benchmark for the subsequent attempts to interpret the asymptotic constants (see, in particular, [S-D], [Pe] and [PT]).

More precisely, we consider a diagonal cubic surface $V \subset \mathbf{P}_\mathbf{Q}^3$ given by an equation of the form

$$aX^3 + bY^3 + cZ^3 + dT^3 = 0.$$

Let H be the height function on $\mathbf{P}^3(\mathbf{Q})$ defined by the formula: for any point $Q = (x : y : z : t)$ in $\mathbf{P}^3(\mathbf{Q})$, one has

$$H(Q) = \max\{|x|,|y|,|z|,|t|\} \text{ if } \begin{cases} (x,y,z,t) \in \mathbf{Z}^4, \\ \gcd(x,y,z,t) = 1. \end{cases}$$

Let U be the complement in V to the 27 lines. We are interested in the asymptotic behavior of the cardinal

$$N_{U,H}(B) = \#\{\, Q \in U(\mathbf{Q}) \mid H(Q) \leqslant B \,\}$$

as B goes to infinity.

Assume that $V(\mathbf{Q})$ is Zariski dense, which by a result of Segre (see [Man2, §29,§30]) is equivalent to $V(\mathbf{Q}) \neq \emptyset$. It is expected that

$$N_{U,H}(B) = BP(\log(B)) + o(B)$$

as B goes to $+\infty$, where P is a polynomial of degree $\operatorname{rk}\operatorname{Pic}(V) - 1$, with leading coefficient $\boldsymbol{\theta}_H(V)$. This constant has a conjectural description. The goal is to compute $\boldsymbol{\theta}_H(V)$ explicitly in the examples at hand and to compare it with the constant obtained from numerical data. Our previous paper [PT] was devoted to surfaces with Picard groups of rank one with or without Brauer-Manin obstruction to weak approximation. In this paper, we consider examples with Picard groups of higher rank.

Note that in these examples the relative error term

$$(N_{U,H}(B) - \boldsymbol{\theta}_H(V)B(\log B)^{\operatorname{rk}\operatorname{Pic}(V)-1})/B(\log B)^{\operatorname{rk}\operatorname{Pic}(V)-1}$$

is expected to decrease more slowly. Indeed, if $\operatorname{rk}\operatorname{Pic}(V) = 1$ this error term is expected to decrease as $1/B^\varepsilon$ for some $\varepsilon > 0$, whereas for higher ranks it should be comparable to $1/\log B$. Thus we decided not only to compare the conjectural constant $\boldsymbol{\theta}_H(V)$ with the quotient

$$N_{U,H}(B)/B(\log B)^{\operatorname{rk}\operatorname{Pic}(V)-1}$$

with $B = 10^5$, but also to take into account that a polynomial P of degree $\operatorname{rk}\operatorname{Pic}(V) - 1$ should appear in the asymptotics and use a naive statistical formula to estimate its leading coefficient $\boldsymbol{\theta}_H^{\operatorname{stat}}(V)$ from the data. We observe a quite good accordance: the difference between $\boldsymbol{\theta}_H(V)$ and $\boldsymbol{\theta}_H^{\text{stat}}(V)$ is less than 6% in the examples. Moreover the fact that $\boldsymbol{\theta}_H^{\text{stat}}(V)$ is nearer to $\boldsymbol{\theta}_H(V)$ than the above quotient is in itself a point in favor of the conjecture: indeed there is no obvious purely statistical reason for which this should be true in general.

The paper is organized as follows: in section 1 we define $\boldsymbol{\theta}_H(V)$. Section 2 contains the description of the Galois action on the geometric Picard group $\text{Pic}(\overline{V})$. In section 3 we compute the Euler product corresponding to good reduction places. In section 4 we explain how to compute the local densities at the places of bad reduction. In section 5 we determine in each case the value of the geometric constant $\alpha(V)$ defined in §1. Section 6 is devoted to the description of statistical tools we used to analyze the numerical data. In section 7 we present the results.

We would like to thank the referee for the improvements he suggested.

1. Description of the conjectural constant

In this section we give a short description of the conjectural asymptotic constant for heights defined by an adelic metrization of the anticanonical line bundle (see [**Pe**] for more details and [**BT**] for a discussion in a more general setting).

Notations 1.1. — For any field E, we denote by $\overline{E}$ an algebraic closure of E. If X is a variety over E, then $X(E)$ denotes the set of rational points of X and $\overline{X}$ the product $X \times_{\text{Spec}(E)} \text{Spec}\,\overline{E}$. The cohomological Brauer group $\text{Br}(X)$ is defined as the étale cohomology group $H^2_{\text{ét}}(X, \mathbf{G}_m)$. For any A in $\text{Br}(X)$, any extension E' of E and any P in $V(E')$, we denote by $A(P)$ the evaluation of A at P.

For a number field F we denote by $\text{Val}(F)$ the set of places of F and by $\text{Val}_f(F)$ the set of finite places. The absolute discriminant of F is denoted by d_F. For any place v of F, let F_v be the v-adic completion of F. If v is finite, then $\mathscr{O}_v$ is the ring of v-adic integers and $\mathbf{F}_v$ the residue field. By global class field theory we have an exact sequence

(1.1) $$0 \to \text{Br}(F) \to \bigoplus_{v \in \text{Val}(F)} \text{Br}(F_v) \xrightarrow{\sum \text{inv}_v} \mathbf{Q}/\mathbf{Z} \to 0.$$

In the following, V is a smooth projective geometrically integral variety over a number field F satisfying the conditions:

(i) $H^i(V, \mathcal{O}_V) = 0$ for $i = 1$ or 2,
(ii) $\operatorname{Pic}(\overline{V})$ has no torsion,
(iii) the anticanonical line bundle ω_V^{-1} belongs to the interior of the cone of classes of effective divisors $\Lambda_{\operatorname{eff}}(V) \subset \operatorname{Pic}(V) \otimes_{\mathbf{Z}} \mathbf{R}$.

The adelic space $V(\mathbf{A}_F)$ of V is the product $\prod_{v \in \operatorname{Val}(F)} V(F_v)$. By [**CT**, lemma 1], for any class A in $\operatorname{Br}(V)$, one has a map ρ_A defined as the composition

$$V(\mathbf{A}_F) \to \bigoplus_{v \in \operatorname{Val}(F)} \operatorname{Br}(F_v) \xrightarrow{\sum \operatorname{inv}_v} \mathbf{Q}/\mathbf{Z}$$
$$(P_v)_{v \in \operatorname{Val}(F)} \mapsto (A(P_v))_{v \in \operatorname{Val}(F)}.$$

Then one defines

$$V(\mathbf{A}_F)^{\operatorname{Br}} = \bigcap_{A \in \operatorname{Br}(V)} \ker(\rho_A) \subset V(\mathbf{A}_F).$$

By the exact sequence (1.1), one has the inclusion

$$\overline{V(F)} \subset V(\mathbf{A}_F)^{\operatorname{Br}}$$

where $\overline{V(F)}$ denotes the topological closure of the set of rational points. Conjecturally both sets coincide for cubic surfaces. (See also the text of Swinnerton-Dyer in this volume). There is a *Brauer-Manin obstruction to weak approximation*, as described by Manin in [**Man1**] and by Colliot-Thélène and Sansuc in [**CTS**], if one has

$$V(\mathbf{A}_F)^{\operatorname{Br}} \neq V(\mathbf{A}_F).$$

Let us assume that the height H on V is defined by an adelic metric $(\|\cdot\|_v)_{v \in \operatorname{Val}(F)}$ on ω_V^{-1}. By definition, this means that we consider ω_V^{-1} as a line bundle, that the functions $\|\cdot\|_v$ are v-adically continuous metrics on $\omega_V^{-1}(F_v)$ which for almost all places v are given by a smooth model of V, and that the height of a rational point x of V is given by the formula

$$\forall y \in \omega_V^{-1}(x) - \{0\}, \quad H(x) = \prod_{v \in \operatorname{Val}(F)} \|y\|_v^{-1}$$

where $\omega_V^{-1}(x)$ is the fiber of ω_V^{-1} at x.

If $v \in \operatorname{Val}(F)$, the Haar measure $\mathrm{d}x_v$ on F_v is normalized as follows:
- $\int_{\mathcal{O}_v} \mathrm{d}x_v = 1$ if v is finite,
- $\mathrm{d}x_v$ is the usual Lebesgue measure if $F_v \xrightarrow{\sim} \mathbf{R}$,
- $\mathrm{d}x_v = \mathrm{d}z\, \mathrm{d}\bar{z} = 2\mathrm{d}x\, \mathrm{d}y$ if $F_v \xrightarrow{\sim} \mathbf{C}$.

The metric $\|\cdot\|_v$ defines a measure $\omega_{H,v}$ on the locally compact space $V(F_v)$. In local v-adic analytic coordinates $x_{1,v}\ldots x_{n,v}$ on $V(F_v)$ this measure is given by the formula

$$\omega_{H,v} = \left\|\frac{\partial}{\partial x_{1,v}} \wedge \cdots \wedge \frac{\partial}{\partial x_{n,v}}\right\|_v \mathrm{d}x_{1,v}\ldots \mathrm{d}x_{n,v}.$$

If M is a discrete representation of $\mathrm{Gal}(\overline{F}/F)$ over $\mathbf{Q}$, then for any finite place $\mathfrak{p}$ of F, the local term of the corresponding Artin L-function is defined as follows: we choose an algebraic closure $\overline{F}_\mathfrak{p}$ of $F_\mathfrak{p}$ containing $\overline{F}$. We get an exact sequence

$$1 \to I_\mathfrak{p} \to D_\mathfrak{p} \to \mathrm{Gal}(\overline{\mathbf{F}}_\mathfrak{p}/\mathbf{F}_\mathfrak{p}) \to 1$$

where $D_\mathfrak{p}$ is the decomposition group and $I_\mathfrak{p}$ the inertia. We denote by $\widetilde{\mathrm{Fr}}_\mathfrak{p}$ a lifting of the Frobenius map to $D_\mathfrak{p} \subset \mathrm{Gal}(\overline{F}/F)$ (which up to conjugation depends only on $\mathfrak{p}$), and put

$$L_\mathfrak{p}(s, M) = \frac{1}{\mathrm{Det}(1 - (\#\mathbf{F}_\mathfrak{p})^{-s}\widetilde{\mathrm{Fr}}_\mathfrak{p} \mid M^{I_\mathfrak{p}})}.$$

We fix a finite set S of bad places containing the archimedean ones so that V admits a smooth projective model $\mathscr{V}$ over the ring of S-integers $\mathscr{O}_S$. For any $\mathfrak{p}$ in $\mathrm{Val}(F) - S$ we consider

$$L_\mathfrak{p}(s, \mathrm{Pic}(\overline{V})) = L_\mathfrak{p}(s, \mathrm{Pic}(\overline{V}) \otimes_\mathbf{Z} \mathbf{Q}).$$

The corresponding global L-function is given by the Euler product

$$L_S(s, \mathrm{Pic}(\overline{V})) = \prod_{\mathfrak{p} \in \mathrm{Val}(F) - S} L_\mathfrak{p}(s, \mathrm{Pic}(\overline{V}))$$

which converges for $\mathrm{Re}\, s > 1$ and has a meromorphic continuation to $\mathbf{C}$ with a pole of order $t = \mathrm{rk}\,\mathrm{Pic}(V)$ at 1. One introduces local convergence factors λ_v given by

$$\lambda_v = \begin{cases} L_v(1, \mathrm{Pic}(\overline{V})) & \text{if } v \in \mathrm{Val}(F) - S, \\ 1 & \text{otherwise.} \end{cases}$$

The Weil conjectures (proved by Deligne) imply that the Tamagawa measure

$$\prod_{v \in \mathrm{Val}(F)} \lambda_v^{-1} \omega_{H,v}$$

converges on $V(\mathbf{A}_F)$ (see [Pe, proposition 2.2.2]).

Definition 1.2. — The *Tamagawa measure* on $V(\mathbf{A}_F)$ corresponding to the adelic metric $(\|\cdot\|_v)_{v\in\mathrm{Val}(F)}$ is defined by

$$\omega_H = \frac{1}{\sqrt{|d_F|}^{\dim V}} \lim_{s\to 1}(s-1)^t L_S(s,\mathrm{Pic}(\overline{V})) \prod_{v\in\mathrm{Val}(F)} \lambda_v^{-1}\omega_{H,v}.$$

From the arithmetic standpoint, it seems more natural to integrate ω_H over the closure $\overline{V(F)} \subset V(\mathbf{A}_F)$ (as in the original approach to the Tamagawa number). However, computationally, it is easier to work with $V(\mathbf{A}_F)^{\mathrm{Br}}$. Therefore, following a suggestion of Salberger, we define here

Definition 1.3. —

(1.2) $$\tau_H(V) = \omega_H(V(\mathbf{A})^{\mathrm{Br}}).$$

Let $\mathrm{Pic}(V)^\vee$ be the dual lattice to $\mathrm{Pic}(V)$. We denote by $d\mathbf{y}$ the corresponding Lebesgue measure on $\mathrm{Pic}(V)^\vee \otimes_\mathbf{Z} \mathbf{R}$ and by

$$\Lambda_{\mathrm{eff}}(V)^\vee = \{\, x \in \mathrm{Pic}(V)^\vee \otimes_\mathbf{Z} \mathbf{R} \mid \forall y \in \Lambda_{\mathrm{eff}}(V),\ \langle x,y\rangle \geq 0 \,\}$$

the dual cone of $\Lambda_{\mathrm{eff}}(V)$.

Definition 1.4. — We define

$$\alpha(V) = \frac{1}{(t-1)!} \int_{\Lambda_{\mathrm{eff}}(V)^\vee} e^{-\langle \omega_V^{-1},\mathbf{y}\rangle} d\mathbf{y}$$

and

$$\beta(V) = \#H^1(k,\mathrm{Pic}(\overline{V})).$$

The theoretical constant attached to V and H is defined as

(1.3) $$\boldsymbol{\theta}_H(V) = \alpha(V)\beta(V)\tau_H(V).$$

In the following sections we compute $\boldsymbol{\theta}_H(V)$ for various diagonal cubic surfaces.

2. The Galois module $\mathrm{Pic}(\overline{V})$

The description of this Galois module is based upon the study of the 27 lines of the cubic. We fix notations for these lines which are slightly different from those given by Colliot-Thélène, Kanevsky and Sansuc in [**CTKS**, p. 9].

Notations 2.1. — From now on V is a diagonal cubic surface V given by an equation of the form

(2.1) $$aX^3 + bY^3 + cZ^3 + dT^3 = 0$$

where a, b, c and d are strictly positive integers with $\gcd(a,b,c,d) = 1$. Let

$$S = \{\infty, 3\} \cup \{p \mid p|abcd\}$$

We fix a cubic root α (resp. α', α'') of b/a (resp. $c/a, d/a$) (which is assumed to be in $\mathbf{Q}$ if b/a (resp. $c/a, d/a$) is a cube in $\mathbf{Q}$) and we put

$$\beta = \frac{\alpha''}{\alpha'} = \sqrt[3]{\frac{d}{c}}, \qquad \beta' = \frac{\alpha}{\alpha''} = \sqrt[3]{\frac{b}{d}} \quad \text{and} \quad \beta'' = \frac{\alpha'}{\alpha} = \sqrt[3]{\frac{c}{b}}.$$

We also consider

$$\gamma = \frac{\alpha}{\alpha'\alpha''} = \sqrt[3]{\frac{ab}{cd}}, \qquad \gamma' = \frac{\alpha'}{\alpha''\alpha} = \sqrt[3]{\frac{ac}{bd}} \quad \text{and} \quad \gamma'' = \frac{\alpha''}{\alpha\alpha'} = \sqrt[3]{\frac{ad}{bc}}.$$

We denote by θ a primitive third root of one. The 27 lines of the cubic surface (2.1) are given by the following equations, where i belongs to $\mathbf{Z}/3\mathbf{Z}$:

$$L(i): \begin{cases} X+\theta^i \alpha Y = 0, \\ Z+\theta^i \beta T = 0. \end{cases} \quad L'(i): \begin{cases} X+\theta^i \alpha Y = 0, \\ Z+\theta^{i+1} \beta T = 0. \end{cases} \quad L''(i): \begin{cases} X+\theta^i \alpha Y = 0, \\ Z+\theta^{i+2} \beta T = 0. \end{cases}$$

$$M(i): \begin{cases} X+\theta^i \alpha' Z = 0, \\ T+\theta^i \beta' Y = 0. \end{cases} \quad M'(i): \begin{cases} X+\theta^i \alpha' Z = 0, \\ T+\theta^{i+1} \beta' Y = 0. \end{cases} \quad M''(i): \begin{cases} X+\theta^i \alpha' Z = 0, \\ T+\theta^{i+2} \beta' Y = 0. \end{cases}$$

$$N(i): \begin{cases} X+\theta^i \alpha'' T = 0, \\ Y+\theta^i \beta'' Z = 0. \end{cases} \quad N'(i): \begin{cases} X+\theta^i \alpha'' T = 0, \\ Y+\theta^{i+1} \beta'' Z = 0. \end{cases} \quad N''(i): \begin{cases} X+\theta^i \alpha'' T = 0, \\ Y+\theta^{i+2} \beta'' Z = 0. \end{cases}$$

Let K be the field $\mathbf{Q}(\theta, \alpha, \alpha', \alpha'')$. It is a Galois extension of $\mathbf{Q}$. In the generic case, K is an extension of degree 54 with a Galois group isomorphic to

$$(\mathbf{Z}/3\mathbf{Z})^3 \rtimes \mathbf{Z}/2\mathbf{Z}.$$

It is generated by the elements c, τ, τ' and τ'' characterized by their action on θ, α, α' and α''.

	θ	α	α'	α''
c	θ^2	α	α'	α''
τ	θ	$\theta\alpha$	α'	α''
τ'	θ	α	$\theta\alpha'$	α''
τ''	θ	α	α'	$\theta\alpha''$

Their action on the 27 lines is given as follows: for τ we have

(2.2)
$$L(i) \longrightarrow L''(i+1) \qquad M(i) \longrightarrow M'(i) \qquad N(i) \longrightarrow N''(i)$$
$$\nwarrow \quad \swarrow \qquad\qquad \nwarrow \quad \swarrow \quad \text{and} \quad \nwarrow \quad \swarrow$$
$$L'(i+2) \qquad\qquad M''(i) \qquad\qquad N'(i)$$

for τ':

(2.3)
$$L(i) \longrightarrow L''(i) \qquad M(i) \longrightarrow M''(i+1) \qquad N(i) \longrightarrow N'(i)$$
$$\nwarrow \quad \swarrow \qquad\qquad \nwarrow \quad \swarrow \quad \text{and} \quad \nwarrow \quad \swarrow$$
$$L'(i) \qquad\qquad M'(i+2) \qquad\qquad N''(i)$$

for τ'':

(2.4)
$$L(i) \longrightarrow L'(i) \qquad M(i) \longrightarrow M''(i) \qquad N(i) \longrightarrow N''(i+1)$$
$$\nwarrow \quad \swarrow \qquad\qquad \nwarrow \quad \swarrow \quad \text{and} \quad \nwarrow \quad \swarrow$$
$$L''(i) \qquad\qquad M'(i) \qquad\qquad N'(i+2)$$

for c:

(2.5)
$$\begin{array}{ccc ccc ccc}
L(0) & L'(0) \leftrightarrow L''(0) & M(0) & M'(0) \leftrightarrow M''(0) & N(0) & N'(0) \leftrightarrow N''(0) \\
L(1) & L'(1) \quad L''(1) & M(1) & M'(1) \quad M''(1) & N(1) & N'(1) \quad N''(1) \\
\downarrow & \times & \downarrow & \times & \downarrow & \times \\
L(2) & L'(2) \quad L''(2) & M(2) & M'(2) \quad M''(2) & N(2) & N'(2) \quad N''(2).
\end{array}$$

To describe the relations between the classes of these divisors in $\mathrm{Pic}(\overline{V})$, which shall be useful for the computation of $\alpha(V)$, we consider $\overline{V}$ as the blow-up of a plane $\mathbf{P}^2_{\mathbf{Q}}$ in six points P_1, P_2, P_3, P_4, P_5 and P_6. The 27 lines may then be described as the 6 exceptional divisors $E_1, \ldots, E_6$, the 15 strict transforms $L_{i,j}$ of the projective lines $(P_i P_j)$ for $1 \leqslant i < j \leqslant 6$ and the 6 strict transforms of the conics Q_i going through all points except P_i. Let Λ be the preimage of a line of $\mathbf{P}^2_{\mathbf{Q}}$ which does not contain any of the points $P_1, \ldots, P_6$. Then

$$([\Lambda], [E_1], [E_2], [E_3], [E_4], [E_5], [E_6])$$

is a basis of $\mathrm{Pic}(\overline{V})$ and we have the following relations in $\mathrm{Pic}(\overline{V})$:

(2.6)
$$[L_{i,j}] = [\Lambda] - [E_i] - [E_j] \qquad \text{for } 1 \leqslant i < j \leqslant 6,$$
$$[Q_i] = 2[\Lambda] - \sum_{j \neq i}[E_j].$$

In the following, we choose the projection of $\overline{V}$ to $\mathbf{P}^2_{\mathbf{Q}}$ so that we have the equalities:

(2.7)
$$\begin{aligned}
E_1 &= L(0), & E_2 &= L(1), & E_3 &= L(2), \\
E_4 &= M(1), & E_5 &= M'(2), & E_6 &= M''(0), \\
Q_1 &= L'(1), & Q_2 &= L'(2), & Q_3 &= L'(0), \\
Q_4 &= M(0), & Q_5 &= M'(1), & Q_6 &= M''(2), \\
L_{1,2} &= L''(1), & L_{2,3} &= L''(2), & L_{3,1} &= L''(0), \\
L_{4,5} &= M''(1), & L_{5,6} &= M(2), & L_{6,4} &= M'(0), \\
L_{1,4} &= N(0), & L_{1,5} &= N(1), & L_{1,6} &= N(2), \\
L_{2,4} &= N'(1), & L_{2,5} &= N'(2), & L_{2,6} &= N'(0), \\
L_{3,4} &= N''(2), & L_{3,5} &= N''(0), & L_{3,6} &= N''(1).
\end{aligned}$$

Notations 2.2. — We consider the étale algebra E_1 over $\mathbf{Q}$ defined as $\mathbf{Q}(\gamma)$ if ab/cd is not a cube in $\mathbf{Q}$ and as $\mathbf{Q}(\theta) \times \mathbf{Q}$ otherwise. Similarly, we define the algebra E_2 (resp. E_3) corresponding to γ' (resp. γ'') and we put

$$E = E_1 \times E_2 \times E_3.$$

We also consider the following elements of $\mathrm{Pic}(\overline{V})$

$$\begin{aligned}
e_0^1 &= [M(0)] + [M(1)] + [M(2)], & e_1^1 &= [M'(0)] + [M'(1)] + [M'(2)], \\
e_2^1 &= [M''(0)] + [M''(1)] + [M''(2)], & e_0^2 &= [N(0)] + [N(1)] + [N(2)], \\
e_1^2 &= [N'(0)] + [N'(1)] + [N'(2)], & e_2^2 &= [N''(0)] + [N''(1)] + [N''(2)], \\
e_0^3 &= [L(0)] + [L(1)] + [L(2)], & e_1^3 &= [L'(0)] + [L'(1)] + [L'(2)], \\
e_2^3 &= [L''(0)] + [L''(1)] + [L''(2)]
\end{aligned}$$

and the sets

$$\mathscr{E}_1 = \{e_0^1, e_1^1, e_2^1\}, \quad \mathscr{E}_2 = \{e_0^2, e_1^2, e_2^2\}, \quad \mathscr{E}_3 = \{e_0^3, e_1^3, e_2^3\},$$

and $\mathscr{E} = \mathscr{E}_1 \cup \mathscr{E}_2 \cup \mathscr{E}_3$.

Lemma 2.3. — *The sets $\mathscr{E}_1$, $\mathscr{E}_2$ and $\mathscr{E}_3$ are globally invariant under the action of $\mathrm{Gal}(K/\mathbf{Q})$ and the étale algebra corresponding to the set $\mathscr{E}_i$ is isomorphic to E_i.*

Proof. — The fact that the sets $\mathscr{E}_1$, $\mathscr{E}_2$ and $\mathscr{E}_3$ are globally invariant follows immediately from the descriptions (2.2)–(2.5). The étale algebra F corresponding to a finite $\mathrm{Gal}(K/\mathbf{Q})$-set $\mathscr{F}$ may be defined as the algebra

$$(K[\mathscr{F}])^{\mathrm{Gal}(K/\mathbf{Q})}$$

where $K[\mathscr{F}]$ is the algebra $K^{\mathscr{F}}$ and where $\mathrm{Gal}(K/\mathbf{Q})$ acts simultaneously on K and $\mathscr{F}$. In the generic case, let us consider

$$\sigma = \tau'\tau'', \quad \sigma' = \tau''\tau \quad \text{and} \quad \sigma'' = \tau\tau'.$$

Then σ sends γ on $\theta\gamma$ and acts trivially on γ', γ'' and θ. We may describe similarly the actions of σ' and σ''. The action of $\mathrm{Gal}(K/\mathbf{Q})$ on $\mathscr{E}_1$ in the generic case is given by the table

	e_0^1	e_1^1	e_2^1
c	e_0^1	e_2^1	e_1^1
σ	e_1^1	e_2^1	e_0^1
σ'	e_0^1	e_1^1	e_2^1
σ''	e_0^1	e_1^1	e_2^1

This implies that if ab/cd is not a cube in $\mathbf{Q}$, then $\mathscr{E}_1$ is isomorphic to

$$\mathrm{Gal}(K/\mathbf{Q})/\mathrm{Gal}(K/\mathbf{Q}(\gamma))$$

as a $\mathrm{Gal}(K/\mathbf{Q})$-set. Then the corresponding étale algebra is

$$(K[\mathrm{Gal}(K/\mathbf{Q})/\mathrm{Gal}(K/\mathbf{Q}(\gamma))])^{\mathrm{Gal}(K/\mathbf{Q})} \xrightarrow{\sim} K^{\mathrm{Gal}(K/\mathbf{Q}(\gamma))} = \mathbf{Q}(\gamma) = E_1.$$

Similarly if ab/cd is a cube in $\mathbf{Q}$, then we may decompose $\mathscr{E}_1$ into two orbits and we see that the corresponding étale algebra is $\mathbf{Q}(\theta) \times \mathbf{Q} = E_1$. The proofs for $\mathscr{E}_2$ and $\mathscr{E}_3$ are similar. □

Lemma 2.4. — *There exists an exact sequence of $\mathrm{Gal}(K/\mathbf{Q})$ modules*

$$0 \to \mathbf{Q}^2 \to \mathbf{Q}[\mathscr{E}] \to \mathrm{Pic}(\overline{V}) \otimes_{\mathbf{Z}} \mathbf{Q} \to 0.$$

Proof. — By (2.6) and (2.7), we have in $\mathrm{Pic}(\overline{V})$ the relations

$$e_0^1 = 3[\Lambda] - [E_1] - [E_2] - [E_3] + [E_4] - 2[E_5] - 2[E_6],$$
$$e_1^1 = 3[\Lambda] - [E_1] - [E_2] - [E_3] - 2[E_4] + [E_5] - 2[E_6],$$
$$e_2^1 = 3[\Lambda] - [E_1] - [E_2] - [E_3] - 2[E_4] - 2[E_5] + [E_6],$$

$$e_0^2 = 3[\Lambda] - 3[E_1] - [E_4] - [E_5] - [E_6],$$
$$e_1^2 = 3[\Lambda] - 3[E_2] - [E_4] - [E_5] - [E_6],$$
$$e_2^2 = 3[\Lambda] - 3[E_3] - [E_4] - [E_5] - [E_6],$$
$$e_0^3 = [E_1] + [E_2] + [E_3],$$
$$e_1^3 = 6[\Lambda] - 2[E_1] - 2[E_2] - 2[E_3] - 3[E_4] - 3[E_5] - 3[E_6],$$
$$e_2^3 = 3[\Lambda] - 2[E_1] - 2[E_2] - 2[E_3]$$

which proves that the natural projection from $\mathbf{Q}[\mathscr{E}]$ to $\mathrm{Pic}(\overline{V}) \otimes_{\mathbf{Z}} \mathbf{Q}$ is surjective. Moreover one has the relations

$$3\omega_V^{-1} = \sum_{x \in \mathscr{E}_1} x = \sum_{x \in \mathscr{E}_2} x = \sum_{x \in \mathscr{E}_3} x,$$

which gives a homomorphism of $\mathrm{Gal}(K/\mathbf{Q})$-modules

$$\mathbf{Q}^2 \to \mathbf{Q}[\mathscr{E}]$$

and the exact sequence of the lemma. □

Notations 2.5. — For any prime p and any finite field extension F of $\mathbf{Q}$, we consider the local factor $\zeta_{F,p}$ of the function ζ_F at p which is defined by

$$\zeta_{F,p}(s) = \prod_{\{v \in \mathrm{Val}(F) | v | p\}} (1 - \#\mathbf{F}_v^{-s})^{-1}.$$

Let F be an étale algebra over $\mathbf{Q}$ and $F = \prod_{i \in I} F_i$ its decomposition in fields. Put

$$\zeta_F(s) = \prod_{i \in I} \zeta_{F_i}(s) \quad \text{and} \quad \zeta_{F,p}(s) = \prod_{i \in I} \zeta_{F_i,p}(s).$$

For any prime p, we denote by $\nu_F(p)$ the number of components of $F \otimes_{\mathbf{Q}} \mathbf{Q}_p$ of degree one over $\mathbf{Q}_p$.

Proposition 2.6. — With notation as above, for any prime p not in S, one has

(i) $L_p(s, \mathrm{Pic}(\overline{V})) = \dfrac{\zeta_{E,p}(s)}{\zeta_{\mathbf{Q},p}(s)^2},$

(ii) $\mathrm{Tr}(\widetilde{\mathrm{Fr}}_p | \mathrm{Pic}(\overline{V})) = \nu_E(p) - 2.$

Proof. — By lemma 2.4, we have

$$L_p(s, \mathrm{Pic}(\overline{V})) = \frac{L_p(s, \mathbf{Q}[\mathscr{E}])}{L_p(s, \mathbf{Q})^2}.$$

Thus it is enough to prove that if E is an arbitrary étale algebra over $\mathbf{Q}$ corresponding to a $\operatorname{Gal}(\overline{\mathbf{Q}}/\mathbf{Q})$-set $\mathscr{E}$ and if p is a prime such that $E/\mathbf{Q}$ is not ramified at p, then

$$\zeta_{E,p}(s) = L_p(s, \mathbf{Q}[\mathscr{E}]).$$

This well-known assertion follows from the fact that the components of $E \otimes \mathbf{Q}_p$ are in bijection with the orbits of $\widetilde{\mathrm{Fr}}_p$ in $\mathscr{E}$, and the degree of each component is the length of the corresponding orbit. This proves (i).

But this also shows that

$$\operatorname{Tr}(\widetilde{\mathrm{Fr}}_p \mid \mathbf{Q}[\mathscr{E}]) = \nu_E(p)$$

which implies (ii). □

Remark 2.7. — Thus the factor λ'_p which was defined in proposition 5.1 in [**PT**] coincides with $L_p(1, \operatorname{Pic}(\overline{V}))$ at the good places (as suggested by the referee of that paper).

3. Euler product for the good places

We need to compute the number of solutions of (2.1) modulo p for all primes not in S.

Proposition 3.1. — *For any prime p not in S, one has*

$$\frac{\#V(\mathbf{F}_p)}{p^2} = 1 + \frac{\nu_E(p) - 2}{p} + \frac{1}{p^2}$$

where E is the étale algebra defined in §2.

Proof. — By a result of Weil (see [**Man2**, theorem 23.1]),

$$\#V(\mathbf{F}_p) = 1 + \operatorname{Tr}(\mathrm{Fr}_p \mid \operatorname{Pic}(\overline{V}))p + p^2.$$

Proposition 2.6 implies that

$$\operatorname{Tr}(\mathrm{Fr}_p \mid \operatorname{Pic}(\overline{V})) = \nu_E(p) - 2. \quad □$$

Remark 3.2. — We could have proved this result directly as in [**PT**]. Let $N(p)$ be the number of solutions of (2.1) in $\mathbf{F}_p^4$. We have

$$\#V(\mathbf{F}_p) = \frac{N(p) - 1}{p - 1}.$$

By [**IR**, §8.7 theorem 5], one has

$$N(p) = p^3 + \sum \overline{\chi}_1(a)\overline{\chi}_2(b)\overline{\chi}_3(c)\overline{\chi}_4(d) J_0(\chi_1, \chi_2, \chi_3, \chi_4),$$

where the sum is taken over all quadruples $(\chi_1, \ldots, \chi_4)$ of nontrivial cubic characters from $\mathbf{F}_p^*$ to $\mathbf{C}^*$ such that $\chi_1\chi_2\chi_3\chi_4 = 1$ and where

$$J_0(\chi_1, \chi_2, \chi_3, \chi_4) = \sum_{t_1+\cdots+t_4=0}^{4} \prod_{i=1}^{4} \chi_i(t_i),$$

the characters being extended by $\chi_i(0) = 0$. For $p \equiv 2 \bmod 3$ there are no nontrivial characters and the formula is obvious. Otherwise there are exactly two nontrivial conjugated characters χ and $\overline{\chi}$. By [**PT**, proof of prop. 4.1], we have

$$J_0(\chi_1, \chi_2, \chi_3, \chi_4) = p(p-1)$$

and

$$\#V(\mathbf{F}_p) = 1 + p(1 + \sum \chi_1(a)\chi_2(b)\chi_3(c)\chi_4(d)) + p^2$$

where the sum is taken over the same quadruples as above. The formula

$$\sum \chi_1(a)\chi_2(b)\chi_3(c)\chi_4(d) =$$
$$\chi\left(\frac{ab}{cd}\right) + \overline{\chi}\left(\frac{ab}{cd}\right) + \chi\left(\frac{ac}{bd}\right) + \overline{\chi}\left(\frac{ac}{bd}\right) + \chi\left(\frac{ad}{bc}\right) + \overline{\chi}\left(\frac{ad}{bc}\right)$$

implies the result.

Notations 3.3. — For any place v of $\mathbf{Q}$, we put

$$\lambda_v = \begin{cases} \frac{\zeta_{E,v}(s)}{\zeta_{\mathbf{Q},v}(s)^2} & \text{if } v \text{ is finite,} \\ 1 & \text{otherwise.} \end{cases}$$

Remark 3.4. — By proposition 2.6, $\lambda_p = L_p(1, \operatorname{Pic}(\overline{V}))$ if $p \in \operatorname{Val}(\mathbf{Q}) - S$. Thus the Tamagawa measure ω_H is given by the formula

$$\omega_H = \lim_{s \to 1} (s-1)^{\operatorname{rk} \operatorname{Pic}(V)} \left(\frac{\zeta_E(s)}{\zeta_{\mathbf{Q}}(s)^2}\right) \times \prod_{v \in \operatorname{Val}(\mathbf{Q})} \lambda_v^{-1} \omega_{H,v}.$$

By lemmata 3.2 and 3.4 in [**PT**] and lemma 5.4.6 in [**Pe**], for any p in $\operatorname{Val}(\mathbf{Q}) - S$ one has

$$\omega_{H,p}(V(\mathbf{Q}_p)) = \frac{\#V(\mathbf{F}_p)}{p^2}$$

(see also [**Pe**, lemma 2.2.1] and [**PT**, remark 5.2]). Therefore, the local factor at a good place p is given by

$$\left(1-\tfrac{1}{p}\right)^7\left(1+\tfrac{7}{p}+\tfrac{1}{p^2}\right) \qquad \text{if } p \equiv 1 \text{ mod } 3 \text{ and } \nu_E(p) = 9$$

$$\left(1-\tfrac{1}{p}\right)^4\left(1-\tfrac{1}{p^3}\right)\left(1+\tfrac{4}{p}+\tfrac{1}{p^2}\right) \qquad \text{if } p \equiv 1 \text{ mod } 3 \text{ and } \nu_E(p) = 6$$

$$\left(1-\tfrac{1}{p}\right)\left(1-\tfrac{1}{p^3}\right)^2\left(1+\tfrac{1}{p}+\tfrac{1}{p^2}\right) \qquad \text{if } p \equiv 1 \text{ mod } 3 \text{ and } \nu_E(p) = 3$$

$$\left(1-\tfrac{1}{p}\right)^{-2}\left(1-\tfrac{1}{p^3}\right)^3\left(1-\tfrac{2}{p}+\tfrac{1}{p^2}\right) \qquad \text{if } p \equiv 1 \text{ mod } 3 \text{ and } \nu_E(p) = 0$$

$$\left(1-\tfrac{1}{p}\right)\left(1-\tfrac{1}{p^2}\right)^3\left(1+\tfrac{1}{p}+\tfrac{1}{p^2}\right) \qquad \text{if } p \equiv 2 \text{ mod } 3.$$

We get (for the good places) the factors C_0, C_1, C_2 and C_3 where

$$C_0 = \prod_{\substack{p \nmid 3abcd, \\ p \equiv 2 \text{ mod } 3.}} \left(1-\tfrac{1}{p^3}\right)\left(1-\tfrac{1}{p^2}\right)^3,$$

$$C_1 = \prod_{\substack{p \nmid 3abcd, \\ p \equiv 1 \text{ mod } 3, \\ \nu_E(p)=9.}} \left(1-\tfrac{1}{p}\right)^7\left(1+\tfrac{7}{p}+\tfrac{1}{p^2}\right),$$

$$C_2 = \prod_{\substack{p \nmid 3abcd, \\ p \equiv 1 \text{ mod } 3, \\ \nu_E(p)=6.}} \left(1-\tfrac{1}{p^3}\right)\left(1-\tfrac{1}{p}\right)^4\left(1+\tfrac{4}{p}+\tfrac{1}{p^2}\right),$$

$$C_3 = \prod_{\substack{p \nmid 3abcd, \\ p \equiv 1 \text{ mod } 3, \\ \nu_E(p)=0 \text{ or } 3.}} \left(1-\tfrac{1}{p^3}\right)^3.$$

These products converge rapidly and are easily approximated.

4. Density at the bad places

In this section we restrict to cubic surfaces with equations of the form

(4.1) $$X^3 + Y^3 + qZ^3 + q^2T^3 = 0$$

with q prime and

(4.2) $$aX^3 + aY^3 + qZ^3 + qT^3 = 0$$

with q prime and a an integer coprime to q.

Notations 4.1. — If V is defined by the equation (2.1), and p is a prime, then we consider the cardinal $N^*(p^r)$ of the set

$$\{(x,y,z,t) \in (\mathbf{Z}/p^r\mathbf{Z})^4 - (p\mathbf{Z}/p^r\mathbf{Z})^4 \mid ax^3+by^3+cz^3+dt^3=0 \text{ in } \mathbf{Z}/p^r\mathbf{Z}\}$$

Remark 4.2. — By [**PT**, lemmata 3.2 and 3.4], there is an explicit integer r_0 such that

$$\omega_{H,p}(V(\mathbf{Q}_p)) = \frac{1}{1-p^{-1}} \times \frac{N^*(p^{r_0})}{p^{3r_0}}.$$

If $p = 3$ and $3 \nmid abcd$, then a direct computation in $(\mathbf{Z}/9\mathbf{Z})^4$ gives the value of $N^*(9)$ and thus of $\omega_{H,p}(V(\mathbf{Q}_p))$. Thus, in the following lemma we restrict to the case when V is given by (4.1) or (4.2) and $p = q$.

Lemma 4.3. — If V is given by the equation

$$X^3 + Y^3 + pZ^3 + p^2T^3 = 0$$

then for $r \geqslant 2$,

$$\frac{N^*(p^r)}{p^{3r}} = \begin{cases} 1 - \frac{1}{p} & \text{if } p \equiv 2 \mod 3, \\ 3\left(1 - \frac{1}{p}\right) & \text{if } p \equiv 1 \mod 3, \\ \frac{2}{3} & \text{if } p = 3. \end{cases}$$

If V is given by the equation

$$aX^3 + aY^3 + pZ^3 + pT^3 = 0,$$

with $p \nmid a$, then for $r \geqslant 3$,

$$\frac{N^*(p^r)}{p^{3r}} = \begin{cases} 1 - \frac{1}{p^2} & \text{if } p \equiv 2 \mod 3, \\ 3\left(1 - \frac{1}{p^2}\right) & \text{if } p \equiv 1 \mod 3, \\ \frac{4}{3} & \text{if } p = 3. \end{cases}$$

Remark 4.4. — This lemma implies that if V is given by the first equation then the local factor at p is given by

$$\lambda_p \omega_{H,p}(V(\mathbf{Q}_p)) = \begin{cases} \left(1 - \frac{1}{p^2}\right)\left(1 - \frac{1}{p}\right) & \text{if } p \equiv 2 \mod 3, \\ 3\left(1 - \frac{1}{p}\right)^3 & \text{if } p \equiv 1 \mod 3, \\ \frac{4}{9} & \text{if } p = 3, \end{cases}$$

and if V is given by the second equation then this factor is

$$\lambda_p \omega_{H,p}(V(\mathbf{Q}_p)) = \begin{cases} \left(1 - \frac{1}{p^2}\right)^3 & \text{if } p \equiv 2 \bmod 3, \\ 3\left(1 - \frac{1}{p}\right)^4\left(1 - \frac{1}{p^2}\right) & \text{if } p \equiv 1 \bmod 3, \\ \frac{16}{27} & \text{if } p = 3. \end{cases}$$

Proof. — We shall now consider the set of quadruples (x, y, z, t) belonging to $(\mathbf{Z}/p^r\mathbf{Z})^4 - (p\mathbf{Z}/p^r\mathbf{Z})^4$ such that

(4.3) $\qquad x^3 + y^3 + pz^3 + p^2 t^3 = 0 \quad \text{in } \mathbf{Z}/p^r\mathbf{Z}.$

If $p|x$ then $p|y$, $p|z$ and $p|t$. Therefore, for any (x, y, z, t) as above, $p \nmid x$ and $p \nmid y$. But for any triple (y, z, t) in $(\mathbf{Z}/p^r\mathbf{Z} - p\mathbf{Z}/p^r\mathbf{Z}) \times (\mathbf{Z}/p^r\mathbf{Z})^2$, there exists exactly one x verifying (4.3) if $p \equiv 2 \bmod 3$ and exactly three of them if $p \equiv 1 \bmod 3$. If $p = 3$ and y belongs to $\mathbf{Z}/3^r\mathbf{Z} - 3\mathbf{Z}/3^r\mathbf{Z}$ then (4.3) implies that $3|z$. For any triple (y, z, t) with y in $(\mathbf{Z}/3^r\mathbf{Z}) - (3\mathbf{Z}/3^r\mathbf{Z})$, z in $(3\mathbf{Z}/3^r\mathbf{Z})$ and t in $(\mathbf{Z}/3^r\mathbf{Z})$ there exist exactly three x in $\mathbf{Z}/3^r\mathbf{Z}$ which satisfy (4.3). We get that

$$\frac{N^*(p^r)}{p^{3r}} = \begin{cases} \frac{(p-1)p^{r-1} \times p^r \times p^r}{p^{3r}} = 1 - \frac{1}{p} & \text{if } p \equiv 2 \bmod 3, \\ 3\frac{(p-1)p^{r-1} \times p^r \times p^r}{p^{3r}} = 3\left(1 - \frac{1}{p}\right) & \text{if } p \equiv 1 \bmod 3, \\ 3\frac{2 \times 3^{r-1} \times 3^{r-1} \times 3^r}{3^{3r}} = \frac{2}{3} & \text{if } p = 3. \end{cases}$$

Let us now turn to the set of (x, y, z, t) in $(\mathbf{Z}/p^r\mathbf{Z})^4 - (p\mathbf{Z}/p^r\mathbf{Z})^4$ such that

$$ax^3 + ay^3 + pz^3 + pt^3 = 0.$$

We decompose this set as follows

$$N_1^*(p^r) = \#\left\{(x, y, z, t) \in (\mathbf{Z}/p^r\mathbf{Z})^4 - (p\mathbf{Z}/p^r\mathbf{Z})^4 \middle| \begin{array}{l} p \nmid x, \\ ax^3 + ay^3 + pz^3 + pt^3 = 0. \end{array}\right\}$$

$$N_2^*(p^r) = \#\left\{(x, y, z, t) \in (\mathbf{Z}/p^r\mathbf{Z})^4 - (p\mathbf{Z}/p^r\mathbf{Z})^4 \middle| \begin{array}{l} p|x, \; p \nmid z, \\ ax^3 + ay^3 + pz^3 + pt^3 = 0. \end{array}\right\}$$

As above we have the formula

$$\frac{N_1^*(p^r)}{p^{3r}} = \begin{cases} \frac{(p-1)p^{r-1} \times p^r \times p^r}{p^{3r}} = 1 - \frac{1}{p} & \text{if } p \equiv 2 \bmod 3, \\ 3 \times \frac{(p-1)p^{r-1} \times p^r \times p^r}{p^{3r}} = 3\left(1 - \frac{1}{p}\right) & \text{if } p \equiv 1 \bmod 3, \\ 3\frac{2 \times 3^{r-1} \times 3^{r-1} \times 3^r}{3^{3r}} = \frac{2}{3} & \text{if } p = 3, \end{cases}$$

where for $p = 3$ we use the equality

$$3^{r-1} \times 3^r = \#\{(z, t) \in (\mathbf{Z}/3^r\mathbf{Z})^2 \mid z^3 \equiv t^3 \bmod 3\}.$$

On the other hand,
$$N_2^*(p^r) = p^2 \left\{ (x,y,z,t) \in (\mathbf{Z}/p^{r-1}\mathbf{Z})^4 \;\middle|\; \begin{matrix} p \nmid z \\ ap^2x^3 + ap^2y^3 + z^3 + t^3 = 0. \end{matrix} \right\}$$

and
$$\frac{N_2^*(p^r)}{p^{3r}} = \frac{p^2}{p^3} \times \begin{cases} \frac{(p-1)p^{r-2} \times p^{r-1} \times p^{r-1}}{p^{3(r-1)}} = 1 - \frac{1}{p} & \text{if } p \equiv 2 \bmod 3, \\ 3\frac{(p-1)p^{r-2} \times p^{r-1} \times p^{r-1}}{p^{3(r-1)}} = 3\left(1 - \frac{1}{p}\right) & \text{if } p \equiv 2 \bmod 3, \\ 3\frac{2 \times 3^{r-2} \times 3^{r-1} \times 3^{r-1}}{3^{3(r-1)}} = 2 & \text{if } p = 3. \end{cases}$$

We conclude:
$$\frac{N^*(p^r)}{p^{3r}} = \begin{cases} 1 - \frac{1}{p} + \frac{1}{p} - \frac{1}{p^2} = 1 - \frac{1}{p^2} & \text{if } p \equiv 2 \bmod 3, \\ 3\left(1 - \frac{1}{p^2}\right) & \text{if } p \equiv 1 \bmod 3, \\ \frac{2}{3} + \frac{2}{3} = \frac{4}{3} & \text{if } p = 3. \end{cases} \qquad \square$$

5. The constant $\alpha(V)$

Since the cubic surfaces we consider in this paper are $\mathbf{Q}$-rational (which implies that $\beta(V) = 1$), it remains to compute the rank t of the Picard group and the value of $\alpha(V)$.

Proposition 5.1. — *If V is given by the equation*
$$\text{(5.1)} \qquad X^3 + Y^3 + aZ^3 + a^2T^3 = 0,$$
where a is not a cube in $\mathbf{Q}$, then $\operatorname{rk} \operatorname{Pic}(V) = 2$ *and* $\alpha(V) = 2$.

If V is given by the equation
$$\text{(5.2)} \qquad aX^3 + aY^3 + bZ^3 + bT^3 = 0,$$
where a and b are strictly positive integers and b/a is not a cube in $\mathbf{Q}$, then $\operatorname{rk} \operatorname{Pic}(V) = 3$ *and* $\alpha(V) = 1$.

If V is given by the equation
$$\text{(5.3)} \qquad X^3 + Y^3 + Z^3 + T^3 = 0$$
then $\operatorname{rk} \operatorname{Pic}(V) = 4$ *and* $\alpha(V) = 7/18$.

Proof. — To compute $\alpha(V)$ we shall use its original definition [Pe, §2]:
$$\alpha(V) = \operatorname{Vol}\{ x \in \Lambda_{\text{eff}}(V)^\vee \mid \langle \omega_V^{-1}, x \rangle = 1 \}$$
where the Lebesgue measure on the affine hyperplane
$$\mathcal{H}(\lambda) = \{ x \in \operatorname{Pic}(V)^\vee \otimes_\mathbf{Z} \mathbf{R} \mid \langle \omega_V^{-1}, x \rangle = \lambda \}$$

is defined by the $(t-1)$-form $d\boldsymbol{x}$ which is characterized by the relation

$$d\boldsymbol{x} \wedge d\omega_V^{-1} = d\boldsymbol{y}$$

(where $d\omega_V^{-1}$ is the linear form defined by ω_V^{-1} on $\mathrm{Pic}(V)^\vee$ and $d\boldsymbol{y}$ is the form corresponding to the natural Lebesgue measure on $\mathrm{Pic}(V)^\vee \otimes_{\mathbf{Z}} \mathbf{R}$). More explicitely, let $(e_1, \ldots, e_t)$ be a basis of $\mathrm{Pic}(V)$ and $(e_1^\vee, \ldots, e_t^\vee)$ be the dual basis. Write

$$\omega_V^{-1} = \sum_{i=1}^{t} \lambda_i e_i$$

with $\lambda_t \neq 0$. Let $f_1, \ldots, f_{t-1}$ be the projection of $e_1^\vee, \ldots, e_{t-1}^\vee$ on $\mathscr{H}(0)$ along $e_t^\vee$. Then

$$d\boldsymbol{x} = \frac{1}{\lambda_t} df_1^\vee \wedge \cdots \wedge df_{t-1}^\vee.$$

By [SK, pages 14 and 55], if $O_1, \ldots, O_m$ are the orbits of the action of $\mathrm{Gal}(K/\mathbf{Q})$ on the 27 lines, then $\Lambda_{\mathrm{eff}}(V)$ is generated by the classes $[O_i] = \sum_{x \in O_i} [x]$.

When V is given by the equation (5.1) the Galois group $\mathrm{Gal}(K/\mathbf{Q})$ is

$$\mathbf{Z}/3\mathbf{Z} \rtimes \mathbf{Z}/2\mathbf{Z}$$

and the orbits of its action on the 27 lines are

$$O_1 = \{L(0), L'(0), L''(0)\},$$
$$O_2 = \{L(1), L(2), L'(1), L'(2), L''(1), L''(2)\},$$
$$O_3 = \{M(0), M(1), M(2)\},$$
$$O_4 = \{M'(0), M'(1), M'(2), M''(0), M''(1), M''(2)\},$$
$$O_5 = \{N(0), N'(1), N''(2)\},$$
$$O_6 = \{N(1), N(2), N'(0), N'(2), N''(0), N''(1)\}.$$

In the basis $([\Lambda], [E_1], \ldots, [E_6])$, a basis of $\mathrm{Pic}(V) = (\mathrm{Pic}\,\overline{V})^{\mathrm{Gal}(K:\mathbf{Q})}$ is given by

$$e_1 = \omega_V^{-1}, \qquad e_2 = -2[E_4] + [E_5] + [E_6].$$

In the basis (e_0, e_1), the effective cone $\Lambda_{\mathrm{eff}}(V)$ is generated by the classes

$$[O_1] = e_1, \qquad [O_2] = 2e_1, \qquad [O_3] = e_1 - e_2,$$
$$[O_4] = 2e_1 + e_2, \qquad [O_5] = e_1 + e_2, \qquad [O_6] = 2e_1 - e_2.$$

Therefore, this cone is generated by the elements $e_1 - e_2$ and $e_1 + e_2$ and $\alpha(V)$ is given as the volume of the domain

$$x = 1, \quad x + y > 0 \quad \text{and} \quad x - y > 0,$$

that is, as the volume of the segment $[-1, 1]$ and $\alpha(V) = 2$.

If V is given by the equation (5.2) then $\text{Gal}(K/\mathbf{Q})$ is isomorphic to

$$\mathbf{Z}/3\mathbf{Z} \rtimes \mathbf{Z}/2\mathbf{Z}$$

and the orbits of the Galois action on the 27 lines are

$$\begin{aligned}
O_1 &= \{L(0)\}, \\
O_2 &= \{L(1), L(2)\}, \\
O_3 &= \{L'(0), L''(0)\}, \\
O_4 &= \{L'(1), L''(2)\}, \\
O_5 &= \{L'(2), L''(1)\}, \\
O_6 &= \{M(0), M'(1), M''(2)\}, \\
O_7 &= \{M(1), M(2), M'(0), M'(2), M''(0), M''(1)\}, \\
O_8 &= \{N(0), N(1), N(2)\}, \\
O_9 &= \{N'(0), N'(1), N'(2), N''(0), N''(1), N''(2)\}.
\end{aligned}$$

A basis of $\text{Pic}(V)$ is given by

$$e_1 = \omega_V^{-1}, \quad e_2 = [E_1], \quad e_3 = [E_2] + [E_3],$$

and the cone $\Lambda_{\text{eff}}(V)$ is generated by

$$\begin{aligned}
&[O_1] = e_2, & &[O_2] = e_3, & &[O_3] = e_1 - e_2, \\
&[O_4] = e_1 + e_2 - e_3, & &[O_5] = e_1 - e_2, & &[O_6] = 2e_1 - e_2 - e_3, \\
&[O_7] = e_1 + e_2 + e_3, & &[O_8] = e_1 - 2e_2 + e_3, & &[O_9] = 2e_1 + 2e_2 - e_3,
\end{aligned}$$

that is, by

$$e_2, \quad e_3, \quad e_1 + e_2 - e_3, \quad 2e_1 - e_2 - e_3, \quad e_1 - 2e_2 + e_3$$

(since $3[O_3] = [O_6] + [O_8]$). Thus $\alpha(V)$ is the volume of the domain given by

$$\begin{cases} x = 1, \ y > 0, \ z > 0, \\ x + y - z > 0, \\ 2x - y - z > 0, \\ x - 2y + z > 0. \end{cases}$$

Using the description above, $\alpha(V)$ is the volume of

$$\begin{cases} 0 < y, \ 0 < z, \\ z - y < 1, \\ y + z < 2, \\ 2y - z < 1. \end{cases}$$

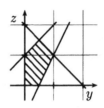

Therefore $\alpha(V) = 1$.

If V is given by the equation (5.3), then $\mathrm{Gal}(K/\mathbf{Q}) = \mathbf{Z}/2\mathbf{Z}$ and the orbits of the Galois action on the 27 lines are given by

$O_1 = \{L(0)\}$, $O_2 = \{L(1), L(2)\}$, $O_3 = \{L'(0), L''(0)\}$,
$O_4 = \{L'(1), L''(2)\}$, $O_5 = \{L'(2), L''(1)\}$,
$O_6 = \{M(0)\}$, $O_7 = \{M(1), M(2)\}$, $O_8 = \{M'(0), M''(0)\}$,
$O_9 = \{M'(1), M''(2)\}$, $O_{10} = \{M'(2), M''(1)\}$,
$O_{11} = \{N(0)\}$, $O_{12} = \{N(1), N(2)\}$, $O_{13} = \{N'(0), N''(0)\}$,
$O_{14} = \{N'(1), N''(2)\}$, $O_{15} = \{N'(2), N''(1)\}$.

A basis of the Picard group is given by

$$e_1 = [\Lambda] - [E_4], \quad e_2 = [E_1], \quad e_3 = [E_2] + [E_3], \quad e_4 = -2[E_4] + [E_5] + [E_6].$$

The effective cone $\Lambda_{\mathrm{eff}}(V)$ is generated by

$[O_1] = e_2$, $[O_2] = e_3$,
$[O_3] = 3e_1 - 2e_2 - e_3 - e_4$, $[O_4] = 3e_1 - 2e_3 - e_4$,
$[O_5] = 3e_1 - 2e_2 - e_3 - e_4$, $[O_6] = 2e_1 - e_2 - e_3 - e_4$,
$[O_7] = e_1 - e_4$, $[O_8] = e_1$,
$[O_9] = 4e_1 - 2e_2 - 2e_3 - e_4$, $[O_{10}] = e_1$,
$[O_{11}] = e_1 - e_2$, $[O_{12}] = 2e_1 - 2e_2 - e_4$,
$[O_{13}] = 2e_1 - e_3 - e_4$, $[O_{14}] = 2e_1 - e_3$,
$[O_{15}] = 2e_1 - e_3 - e_4$.

Since $[O_3] = [O_5] = [O_6] + [O_{11}]$ and $[O_{13}] = [O_{15}] = [O_6] + [O_2]$, we get that $\Lambda_{\mathrm{eff}}(V)$ is generated by

$$e_2, \quad e_3, \quad 3e_1 - 2e_3 - e_4, \quad 2e_1 - e_2 - e_3 - e_4, \quad e_1 - e_4,$$
$$4e_1 - 2e_2 - 2e_3 - e_4, \quad e_1 - e_2, \quad 2e_1 - 2e_2 - e_4, \quad 2e_1 - e_3.$$

The anticanonical class is given by
$$\omega_V^{-1} = 3e_1 - e_2 - e_3 - e_4.$$
Thus $\alpha(V)$ is the volume of the domain
$$\begin{cases} 3x - y - z - t = 1, \\ y > 0, \quad z > 0, \\ x - y > 0, \\ 2x - z > 0, \\ x - t > 0, \\ 3x - 2z - t > 0, \\ 2x - y - z - t > 0, \\ 4x - 2y - 2z - t > 0, \\ 2x - 2y - t > 0, \end{cases}$$
that is, of the domain P in $\mathbf{R}^3$ given by
$$\begin{cases} y > 0, \quad z > 0, \\ x - y > 0, \\ 2x - z > 0, \\ 1 - 2x + y + z > 0, \\ 1 + y - z > 0, \\ 1 - x > 0, \\ 1 + x - y - z > 0, \\ 1 - x - y + z > 0. \end{cases}$$

We compute its volume as follows: decompose P into cones with apex $(0,0,0)$ and supported by the faces not containing this point. Thus we consider the following faces of P:

$F_1:$ $1 - x = 0,$ $\qquad$ $F_2:$ $1 - 2x + y + z = 0,$
$F_3:$ $1 + y - z = 0,$ $\qquad$ $F_4:$ $1 + x - y - z = 0,$
$F_5:$ $1 - x - y + z = 0.$

One has
$$\alpha(V) = \mathrm{Vol}(P) = \frac{1}{3} \sum_{i=1}^{5} \mathrm{Area}(F_i).$$

The area of F_1 is the volume of the domain

$$\begin{cases} y > 0, \quad z > 0, \\ 1 - y > 0, \\ 2 - z > 0, \\ -1 + y + z > 0, \\ 1 + y - z > 0, \\ 2 - y - z > 0, \\ z - y > 0, \end{cases}$$

and we get $\text{Area}(F_1) = \frac{1}{2}$. For F_2 we have the equations

$$\begin{cases} y > 0, \\ -1 + 2x - y > 0, \\ x - y > 0, \\ 1 + y > 0, \\ 2 - 2x + 2y > 0, \\ 1 - x > 0, \\ 2 - x > 0, \\ x - 2y > 0. \end{cases}$$

We get $\text{Area}(F_2) = \frac{1}{6}$. For F_3 we have the same equations and the same area. For F_4 we have the equations

$$\begin{cases} y > 0, \\ 1 + x - y > 0, \\ x - y > 0, \\ -1 + x + y > 0, \\ 2 - x > 0, \\ -x + 2y > 0, \\ 1 - x > 0, \\ 2 - 2y > 0. \end{cases}$$

We find $\text{Area}(F_4) = 1/8 + 1/24 = 1/6$. The face F_5 is given by the same equations and $\text{Area}(F_5) = 1/6$. Finally

$$\alpha(V) = \frac{1}{3}\left(\frac{1}{2} + \frac{4}{6}\right) = \frac{7}{18}. \quad \square$$

6. Some statistical formulae

The most naive way to test the conjecture is to compute the quotient

(6.1) $$N_{U,H}(B)/\theta_H(V)B(\log B)^{t-1}$$

for large values of B. However, as explained in the introduction, the relative error term is expected to decrease slowly. Therefore it is natural to use the fact that we expect an asymptotic of the form

$$N_{U,H}(B) = BP(\log(B)) + o(B),$$

where P is a polynomial of degree $t-1$ with a dominant coefficient equal to $\theta_H(V)$. With the program of D. J. Bernstein [Be], we can get a family of pairs $(B_i, N_{U,H}(B_i))_{1 \leq i \leq N}$. In the examples below we took for B_i successive powers of $6/5$ between 200 and 10^5. For any i between 1 and N, let

$$x_i = \log(B_i) \quad \text{and} \quad y_i = N_{U,H}(B_i)/B_i.$$

The simplest statistical tool in this setting is to look for a polynomial Q of degree $t-1$ such that

$$\sum_{i=1}^{N}(Q(x_i) - y_i)^2$$

is minimal and to compute its leading coefficient A_{t-1}. We then test the conjecture using the quotient

(6.2) $$A_{t-1}/\theta_H(V).$$

The advantage of this method is that, if the expected formula is correct, and if we take for B_i successive powers of a fixed real number λ between B_1 and B_N, then the relative error term for (6.2) should at least decrease as

$$C/(\log(B_N) - \log(B_1))^{t-1}$$

for $\log(B_N)/\log(B_1)$ large enough with a constant C going to 0 as B_1 goes to infinity.

Of course, due to the arithmetic nature of $N_{U,H}(B)$, the errors are not as independent as one would need for a clean statistical treatment of the data. Also, since we do not have a good understanding of the difference

$$N_{U,H}(B) - BP(\log(B)),$$

and in order to limit the number of arbitrary parameters involved in the statistical computation, we prefered not to weight the points.

Notations 6.1. — Let $R(X,Y)$ be a polynomial in $\mathbf{Q}[X,Y]$ and denote by $\langle R(X,Y)\rangle$ the mean value of $(R(x_i,y_i))_{1\leqslant i\leqslant N}$, that is,

$$\langle R(X,Y)\rangle = \frac{1}{N}\sum_{i=1}^{N} R(x_i,y_i).$$

If $t = 2$ the leading coefficient of Q (if it is uniquely defined) is given by

$$A_1 = \frac{\langle XY\rangle - \langle Y\rangle\langle X\rangle}{\langle X^2\rangle - \langle X\rangle^2}.$$

If $t = 3$ the leading coefficient is

$$A_2 = \frac{\langle YX^2\rangle - \langle Y\rangle\langle X^2\rangle - \frac{(\langle X^3\rangle - \langle X\rangle\langle X^2\rangle)(\langle YX\rangle - \langle Y\rangle\langle X\rangle)}{\langle X^2\rangle - \langle X\rangle^2}}{\langle X^4\rangle - \langle X^2\rangle^2 - \frac{(\langle X^3\rangle - \langle X\rangle\langle X^2\rangle)^2}{\langle X^2\rangle - \langle X\rangle^2}}.$$

If $t = 4$, the leading coefficient is

$$A_3 = \frac{\langle YX^3\rangle - \langle Y\rangle\langle X^3\rangle - \frac{(\langle X^4\rangle - \langle X\rangle\langle X^3\rangle)(\langle YX\rangle - \langle Y\rangle\langle X\rangle)}{\langle X^2\rangle - \langle X\rangle^2} - \frac{\beta\delta}{\gamma}}{\langle X^6\rangle - \langle X^3\rangle^2 - \frac{(\langle X^4\rangle - \langle X\rangle\langle X^3\rangle)^2}{\langle X^2\rangle - \langle X\rangle^2} - \frac{\beta^2}{\gamma}},$$

with

$$\beta = \langle X^5\rangle - \langle X^3\rangle\langle X^2\rangle - \frac{\langle X^3\rangle - \langle X\rangle\langle X^2\rangle}{\langle X^2\rangle - \langle X\rangle^2}(\langle X^4\rangle - \langle X^3\rangle\langle X\rangle),$$

$$\gamma = \langle X^4\rangle - \langle X^2\rangle^2 - \frac{(\langle X^3\rangle - \langle X\rangle\langle X^2\rangle)^2}{\langle X^2\rangle - \langle X\rangle^2},$$

$$\delta = \langle YX^2\rangle - \langle Y\rangle\langle X^2\rangle - \frac{\langle X^3\rangle - \langle X\rangle\langle X^2\rangle}{\langle X^2\rangle - \langle X\rangle^2}(\langle YX\rangle - \langle Y\rangle\langle X\rangle).$$

In the next section, we denote by $\theta_H^{\mathrm{stat}}(V)$ the leading coefficient A_{t-1}.

7. Presentation of the results

We consider only cubic surfaces of the form (5.1), (5.2), or (5.3). By [CTKS, Lemme 1], the corresponding surface V is $\mathbf{Q}$-rational and, in particular, $\mathrm{Br}(V) = \mathrm{Br}(\mathbf{Q})$. Thus the Brauer-Manin obstruction to weak approximation is void and

$$V(\mathbf{A_Q})^{\mathrm{Br}} = V(\mathbf{A_Q}) = \prod_{v\in\mathrm{Val}(\mathbf{Q})} V(\mathbf{Q}_v).$$

Moreover,

$$\beta(V) = \#H^1(\mathbf{Q},\mathrm{Pic}(\overline{V})) = 1.$$

By (1.2) and (1.3), the constant $\theta_H(V)$ may be written as
$$\theta_H(V) = \alpha(V)\omega_H(V(\mathbf{A_Q})).$$
Using remark 3.4 we get
$$\theta_H(V) = \alpha(V)\lim_{s\to 1}(s-1)^{t+2}\zeta_E(s) \times \omega_{H,\infty}(V(\mathbf{R}))$$
$$\times \prod_{p\nmid 3abcd} \lambda_p\omega_{H,p}(V(\mathbf{Q}_p)) \times \prod_{i=0}^{3} C_i,$$
where E is the étale algebra defined in 2.2. The residue of the zeta function could have been computed directly (see, for example, [Co, chapter 4]), but instead we used PARI. The volume at the real place is given by the formula
$$\frac{1}{2}\int_{\{(x,y,z,t)\}\left|\begin{matrix}ax^3+by^3+cz^3+dt^3=0\\ \sup(|x|,|y|,|z|,|t|)\leq 1\end{matrix}\right.} \omega_L(x,y,z,t),$$
where ω_L is the Leray form
$$\omega_L(x,y,z,t) = \frac{\sqrt[3]{d}^{-1}}{3(ax^3+by^3+cz^3)^{2/3}} dx\,dy\,dz.$$
Decomposing the domain of integration (and using the various expressions of the Leray form) it is possible to remove the singularities of this integral which is then easily estimated on a computer. The factors corresponding to the bad places have been described in section 4 and the constants C_0, C_1, C_2, and C_3 may be computed directly as in section 3.

We considered the following examples: for the cubic surfaces with a Picard group of rank 2 we used

(S_1) $X^3 + Y^3 + 2Z^3 + 4T^3 = 0,$
(S_2) $X^3 + Y^3 + 5Z^3 + 25T^3 = 0,$
(S_3) $X^3 + Y^3 + 3Z^3 + 9T^3 = 0.$

For the rank 3 case:

(S_4) $X^3 + Y^3 + 2Z^3 + 2T^3 = 0,$
(S_5) $X^3 + Y^3 + 5Z^3 + 5T^3 = 0,$
(S_6) $X^3 + Y^3 + 7Z^3 + 7T^3 = 0,$
(S_7) $2X^3 + 2Y^3 + 3Z^3 + 3T^3 = 0,$

and for rank 4:

(S_8) $X^3 + Y^3 + Z^3 + T^3 = 0.$

We draw below the corresponding experimental curves in which we compare the value of $N_{U,H}(B)/(B(\log B)^{t-1})$ with $\boldsymbol{\theta}_H(V)$. On each drawing, only the points on the right of the vertical line have been used for the computation of $\boldsymbol{\theta}_H^{\mathrm{stat}}(V)$.

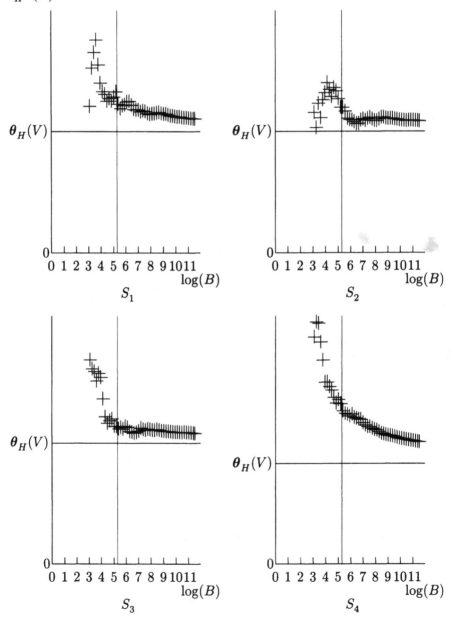

TAMAGAWA NUMBERS OF CUBIC SURFACES

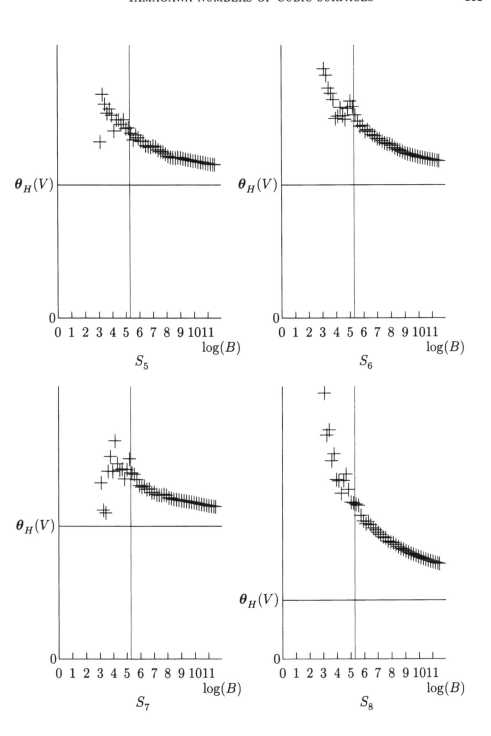

We finish with tables of numerical results. The value of $\theta_H^{\text{stat}}(V)$ is obtained from the pairs $(B_i, N_{U,H}(B_i))$ as described in section 6. We denote by $\theta_H^{\text{exp}}(V)$ the quotient $N_{U,H}(B)/B\log(B)^{t-1}$ and by $\zeta_{E_i}^*(1)$ the limit

$$\zeta_{E_i}^*(1) = \lim_{s \to 1}(s-1)^{t_i}\zeta_{E_i}^*(s),$$

where t_i is the number of components of E_i. Note that for the examples with a Picard group of rank 2, C_2 is equal to 1.

Surface	S_1	S_2	S_3
B	100000	100000	100000
$N_{U,H}(B)$	433526	286040	455164
$\alpha(V)$	2	2	2
ab/cd	1/8	1/125	1/27
$\zeta_{E_0}^*(1)$	0.6045998	0.6045998	0.6045998
ac/bd	1/2	1/5	1/3
$\zeta_{E_1}^*(1)$	0.8146241	1.163730	1.017615
ad/bc	2	5	3
$\zeta_{E_2}^*(1)$	0.8146241	1.163730	1.017615
$\lambda_3' \omega_H(V(\mathbf{Q}_3))$	4/9	4/9	4/9
p_0	2	5	
$\lambda_{p_0}' \omega_H(V(\mathbf{Q}_{p_0}))$	3/8	96/125	
C_0	0.8306815	0.3493824	0.3066383
C_1	0.9540383	0.8704106	0.9762028
C_3	0.9893865	0.9906098	0.9892790
$\omega_H(V(\mathbf{R}))$	3.255161	1.360417	2.221359
$\theta_H(V)$	0.3413500	0.2290769	0.3660885
$\theta_H^{\text{exp}}(V)/\theta_H(V)$	1.103137	1.084575	1.079931
$\theta_H^{\text{stat}}(V)/\theta_H(V)$	0.988687	1.067208	1.051041

For the examples with a Picard group of rank 3, C_3 is equal to 1.

Surface	S_4	S_5	S_6	S_7
B	100000	100000	100000	100000
$N_{U,H}(B)$	3051198	1976482	3420784	1966160
$\alpha(V)$	1	1	1	1
ab/cd	1/4	1/25	1/49	4/9
$\zeta^*_{E_0}(1)$	0.8146241	1.163730	1.265025	1.028996
ac/bd	1	1	1	1
$\zeta^*_{E_1}(1)$	0.6045998	0.6045998	0.6045998	0.6045998
ad/bc	1	1	1	1
$\zeta^*_{E_2}(1)$	0.6045998	0.6045998	0.6045998	0.6045998
$\lambda'_3 \omega_H(V(\mathbf{Q}_3))$	16/27	16/27	16/27	16/27
p_0	2	5	7	2
$\lambda'_{p_0} \omega_H(V(\mathbf{Q}_{p_0}))$	27/64	13824/15625	186624/117649	27/64
C_0	0.8306815	0.3493824	0.3066383	0.8306815
C_1	0.9540383	0.8704106	0.9297617	0.8196347
C_2	0.7827314	0.8112747	0.9228033	0.8294515
$\omega_H(V(\mathbf{R}))$	4.105301	2.347970	1.910125	2.430506
$\theta_H(V)$	0.1895795	0.1291945	0.2184437	0.1290720
$\theta_H^{\exp}(V)/\theta_H(V)$	1.214249	1.154191	1.181448	1.149252
$\theta_H^{\mathrm{stat}}(V)/\theta_H(V)$	0.981952	1.035070	0.999247	1.063376

For the last example we have $C_2 = C_3 = 1$ and $E_1 = E_2 = E_3$ and we get

Surface	S_8
B	100000
$N_{U,H}(B)$	12137664
$\alpha(V)$	7/18
$\zeta^*_{E_i}(1)$	6.045998×10^{-1}
$\lambda'_3 \omega_H(V(\mathbf{Q}_3))$	16/27
C_0	3.066383×10^{-1}
C_1	5.129319×10^{-1}
$\omega_H(V(\mathbf{R}))$	6.121864
$\boldsymbol{\theta}_H(V)$	4.904057×10^{-2}
$\boldsymbol{\theta}_H^{\exp}(V)/\boldsymbol{\theta}_H(V)$	1.621894
$\boldsymbol{\theta}_H^{\text{stat}}(V)/\boldsymbol{\theta}_H(V)$	1.012304

References

[BM] V. V. Batyrev et Y. I. Manin, *Sur le nombre des points rationnels de hauteur bornée des variétés algébriques*, Math. Ann. **286** (1990), 27–43.

[BT] V. V. Batyrev and Y. Tschinkel, *Tamagawa numbers of polarized algebraic varieties*, Nombre et répartition de points de hauteur bornée, Astérisque, vol. 251, SMF, Paris, 1998, 299–340.

[Be] D. J. Bernstein, *Enumerating solutions to $p(a) + q(b) = r(c) + s(d)$*, Math. Comp. **70** (2001), 389–394.

[Co] H. Cohen, *A course in computational algebraic number theory*, Graduate Texts in Math., vol. 138, Springer-Verlag, Berlin, Heidelberg and New York, 1993.

[CT] J.-L. Colliot-Thélène, *The Hasse principle in a pencil of algebraic varieties*, Number theory (Tiruchirapalli, 1996), Contemp. Math., vol. 210, Amer. Math. Soc., Providence, 1998, 19–39.

[CTKS] J.-L. Colliot-Thélène, D. Kanevsky, et J.-J. Sansuc, *Arithmétique des surfaces cubiques diagonales*, Diophantine approximation and transcendence theory (Bonn, 1985), Lecture Notes in Math., vol. 1290, Springer-Verlag, Berlin, Heidelberg and New York, 1987, 1–108.

[CTS]   J.-L. Colliot-Thélène et J.-J. Sansuc, *La descente sur une variété rationnelle définie sur un corps de nombres*, C. R. Acad. Sci. Paris Sér. I Math. **284** (1977), 1215–1218.

[FMT]   J. Franke, Y. I. Manin, and Y. Tschinkel, *Rational points of bounded height on Fano varieties*, Invent. Math. **95** (1989), 421–435.

[H-B]   D. R. Heath-Brown, *The density of zeros of forms for which weak approximation fails*, Math. Comp. **59** (1992), 613–623.

[IR]   K. Ireland and M. Rosen, *A classical introduction to modern number theory (second edition)*, Graduate texts in Math., vol. 84, Springer-Verlag, Berlin, Heidelberg and New York, 1990.

[Man1]   Y. I. Manin, *Le groupe de Brauer-Grothendieck en géométrie diophantienne*, Actes Congrès intern. math., Tome 1 (Nice, 1970), Gauthiers-Villars, Paris, 1971, 401–411.

[Man2]   _____, *Cubic forms (second edition)*, North-Holland Math. Library, vol. 4, North-Holland, Amsterdam, New York and Oxford, 1986.

[Pe]   E. Peyre, *Hauteurs et mesures de Tamagawa sur les variétés de Fano*, Duke Math. J. **79** (1995), n° 1, 101–218.

[PT]   E. Peyre and Y. Tschinkel, *Tamagawa numbers of diagonal cubic surfaces, numerical evidence*, Math. Comp. **70** (2001), 367–387.

[SK]   K. E. Smith, *Rational and non-rational algebraic varieties: lectures of János Kollár*, http://xxx.lanl.gov/abs/alg-geom/9707013 (1997).

[S-D]   P. Swinnerton-Dyer, *Counting rational points on cubic surfaces*, (L'Aquila, 1992) (C. Ciliberto, E. L. Livorni, and A. J. Sommese, eds.), Contemp. Math., vol. 162, AMS, Providence, 1994, 371–379.

Rational points on algebraic varieties
(E. PEYRE, Y. TSCHINKEL, ed.), p. 307–311

THE HASSE PRINCIPLE FOR COMPLETE INTERSECTIONS IN PROJECTIVE SPACE

Bjorn Poonen
   Department of Mathematics, University of California, Berkeley, CA 94720-3840, USA • E-mail : poonen@math.berkeley.edu

Abstract. — Assuming the existence of a smooth geometrically integral complete intersection X of dimension ≥ 3 in $\mathbf{P}^n$ over a number field K, such that the Zariski closure of $X(K)$ is nonempty but of codimension ≥ 2 in X, we construct a 3-dimensional smooth geometrically integral complete intersection X' in $\mathbf{P}^n_K$ that violates the Hasse principle. Such a violation could not be explained by the Brauer-Manin obstruction or Skorobogatov's generalization thereof.

In [**SW**], Sarnak and Wang showed that the finiteness of $X(\mathbf{Q})$ for a certain nonsingular hypersurface X in $\mathbf{P}^5$ over $\mathbf{Q}$ would imply the existence of a nonsingular hypersurface Y in $\mathbf{P}^4$ that violated the Hasse principle. They chose X so that $X(\mathbf{C})$ was hyperbolic, so that the finiteness of $X(\mathbf{Q})$ would follow from a conjecture of Lang. Colliot-Thélène noted that the Brauer-Manin obstruction for a nonsingular complete intersection of dimension ≥ 3 in projective space vanishes, so Y then would be an example for which the Brauer-Manin obstruction to the Hasse principle was not the only one. Later, Skorobogatov [**Sk**] constructed a different sort of variety over $\mathbf{Q}$ (a bielliptic surface) for which the

---

Much of this research was done while the author was at Princeton University supported by an NSF Mathematical Sciences Postdoctoral Research Fellowship. The author is currently supported by NSF grant DMS-9801104, a Sloan Fellowship, and a Packard Fellowship.

Brauer-Manin obstruction is not the only one, by applying the Brauer-Manin obstruction to unramified covers. A complete intersection Y as above is simply connected, however, so its lack of rational points would not be explainable by Skorobogatov's methods.

The construction of Sarnak and Wang involved slicing X with hyperplanes, and performing various calculations involving special properties of X, which had to be carefully chosen. The purpose of this note is to show that using a similarly elementary approach, but with *ramified coverings* instead of slices, one can show that the finiteness and non-emptiness of the rational points in *any* nonsingular hypersurface in $\mathbf{P}^n$ over a number field K implies the existence of a nonsingular hypersurface in $\mathbf{P}^n$ over K that violates the Hasse principle. In particular, it is no longer necessary to start with a variety of dimension one higher in order to obtain a Hasse principle counterexample of desired dimension. More generally, we prove the following.

Theorem 1. — *Fix a number field K and integers $d \geq 2$, $n \geq 3$. If there exists a d-dimensional smooth geometrically integral complete intersection X in $\mathbf{P}^n$ over K such that the Zariski closure of $X(K)$ has dimension at most 1, but $X(K)$ is not empty, then there exists also a d-dimensional smooth geometrically integral complete intersection X' in $\mathbf{P}^n$ over K such that $X'(K_v)$ is nonempty for all completions K_v of K, but $X'(K)$ is empty.*

Proof. — Suppose that X exists. It follows from Bertini's theorem (see Theorem II.8.18 of [**Ha**] and its proof) that after a linear change of variables on $\mathbf{P}^n$ over K, we may assume both of the following:

(1) Each coordinate subspace meets X transversely. (By "coordinate subspace," we mean a linear subspace of $\mathbf{P}^n$ of the form $x_{i_1} = x_{i_2} = \cdots = x_{i_q} = 0$.)

(2) If Z is the Zariski closure of $X(K)$ in X, and R is any 1-dimensional irreducible component of $Z \times_K \overline{K}$, then each coordinate subspace meets R transversely (i.e., the hyperplane $x_i = 0$ meets R in r distinct points where r is the degree of R in $\mathbf{P}^n$, and the subspace $x_i = x_j = 0$ for $i < j$ does not meet R).

Let $X' = \varphi^{-1}(X)$ where the morphism $\varphi : \mathbf{P}^n \to \mathbf{P}^n$ is given in homogeneous coordinates by

$$(x_0 : x_1 : \cdots : x_n) \mapsto (c_0 x_0^p : c_1 x_1^p : \cdots : c_n x_n^p)$$

for some large prime p and for some $c_0, c_1, \ldots, c_n \in K^*$ to be specified later. Since the coordinate subspaces meet X transversely, X' is smooth. If X is given by the homogeneous equations

$$f_1(x_0, \ldots, x_n) = \cdots = f_{n-d}(x_0, \ldots, x_n) = 0,$$

then X' is a complete intersection of dimension d as well, given by the equations
$$f_1(c_0x_0^p, \ldots, c_nx_n^p) = \cdots = f_{n-d}(c_0x_0^p, \ldots, c_nx_n^p) = 0.$$
If one successively adjoins to the function field $\overline{K}(X)$ the p-th roots of the functions x_i/x_0 on X, each such extension is totally ramified along the poles and zeros of these functions, so X' (which is of pure dimension d) is geometrically integral.

Suppose that R is a 1-dimensional irreducible component of $Z \times_K \overline{K}$ as above. Let $S = \varphi^{-1}(R)$, which is a curve over $\overline{K}$. The same proof that showed that X' was geometrically integral shows that S is irreducible. We will prove next that the genus g_S of S is at least 2 if $p > 2$. (By genus, we mean geometric genus, i.e., the genus of a smooth projective desingularization.) The Riemann-Hurwitz formula yields
$$2g_S - 2 = (\deg \iota)(2g_R - 2) + \sum(e_i - 1).$$
Let r be the degree of R in $\mathbf{P}^n$. Then each coordinate hyperplane meets R transversely in exactly r points, each belonging to only one coordinate hyperplane. These are the only points on R which ramify, and above each there are p^{n-1} points of S, each with ramification index p. Thus
$$\begin{aligned}2g_S - 2 &= p^n(2g_R - 2) + (n+1)rp^{n-1}(p-1) \\ &\geq -2p^n + 4(p^n - p^{n-1}).\end{aligned}$$
If $p > 2$, then $g_S \geq 2$.

Thus $\varphi^{-1}(Z) \times_K \overline{K}$ consists of isolated points and curves over $\overline{K}$ of genus ≥ 2. Applying Faltings' theorem to each curve (over a finite extension of K over which it is defined) shows that $\varphi^{-1}(Z)(K)$ is finite. But $X'(K) \subseteq \varphi^{-1}(Z)$, so $X'(K)$ is finite too. On the other hand, we are free to choose the c_i so that at least one point in $X(K)$ has a K-rational preimage. Thus by replacing X by X', we have reduced the problem to the case where $X(K)$ is finite (and nonempty).

The next step is to reduce to the case where $X(K)$ consists of a single point. Suppose instead that there exist distinct points $P_1, P_2 \in X(K)$. We construct a new covering $X' \to X$ exactly as before, choosing p large enough so that K has no nontrivial p-th roots of unity. Because we are allowed a generic change of variables on the projective space containing X, we may assume that P_1 and P_2 are outside all coordinate hyperplanes and that the values taken by the rational function x_1/x_0 at P_1 and P_2 are distinct in K^*/K^{*p}. Then we may pick the c_i so that P_1 has a K-rational preimage in X' but P_2 does not. Moreover, the choice of p guarantees that each K-rational point on X has at most one K-rational preimage on X', so we have succeeded in constructing a cover X' with fewer K-rational points than X, but still with at least one K-rational point. Iterating such covers eventually leads to a cover with exactly one K-rational point.

Therefore we now assume $X(K)$ consists of a single point P. Again perform a generic change of variables and define a covering X' as before, using $c_i = 1$ for $i \neq 0$ ($c_0 \in K^*$ will be chosen later). By Bertini's theorem, we may assume that the intersection Y of X' with the hyperplane $x_0 = 0$ is smooth and geometrically integral, and we may also assume that $P = (1 : 1 : \cdots : 1)$. Over all but finitely many completions of K, Y will have a point, and hence so will X'. Note that Y does not depend on c_0, so we are free to apply weak approximation to choose $c_0 \in K^*$ such that $c_0 \notin K^{*p}$ but such that $c_0 \in K_v^{*p}$ for each of the finitely many completions K_v of K for which $Y(K_v) = \emptyset$. This ensures that X' has a point (namely a preimage of P) over each of the finitely many completions remaining. On the other hand, $X'(K)$ is empty, since any K-rational point of X' would have to lie above P. □

Corollary 2. — *If there exists a smooth geometrically integral complete intersection X of dimension ≥ 3 in $\mathbf{P}^n$ over a number field K, such that the Zariski closure of $X(K)$ has codimension ≥ 2 in X, but $X(K)$ is not empty, then there exists also a 3-dimensional smooth geometrically integral complete intersection X' in projective space over K such that $X'(K)$ is empty, but such that the emptiness of $X'(K)$ cannot be explained by the Brauer-Manin obstruction or Skorobogatov's generalization thereof.*

Proof. — Choose $P \in X(K)$. If $\dim X > 3$, then we may repeatedly use the Bertini-type theorems of [**KA**] to replace X by its intersection with a sufficiently general K-rational hypersurface of large degree m through P, in order to reduce to the case where $\dim X = 3$. Now apply Theorem 1. □

Acknowledgement.
I thank the referee for several useful comments.

References

[Ha] HARTSHORNE, R., *Algebraic geometry*, Springer-Verlag, New York, 1977.
[KA] KLEIMAN, S. AND ALTMAN, A., Bertini theorems for hypersurface sections containing a subscheme, *Comm. Algebra* **7** (1979), no. 8, 775–790.
[SW] SARNAK, P. AND WANG, L., Some hypersurfaces in $\mathbf{P}^4$ and the Hasse-principle, *C. R. Acad. Sci. Paris* **321** (1995), 319–322.

[Sk] SKOROBOGATOV, A. N., Beyond the Manin obstruction, *Invent. Math.* **135** (1999), no. 2, 399–424.

Rational points on algebraic varieties
(E. PEYRE, Y. TSCHINKEL, ed.), p. 313–334

UNE CONSTRUCTION DE COURBES k-RATIONNELLES SUR LES SURFACES DE KUMMER D'UN PRODUIT DE COURBES DE GENRE 1.

Philippe Satgé

Département de Mathématiques, Université de Caen, Campus II, B.P. 5186, 14032 CAEN Cedex, France • *E-mail* : Philippe.Satge@math.unicaen.fr

Résumé. — k étant un corps de caractéristique différente de 2, nous décrivons une méthode permettant de construire des courbes k-rationnelles (i.e. k-birationnellement équivalentes à la droite projective) sur les surfaces de Kummer associées à un produit de courbes de genre 1 munies d'involutions k-hyperelliptiques. Nous ramenons ce problème à un problème de géométrie énumérative sur le produit $\mathbf{P}_{1,k} \times \mathbf{P}_{1,k}$ de la droite projective par elle même. Bien que la résolution générale du problème de géométrie énumérative auquel nous arrivons soit hors de portée des méthodes que nous connaissons, la recherche de solutions particulières dans des systèmes linéaires convenablement choisis permet d'obtenir des exemples interessants. On constate par exemple que l'on retrouve ainsi, de manière assez systématique, plusieurs résultats qui apparaissent de manière isolée dans la littérature.

Introduction

On fixe un corps k de caractéristique différente de 2. Le but de ce papier est la construction de courbes k-rationnelles (i.e. k-birationnellement équivalentes à la droite projective) sur la surface de Kummer $X = X_1 \times X_2/(\iota_1 \times \iota_2)$ où X_1 et X_2 sont deux k-courbes de genre 1, et ι_1 et ι_2 deux involutions k-hyperelliptiques de X_1 et X_2, c'est à dire deux k-involutions de X_1 et X_2 telles que les quotients X_1/ι_1 et X_2/ι_2 sont k-isomorphes à la droite projective. Il est facile de calculer des équations locales de la surface X (ce calcul est donné au début du §2) et cela permet de reformuler le problème qui nous interesse de la manière très explicite suivante : étant donnés deux polynômes $P_1(x_1)$ et $P_2(x_2)$ de degré 3 ou 4, à coefficients dans k et sans racines multiples, nous cherchons à construire des solutions, dans le corps $k(u)$ des fractions rationnelles à coefficients dans k, à l'équation diophantienne $P_2(x_2(u))y(u)^2 = P_1(x_1(u))$.

Les raisons de s'interesser aux courbes k-rationnelles de la surface de Kummer X sont nombreuses et classiques ; rappelons en quelques unes. On montre facilement que l'image réciproque d'une courbe k-rationnelle de X par la projection canonique de $X_1 \times X_2$ sur X est une courbe k-hyperelliptique ; la construction de courbes k-rationnelles sur X est donc équivalente à la construction de courbes k-hyperelliptiques dont la jacobienne contient deux facteurs elliptiques. Lorsque le corps k est le corps des complexes, c'est le très classique problème de la réduction des intégrales hyperelliptiques à des intégrales elliptiques. Le cas particulier des courbes hyperelliptiques de genre 2 est particulièrement intéressant et est discuté dans deux des exemples que nous traitons ; on notera que nous ne supposons pas le corps de base algébriquement clôs et que le contrôle des corps de rationalité est un point important de notre approche. On sait aussi que la surface X possède plusieurs fibrations en courbes de genre 1 sur la droite projective ; on peut utiliser les courbes k-rationnelles de X qui ne sont pas contenues dans des fibres pour construire des changements de base qui augmentent le rang générique des fibrations ; lorsque le corps k est un corps de nombres, ce procédé est (au langage près) le procédé standard utilisé pour construire par spécialisation des familles de courbes dont le rang est strictement plus grand que le rang générique. Ce point de vue est développé dans [**M-S**]. Enfin, toujours dans le cas où k est un corps de nombres, rappelons que les conjectures de Batyrev-Manin ([**B-M**], §3.5) prédisent que les points rationnels de X sont presque tous situés sur des courbes k-rationnelles, et qu'il semble important de comprendre comment ils s'accumulent sur ces courbes. McKinnon ([**McK**]) a montré que les courbes k-rationnelles qui forment le premier ensemble d'accumulation des points rationnels sur certaines surfaces du type de X sont, pour des hauteurs bien choisies, les courbes k-rationnelles dont la construction est évidente (ce sont les courbes appellées triviales dans le §1.1 de ce papier). On sait très peu de choses sur les ensembles d'accumulations suivants. Bien que nous ne soyons rien capable de prouver dans cette direction à l'heure actuelle, il est raisonnable de penser que certaines des courbes construites dans ce papier appartiennent au second de ces ensembles. La production un peu systématique d'exemples de courbes k-rationnelles sur la surface X nous semble interessante de ce point de vue.

Dans tout le papier k est un corps de caractéristique différente de 2 et $\overline{k}$ une clôture algébrique de k. On fixe deux couples (X_1, ι_1) et (X_2, ι_2) où X_1

et X_2 sont deux k-courbes lisses, géométriquement intègres de genre 1, et où ι_1 et ι_2 sont deux involutions k-hyperelliptiques sur X_1 et X_2, c'est à dire des k-involutions telles que les quotients X_1/ι_1 et X_2/ι_2 sont k-isomorphes à la droite projective $\mathbf{P}_{1,k}$. Le quotient $X = X_1 \times X_2/(\iota_1 \times \iota_2)$ est une k-surface géométriquement intègre, possédant 16 points singuliers, et dont le modèle lisse est une surface de Kummer ; c'est sur cette surface que nous voulons construire des courbes k-rationnelles. Pour $i = 1,2$, on note $\pi_i : X_i \to X_i/\iota_i = \mathbf{P}_{1,k}$ la projection canonique ; le produit $\pi_1 \times \pi_2 : X_1 \times X_2 \to \mathbf{P}_{1,k} \times \mathbf{P}_{1,k}$ se factorise naturellement par la projection canonique $X_1 \times X_2 \to X_1 \times X_2/(\iota_1 \times \iota_2) = X$ et on note $\pi : X \to \mathbf{P}_{1,k} \times \mathbf{P}_{1,k}$ le k-morphisme de factorisation. Le principe de notre construction consiste à construire des courbes k-rationnelles sur X en remontant des courbes k-rationnelles de $\mathbf{P}_{1,k} \times \mathbf{P}_{1,k}$; l'efficacité du procédé repose sur le fait que la géométrie de $\mathbf{P}_{1,k} \times \mathbf{P}_{1,k}$ est plus simple que celle de X.

Nous avons divisé ce travail en deux paragraphes. Dans le premier, on étudie les composantes irréductibles de l'image réciproque d'une courbe de $\mathbf{P}_{1,k} \times \mathbf{P}_{1,k}$ par le morphisme canonique $\pi : X \to \mathbf{P}_{1,k} \times \mathbf{P}_{1,k}$. On montre (corollaire 1.2.4) que les courbes de $\mathbf{P}_{1,k} \times \mathbf{P}_{1,k}$ dont les images réciproques par π ont des composantes k-rationnelles sont les courbes k-rationnelles de $\mathbf{P}_{1,k} \times \mathbf{P}_{1,k}$ qui rencontrent la courbe de ramification Z du k-morphisme π d'une manière imposée ; on ramène ainsi le problème de la recherche des courbes k-rationnelles de X à un problème de géométrie énumérative dans $\mathbf{P}_{1,k} \times \mathbf{P}_{1,k}$. On sait que c'est la un problème difficile en général ; on montre dans le second paragraphe qu'il est abordable pour des courbes dont le bidegré reste raisonable (par rapport aux moyens de calculs dont on dispose) et donne des résultats non triviaux. Dans le second paragraphe nous traitons quatres exemples. Dans le premier exemple nous traitons le cas le plus simple qui est celui des courbes de bidegré $(1,1)$. On constate que l'on retrouve de cette manière des formules classiques attribuées à Legendre et Jacobi. Le deuxième exemple semble moins classique ; il illustre le fait que les couples (X_1, ι_1) et (X_2, ι_2) ne sont pas nécessairement formés d'une courbe elliptique et de la multiplication par -1 ; dans le point de vue des équations diophantiennes, cela se traduit par le fait que les polynômes $P_1(x_1)$ et $P_2(x_2)$ sont de degré 4 et n'ont pas nécessairement de racines dans le corps de base k. Le troisième exemple est choisi de manière à retrouver un résultat établi, dans un autre contexte, par J.-F. Mestre. Enfin, dans le quatrième exemple, on s'intéresse aux courbes k-rationnelles de X qui correspondent aux courbes hyperelliptique de genre 2 dont la jacobienne est décomposée. Cela nous amène à rechercher dans $\mathbf{P}_{1,k} \times \mathbf{P}_{1,k}$ des courbes k-rationnelles de bidegré (n,n), très singulières, et rencontrant la courbe de ramification Z de π de manière imposée. Le cas $n = 3$ est facile à traiter et redonne des formules classiques. Peu de choses sont connus pour les valeurs de $n > 3$. La méthode présentée ici permet d'aborder les cas $n = 4$ et $n = 5$; chaque fois que l'on a trouvé un exemple dans la littérature (on en signale un certains nombre à la fin du papier) il nous a été relativement facile de le retrouver par notre procédé. Une étude plus générale nécessiterait d'avoir des informations complémentaires sur les singularités des courbes de bidegré (n,n) que l'on cherche à contruire. Cette étude est abordée dans [Sat] et mérite sans doute d'être poussée plus loin.

1. Relèvement des courbes de $\mathbf{P}_{1,k} \times \mathbf{P}_{1,k}$ sur la surface de Kummer

Dans ce premier paragraphe nous expliquons comment calculer le genre géométrique des composantes irréductibles des images réciproques par π des courbes de $\mathbf{P}_{1,k} \times \mathbf{P}_{1,k}$ et nous en déduisons la caractérisation des courbes de $\mathbf{P}_{1,k} \times \mathbf{P}_{1,k}$ dont l'image réciproque par π contient des courbes k-rationnelles en utilisant le fait que les courbes k-rationnelles sont les courbes de genre géométrique 0 qui possèdent des points rationnels.

1.1. Notations.— Pour $i = 1, 2$, on désigne par $R_0^{(i)}, \ldots, R_3^{(i)}$ les quatre $\overline{k}$-points fixes de l'involution ι_i, et par $r_0^{(i)}, \ldots, r_3^{(i)}$ leurs images respectives par la projection canonique $\pi_i : X_i \to X_i/\iota_i = \mathbf{P}_{1,k}$; on note $\mathscr{P}$ l'ensemble des seize $\overline{k}$-points de $\mathbf{P}_{1,k} \times \mathbf{P}_{1,k}$ formé des $r_{m,n} = (r_m^{(1)}, r_n^{(2)})$ où $m, n = 0, \ldots, 3$. Pour tout $\overline{k}$-point r de $\mathbf{P}_{1,k}$, on note $L_1(r)$ (resp. $L_2(r)$) la $\overline{k}$-courbe $\mathbf{P}_{1,k} \times \{r\}$ (resp. $\{r\} \times \mathbf{P}_{1,k}$) de $\mathbf{P}_{1,k} \times \mathbf{P}_{1,k}$. Pour $n, m \in \{0, \ldots, 3\}$, on pose $L_1(n) = L_1(r_n^{(2)})$ et $L_2(m) = L_2(r_m^{(1)})$; les courbes $L_1(n)$ et $L_2(m)$ se coupent donc en $r_{m,n}$; on note $\mathscr{L}$ l'ensemble des huit courbes $\{L_i(n),\ i = 1, 2,\ n = 0, \ldots, 3\}$. On désigne par Z la courbe (réduite) de $\mathbf{P}_{1,k} \times \mathbf{P}_{1,k}$ dont le support est la réunion des huit courbes de $\mathscr{L}$; c'est la courbe de ramification du morphisme π; ses points singuliers sont les seize points de $\mathscr{P}$ et sont aussi les images par π des seize points singuliers de X.

Toute courbe $L_i(n) \in \mathscr{L}$ est une courbe $\overline{k}$-rationnelle de $\mathbf{P}_{1,k} \times \mathbf{P}_{1,k}$; elle est k-rationnelle si et seulement si le point $r_n^{(i)}$ est défini sur k; comme $L_i(n)$ est dans le lieu de ramification de π, son image réciproque, munie de sa structure de schéma réduit, est une courbe $\overline{k}$-rationnelle de X qui est k-rationnelle si et seulement si $L_i(n)$ est k-rationnelle, c'est à dire si et seulement si $r_n^{(i)}$ est défini sur k. Dans la suite ces courbes k-rationnelles seront appelées les courbes k-rationnelles triviales de X.

Si D est une k-courbe géométriquement intègre, on note $\overline{k}(D)$ le corps des $\overline{k}$-fonctions rationnelles sur D ; les valuations discrètes de $\overline{k}(D)$ qui sont triviales sur $\overline{k}$ et normalisées par le fait que leurs groupes des valeurs est $\mathbf{Z}$ tout entier seront appelées les branches de D. On note $\mathscr{V}_D$ l'ensemble des branches de D ; le corps de définition de la branche $v \in \mathscr{V}_D$ est le corps résiduel de l'anneau de la restriction de v au sous corps $k(D)$ de $\overline{k}(D)$ formé des k-fonctions rationnelles. S'il existe un $\overline{k}$-point de D dont l'anneau local dans $\overline{k}(D)$ est dominé par l'anneau de la valuation v, ce $\overline{k}$-point est unique (les k-courbes sont, par hypothèse, des k-schémas séparés) et est appelé le centre de v. Toutes les courbes considérées dans ce papier sont des courbes projectives, donc toutes les branches ont un centre.

1.2. Le calcul du genre.— Pour $i \in \{1, 2\}$, choisissons des coordonnées $(u_{i,0} : u_{i,1})$ sur la droite projective $\mathbf{P}_{1,k} = X_i/\iota_i$ et notons U_i et U'_i les k-ouverts $\mathbf{P}_{1,k} \setminus \{(0 : 1)\}$ et $\mathbf{P}_{1,k} \setminus \{(1 : 0)\}$ de $\mathbf{P}_{1,k}$. On sait que le couple (X_i, ι_i) se décrit de la manière suivante : il existe une forme $F_i(u_{i,0}, u_{i,1})$ de degré 4, bien définie à multiplication par le carré d'un élément non nul de k près, telle que les ouverts $\pi^{-1}(U_i)$ et $\pi^{-1}(U'_i)$ de X_i qui recouvrent X sont les k-courbes affines d'équations respectives $v_i^2 = P_i(u_i)$ et $v'^2_i = P'_i(u'_i)$ où $P_i(u_i) = F_i(1, u_i)$ et $P'_i(u'_i) = F_i(1, u'_i)$; ces deux ouverts se recollent par $u'_i = 1/u_i$, $v'_i = v_i/u_i^2$, et ι_i envoie le point de coordonnées (u_i, v_i) (resp. (u'_i, v'_i)) sur le point de coordonnées $(u_i, -v_i)$ (resp. $(u'_i, -v'_i)$). Un calcul élémentaire montre que l'ouvert $\pi^{-1}(U_1 \times U_2)$ de X est la k-surface affine de l'espace affine $\mathbf{A}^3_k$ d'équation $v^2 = P_1(u_1)P_2(u_2)$; notons que $P_1(u_1)P_2(u_2) = 0$ est une équation de la courbe Z sur $U_1 \times U_2$, donc on obtient une équation de l'ouvert $\pi^{-1}(U_1 \times U_2)$ de X en extrayant la racine carré d'une équation locale de Z bien normalisée. On décrit bien sûr de manière analogue $\pi^{-1}(U_1 \times U'_2)$, $\pi^{-1}(U'_1 \times U_2)$, et $\pi^{-1}(U'_1 \times U'_2)$ qui, avec $\pi^{-1}(U_1 \times U_2)$, recouvrent X ; les formules de recollement sont immédiates à écrire. On introduit la définition suivante :

Définition 1.2.1. — *Une bonne k-équation locale de Z est une k-équation locale (U, f) de Z (i.e. un k-ouvert affine $U \subset \mathbf{P}_{1,k} \times \mathbf{P}_{1,k}$ et une k-fonction régulière f sur U dont le schéma des zéros est $Z \cap U$) telle que l'anneau des coordonnées du k-ouvert affine $\pi^{-1}(U)$ de X est $A[\sqrt{f}]$ où A désigne l'anneau des coordonnées de U.*

On considère maintenant une k-courbe D géométriquement intègre de $\mathbf{P}_{1,k} \times \mathbf{P}_{1,k}$ qui n'est pas contenue dans Z. Soit v une branche de D et (U, f) une bonne k-équation locale de Z telle que le centre de v appartient à U (on peut par exemple prendre pour U l'un des quatres ouverts $U_1 \times U'_1$, $U_1 \times U'_2$, $U_2 \times U'_1$

ou $U_2 \times U_2'$ qui recouvrent $\mathbf{P}_{1,k} \times \mathbf{P}_{1,k}$ et pour f l'équation locale que l'on a explicitée plus haut). Comme D n'est pas inclue dans Z, la restriction de f à D n'est pas nulle et définit donc un élément non nul du corps $k(D)$ que l'on note f_D. L'entier $v(f_D)$ ne dépend pas du choix de (U, f) et est, par définition, la multiplicité d'intersection de la branche v avec Z. Si P est un $\overline{k}$-point de D, la multiplicité d'intersection de D et Z en P est donc la somme des multiplicités des branches de D centrées en P avec Z. On pose :

Définition 1.2.2. — *Soit D une k-courbe de $\mathbf{P}_{1,k} \times \mathbf{P}_{1,k}$ qui est géométriquement intègre et qui n'est pas contenue dans la courbe Z. Pour toute branche v de D, on note $m(v)$ la multiplicité d'intersection de la branche v avec Z. On note $\mathscr{R}_D$ l'ensemble des branches de D pour lesquelles $m(v)$ est impair.*

On rappelle que le genre géométrique d'une k-courbe géométriquement intègre est, par définition, le genre d'un $\overline{k}$ modèle propre et lisse de cette courbe ; on a :

Proposition 1.2.3. — *Soit D une k-courbe de $\mathbf{P}_{1,k} \times \mathbf{P}_{1,k}$ qui est géométriquement intègre et qui n'est pas contenue dans Z. Si g_D est le genre géométrique de D et si r_D est le cardinal de l'ensemble $\mathscr{R}_D$ défini en 1.2.2, on a :*
i) Si $\mathscr{R}_D$ est non vide, l'image réciproque $\pi^{-1}(D)$ de D par π est une k-courbe de X qui est géométriquement intègre et dont le genre géométrique est $g = 1 + 2(g_D - 1) + \frac{r_D}{2}$ (et donc r_D est pair).
ii) Si $\mathscr{R}_D$ est vide, si (U, f) est une bonne k-équation locale de Z et si $f_D \in k(D)$ est la restriction de f à D, alors :
  ii)$_1$ si f_D n'est pas un carré dans le corps $\overline{k}(D)$, l'image réciproque $\pi^{-1}(D)$ de D par π est une k-courbe de X qui est géométriquement intègre et dont le genre géométrique est $g = 1 + 2(g_D - 1)$;
  ii)$_2$ si f_D est un carré dans le corps $\overline{k}(D)$, l'image réciproque $\pi^{-1}(D)$ de D par π est la réunion de deux $\overline{k}$-courbes de X qui sont $\overline{k}$-birationnellement équivalentes à D ; ces deux $\overline{k}$-courbes sont des k-courbes de X si et seulement si f_D est un carré dans le corps $k(D)$ et, dans ce cas, elles sont k-birationnellement équivalentes à D.

Démonstration: Fixons une famille finie $(U^{(i)}, f_i)_{i \in I}$ de bonnes k-équations locales de Z dans laquelle $(U^{(i)})_{i \in I}$ est un recouvrement ouvert de $\mathbf{P}_{1,k} \times \mathbf{P}_{1,k}$ et les $U^{(i)}$ sont des k-ouverts affines d'anneau de coordonnées A_i. La famille $(U^{(i)} \cap D)_{i \in I}$ est un recouvrement ouvert de D et les $U^{(i)} \cap D$ sont des k-ouverts affines dont les anneaux de coordonnées sont notés $A_{i,D}$; pour $i \in I$, on note $f_{i,D} \in A_{i,D}$ la restriction de f_i à D. Par définition $\pi^{-1}(U^{(i)})$ est un k-ouvert affine de X d'anneau de coordonnées $A_i[\sqrt{f_i}]$, donc $\pi^{-1}(U^{(i)}) \cap \pi^{-1}(D)$ est un le k-schéma affine d'anneau des coordonnées $A_i[\sqrt{f_i}] \otimes_{A_i} A_{i,D} = A_{i,D}[\sqrt{f_{i,D}}]$. Ainsi, $\pi^{-1}(D)$ est géométriquement intègre si et seulement si, quelque soit $i \in I$, $f_{i,D}$ n'est pas un carré dans le corps des fractions de $A_{i,D} \otimes_k \overline{k}$ qui est $\overline{k}(D)$. Soit maintenant (U, f) une bonne k-équation locale de Z; il résulte immédiatement des définitions que, pour tout $i \in I$, il existe un $g_i \in k(\mathbf{P}_{1,k} \times \mathbf{P}_{1,k})$ tel que $f = f_i g_i^2$; en conséquence la restriction f_D de f à D est $f_D = f_{i,D} g_{i,D}^2$ où $g_{i,D}$ est la restriction de g_i à D. Ainsi f_D n'est pas un carré dans $\overline{k}(D)$ si et seulement si aucun des $f_{i,D}$ n'est un carré dans $\overline{k}(D)$, et donc si et seulement si la courbe $\pi^{-1}(D)$ est géométriquement intègre.

Montrons $i)$: Par hypothèse $\mathscr{R}_D$ n'est pas vide, c'est à dire qu'il existe au moins une valuation discrète v de $\overline{k}(D)$ triviale sur $\overline{k}$ telle que $v(f_D)$ est impair, donc f_D n'est pas un carré dans $\overline{k}(D)$; on vient de voir que cela implique que $\pi^{-1}(D)$ est géométriquement intègre. D'autre part, la théorie de Kummer montre qu'une valuation discrète w de $\overline{k}(D)$ triviale sur $\overline{k}$ est ramifiée dans l'extension $\overline{k}(D)(\sqrt{f_D})$ si et seulement si $w(f_D)$ est impair, c'est à dire si et seulement si $w \in \mathscr{R}_D$. Il résulte alors de la formule de Riemann-Hurwitz que le genre du corps $\overline{k}(D)(\sqrt{f_D})$, qui est le genre géométrique de $\pi^{-1}(D)$, est $1 + 2(g_D - 1) + \frac{r_D}{2}$, ce qu'on voulait.

Montrons $ii)$: Si f_D n'est pas un carré dans $\overline{k}(D)$, on raisonne comme dans le cas $i)$. Sinon, les assertions résultent immédiatement de la description locale de $\pi^{-1}(D)$ que l'on a explicité dans la première partie de cette démonstration.□

Remarque: Les équivalences birationnelles du cas $ii)_2$ de la proposition précédente ne sont pas nécessairement des isomorphismes (on fabrique facilement des exemples où D est une courbe singulière, et où ces équivalences birationnelles sont des désingularisations de $D \times_k \overline{k}$.

Dans ce papier nous utiliserons le corollaire suivant de la proposition 1.2.3 :

***Corollaire* 1.2.4.** — *Soit D une courbe k-rationnelle de $\mathbf{P}_{1,k} \times \mathbf{P}_{1,k}$ qui n'est pas contenue dans Z, et soit r_D le cardinal de $\mathscr{R}_D$ (définition 1.2.2). Alors*
1) Les deux assertions suivantes sont équivalentes :
  (i) l'image réciproque $\pi^{-1}(D)$ de D dans X est une k-courbe géométriquement intègre de genre géométrique zéro;
  (ii) $r_D = 2$.
2) Les deux assertions suivantes sont équivalentes :
  (i) l'image réciproque $\pi^{-1}(D)$ de D dans X possède deux $\overline{k}$-composantes irréductibles C_1 et C_2 qui, si on les munit de leurs structures de schémas réduits, sont de genre géométrique zéro;
  (ii) $r_D = 0$.

3) Si $r_D \neq 0, 2$, alors aucune composante de $\pi^{-1}(D)$ n'est $\overline{k}$-rationnelle.

Démonstration: La courbe D étant de genre géométrique 0, elle n'admet pas de revêtement non ramifié de degré strictement positif, donc si $\mathscr{R}_D$ est vide i.e. si $r_D = 0$, l'image réciproque $\pi^{-1}(D)$ de D n'est pas géométriquement intègre. Compte tenu de cette remarque, nos assertions résultent immédiatement de la proposition 1.2.3. □

Comme toute courbe k-rationnelle de X est une composante de l'image réciproque par π de son image dans $\mathbf{P}_{1,k} \times \mathbf{P}_{1,k}$, et que l'image d'une courbe k-rationnelle par un k-morphisme est une courbe k-rationnelle, le corollaire précédent ramène la recherche des courbes k-rationnelles sur la surface de Kummer X aux deux problèmes suivants : d'une part le problème géométrique de trouver les courbes k-rationnelles D de $\mathbf{P}_{1,k} \times \mathbf{P}_{1,k}$ avec $r_D = 0$ ou 2, et d'autre part le problème arithmétique de décider, dans le cas $r_D = 0$ si les deux $\overline{k}$-composantes de $\pi^{-1}(D)$ sont définies sur k (ce qui implique qu'elles sont k-rationnelles puisqu'elles sont alors k-birationnellement équivalentes à D), et dans le cas $r_D = 2$ si la k-courbe géométriquement intègre $\pi^{-1}(D)$ qui est de genre géométrique 0 possède des points rationnels sur k.

2. Exemples

Dans la présentation des exemples nous utiliserons les notations introduites au début du §1.2. Les coordonnées $(u_{i,0} : u_{i,1})$ sur la droite projective $\mathbf{P}_{1,k} = X_i/\iota_i$ seront choisis de manière adaptée à l'exemple traité. Rappelons que le k-ouvert affine $\pi^{-1}(U_1 \times U_2)$ de X est k-isomorphe à la k-surface affine de $\mathbf{A}_k^3$ d'équation $v^2 = P_1(u_1)P_2(u_2)$. Les points singuliers de cet ouvert sont les points dont la coordonnée v est 0 ; le morphisme d'éclatement de ces points singuliers est le morphisme de la k-surface $\mathbf{A}_k^2 \times \mathbf{P}_{1,k}$ d'équation $P_2(x_2)y_1^2 = P_1(x_1)y_0^2$ sur la k-surface de l'espace affine $\mathbf{A}_k^3$ d'équation $v^2 = P_1(u_1)P_2(u_2)$ qui envoie le point de coordonnées $(x_1, x_2, (y_0 : y_1))$ sur $(u_1 = x_1, u_2 = x_2, v = P_2(x_2)y_1/y_0)$ si $y_0 \neq 0$ et sur $(u_1 = x_1, u_2 = x_2, v = P_1(x_1)y_0/y_1)$ si si $y_1 \neq 0$. La partie $y_0 \neq 0$ de cet éclaté est donc un k-ouvert affine du modèle lisse X^{lisse} de X qui est k-isomorphe à la k-surface affine de $\mathbf{A}_k^3$ d'équation $P_2(x_2)y^2 = P_1(x_1)$. Nous explicitons les paramétrisations des courbes k-rationnelles que nous construisons dans cet ouvert affine de X^{lisse} ; on trouve ainsi les solutions de l'équation diophantienne qui a été mentionnée dans l'introduction.

Pour toute k-courbe D de $\mathbf{P}_{1,k} \times \mathbf{P}_{1,k}$ non contenue dans Z, toute branche v de D, et tout couple (i, n) avec $i = 1, 2$ et $n = 0, \ldots, 3$, on note $m(i, n; v)$

la multiplicité d'intersection de la branche v avec la droite $L_i(n)$; on a donc $v \in \mathscr{R}_D$ si et seulement si $\sum_{(i,n)} m(i,n;v)$ est impair.

2.1. Exemple 1.— Relèvement des courbes de bidegré $(1,1)$

Une k-courbe D intègre et de bidegré $(1,1)$ est propre, lisse et k-rationnelle; l'application qui envoie une branche de D sur son centre permet donc d'identifier l'ensemble des branches de D avec l'ensemble des $\overline{k}$-points de D. Pour tout couple (j,m), la courbe D coupe $L_j(m)$ en un unique point $p_{j,m}$ et avec multiplicité 1; ainsi, si $v^{(j,m)}$ est la branche de D dont le centre est $p_{j,m}$, on a $m(j,m;v^{(j,m)}) = 1$. Si $p_{j,m}$ n'est sur aucune droite de $\mathscr{L}$ distincte de $L_j(m)$, on a $m(i,n;v^{(j,m)}) = 0$ pour tout $(i,n) \neq (j,m)$, et donc $v^{(j,m)} \in \mathscr{R}_D$. Sinon il existe un $n \in \{0,\ldots 3\}$ et un seul tel que $p_{j,m}$ est le point d'intersection de la droite $L_j(m)$ avec la droite $L_i(n)$ où i est défini par $\{i,j\} = \{1,2\}$; on a alors $p_{j,m} = r_{m,n}$ (resp. $r_{n,m}$) si $j = 2$ donc $i = 1$ (resp. $j = 1$ donc $i = 2$). Dans ce cas on a $m(j,m;v^{(j,m)}) = m(i,n;v^{(j,m)}) = 1$ et, pour tout $(i',n') \neq (j,m)$, (i,n), on a $m(i',n';v^{(j,m)}) = 0$; on a donc $v^{(j,m)} \notin \mathscr{R}_D$.

Le cas $\mathbf{r_D = 0}$. Supposons qu'il existe une k-courbe D intègre et de bidegré $(1,1)$ avec $r_D = 0$; alors, pour tout $m = 0,\ldots,3$, il existe un et un seul $n = \alpha(m) \in \{0,\ldots,3\}$ tel que $p_{2,m} = p_{1,\alpha(m)}$ est le point d'intersection $r_{m,\alpha(m)}$ de $L_2(m)$ avec $L_1(\alpha(m))$. L'application α est une bijection de l'ensemble $\{0,\ldots,3\}$ sur lui même (si $\alpha(m) = \alpha(m')$ avec $m \neq m'$, la courbe D contient $L_1(\alpha(m))$, donc n'est pas intègre) et le k-isomorphisme de $\mathbf{P}_{1,k}$ dont le graphe est D envoie les quatre $\overline{k}$-points $r_0^{(1)},\ldots,r_3^{(1)}$ respectivement sur $r_{\alpha(0)}^{(2)},\ldots,r_{\alpha(3)}^{(2)}$; notons ξ cet isomorphisme. Le k-isomorphisme $\xi : \mathbf{P}_{1,k} \to \mathbf{P}_{1,k}$ se relève en un $\overline{k}$ isomorphisme $\tilde{\xi} : X_1 \to X_2$ qui est compatible avec les involutions. Il est alors facile de vérifier que les deux $\overline{k}$-composantes de $\pi^{-1}(D)$ sont les images par $\theta : X_1 \times X_2 \to X$ des deux courbes images de $(id,\tilde{\xi}) : X_1 \to X_1 \times X_2$ et de $(\iota_1,\tilde{\xi}) : X_1 \to X_1 \times X_2$. Ces composante sont définies sur k si et seulement si $\tilde{\xi}$ est défini sur k. Bien entendu les courbes k-rationnelles ainsi produites sont évidentes à trouver directement, et nous n'avons traité ce cas que pour être complet.

Le cas $\mathbf{r_D = 2}$. Supposons qu'il existe une k-courbe D intègre et de bidegré $(1,1)$ avec $r_D = 2$; alors, il existe au moins un $m(D) \in \{0,1,2,3\}$ tel que $p_{2,m(D)}$ n'est sur aucune des quatre droites $L_1(n)$ pour $n = 0,\ldots,3$ puisque sinon, comme on vient de le voir, on aurait $r_D = 0$. Pour la même raison, il existe au moins un $n(D) \in \{0,1,2,3\}$ tel que $p_{1,n(D)}$ n'est sur aucune des quatre droites $L_2(m)$ pour $m = 0,\ldots,3$. Ainsi les deux branches de D centrées respectivement en $p_{2,m(D)}$ et en $p_{1,n(D)}$ sont dans $\mathscr{R}_D$, et donc aucune autre branche de D n'est dans $\mathscr{R}_D$. Comme dans le cas précédent, on en déduit que si $m \in \{0,1,2,3\} \setminus \{m(D)\}$, il existe un et un seul $n = \alpha(m) \in \{0,\ldots,3\} \setminus \{n_1(D)\}$ tel que $p_{2,m} = p_{1,\alpha(m)}$ est le point d'intersection $r_{m,\alpha(m)}$ de $L_2(m)$ avec $L_1(\alpha(m))$, et que α est une bijection de $\{0,1,2,3\} \setminus \{m(D)\}$ sur $\{0,1,2,3\} \setminus \{n(D)\}$. Quitte à changer les indices nous pouvons supposer que $m(D) = n(D) = 0$, de sorte que la situation se décrit par la figure suivante où les obliques représentent les intersections de la courbe D avec les $L_i(n)$ (dans l'exemple ci-dessous, on respecte le choix $m(D) = n(D) = 0$, et on a $\alpha(1) = 1$, $\alpha(2) = 3$ et $\alpha(3) = 2$) :

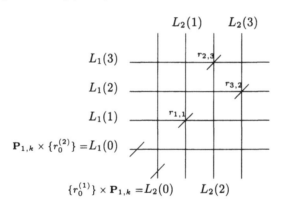

Comme la courbe D est définie sur k, les deux droites $L_2(0)$ et $L_1(0)$ sont définies sur k, donc les deux points $r_0^{(1)}$ et $r_0^{(2)}$ sont définis sur k et le zéro cycle $(r_{1,\alpha(1)}) + (r_{2,\alpha(2)}) + (r_{3,\alpha(3)})$ de $\mathbf{P}_{1,k} \times \mathbf{P}_{1,k}$ est rationnel sur k. Notons (*) la condition suivante :

(*) Il existe une bijection α de l'ensemble $\{1,2,3\}$ sur lui même telle que le zéro cycle $(r_{1,\alpha(1)}) + (r_{2,\alpha(2)}) + (r_{3,\alpha(3)})$ de $\mathbf{P}_{1,k} \times \mathbf{P}_{1,k}$ est rationnel sur k.

La condition (*) est une partie des conditions nécessaires que l'on vient de mettre en évidence, à l'existence d'une k-courbe D intègre de bidegré $(1,1)$ avec $r_D = 2$. Notons qu'elle implique que chacun des deux points $r_0^{(1)}$ et $r_0^{(2)}$ est défini sur k, et donc que chacune des deux droites $L_2(0)$ et $L_1(0)$ est aussi définie sur k.

Supposons la condition (*) vérifiée ; notons qu'elle implique que les trois $\overline{k}$-points $r_{1,\alpha(1)}$, $r_{2,\alpha(2)}$, et $r_{3,\alpha(3)}$ ne sont pas situés sur une réunion de courbes de bidegré $(0,1)$ ou $(1,0)$). Il en résulte qu'il existe une et une seule courbe D_α de bidegré $(1,1)$ passant par les trois $\overline{k}$-points $r_{1,\alpha(1)}$, $r_{2,\alpha(2)}$, et $r_{3,\alpha(3)}$ de $\mathbf{P}_{1,k} \times \mathbf{P}_{1,k}$, et que cette courbe est intègre ; de plus elle est définie sur k puisque le zéro cycle $(r_{1,\alpha(1)}) + (r_{2,\alpha(2)}) + (r_{3,\alpha(3)})$ de $\mathbf{P}_{1,k} \times \mathbf{P}_{1,k}$ est, par hypothèse, rationnel sur k. On a $r_{D_\alpha} = 2$ si et seulement si les points d'intersection $p_{1,0}$ et $p_{2,0}$ de D_α avec $L_1(0)$ et $L_2(0)$ sont distincts, c'est à dire si D_α ne passe pas par le point $r_{0,0}$.

Notons enfin que, lorsque D_α vérifie toute ces conditions, la partie arithmétique est automatiquement résolue : en effet, les images réciproques des deux points rationnels $p_{1,0}$ et $p_{2,0}$ par π sont deux points de $\pi^{-1}(D_\alpha)$ qui sont aussi rationnels puisque ce sont des points de ramification de π. La courbe $\pi^{-1}(D_\alpha)$ est donc k-rationnelle.

Plaçons nous dans des coordonnées adaptées à notre situation : pour $i = 1, 2$, on choisit les coordonnées $(u_{i,0} : u_{i,1})$ sur la droite projective de sorte que $r_0^{(i)}$ est le point de coordonnées $(0 : 1)$ et que la $i^{\text{ème}}$ projection du point rationnel $p_{i,0}$ est le point de coordonnées $(0 : 1)$. Avec ces choix, les points $p_{1,0}$ et $p_{2,0}$ sont les points de $\mathbf{P}_{1,k} \times \mathbf{P}_{1,k}$ de coordonnées $((1 : 0), (0 : 1))$ et $((0 : 1), (1 : 0))$; la courbe D_α passe par ces deux points, est intègre, et est définie sur k, donc elle a pour équation $u_{1,1} u_{2,1} - \lambda u_{1,0} u_{2,0}$ où λ est un élément non nul de k. D'autre part, le point $r_0^{(i)}$ étant défini sur k, le point $R_0^{(i)}$ est aussi défini sur k ; on munit X_i de la structure de k-courbe elliptique pour laquelle $R_0^{(i)}$ est l'origine ; l'involution ι_i est alors la multiplication par -1 et on pose $X_i = E_i$ pour rappeler ce choix. On prend $u_i = u_{i,1}/u_{i,0}$ comme coordonnée dans l'ouvert affine U_i défini par $u_{i,0} \neq 0$ et on note $P_i(u_i)$ le polynôme unitaire dont les racines sont les coordonnées des trois points $r_1^{(i)}$, $r_2^{(i)}$ et $r_3^{(i)}$ de U_i. Ainsi $v_i^2 = P_i(u_i)$ est un modèle de Weierstrass de E_i ; comme la $i^{\text{ème}}$ projection de $p_{i,0}$ est le point de coordonnées $u_i = 0$ dans U_i, et n'est ni $r_1^{(i)}$, ni $r_2^{(i)}$, ni $r_3^{(i)}$, on a $P_i(0) \neq 0$. Enfin, pour $m = 1, 2, 3$, la courbe D_α passe par le point $(r_m^{(1)}, r_{\alpha(m)}^{(2)})$ de $U_1 \times U_2$, donc le polynôme $P_2(u_2)$ et la fonction rationnelle $P_1(\lambda/u_2)$ ont les mêmes zéros ; si l'on pose $P_i(u_i) = u_i^3 + a_i u_i^2 + b_i u_i + c_i$ cela se traduit par $a_2 = b_1 \lambda/c_1, b_2 = a_1 \lambda^2/c_1$ et $c_2 = \lambda^3/c_1$. Si toutes ces conditions sont vérifiées, la courbe D_α d'équation $u_{1,1} u_{2,1} - \lambda u_{1,0} u_{2,0}$ est une k-courbe intègre de bidegré $(1,1)$ pour laquelle $r_D = 2$. Il est facile de calculer et de paramétriser l'image réciproque de D_α ; en posant $w = \lambda/c_1$ on trouve :

***Proposition* 2.1.1**. — *La surface X contient une courbe k-rationnelle C dont l'image dans $\mathbf{P}_{1,k} \times \mathbf{P}_{1,k}$ est une courbe D de bidegré $(1,1)$ avec $r_D = 2$ si et seulement si X_1 et X_2 sont deux k-courbes elliptiques E_1 et E_2 de modèle de Weirstrass $v_1^2 = u_1^3 + au_1^2 + bu_1 + c$ et $v_2^2 = u_2^3 + bwu_2^2 + acw^2 u_2 + c^2 w^3$ où a, b, c et w sont dans k, et où c et w non nuls, les involutions ι_1 et ι_2 étant la multiplication par -1.*

Dans l'ouvert affine de X^{lisse} d'équation $(x_2^3 + bwx_2^2 + acw^2x_2 + c^2w^3)y^2 = (x_1^3 + ax_1^2 + bx_1 + c)$ dans $\mathbf{A}_k^3$, la courbe k-rationnelle C admet la paramétrisation $z \to (x_1 = \xi_1(z), x_2 = \xi_2(z), y = \xi_3(z))$ où

$$\xi_1(z) = wz^2, \quad \xi_2(z) = \frac{c}{z^2}, \quad et \quad \xi_3(z) = \frac{z^3}{c}.$$

On peut interpréter la condition (*) en introduisant les modules galoisiens $E_1[2]$ et $E_2[2]$ des $\overline{k}$-points tués par 2 des courbes elliptiques E_1 et E_2. La bijection ensembliste de $E_1[2] = \{R_0^{(1)}, \ldots, R_3^{(1)}\}$ sur $E_2[2] = \{R_0^{(2)}, \ldots, R_3^{(2)}\}$ qui envoie $R_0^{(1)}$ sur $R_0^{(2)}$ et $R_m^{(1)}$ sur $R_{\alpha(m)}^{(2)}$ pour $m = 1, 2, 3$ est un isomorphisme de groupe et la condition (*) est équivalente au fait que cet isomorphisme de groupe est un isomorphisme de $Gal(\overline{k}/k)$ module. Il est immédiat de vérifier que la condition pour que la courbe D_α ne passe pas par le point $r_{0,0}$, c'est à dire pour que $r_{D_\alpha} = 2$, est que ce k-isomorphisme de $Gal(\overline{k}/k)$-module n'est pas induit par un isomorphisme de courbes elliptiques. En utilisant, dans le cas particulier qui nous intéresse ici, les techniques de [**F-K**] ou [**Sat**], on peut vérifier que cette condition est la condition pour qu'il existe une k-courbe $\mathscr{C}$ de genre 2 et deux k-morphismes $\varphi_1 : \mathscr{C} \to E_1$ et $\varphi_2 : \mathscr{C} \to E_2$ qui sont de degré 2 et indépendants. On retrouve ce résultat en calculant l'image réciproque de la courbe k-rationnelle C de X par le morphisme $\theta : E_1 \times E_2 \to X$; on trouve, avec les notations introduites dans la proposition 2.1.1, que cette image réciproque est la courbe plane d'équation $y^2 = w^3x^6 + aw^2x^4 + bwx^2 + c$, et que la restriction à cette courbe des deux projections q_1 et q_2 de $E_1 \times E_2$ sur E_1 et E_2 sont les morphismes φ_1 et φ_2 qui envoient le point de coordonnées (x, y) sur le point de E_1 de coordonnées $(u_1, v_1) = (wx^2, y)$ et sur le point de E_2 de coordonnées $(u_2, v_2) = (c/x^2, cy/x^3)$.

A la fin du dix neuvième siècle, la forme générale de l'intégrale hyperelliptique de genre 2 qui se ramène a des intégrales elliptiques par des transformations rationnelles de degré 2 a été donnée par Legendre ([**Leg**]) et Jacobi ([**Jac**]), et beaucoup discutée ([**Kra**], Chapitre X1, [**Pic**] par exemple). Comme c'était l'habitude au XIXième siècle, Legendre et Jacobi énoncent leurs résultats à partir d'une forme de Rosenhain de la courbe hyperelliptique de genre 2, c'est à dire pour la courbe mise sous la forme $Y^2 = X(X-1)(X-\alpha)(X-\beta)(X-\gamma)$. Ils montrent que les intégrales des formes holomorphes de cette courbe se ramènent à des intégrales elliptiques par des transformations rationnelles de degré 2 si et seulement si $\gamma = \alpha\beta$. On retrouve ce résultat à partir du nôtre de la manière suivante : on note t_1, t_2, t_3 les trois racines de l'équation polynômiale $w^3T^3 + aw^2T^2 + bwT + c = 0$ et on calcule la forme de Rosenhain de la courbe d'équation $y^2 = w^3x^6 + aw^2x^4 + bwx^2 + c$ pour laquelle les trois points de Weierstrass de coordonnées $(t_1, 0)$, $(-t_1, 0)$ et $(t_2, 0)$ sont envoyés respectivement sur les points d'abscisses 0, ∞ et 1. Les trois points de Weierstrass restants qui sont les points de coordonnées $(-t_2, 0)$, $(t_3, 0)$ et $(-t_3, 0)$ sont alors respectivement envoyés sur les points d'abscisses $\gamma = (\frac{t_2+t_1}{t_2-t_1})^2$, $\alpha = \frac{t_2+t_1}{t_2-t_1}\frac{t_3+t_1}{t_3-t_1}$ et $\beta = \frac{t_2+t_1}{t_2-t_1}\frac{-t_3+t_1}{-t_3-t_1}$; on a bien $\gamma = \alpha\beta$. Bien entendu, la forme de Rosenhain n'est pas adaptée aux questions de rationalité puisque les points de Weierstrass ne sont pas en général définis sur le corps de base de la courbe, ni même d'ailleurs sur le corps de définition des quotients elliptiques de la Jacobienne de cette courbe.

2.2. Exemple 2.— Relèvement non ramifié de certaines courbes de bidegré $(2,2)$.

Dans cet exemple on cherche les courbes k-rationnelles de X dont l'image dans $\mathbf{P}_{1,k} \times \mathbf{P}_{1,k}$ est une courbe D de bidegré $(2,2)$ qui rencontre les huit droites $L_i(n)$ de $\mathscr{L}$ de la manière décrite dans la figure ci-dessous :

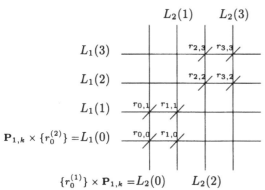

où les obliques représentent les intersections de la courbe D avec les $L_i(n)$. Autrement dit, la courbe D de bidegré $(2,2)$ passe par le sous-ensemble $\mathscr{P}_0$ de $\mathscr{P}$ formé des huit points $r_{0,0}$, $r_{0,1}$, $r_{1,0}$, $r_{1,1}$, $r_{2,2}$, $r_{2,3}$, $r_{3,2}$, et $r_{3,3}$. Les seules branches d'une telle courbe qui rencontrent les $L_i(n)$ sont les huit $v^{(n,m)}$ centrées en $r_{n,m} \in \mathscr{P}_0$. Pour ces huit branches, on a

$$m(1, n; v^{(n,m)}) = m(2, m; v^{(n,m)}) = 1$$

et tous les autres $m(i', n'; v^{(n,m)})$ sont nuls, donc $v^{(n,m)} \notin \mathscr{R}_D$, et donc $r_D = 0$.

Supposons qu'il existe une courbe k-rationnelle D de bidegré $(2,2)$ qui coupe les $L_i(n)$ comme indiqué. Le fait que la courbe D est définie sur k implique que les deux droites $L_1(0)$ et $L_1(1)$ (resp. $L_1(2)$ et $L_1(3)$, resp. $L_2(0)$ et $L_2(1)$, resp. $L_2(2)$ et $L_2(3)$) sont soit définies sur k, soit définies sur une extension quadratique de k et conjuguées ; les deux points $r_0^{(2)}$ et $r_1^{(2)}$ (resp. $r_2^{(2)}$ et $r_3^{(2)}$, resp. $r_0^{(1)}$ et $r_1^{(1)}$, resp. $r_2^{(1)}$ et $r_3^{(1)}$) ont donc la même propriété. D'autre part, comme une courbe de bidegré $(2,2)$ est de genre arithmétique 1, la courbe D a un point singulier et un seul, et donc ce point singulier est un point de $\mathbf{P}_{1,k} \times \mathbf{P}_{1,k}$ rationnel sur k. Enfin, on suppose que la courbe k-rationnelle D a des points lisses rationnels en dehors de l'ensemble $\mathscr{P}$ (ce qui est toujours le cas si le cardinal de k est plus grand ou égal à 11) et on fixe un tel point. Pour $i = 1, 2$, on choisit alors les coordonnées $(u_{i,0} : u_{i,1})$ de sorte que les points de $\mathbf{P}_{1,k} \times \mathbf{P}_{1,k}$ de coordonnées $((0:1),(0:1))$ et $((1:0),(1:0))$ sont respectivement le point rationnel lisse que l'on a fixé et le point singulier de D. Prenons $u_i = u_{i,1}/u_{i,0}$ comme coordonnée dans l'ouvert affine U_i défini par $u_{i,0} \neq 0$. L'ouvert affine $\pi_i^{-1}(U_i)$ de X_i a un modèle plan d'équation $v_i^2 = P_i(u_i)$ où P_i est un polynôme de degré 4 dont les racines sont les coordonnées des quatre points $r_0^{(i)}, \ldots, r_3^{(i)}$ de U_i. Comme les deux points $r_0^{(i)}$ et $r_1^{(i)}$ (resp. $r_2^{(i)}$ et $r_3^{(i)}$) sont soit définis sur k, soit définis sur une extension quadratique de k et conjugués, on a $P_i(u_i) = e_i(u_i^2 + a_i u_i + b_i)(u_i^2 + c_i u_i + d_i)$ où a_i, b_i, c_i, d_i et e_i sont dans k ; on peut noter que b_i et d_i sont non nuls puisque le point singulier de D est le point de coordonnées $(0,0)$ dans $U_1 \times U_2$ et que ce point singulier n'est sur aucune des $L_i(n)$ par hypothèse.

La courbe D appartient au système linéaire des courbes de $\mathbf{P}_{1,k} \times \mathbf{P}_{1,k}$ de bidegré $(2,2)$ passant par les huit points de $\mathscr{P}_0$. Ce système linéaire est de dimension 1 puisque le système linéaire de toutes les courbes de bidegré $(2,2)$ est de dimension 9 ; il est formé des courbes $\mathscr{D}_{(\lambda:\mu)}$ dont trace dans $U_1 \times U_2$ a pour équation

$(\lambda(a_2 - c_2))u_1^2 u_2^2 + (\lambda(a_1 a_2 - c_1 c_2) + \mu(c_1 - a_1))u_1 u_2^2 + \mu(a_2 - c_2)u_1^2 u_2 + (\lambda(a_2 b_1 - d_1 c_2) + \mu(d_1 - b_1))u_2^2 + (\lambda(a_2 d_2 - b_2 c_2) + \mu(b_2 - d_2))u_1^2 + (\lambda a_2 c_2(a_1 - c_1) + \mu(a_2 c_1 - a_1 c_2))u_1 u_2 + (\lambda a_2 c_2(b_1 - d_1) + \mu(d_1 a_2 - b_1 c_2))u_2 + (\lambda(d_2 a_2 a_1 - c_2 b_2 c_1) + \mu(b_2 c_1 - d_2 a_1))u_1 + \lambda(b_1 a_2 d_2 - b_2 c_2 d_1) + \mu(d_1 b_2 - b_1 d_2) = 0$.

Conformément à l'habitude, lorsque nous travaillons dans $U_1 \times U_2$ nous notons (∞, ∞) le point de $\mathbf{P}_{1,k} \times \mathbf{P}_{1,k}$ de coordonnées $((0:1),(0:1))$. La seule courbe de cette famille qui peut être géométriquement intègre et passer par le point de coordonnées $(u_1, u_2) = (0,0)$ avec multiplicité 2 et par le point (∞, ∞) avec multiplicité 1, est la courbe $D = D_{(0:1)}$. Cette courbe passe par $(0,0)$ et (∞, ∞) avec les multiplicités voulues si et seulement si $d_1 a_2 - b_1 c_2 = 0$, $c_1 b_2 - d_2 a_1 = 0$ et $d_1 b_2 - b_1 d_2 = 0$; elle est géométriquement intègre si et seulement si $d_2 c_1^2 - d_1 c_2^2 \neq 0$ comme on le voit sur l'équation de sa trace dans $U_1 \times U_2$ qui est :

$$c_2 u_1^2 u_2 - c_1 u_2^2 u_1 + d_2 u_1^2 - d_1 u_2^2 = 0.$$

En remontant cette courbe D, on trouve que les deux composantes de $\pi^{-1}(D)$ sont définies sur l'extension quadratique $k(\sqrt{e_1/e_2})$ et un calcul facile donne :

Proposition 2.2.1. — *Pour $i = 1, 2$ on désigne par X_i la complétée de la courbe affine plane lisse d'équation $v_i^2 = e_i(u_i^2 + a_i u_i + b_i)(u_i^2 + c_i u_i + d_i)$ où a_i, b_i, c_i, d_i et e_i sont dans k, et où e_i, b_i et d_i sont non nuls; on munit X_i de l'involution ι_i qui envoie le point de coordonnées (u_i, v_i) sur le point de coordonnées $(u_i, -v_i)$. On suppose que $d_1 a_2 - b_1 c_2 = 0$, $c_1 b_2 - d_2 a_1 = 0$, $d_1 b_2 - b_1 d_2 = 0$, que $d_2 c_1^2 - d_1 c_2^2 \neq 0$, et que $e_1/e_2 = \alpha^2$ pour un $\alpha \in k$. Alors la surface de Kummer X contient deux courbes k-rationnelles C_1 et C_2 dont les images dans $\mathbf{P}_{1,k} \times \mathbf{P}_{1,k}$ sont de bidegré $(2, 2)$ et rencontrent les $L_i(n)$ comme indiqué sur la figure.*

Dans l'ouvert affine de X^{lisse} d'équation $e_2(x_2^2 + a_2 x_2 + b_2)(x_2^2 + c_2 x_2 + d_2)y^2 = e_1(x_1^2 + a_1 x_1 + b_1)(x_1^2 + c_1 x_1 + d_1)$ dans $\mathbf{A}_k^3$, les courbes k-rationnelle C_1 et C_2 admettent respectivement les paramétrisations $z \to (x_1 = \xi_1(z), x_2 = \xi_2(z), y = \xi_3(z))$ et $z \to (x_1 = \xi_1(z), x_2 = \xi_2(z), y = -\xi_3(z))$ où

$$\xi_1(z) = -\frac{d_2 z^2 - d_1}{c_2 z - c_1}$$
$$\xi_2(z) = -\frac{d_2 z^2 - d_1}{z(c_2 z - c_1)}$$
$$\xi_3(z) = \alpha z^2.$$

2.3. Exemple 3.— L'exemple de Mestre.

Deux k-courbes elliptiques E_1 et E_2 étant fixées, J.-F. Mestre ([Mes]) montre qu'on peut toujours construire un k-revêtement $\mathscr{C}$ de degré 2 de $\mathbf{P}_{1,k}$ (c'est à dire une courbe k-hyperelliptique) et deux k-morphismes φ_1 et φ_2 de $\mathscr{C}$ vers E_1 et E_2 qui vérifient $\varphi_1 \circ \iota_{\mathscr{C}} = -\varphi_1$ et $\varphi_2 \circ \iota_{\mathscr{C}} = -\varphi_2$ où $\iota_{\mathscr{C}}$ est l'involution hyperelliptique de $\mathscr{C}$, et qui sont indépendants. La construction du triplet $(\mathscr{C}, \varphi_1, \varphi_2)$ est essentiellement équivalente à la construction de la courbe k-rationnelle de la surface de Kummer $X = E_1 \times E_2 / \{\pm id\}$ qui est l'image de $\mathscr{C}$ par le composé de $(\varphi_1, \varphi_2) : \mathscr{C} \to E_1 \times E_2$ avec la projection canonique de $E_1 \times E_2$ sur X.

On retrouve le résultat de Mestre en prenant $(X_1, \iota_1) = (E_1, -id)$, $(X_2, \iota_2) = (E_2, -id)$, et en relevant une courbe D de $\mathbf{P}_{1,k} \times \mathbf{P}_{1,k}$ de bidegré $(3, 3)$ dont les intersections avec les $L_i(n)$ sont décrites par le diagramme suivant :

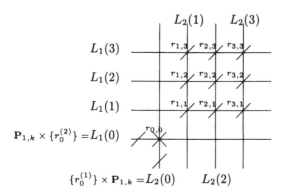

où, comme plus haut, les obliques représentent les intersections de la courbe D avec les $L_i(n)$. Autrement dit, la courbe D de bidegré $(3,3)$ passe par les neuf points $r_{1,1}$, $r_{1,2}$, $r_{1,3}$, $r_{2,1}$, $r_{2,2}$, $r_{2,3}$, $r_{3,1}$, $r_{3,2}$, et $r_{3,3}$ avec multiplicité 1 et par le point $r_{0,0}$ avec multiplicité 2 ; notons que cela implique que le troisième point d'intersection de D avec $L_1(0)$ (resp. $L_2(0)$) n'est ni $r_{1,0}$, ni $r_{2,0}$, ni $r_{3,0}$ (resp. ni $r_{0,1}$, ni $r_{0,2}$, ni $r_{0,3}$). Comme dans le travail de Mestre nous obtenons comme cela le résultat cherché génériquement ; les cas particuliers restants se traitent après coup.

On vérifie (en reprenant les arguments explicités dans les exemples précédents) que $r_D = 2$ donc, si une telle courbe existe, son image réciproque $\pi^{-1}(D)$ est une k-courbe géométriquement intègre de genre géométrique 0 de X. Notons $p_{1,0}$ et $p_{2,0}$ les points d'intersection de D avec $L_1(0)$ et $L_2(0)$ distincts de $r_{0,0}$; comme dans l'exemple 1, ces points sont des points de D qui sont lisses et rationnels sur k, et au dessus desquels π est ramifié ; leurs images réciproques sont donc deux points rationnels de $\pi^{-1}(D)$, et donc $\pi^{-1}(D)$ est une courbe k-rationnelle. Le genre arithmétique d'une courbe de bidegré $(3,3)$ étant égal à 4, il faut imposer d'autres singularités que le point double en $r_{0,0}$ pour assurer que le genre géométrique de D est 0. Pour retrouver les équations de [Mes], nous imposons dans la suite que D possède un point triple ; cela implique que le genre arithmétique est 0, donc aussi que le genre géométrique est 0 (mais ce n'est pas la seule manière, par exemple imposer trois points doubles ordinaires en plus de $r_{0,0}$ marcherait aussi). Notons qu'alors le point double en $r_{0,0}$ et le point triple sont les seuls points singuliers de D, donc le point triple est un point de $\mathbf{P}_{1,k} \times \mathbf{P}_{1,k}$ qui est rationnel sur k ; de plus il n'est sur aucune des $L_i(n)$ comme on le voit sur la figure représentant l'intersection de D avec les $L_i(n)$.

Pour prouver l'existence d'une courbe k-rationnelle D rencontrant les $L_i(n)$ comme sur la figure, nous choisissons des coordonnées adaptées à la situation.

Pour $i = 1, 2$, on choisit les coordonnées $(u_{i,0} : u_{i:1})$ sur la droite projective de sorte que $r_0^{(i)}$ est le point à l'infini i.e. le point de coordonnées $(0 : 1)$, et que la $i^{\text{ème}}$ projection du point triple de D est le point zéro de coordonnées $(1 : 0)$. On prend $u_i = u_{i,1}/u_{i,0}$ comme coordonnée sur l'ouvert U_i défini par $u_{i,0} \neq 0$, et on note $P_i(u_i)$ le polynôme unitaire de degré 3 à coefficients dans k dont les racines sont les coordonnées des trois points $r_1^{(i)}$, $r_2^{(i)}$ et $r_3^{(i)}$ dans U_i. Ainsi $v_i^2 = P_i(u_i)$ est un modèle de Weierstrass de E_i; notons que $P_i(0) \neq 0$ puisque le point triple de D est le point de $U_1 \times U_2$ de coordonnées $(0,0)$ et n'est sur aucune des $L_i(n)$.

Une courbe D qui rencontre les $L_i(n)$ comme indiqué sur la figure, appartient au système linéaire des courbes de $\mathbf{P}_{1,k} \times \mathbf{P}_{1,k}$ de bidegré $(3,3)$ qui passent par les neufs points $r_{n,m}$ pour $n, m = 1, 2, 3$ avec multiplicité 1 et par le point $r_{0,0}$ avec multiplicité 2. Ce système linéaire est de dimension 3 puisque le système linéaire de toutes les courbes de bidegré $(3,3)$ est de dimension 14; il est formé des courbes $D_{(\lambda_0:\lambda_1:\lambda_2:\lambda_3)}$ dont la trace dans l'ouvert $U_1 \times U_2$ de $\mathbf{P}_{1,k} \times \mathbf{P}_{1,k}$ a pour équation

$$P_1(u_1)(\lambda_0 u_2 + \lambda_1) + P_2(u_2)(\lambda_2 u_1 + \lambda_3) = 0.$$

Pour $i = 1, 2$ posons $P_i(u_i) = u_i^3 + a_i u_i^2 + b_i u_i + c_i$ de sorte que a_i, b_i et c_i sont des éléments de k avec $c_i \neq 0$ puisque $P_i(0) \neq 0$. L'équation de la courbe $D_{(\lambda_0:\lambda_1:\lambda_2:\lambda_3)}$ est $(u_1^3 + a_1 u_1^2 + b_1 u_1 + c_1)(\lambda_0 u_2 + \lambda_1) + (u_2^3 + a_2 u_2^2 + b_2 u_2 + c_2)(\lambda_2 u_1 + \lambda_3) = 0$; le point $(u_1, u_2) = (0, 0)$ est un point de multiplicité au moins égale à trois si et seulement si on a $a_1 \lambda_1 = 0$, $b_2 \lambda_3 = 0$, $c_1 \lambda_1 + c_2 \lambda_3 = 0$, $c_1 \lambda_0 + b_2 \lambda_3 = 0$, $b_1 \lambda_1 + c_2 \lambda_2 = 0$, et $b_1 \lambda_0 + b_2 \lambda_2 = 0$; comme $a_1 c_2 \neq 0$, on a necessairement $a_1 = a_2 = 0$ et $(\lambda_0 : \lambda_1 : \lambda_2 : \lambda_3) = (b_2 : c_2 : -b_1 : -c_1)$. L'équation de la trace de $D_{(b_2:c_2:-b_1:-c_1)}$ sur $U_1 \times U_2$ est $b_2 u_1^3 u_2 - b_1 u_1 u_2^3 + c_2 u_1^3 - c_1 u_2^3$. Le point $r_{0,0} = (\infty, \infty)$ est un point double sur cette courbe si l'un au moins des coefficients b_1 ou b_2 est non nul. Enfin, si b_1 ou b_2 est non nul, on voit en tenant compte du fait que c_1 et c_2 sont non nuls, que $D_{(b_2:c_2:-b_1:-c_1)}$ est géométriquement intègre si et seulement si $b_1^3 c_2^2 - b_2^3 c_1^2 \neq 0$. Sous ces hypothèses, on calcule et on paramétrise facilement l'image réciproque de $D_{(b_2:c_2:-b_1:-c_1)}$ dans X et on trouve :

Proposition 2.3.1. — *Soient E_1 et E_2 deux k-courbes elliptiques qui admettent des modèles de Weierstrass $v_1^2 = u_1^3 + b_1 u_1 + c_1$ et $v_2^2 = u_2^3 + b_2 u_2 + c_2$. On suppose c_1 et c_2 non nuls, b_1 ou b_2 non nul, et $b_1^3 c_2^2 - b_2^3 c_1^2 \neq 0$. Alors la surface de Kummer $X = E_1 \times E_2/(\pm id)$ contient une courbe k-rationnelle C dont l'image dans $\mathbf{P}_{1,k} \times \mathbf{P}_{1,k}$ est de bidegré $(3,3)$ et rencontre les $L_i(n)$ comme indiquée sur la figure.*

Dans l'ouvert affine de X^{lisse} d'équation $(x_2^3 + b_2x_2 + c_2)y^2 = (x_1^3 + b_1x_1 + c_1)$ dans $\mathbf{A}_k^3$ la courbe C admet la paramétrisation $z \to (x_1 = \frac{c_2z^6-c_1}{b_1-b_2z^4}, x_2 = \frac{c_2z^6-c_1}{z^2(b_1-b_2z^4)}, y = z^3)$.

On reconnait bien dans cette proposition les formules de Mestre. Les cas exclus dans la proposition précédente se traitent par une étude cas par cas. Traitons par exemple le cas suivant : on conserve les hypothèses c_1 et c_2 non nuls, b_1 ou b_2 non nul, mais on suppose $b_1^3c_2^2 - b_2^3c_1^2 = 0$. On a alors b_1 et b_2 non nuls et on pose $\lambda = b_1c_2/b_2c_1$ de sorte que $b_1 = \lambda^2 b_2$ et $c_2 = \lambda^3 c_1$ (l'hypothèse est donc que les deux courbes elliptiques E_1 et E_2 sont isomorphes sur le corps $k(\sqrt{\lambda})$ et que leurs invariants modulaires sont différents de 0 et 1728). L'équation de la trace de $D_{(b_2:c_2:-b_1:-c_1)}$ dans $U_1 \times U_2$ devient $(\lambda u_1 - u_2)(b_1\lambda u_1^2 u_2 + b_1 u_1 u_2^2 + c_1 \lambda^2 u_1^2 + c_1 \lambda u_1 u_2 + c_1 u_2^2)$; ainsi $D_{(b_2:c_2:-b_1:-c_1)}$ est la réunion d'une courbe de bidegré $(1,1)$ et d'une courbe de bidegré $(2,2)$; on vérifie que l'image réciproque de la courbe de bidegré $(2,2)$ est une courbe k-rationnelle sur X qui est non triviale, ce qui règle ce cas. Dans l'ouvert affine de X^{lisse} d'équation $(x_2^3 + \lambda^2 b_1 x_2 + \lambda^3 c_1)y^2 = (x_1^3 + b_1 x_1 + c_1)$ dans $\mathbf{A}_k^3$ cette courbe k-rationnelle admet la paramétrisation $z \to (x_1 = \frac{-c_1(\lambda^2 z^4 + \lambda z^2 + 1)}{b_1(\lambda z^2 + 1)}, x_2 = \frac{-c_1(\lambda^2 z^4 + \lambda z^2 + 1)}{b_1 z^2(\lambda z^2 + 1)}, y = z^2)$. Notons que l'image réciproque de la courbe de bidegré $(1,1)$ se calcule comme on l'a fait dans l'exemple 1 ; elle se compose de deux composantes $\overline{k}$-rationnelles conjuguées sur $k(\sqrt{\lambda})$ et donc ne permet de construire une courbe k-rationnelle sur X que dans le cas particulier où λ est un carré dans k, i.e. lorsque E_1 et E_2 sont isomorphes sur k.

Les cas restants sont essentiellement ceux pour lesquels les invariants modulaires des courbes E_1 et E_2 sont 0 ou 1728. Les choix de coordonnées que nous avons fait ne sont alors pas adaptés à la situation (essentiellement on ne peut plus demander dans ces cas que le point triple soit le point de coordonnées $((0:1),(0:1))$ dans $\mathbf{P}_{1,k} \times \mathbf{P}_{1,k}$). Les changements à faire pour traiter ces cas ne posent pas de difficultés.

2.4. Exemple 4.— Courbes k-rationnelles associées aux courbes de genre 2.

On suppose encore que $(X_1, \iota_1) = (E_1, -id)$ et $(X_2, \iota_2) = (E_2, -id)$ où E_1 et E_2 sont deux k-courbes elliptiques. On fixe un entier naturel n.

Si l'entier n est impair et strictement plus grand que 1, on considère les courbes k-rationnelles D de $\mathbf{P}_{1,k} \times \mathbf{P}_{1,k}$ de bidegré (n,n) qui rencontrent les $L_i(n)$ de la manière suivante : pour $n = 1, 2, 3$ (resp. $m = 1, 2, 3$) la courbe D rencontre la droite $L_1(n)$ (resp. $L_2(m)$) en le point $r_{0,n}$ (resp. $r_{m,0}$) avec multiplicité 1, et en $(n-1)/2$ autres points qui sont des points lisses de D, qui ne sont situés sur aucune des $L_2(m)$ (resp. $L_1(n)$), avec multiplicité 2; la courbe D rencontre $L_1(0)$ (resp. $L_2(0)$) en les trois points $r_{1,0}$, $r_{2,0}$ et $r_{3,0}$ (resp. $r_{0,1}$, $r_{0,2}$ et $r_{0,3}$) avec multiplicité 1, et en $(n-3)/2$ autres points qui sont des points lisses de D, qui ne sont situés sur aucune des $L_2(m)$ (resp. $L_1(n)$), avec multiplicité 2. Par exemple, pour $n = 3$, cette situation se représente par la figure suivante :

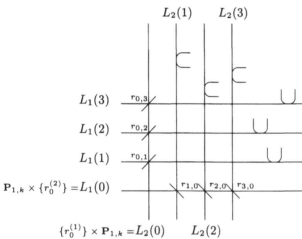

où les intersections de D avec les $L_i(n)$ sont représentées par des obliques lorsqu'elles sont de multiplicité 1 et par des ovales lorsqu'elles sont de multiplicité 2. Une branche d'une courbe D qui vérifie ces conditions rencontre soit deux des droites $L_i(n)$ avec multiplicité 1, soit une des $L_i(n)$ avec multiplicité 2 de sorte que $r_D = 0$. L'image réciproque $\pi^{-1}(D)$ de D sur X est donc formée de deux composantes $\bar{k}$-rationnelles. Supposons trouvée une telle courbe D, et notons φ une paramétrisation de D, i.e. un k-morphisme $\varphi : \mathbf{P}_{1,k} \to \mathbf{P}_{1,k} \times \mathbf{P}_{1,k}$ dont l'image est D; notons enfin C_1 et C_2 les deux composantes géométriques de $\pi^{-1}(D)$. Choisissons des coordonnées $(z_0 : z_1)$ sur la droite projective source de φ, posons $z = z_1/z_0$, et notons $z_{1,0}$, $z_{2,0}$, $z_{3,0}$, $z_{0,1}$, $z_{2,0}$ et $z_{0,3}$ les paramètres des six points $r_{1,0}, \ldots, r_{0,3}$, c'est à dire les six valeurs telles que $\varphi(z_{1,0}) = r_{1,0}, \ldots, \varphi(z_{0,3}) = r_{0,3}$. Un calcul facile montre que, pour $i = 1, 2$, l'image réciproque $\mathscr{C}_i = \theta^{-1}(C_i)$ de C_i par le morphisme canonique $\theta : E_1 \times E_2 \to X$ est une courbe de genre 2 sur $E_1 \times E_2$ qui est $\bar{k}$-birationnellement équivalentes au revêtement de degré 2 de $\mathbf{P}_{1,k}$ ramifié en les six points $z_{1,0}$, $z_{2,0}$, $z_{3,0}$, $z_{0,1}$, $z_{2,0}$ et $z_{0,3}$; si C_1 et C_2 sont k-rationnelles, les courbes $\mathscr{C}_1$ et $\mathscr{C}_2$ sont définies sur k et sont k-birationnellement équivalentes au revêtement de degré 2 de $\mathbf{P}_{1,k}$ ramifié en ces six points. Le k-isomorphisme $(-id_{E_1}) \times id_{E_2}$ de $E_1 \times E_2$ échange ces deux courbes.

Si l'entier n est pair, on considère les courbes k-rationnelles D de $\mathbf{P}_{1,k} \times \mathbf{P}_{1,k}$ de bidegré (n,n) qui rencontrent les $L_i(n)$ de la manière suivante : la courbe D rencontre $L_1(0)$ (resp. $L_2(0)$) en $n/2$ points qui sont des points lisses de D, qui ne sont situés sur aucune des $L_2(m)$ (resp. $L_1(n)$), avec multiplicité 2 ; les trois points $r_{1,1}$, $r_{2,2}$ et $r_{3,3}$ sont trois points doubles ordinaires de D et chacune des six branches de D centrées en l'un de ces trois points rencontre les $L_i(n)$ passant par ce point avec multiplicité 1 ; enfin, pour $m = 1, 2, 3$ (resp. $n = 1, 2, 3$) la courbe D rencontre la droite $L_2(m)$ (resp. $L_1(n)$) en $(n-2)/2$ points qui sont des points lisses de D, qui ne sont situés sur aucune des $L_1(n)$ (resp. $L_2(m)$), avec multiplicité 2, et en le point double ordinaire $r_{m,m}$ (resp. $r_{n,n}$) comme on vient de le décrire. Pour $n = 4$, cette situation se représente par la figure suivante :

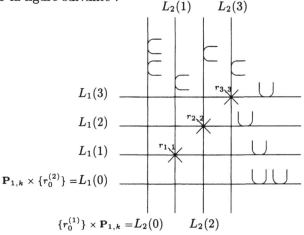

où, comme plus haut, les obliques et les ovales représentent respectivement les branches de D qui coupent les $L_i(n)$ avec multiplicités 1 et 2 ; on vérifie immédiatement que, dans ce cas encore, on a $r_D = 0$. Supposons trouvée une telle courbe D, et notons φ une paramétrisation de D, i.e. un k-morphisme $\varphi : \mathbf{P}_{1,k} \to \mathbf{P}_{1,k} \times \mathbf{P}_{1,k}$ dont l'image est D ; notons enfin C_1 et C_2 les deux composantes géométriques de $\pi^{-1}(D)$. Choisissons des coordonnées $(z_0 : z_1)$ sur la droite projective source de φ, posons $z = z_1/z_0$ et, pour $n = 1, 2, 3$, notons $z_{n,1}$, $z_{n,2}$ les deux paramètres du point double ordinaire $r_{n,n}$ de D, c'est à dire les deux valeurs telles que $\varphi(z_{n,1}) = \varphi(z_{n,2}) = r_{n,n}$. Un calcul facile montre que, pour $i = 1, 2$, l'image réciproque $\mathscr{C}_i = \theta^{-1}(C_i)$ de C_i par le morphisme canonique $\theta : E_1 \times E_2 \to X$ est une courbe de genre 2 sur $E_1 \times E_2$ qui est $\bar{k}$-birationnellement équivalente au revêtement de degré 2 de $\mathbf{P}_{1,k}$ ramifié en les six points $z_{1,1}$, $z_{1,2}$, $z_{2,1}$, $z_{2,2}$, $z_{3,1}$ et $z_{3,2}$; si C_1 et C_2 sont k-rationnelles, les courbes $\mathscr{C}_1$ et $\mathscr{C}_2$ sont définies sur k et sont k-birationnellement équivalentes au revêtement de degré 2 de $\mathbf{P}_{1,k}$ ramifié en ces six points. Le k-isomorphisme $(-id_{E_1}) \times id_{E_2}$ de $E_1 \times E_2$ échange ces deux courbes.

Chaque fois que l'on trouve un courbe k-rationnelle D dans $\mathbf{P}_{1,k} \times \mathbf{P}_{1,k}$ qui vérifie les conditions que l'on vient de décrire, on note $\mathscr{C}$ l'une des deux courbes $\mathscr{C}_1$ ou $\mathscr{C}_2$. On vérifie immédiatement que les deux projections q_1 et q_2 du produit $E_1 \times E_2$ sur E_1 et E_2 induisent deux morphismes de degré n de $\mathscr{C}$ sur E_1 et E_2. On montre dans [**Sat**] que toute courbe k-hyperelliptique de genre 2 dont la jacobienne est k-isogène à un produit de k-courbes elliptiques est obtenue de cette manière avec un bon choix de n.

Comme on l'a rappelé dans l'introduction, lorsque le corps k est le corps de complexes, la recherche des courbes hyperelliptiques dont la jacobienne contient des courbes elliptiques est le problème classique de la recherche d'intégrales hyperelliptiques qui se réduisent à des intégrales elliptiques. Dans le cas des intégrales hyperelliptiques de genre 2 qui nous interessent dans cet exemple, c'est un execice amusant de retrouver les formules classiques que l'on trouve par exemple dans [**Bol**], dans le chapitre XI de [**Kra**], ou encore dans [**Pic**] en construisant les courbes D correspondantes. Terminons ce travail en revenant au cas où le corps de base est quelconque, et en signalant deux travaux qui, au point de vue près, traitent des cas particuliers de la situation que l'on décrit ici. Tout d'abord, les calculs du §6 de [**Kuh**] s'interprètent immédiatement comme l'explicitation de la paramétrisation de φ dans le cas $n = 3$. D'autre part, notons $E_1[n]$ et $E_2[n]$ les $Gal(\overline{k}/k)$-modules galoisiens des points de E_1 et E_2 qui sont tués par n. On sait ([**F-K**] ou [**Sat**] par exemple) que les courbes k-hyperelliptiques de genre 2 tracées sur $E_1 \times E_2$ dont l'image dans $\mathbf{P}_{1,k} \times \mathbf{P}_{1,k}$ est une courbe k-rationnelle de type (n,n) qui coupent les $L_i(n)$ comme indiqué plus haut correspondent aux $Gal(\overline{k}/k)$-isomorphismes de $E_1[n]$ sur $E_2[n]$ qui sont anti-symplectique (pour le pairing de Weil). Ainsi, les calculs de [**R-S**] donnent des exemples de la situation décrite ici avec $n = 3$ et $n = 5$. Notons que les calculs présentés dans les deux travaux que nous venons de citer nécessitent l'emploi d'un logiciel de calcul formel. Nous avons recalculé ces exemples sans difficulté dans le cadre exposé ici en utilisant Maple.

Références

[B-M]   V.-V. Batyrev and Yu. I. Manin, *Sur le nombre de points rationnels de hauteur borné des variétés algébriques*, Math. Ann. **286** (1990), 27–43.

[Bol]   O. Bolza, *Ueber die Reduction hyperelliptischer Integrale erster ordnung und erster Gattung auf elliptische durch eine transformation vierten Grades*, Math. Ann. **XXVIII** (1886), 447–456.

[F-K] G. Frey and E. Kani, *Curves of genus 2 covering elliptic curves*, Arithmetic Algebraic Geometry, Prog. in Math. 89, Birkhäuser (1991), 153–175.

[Jac] C.G. Jacobi, *Anzeige von Legendre, Théorie des fonctions elliptiques, troisième supplément (1832)*, Journ. reine und angew. Math., Bd. **8**, reproduit dans les Gesammelte Werke, Bd. 1, Berlin (1881).

[Kra] A. Krazer, *Lehrbuch der Thetafunctionen*, Teubner, Leipzig 1903 (réédité par Chelsea, New-York 1970).

[Kuh] R.-M. Kuhn, *Curves of genus 2 with split Jacobian*, Trans. of the Amer. Math. Society, Vol. **307** (1988), 41–49.

[Leg] A.-M. Legendre, *Traité des fonctions elliptiques et des intégrales eulériennes*, 3ième supplément, (1832).

[M-S] N. Maïer et Ph. Satgé, *Fibrations en courbes de genre 1 sur la surface de Kummer associée au produit de deux courbes de genre 1*, prépublications de l'Université de Caen.

[McK] D. McKinnon, *Counting rational points of bounded height on K3 surfaces*, J. Number Theory **84** (2000), no. 1, 49–62.

[Mes] J.-F. Mestre, *Rang de courbes elliptiques d'invariant donné*, C.R. Acad. Sci. Paris Sér. I Math. **314** (1992), 919-922.

[Pic] E. Picard, *Sur quelques exemples de réduction d'intégrales abéliennes aux intégrales elliptiques*, C.R. **93**, Séance du 16 décembre 1881, 1126-11288; *Sur la réduction des intégrales abéliennes aux intégrales elliptiques*, C.R. **94**, Séance du 26 juin 1882, 1704-1707; reproduits dans le volume III des oeuvres complètes, Editions du CNRS, (1980).

[R-S] K. Rubin and A. Silverberg, *Families of elliptic curves with constant mod. p representations*, Elliptic curves, modular forms, and Fermat's last theorem (Hong Kong, 1993), 148–161, Ser. Number Theory, I, Intern. Press, Cambridge, MA, 1995.

[Sat] Ph. Satgé, *Le morphisme complémentaire d'un morphisme d'une courbe de genre 2 vers une courbe de genre 1*, soumis aux actes des journées revêtements de St Etienne (Mars 2000).

Rational points on algebraic varieties
(E. PEYRE, Y. TSCHINKEL, ed.), p. 335–355
Progress in Mathematics, Vol. 199, © 2001 Birkhäuser Verlag Basel/Switzerland

ARITHMETIC STRATIFICATIONS AND PARTIAL EISENSTEIN SERIES

Matthias Strauch

Mathematisches Institut der Westfälischen Wilhelms-Universität Münster
E-mail : straucm@math.uni-muenster.de

Abstract. — Let $P\backslash G$ and $Q\backslash H$ be generalized flag varieties over a number field F. In this paper we study certain locally trivial fibre bundles Y_η over $P\backslash G$ having $Q\backslash H$ as general fibre, and determine the arithmetic stratification of Y_η with respect to a line bundle. The arithmetic stratification is defined in terms of height zeta functions and the height zeta function of a stratum is of the form

$$\sum_{\gamma \in P(F)\backslash G(F)} e^{\langle s\lambda, H_P(\gamma)\rangle} E_Q^{Qw^{-1}Q_0}(s\mu, \eta(p_\gamma)),$$

where $E_Q^{Qw^{-1}Q_0}$ is a "partial Eisenstein series" associated to the Schubert cell $Q\backslash Qw^{-1}Q_0$. The computation of the constant term of these gives estimates that allow one to determine the abcissa of convergence of the height zeta function of the stratum.

Introduction

Consider a projective variety Y over a number field F and a metrized line bundle $\mathbf{L} = (L, (\|\ \|_v)_v)$ on Y. For any subvariety $U \subset Y$ there is a counting function

$$N_U(\mathbf{L}, H) = \#\{y \in U(F) \mid H_\mathbf{L}(y) \le H\}.$$

Suppose $N_U(\mathbf{L}, H)$ is finite for all H. Then one would like to understand the asymptotic behaviour of the counting function $N_U(\mathbf{L}, H)$ as H tends to infinity. A useful tool when studying this question is the (formal) Dirichlet series

$$Z_U(\mathbf{L}, s) = \sum_{y \in U(F)} \frac{1}{H_\mathbf{L}(y)^s},$$

called a *height zeta function*. Whenever this series converges for all $s \gg 0$ let $\sigma_U(L)$ be its abcissa of convergence:

$$\sigma_U(L) = \inf\{s \in \mathbf{R} \mid Z_U(\mathbf{L}, s) \text{ converges}\}.$$

$\sigma_U(L)$ depends only on the line bundle L underlying $\mathbf{L}$ and not on the metric. $\sigma_U(L) = -\infty$ if and only if $U(F)$ is finite; otherwise

$$\sigma_U(L) = \limsup \frac{\log N_U(\mathbf{L}, H)}{\log H}.$$

Sometimes one can go much further: namely, establish analytic continuation of the height zeta function. Then one even gets an asymptotic formula for the counting function, using a Tauberian Theorem (provided an assumption on the poles is satisfied). However, one encounters some features that should be considered beforehand.

A general phenomenon which may occur is this: there is a proper subvariety Y' in Y, such that $\sigma_Y(L) = \sigma_{Y'}(L) > \sigma_{Y-Y'}(L)$. Such a subvariety is called *accumulating*. In this case, counting rational points on Y' is asymptotically the same as counting rational points on Y, and therefore one should study Y' and $Y - Y'$ separately. It may happen that $Y - Y'$ contains again an accumulating subvariety. This leads us to the notion of the *arithmetic stratification* of Y with respect to L, introduced by V.V. Batyrev and Yu.I. Manin in [**BM**]. Suppose that $Y(F)$ is dense in Y and that there is an open subset Y' such that $Z_{Y'(F)}(\mathbf{L}, s)$ converges for all $s \gg 0$. The arithmetic stratification of Y with respect to L is the descending sequence of Zariski open subvarieties

$$Y \supset Y_0 \supset Y_1 \supset Y_2 \ldots$$

where Y_0 is the maximal Zariski open subset of Y for which the height zeta function $Z_{Y_0}(\mathbf{L}, s)$ converges for all $s \gg 0$, and $Y_{i+1} \subset Y_i$ is the maximal open subset of Y_i such that $\sigma_{Y_{i+1}}(L)$ is strictly smaller than $\sigma_{Y_i}(L)$. Note that the arithmetic stratification depends on the line bundle L. As an example, consider the variety obtained by blowing up the projective plane over F in a rational point. The projective line replacing this point will be an accumulating subvariety with respect to the anticanonical line bundle. For a further discussion cf. the introduction of [**FMT**] and [**BM**].

In this paper we study a class of smooth projective varieties, give a description of their Picard groups, and determine the arithmetic stratification with respect to an arbitrary line bundle. Here is a brief description of the varieties and the methods employed.

The varieties are locally trivial fibre bundles $Y_\eta \to W = P\backslash G$ over a generalized flag variety $P\backslash G$. The fibres are isomorphic to another flag variety $Q\backslash H$, and Y_η is constructed via a homomorphism $\eta : P \longrightarrow H$ which factors through a split torus. The Picard group is up to a finite cokernel given by the direct sum of character groups $X^*(Q) \oplus X^*(P)$. In $Pic(Y_\eta)_\mathbf{R}$ there is the closed cone $\Lambda(Y_\eta)_{\text{eff}}$ generated by the line bundles having non-zero global sections, and we denote the interior of this cone by $\Lambda(Y_\eta)_{\text{eff}}^\circ$. From now on we assume that η satisfies assumption 1.4 below. Let L be a line bundle on Y corresponding to the pair $(\mu, \lambda) \in X^*(Q) \oplus X^*(P)$. Suppose L lies in $\Lambda(Y_\eta)_{\text{eff}}^\circ$, and consider the arithmetic stratification $Y_\eta \supset Y_0 \supset Y_1 \supset Y_2 \ldots$ of Y_η with respect to L. It turns out that the strata $Y_i - Y_{i+1}$ are always unions of twisted products $Y^w = X^w \times^P G$ over $P\backslash G$. Here $X^w = Q\backslash Qw^{-1}Q_0$ is the Schubert cell in X corresponding to the Weyl group element w. The height zeta function of Y^w (with respect to a certain metric on L) is a double sum

$$\sum_{\gamma \in P(F)\backslash G(F)} e^{\langle s\lambda, H_P(\gamma) \rangle} E_Q^{Qw^{-1}Q_0}(s\mu, \eta(p_\gamma)).$$

We call $E_Q^{Qw^{-1}Q_0}$ a "partial Eisenstein series", because it is a subseries of a Langlands-Eisenstein series: the summation runs only over the quotient $Q(F)\backslash Q(F)w^{-1}Q_0(F)$. The partial Eisenstein series can be bounded from above and from below by its constant term $E_{Q,Q_0}^{Qw^{-1}Q_0}$. The constant term in turn is equal to

$$e^{\langle s\check{\eta}(w\mu) - \check{\eta}(w\rho_0 - \rho_0), H_P(\gamma) \rangle} E_{Q,Q_0}^{Qw^{-1}Q_0}(s\mu, 1_H).$$

Therefore, replacing the partial Eisenstein series by its constant term, one gets

$$E_{Q,Q_0}^{Qw^{-1}Q_0}(s\mu, 1_H) \cdot E_P^G(s(\lambda + \check{\eta}(w\mu)) - \check{\eta}(w\rho_0 - \rho_0), 1_G).$$

This suffices to determine the abcissa of convergence of the height zeta functions of the strata. The arithmetic stratification with respect to L can be described purely in terms of a finite number of explicitely computable constants involving the parameters μ, λ and the Weyl group elements.

The results in this paper are part of the authors Ph.D. thesis, whose second part concerning asymptotic formulas shall appear in [St]. It is a pleasure to thank Jens Franke for introducing me to the methods used in this paper and for his constant support while working on these questions.

1. The fibre bundles: geometric-arithmetic preliminaries

1.1. Let G and H be two semi-simple linear algebraic groups over a number field F, and fix minimal F-rational parabolic subgroups $P_0 \subset G$ and $Q_0 \subset H$ with (F-rational) Levi decompositions $P_0 = L_0 U_0$, $Q_0 = M_0 V_0$. The quotients of G, respectively H, by parabolic subgroups $P \supset P_0$, $Q \supset Q_0$ are (by definition) generalized flag varieties:

$$W = P\backslash G \quad \text{and} \quad X = Q\backslash H.$$

Consider a homomorphism $\eta : P \to T_0$, where T_0 is the maximal split torus in the center of M_0. One has a right action of P on $X \times G$, via η: $(x, g) \cdot p = (x\eta(p), p^{-1}g)$, and there is a canonical morphism π from

$$Y = Y_\eta = (X \times G)/P$$

to W, giving Y the structure of a locally trivial fibre bundle over W with general fibre X, cf. [J], 5.14.

1.2. Let $\mu \in X^*(Q) = \mathrm{Hom}_{F\text{-}gps}(Q, \mathbf{G}_{m,F})$ be a character of Q and define an action of Q on $H \times \mathbf{A}^1$ by $q \cdot (h, a) = (qh, \mu(q)^{-1}a)$. The quotient of $H \times \mathbf{A}^1$ by Q is a line bundle over $X = Q\backslash H$, to be denoted by L_μ:

$$L_\mu = Q\backslash(H \times \mathbf{A}^1),$$

This gives an injection

$$X^*(Q) \longrightarrow Pic(X)$$

with finite cokernel, cf. for instance [Sa], Prop. 6.10, Lemma 6.9. The same applies to $X^*(P)$ and $Pic(W)$. The action of P on $H \times \mathbf{A}^1$, given by $(h, a) \cdot p = (h\eta(p), a)$, descends to an action of P on L_μ, and the quotient

$$L_\mu^Y = (L_\mu \times G)/P,$$

P acting on $L_\mu \times G$ by $(l, g) \cdot p = (l \cdot p, p^{-1}g)$, is a line bundle over Y. Using the exact sequence (6.10.2) in [Sa], it is not difficult to show that

(1.2.1) $\qquad\qquad (\mu, \lambda) \longmapsto$ class of $L_\mu^Y \otimes \pi^* L_\lambda$

is a monomorphism with finite cokernel

$$X^*(Q) \oplus X^*(P) \longrightarrow Pic(Y).$$

1.3. The choice of Q_0 determines a basis Δ_0^H of the root system of H with respect to T_0. The roots of T_0 occuring in the Lie algebra of the unipotent radical of Q_0 are defined to be positive and Δ_0^H is the basis of simple positive roots with respect to this set of positive roots. Let

$$\mathscr{W}_H = N_{H(F)}(T_0(F))/Z_{H(F)}(T_0(F))$$

be the Weyl group of H ($N_{H(F)}$ and $Z_{H(F)}$ denoting the normalizer, respectively centralizer in $H(F)$), and let $(.,.)$ be a $\mathscr{W}_H$-invariant positive definite form on $X^*(T_0)_{\mathbf{R}}$. Define the set

$$X^*(T_0)_{\mathbf{R}}^+ = \{x \in X^*(T_0)_{\mathbf{R}} \,|\, \text{for all } \beta \in \Delta_0^H : (x,\beta) > 0\}.$$

This is the positive Weyl chamber. Restriction of characters gives an injection and an isomorphism

$$X^*(Q)_{\mathbf{R}} \hookrightarrow X^*(Q_0)_{\mathbf{R}} \xrightarrow{\sim} X^*(T_0)_{\mathbf{R}}.$$

In the sequel we will always consider $X^*(Q)_{\mathbf{R}}$ as a subspace of $X^*(Q_0)_{\mathbf{R}}$ which we will identify with $X^*(T_0)$. Let M denote the unique Levi component of Q which contains M_0. There is a subset Δ_0^M of Δ_0^H such that

$$X^*(Q)_{\mathbf{R}} = \{x \in X^*(T_0)_{\mathbf{R}} \,|\, \text{for all } \beta \in \Delta_0^M : (x,\beta) = 0\}.$$

Denote by $X^*(Q)_{\mathbf{R}}^+$ the interior (in $X^*(Q)_{\mathbf{R}}$) of the intersection of $X^*(Q)_{\mathbf{R}}$ with the closure of $X^*(T_0)_{\mathbf{R}}^+$. Let Δ_Q^H be the complement of Δ_0^M in Δ_0^H. Then we have

$$X^*(Q)_{\mathbf{R}}^+ = \{x \in X^*(Q)_{\mathbf{R}} \,|\, \text{for all } \beta \in \Delta_Q^H : (x,\beta) > 0\}.$$

It is well known, that under the isomorphism $X^*(Q)_{\mathbf{R}} \to Pic(X)_{\mathbf{R}}$, the closure of $X^*(Q)_{\mathbf{R}}^+$ is mapped onto the cone of effective divisors $\Lambda(X)_{\text{eff}}$, i.e., the closed cone that is generated by those line bundles having non-zero global sections (cf. [J], II, 2.6).

The definitions depend on the once and for all chosen minimal parabolic subgroup and the fixed Levi component M_0. Define $X^*(P)_{\mathbf{R}}^+$ by analogy (depending on P_0 and L_0).

The following assumption will be used to describe $\Lambda(Y)_{\text{eff}}$ and to determine the abcissa of convergence of the height zeta functions.

Assumption 1.4. — For all $\beta \in \Delta_0^H$ the element $-\beta \circ \eta$ of $X^*(P)$ lies in the closure of $X^*(P)_{\mathbf{R}}^+$.

Geometrically speaking, we consider those η which map the cone in $X^*(T_0)_{\mathbf{R}}$ generated by $-\Delta_0^H$ into the cone of effective divisors in $X^*(P)_{\mathbf{R}}$. This subset of $Hom_{F\text{-}gps}(P,T_0)$ generates a sublattice of finite index.
Moreover, it is easy to see that if one replaces η by a conjugate $w\eta w^{-1}$, w

being an element of the Weyl group $\mathscr{W}_H$, the variety $Y_{w\eta w^{-1}}$ is isomorphic to Y_η. Sometimes such a conjugate of η satisfies assumption 1.4. This is always the case if P is a maximal proper parabolic subgroup of G. Indeed, the group $X^*(P)$ is in this case of rank one, and if η is not trivial, the kernel of the induced homomorphism $\check{\eta} : X^*(T_0)_{\mathbf{R}} \to X^*(P)_{\mathbf{R}}$ is a hyperplane, consisting of elements which are orthogonal to some $\chi \in X^*(T_0)_{\mathbf{R}}$, orthogonal with respect to the $\mathscr{W}_H$-invariant positive definite form $(.,.)$ on $X^*(T_0)_{\mathbf{R}}$. Replacing χ by $-\chi$ if necessary, we can assume that the half-space of $\nu \in X^*(T_0)_{\mathbf{R}}$ with $(\nu,\chi) < 0$ is mapped to $X^*(P)_{\mathbf{R}}^+$ by $\check{\eta}$. Consider the (closure of the) chamber that contains χ and the base Δ of the root system that corresponds to that chamber. For $\beta \in \Delta$ one has $-\beta \circ \eta \in \overline{X^*(P)_{\mathbf{R}}^+}$, and this shows that a conjugate of η by an element of the Weyl group fulfills 1.4.

Identify $\mathscr{W}_H$ with its image in $Aut(X^*(T_0)_{\mathbf{R}})$. The reflections along the elements of Δ_0^H generate $\mathscr{W}_H$, and define the Bruhat ordering on $\mathscr{W}_H$ and a length function. Let w_H be the longest element of $\mathscr{W}$. Define

$$\Lambda_\eta = \{(\mu,\lambda) \in X^*(Q)_{\mathbf{R}} \oplus X^*(P)_{\mathbf{R}} \mid \mu \in \overline{X^*(Q)_{\mathbf{R}}^+} \text{ and } \check{\eta}(w_H\mu)+\lambda \in \overline{X^*(P)_{\mathbf{R}}^+}\}.$$

This is a closed cone in $X^*(Q)_{\mathbf{R}} \oplus X^*(P)_{\mathbf{R}}$.

Proposition 1.5. — *Suppose η fulfills assumption 1.4. Then the line bundle $L_\mu^Y \otimes \pi^*L_\lambda$ has non-zero global sections if and only if (μ,λ) lies in Λ_η.*

Proof. — Suppose $H^0(Y, L_\mu^Y \otimes \pi^*L_\lambda) \neq 0$. The pull-back of $L_\mu^Y \otimes \pi^*L_\lambda$ by the quotient map $X \times G \longrightarrow Y$ is isomorphic to the pull-back of L_μ by the projection onto the first factor $X \times G \longrightarrow X$. It follows that L_μ has a non-zero global section, hence μ is an element of $\overline{X^*(Q)_{\mathbf{R}}^+}$ (cf. [**J**], II, 2.6). Let T_0 act on $L_\mu = Q \backslash (H \times \mathbf{A}^1)$ by $[h,a]t = [ht,a]$ and on $H^0(X, L_\mu)$ by $(t \cdot s)(x) = s(xt)t^{-1}$. Because T_0 is split, $H^0(X, L_\mu)$ decomposes as a direct sum of one-dimensional representations. For each $\chi \in X^*(T_0)$ let $m(\chi)$ be the multiplicity of the corresponding weight space. Then one has

$$\pi_*(L_\mu^Y \otimes \pi^*L_\lambda) \simeq \bigoplus_{\chi \in X^*(Q)} L_{-\chi \circ \eta + \lambda}^{\oplus m(\chi)}.$$

Hence, there is a χ with $m(\chi) > 0$ and $-\check{\eta}(\chi) + \lambda \in \overline{X^*(P)_{\mathbf{R}}^+}$. This implies

$$\check{\eta}(w_H\mu) + \lambda = -\check{\eta}(-w_H\mu - \chi) - \check{\eta}(\chi) + \lambda \in \overline{X^*(P)_{\mathbf{R}}^+},$$

because $-w_H\mu - \chi \in \sum_{\beta \in \Delta_0^H} \mathbf{Z}_{\geq 0}\beta$.

Now suppose that (μ, λ) lies in Λ_η. Then $-w_H\mu$ is a weight of the representation of T_0 on $H^0(X, L_\mu)$. Therefore,

$$\pi_*(L_\mu^Y \otimes \pi^* L_\lambda) \simeq L_{(w_H\mu)\circ\eta+\lambda}^{\oplus m(-w_H\mu)} \oplus \bigoplus_{\chi \neq -w_H\mu} L_{-\chi\circ\eta+\lambda}^{\oplus m(\chi)}$$

has a non-zero global section, and so does $L_\mu^Y \otimes \pi^* L_\lambda$. $\square$

By Proposition 1.5, under the isomorphism $X^*(Q)_\mathbf{R} \oplus X^*(P)_\mathbf{R} \longrightarrow Pic(Y)_\mathbf{R}$ induced by (1.2.1), the cone of effective divisors $\Lambda(Y_\eta)_{\text{eff}}$ is the image of Λ_η, and the interior Λ_η° of Λ_η is mapped onto $\Lambda(Y_\eta)_{\text{eff}}^\circ$.

1.6. Next we discuss metrics on the line bundles introduced above. To do this, choose a maximal compact subgroup $\mathbf{K}_G = \prod_v \mathbf{K}_{G,v} \subset G(\mathbf{A})$, such that

$$G(\mathbf{A}) = P_0(\mathbf{A})\mathbf{K}_G,$$

where v runs over all places of F and $\mathbf{A}$ denotes the ring of adeles of F. For a place v and $w \in W(F_v)$ choose $k \in \mathbf{K}_{G,v}$ which is mapped to w. If $[g, a] \in L_\lambda(F_v)$ is an F_v-valued point of the line bundle L_λ over w, define its norm by

$$\|[g, a]\|_w = |\lambda(p)a|_v,$$

where $p \in P(F_v)$ is determined by $g = pk$ and $|\cdot|_v$ denotes the local absolute value that is the Haar multiplier. The family $\|\cdot\|_v = (\|\cdot\|_w)_{w \in W(F_v)}$ is a v-adic metric on L_λ and $\mathbf{L}_\lambda = (L_\lambda, (\|\cdot\|_v)_v)$ is a metrization of L_λ, cf. [P], Lemme 6.2.3. Define a map

$$H_P = H_{P,\mathbf{K}_G} : G(\mathbf{A}) \to Hom_\mathbf{R}(X^*(P) \otimes \mathbf{R}, \mathbf{R})$$

by $\langle \lambda, H_P(pk)\rangle = \log(\prod_v |\lambda(p_v)|_v)$, for $\lambda \in X^*(P)$, $p = (p_v)_v \in P(\mathbf{A})$, $k \in \mathbf{K}_G$ and extend linearly. An easy computation shows (cf. [FMT], p. 428) that for the (exponential) height function we have

(1.6.1) $$H_{\mathbf{L}_\lambda}(w) = e^{-\langle\lambda, H_P(\gamma)\rangle},$$

whenever $\gamma \in G(F)$ maps to $w \in W(F) = P(F)\backslash G(F)$.

Similarly, we fix a maximal compact subgroup $\mathbf{K}_H = \prod_v \mathbf{K}_{H,v} \subset H(\mathbf{A})$ such that $H(\mathbf{A}) = Q_0(\mathbf{A})\mathbf{K}_H$, and provide all line bundles L_μ, $\mu \in X^*(Q)$, with the metrics as defined above. Furthermore, $\mathbf{K}_H$ is supposed to induce maximal compact subgroups of all the standard Levi subgroups. More precisely, we assume that

for each parabolic subgroup $Q' \supset Q_0$ of H with Levi decomposition $Q' = M'V'$, $M' \supset M_0$, one has

- $Q'(\mathbf{A}) \cap \mathbf{K}_H = (M'(\mathbf{A}) \cap \mathbf{K}_H)(V'(\mathbf{A}) \cap \mathbf{K}_H)$,
- $M'(\mathbf{A}) \cap \mathbf{K}_H$ is a maximal compact subgroup of $M'(\mathbf{A})$.

By [**MW**], I.1.4., such $\mathbf{K}_H$ exist. This implies in particular that $\mathbf{K}_H$ contains the maximal compact subgroup of $T_0(\mathbf{A})$.

Finally, we define metrics on the line bundles $L_\mu^Y = (L_\mu \times G)/P$. Let $\|\cdot\|_v$ be the v-adic metric on L_μ defined by means of $\mathbf{K}_{H,v}$. Recall the action $L_\mu \times P \to L_\mu = Q \backslash (H \times \mathbf{A}^1)$, given by $l \cdot p = l \cdot \eta(p) = [h\eta(p), a]$ if $l = [h, a]$. For an element $[l, g] \in L_\mu(F_v)$ mapping to $y = [x, g] \in Y(F_v)$ define

$$\| [l, g] \|_y = \| l \cdot \eta(p) \|_{x\eta(p)},$$

where $g = pk$, $p \in P(F_v)$, $k \in \mathbf{K}_{H,v}$. It follows from the remark above, that this is well-defined (because $P(F_v) \cap \mathbf{K}_{G,v}$ is mapped into $\mathbf{K}_{H,v}$). The family $\|\cdot\|_v = (\|\cdot\|_y)_{y \in Y(F_v)}$ is a v-adic metric on L_μ^Y and $\mathbf{L}_\mu^Y = (L_\mu^Y, (\|\cdot\|_v)_v)$ is a metrization of L_μ^Y.

For $\gamma \in G(F)$ and $\delta \in H(F)$ mapping to $x \in X(F)$, the pair (x, γ) maps to an element $y \in Y(F)$, and

(1.6.2) $$H_{\mathbf{L}_\mu^Y}(y) = e^{-\langle \mu, H_Q(\delta \eta(p)) \rangle}$$

with $g = pk$, $p \in P(\mathbf{A})$, $k \in \mathbf{K}_G$.

2. Height zeta functions

Denote by s a variable (taking values in $\mathbf{R}$). By (1.6.1) the height zeta function of the generalized flag variety $X = Q \backslash H$ corresponding to the metrized line bundle $\mathbf{L}_\mu$ is equal to (the value of) an *Langlands-Eisenstein series*, whenever the series converges:

$$Z_X(\mathbf{L}_\mu, s) = \sum_{x \in X(F)} H_{\mathbf{L}_\mu}(x)^{-s} = \sum_{\delta \in Q(F) \backslash H(F)} e^{\langle s\mu, H_Q(\delta) \rangle} = E_Q^H(s\mu, 1_H).$$

This crucial observation is due to J. Franke. In [**FMT**] all analytic properties of these Eisenstein series that are needed for applications concerning rational points on X are proved. In this paper we only need results on the domain of convergence, and so we found it appropriate to prove these more accessible statements here again instead of deducing them from much deeper theorems. Hence our first task is to determine the domain of convergence of these Eisenstein series. Let ρ_Q be half the sum of the roots of T_0 that occur in the Lie algebra of V (= unipotent radical of Q), counted with multiplicity. For convenience, in the definition of E_Q^H, we shifted the coordinate on $X^*(Q) \otimes \mathbf{R}$ by $-\rho_Q$, compared with the usual convention.

The following elementary lemma will turn out to be very useful.

Lemma 2.1. — *Let $\mathscr{F} \subset H(\mathbf{A})$ and $\mathscr{C} \subset X^*(Q)_{\mathbf{R}}$ be relatively compact subsets. Then there is a constant $c > 1$ such that for all $v \in \mathscr{F}$, $\mu \in \mathscr{C}$ and $h \in H(\mathbf{A})$*

$$c^{-1} e^{\langle \mu, H_Q(h) \rangle} \leq e^{\langle \mu, H_Q(hv) \rangle} \leq c e^{\langle \mu, H_Q(h) \rangle}.$$

Proof. — Put $h = qk$, $hv = q_1 k_1$, $q, q_1 \in Q(\mathbf{A})$, $k, k_1 \in \mathbf{K}_H$. Note that $q^{-1} q_1 = kv k_1^{-1} \in \mathbf{K}_H \mathscr{F} \mathbf{K}_H \cap Q(\mathbf{A})$ and the latter is a compact subset of $Q(\mathbf{A})$. Hence there is $c > 1$ such that for all $v_1 \in \mathbf{K}_H \mathscr{F} \mathbf{K}_H$ and $\mu \in \mathscr{C}$

$$c^{-1} \leq e^{\langle \mu, H_Q(v_1) \rangle} \leq c.$$

Hence we have for any triple $(\mu, v, h) \in \mathscr{C} \times \mathscr{F} \times H(\mathbf{A})$

$$\begin{aligned} c^{-1} e^{\langle \mu, H_Q(h) \rangle} &\leq e^{\langle \mu, H_Q(q^{-1} q_1) \rangle} e^{\langle \mu, H_Q(q) \rangle} = e^{\langle \mu, H_Q(hv) \rangle}, \\ e^{\langle \mu, H_Q(hv) \rangle} &= e^{\langle \mu, H_Q(q^{-1} q_1) \rangle} e^{\langle \mu, H_Q(q) \rangle} \leq c e^{\langle \mu, H_Q(hv) \rangle}. \quad \square \end{aligned}$$

2.2. The most important tool for us to investigate the Eisenstein series are Siegel domains. To introduce these, consider the homomorphism

$$\begin{aligned} \bar{\beta} = (\beta)_{\beta \in \Delta_0^H} : T_0 &\to \mathbf{G}_{m,F}^d, \\ t &\mapsto (\beta(t))_{\beta \in \Delta_0^H}. \end{aligned}$$

It is surjective with finite kernel ($d = dim(T)$). Embed $\mathbf{R}_{>0}$ into $\mathbf{A}^*$ by mapping $x \in \mathbf{R}$ to the idèle $(x_v)_v$ with archimedean components $x_v = x$ and non-archimedean components $x_v = 1$. This way we consider $\mathbf{R}_{>0}^d$ as a subgroup of $\mathbf{G}_{m,F}^d(\mathbf{A})$. Let A_{M_0} the connected component of the preimage of $\mathbf{R}_{>0}^d$ under $\bar{\beta}$, which contains the trivial element. For all $w \in \mathscr{W}$ we have $w A_{M_0} w^{-1} = A_{M_0}$. One has $M_0(\mathbf{A}) = A_{M_0} M_0(\mathbf{A})^1$ with

$$M_0(\mathbf{A})^1 = \bigcap_{\chi \in X^*(M_0)} ker(|\chi| : M_0(\mathbf{A}) \to \mathbf{R}_{>0}),$$

where $|\chi|((m_v)_v) := \prod_v |\chi(m_v)|_v$. For $t_0 \in A_{M_0}$ we define

$$A_{M_0}(t_0) = \{ t \in A_{M_0} \mid \text{for all } \beta \in \Delta_0^H : \beta(t) > \beta(t_0) \}.$$

Recall the Levi decomposition $Q = MV$, M being the unique Levi component containing M_0. For a compact subset $\Omega \subset Q_0(\mathbf{A})$ put

$$\mathscr{S}_{Q,\Omega} = \Omega A_M A_{M_0}(t_0) \mathbf{K}_H,$$

where $A_M = A_{M_0} \cap Z(M(\mathbf{A}))$, $Z(M(\mathbf{A}))$ denoting the center of $M(\mathbf{A})$. If Ω is sufficiently big and t_0 is sufficiently small (i.e., $\beta(t_0)$ is sufficiently small for all $\beta \in \Delta_0^H$) then the following holds, c.f. [MW], I.2.1:

$$H(\mathbf{A}) = Q(F) \mathscr{S}_{Q,\Omega}.$$

In this case $\mathscr{S}_{Q,\Omega}$ is called a *Siegel domain*. With these preparations we can determine the abcissa of convergence of the Eisenstein series.

Proposition 2.3. — *a) The series*

$$E_Q^H(\mu + 2\rho_Q, h) = \sum_{\delta \in Q(F)\backslash H(F)} e^{\langle \mu + 2\rho_Q, H_Q(\delta h)\rangle},$$

$\mu \in X^*(Q) \otimes \mathbf{C}$, $h \in H(\mathbf{A})$, *converges uniformly for* $(\Re(\mu), h)$ *in any compact subset of* $X^*(Q)_{\mathbf{R}}^+ \times H(\mathbf{A})$.
b) For any $\mu \in X^*(Q)_{\mathbf{R}} - X^*(Q)_{\mathbf{R}}^+$ *and any* $h \in H(\mathbf{A})$ *the series* $E_Q^H(\mu + 2\rho_Q, h)$ *does not converge.*

Proof. — a) (This part is extracted from [Go].) For each $\beta \in \Delta_0^H$ there is a maximal proper parabolic subgroup Q_β containing Q_0 and a generator μ_β of $X^*(Q_\beta)$ lying in $X^*(Q_\beta)_{\mathbf{R}}^+$ such that

$$X^*(Q_0)_{\mathbf{R}}^+ = \sum_{\beta \in \Delta_0^H} \mathbf{R}_{>0}\mu_\beta.$$

Attached to μ_β there is an ample line bundle L_{μ_β} on $Q_\beta \backslash H$. Hence, there exists $c_\beta > 0$ such that for all $\delta \in H(F)$

$$e^{\langle \mu_\beta, H_{Q_\beta}(\delta) \rangle} = H_{L_{\mu_\beta}}(Q_\beta(F)\delta)^{-1} \leq c_\beta.$$

Now let $U \subset H(\mathbf{A})$ be an open and relatively compact subset. By Lemma 2.1 there is a constant $c > 0$ such that for all $\beta \in \Delta_Q^H$, $\delta \in H(F)$ and $h \in U$

$$e^{\langle \mu_\beta, H_{Q_0}(\delta h) \rangle} \leq c.$$

Therefore, $H(F)U$ is contained in $V_0(\mathbf{A})M_0(\mathbf{A})^1 A_{M_0}^-(t_0^-)\mathbf{K}_H$ for some $t_0^- \in A_{M_0}$, where

$$A_{M_0}^-(t_0^-) = \{t \in A_{M_0} \mid \text{for all } \beta \in \Delta_0^H : \mu_\beta(t) < \mu_\beta(t_0^-)\}.$$

Next we use the fact that $H(\mathbf{A}) = Q(F)\mathscr{S}_{Q,\Omega}$ with some Siegel domain $\mathscr{S}_{Q,\Omega}$. By the definition of $\mathscr{S}_{Q,\Omega}$ we see that there is a fundamental domain for the action of $Q(F)$ on $H(F)U$ which is contained in

$$\Omega'(A_M \cap A_{M_0}^-(t_0^-))\mathbf{K}_H$$

for some compact subset $\Omega' \subset Q_0(\mathbf{A})$ which contains Ω. By Lemma 2.1, for h in a compact subset of $H(\mathbf{A})$, we can bound $E_Q^H(\mu + 2\rho_Q, h)$ by the integral

$$\int_{Q(F)\backslash H(F)U} e^{\langle \mu + 2\rho_Q, H_Q(h_1)\rangle} dh_1,$$

with some Haar measure dh_1 on $H(\mathbf{A})$. Now, $dh_1 = e^{\langle -2\rho_Q, H_Q(m)\rangle} dv\, dm\, dk$ with Haar measures dv on $V(\mathbf{A})$, dm on $M(\mathbf{A})$, dk on $\mathbf{K}_H$ (recall $Q = MV$ and $H(\mathbf{A}) = Q(\mathbf{A})\mathbf{K}_H$). Therefore, the integral above can be bounded by

$$\int_{A_M \cap A_{M_0}^-(t_0^-)} e^{\langle \mu, H_Q(a)\rangle} da,$$

with some Haar measure da on A_M. Writing $\mu = \sum_{\beta \in \Delta_Q^H} s_\beta \mu_\beta$ with $s_\beta \in \mathbf{R}_{>0}$, we see that this integral is equal to

$$(\text{constant depending on } t_0^- \text{ and } da) \cdot \prod_{\beta \in \Delta_Q^H} \frac{1}{s_\beta}.$$

Hence we have shown that $E_Q^H(\mu + 2\rho_Q, h)$ converges uniformly for (μ, h) in compact subsets of $X^*(Q)_{\mathbf{R}}^+ \times H(\mathbf{A})$.

b) To prove the second assertion, it is enough to show that $E_Q^H(\mu + 2\rho_Q, 1_H)$ tends to infinity as μ approaches the boundary of $X^*(Q)_{\mathbf{R}}^+$.
Let $Q' \supset Q$ be a parabolic subgroup with Levi component $M' \supset M$ such that $Q \cap M'$ is a maximal proper parabolic subgroup of M'. Decompose $X^*(Q)_{\mathbf{R}}$ as follows:

$$X^*(Q)_{\mathbf{R}} = X^*(Q')_{\mathbf{R}} \oplus X^*(T_M/T_{M'})_{\mathbf{R}},$$

where T_M (resp. $T_{M'}$) is the maximal split torus in the center of M (resp. M'). With respect to this decomposition we have $2\rho_Q = 2\rho_{Q'} + 2\rho_{Q\cap M'}^{M'}$, where $2\rho_{Q\cap M'}^{M'}$ is the sum of roots of T_0 (counted with multiplicity) which occur in the Lie algebra of the unipotent radical of $Q \cap M'$. $X^*(T_M/T_{M'})_{\mathbf{R}}$ is a one dimensional $\mathbf{R}$-vector space and the image of $X^*(Q)_{\mathbf{R}}^+$ under the projection onto this space is the open half-line generated by $\rho_{Q\cap M'}^{M'}$. Now let $\mu \in X^*(Q)_{\mathbf{R}}^+$ tend to the boundary of $X^*(Q)_{\mathbf{R}}^+$. Call μ' the image of μ under the projection onto $X^*(T_M/T_{M'})_{\mathbf{R}}$, and assume that μ' tends to zero. Next we note that

$$E_Q^H(\mu + 2\rho_Q, 1_H) = \sum_{\delta \in Q'(F) \backslash H(F)} e^{\langle \mu - \mu' + 2\rho_{Q'}, H_{Q'}(\delta)\rangle} E_Q^{Q'}(\mu' + 2\rho_{Q\cap M'}^{M'}, \delta).$$

Hence it suffices to show that $E_{Q\cap M'}^{M'}(\mu' + 2\rho_{Q\cap M'}^{M'}, 1_H)$ tends to infinity as μ' tends to zero. In other words, it suffices to consider the case when Q is a maximal proper parabolic subgroup. Let Q^- be the opposite parabolic subgroup with Levi component V^-. The map from H to $Q \backslash H$ maps V^- isomorphically onto an open subset of $Q \backslash H$. On $V^-(\mathbf{A})$ the function $h \mapsto e^{\langle 2\rho_Q, H_Q(h)\rangle}$ is bounded from above but assumes every sufficiently small positive value. Let $\mathscr{F} \subset V^-(\mathbf{A})$ be an open and relatively compact subset such that $V^-(F)\mathscr{F} = V^-(\mathbf{A})$. Then, with the notations from part a) of the proof, there is some open and relatively compact subset $\omega \subset Q_0(\mathbf{A})$ such that a fundamental domain for the action of $Q(F)$ on $H(F)\mathscr{F}\mathbf{K}_H$ contains

$$\omega(A_M \cap A_{M_0}^-(t_0^-))\mathbf{K}_H.$$

Here we used the fact that A_M is one-dimensional. By Lemma 2.1, $E_Q^H(\mu + 2\rho_Q, 1_H)$ is bounded from below by the integral

$$\int_{Q(F)\backslash H(F)\mathscr{F}\mathbf{K}_H} e^{\langle \mu+2\rho_Q, H_Q(h)\rangle} dh,$$

with some Haar measure dh on $H(\mathbf{A})$, and this in turn can be bounded from below by

$$\int_{A_M \cap A_{M_0}^-(t_0^-)} e^{\langle \mu, H_Q(a)\rangle} da,$$

with some appropriate Haar measure da on A_M. This integral tends to infinity as μ tends to zero, and this completes the proof. $\square$

2.4. Of course, the Proposition 2.3 applies equally to the Eisenstein series

$$E_P^G(\lambda, G) = \sum_{\gamma \in P(F)\backslash G(F)} e^{\langle \lambda, H_P(\gamma g)\rangle}.$$

For $\mu \in X^*(Q)_{\mathbf{R}}^+$ denote by a_μ the abcissa of convergence of $E_Q^H(s\mu, h)$. By Prop. 2.3 we have

$$a_\mu = \inf\{a \,|\, a\mu \in 2\rho_Q + X^*(Q)_{\mathbf{R}}^+\,\}.$$

For $\mu \notin X^*(Q)_{\mathbf{R}}^+$ we put $a_\mu = \infty$. The height zeta function of the variety Y with respect to the metrized line bundle $\mathbf{L}_\mu^Y \otimes \pi^* \mathbf{L}_\lambda$ has, by (1.6.1) and (1.6.2), the following formal expression:

$$\sum_{y \in Y(F)} H_{\mathbf{L}_\mu^Y \otimes \pi^* \mathbf{L}_\lambda}(y)^{-s}$$

$$= \sum_{\gamma \in P(F)\backslash G(F)} e^{\langle s\lambda, H_P(\gamma)\rangle} \sum_{\delta \in Q(F)\backslash H(F)} e^{\langle s\mu, H_Q(\delta\eta(p_\gamma))\rangle}$$

$$= \sum_{\gamma \in P(F)\backslash G(F)} e^{\langle s\lambda, H_P(\gamma)\rangle} E_Q^H(s\mu, \eta(p_\gamma))$$

where $\gamma = p_\gamma k_\gamma$ with $p_\gamma \in P(\mathbf{A})$, $k_\gamma \in \mathbf{K}_G$. Even if $\mathbf{L}_\mu^Y \otimes \pi^* \mathbf{L}_\lambda$ lies in the interior of the cone of effective divisors $\Lambda(Y)_{\text{eff}}$, this series will in general not converge for any s. The reason for this is the occurrence of accumulating subvarieties. A posteriori it turns out that the right approach to find these is to decompose the fibre X into locally closed subvarieties, the images of the Bruhat cells:

$$X = \coprod_{w \in \mathscr{W}^M} X^w,$$

where $\mathscr{W}^M \subset \mathscr{W}_H$ consists of the elements of minimal length in the classes $w\mathscr{W}_M$, w running through $\mathscr{W}_H$, and $\mathscr{W}_M$ is the Weyl group of M with respect to T_0. X^w is the image of the generalized Bruhat cell $Qw^{-1}Q_0 \subset H$. Denote by w_H^M the element of minimal length in $w_H \mathscr{W}_M$. Then $X^{w_H^M}$ is the open stratum, which we denote by X°. The *partial Eisenstein series* are defined by summation over the rational points of X^w:

$$E_Q^{Qw^{-1}Q_0}(\mu, h) = \sum_{\delta \in Q(F) \backslash Q(F) w^{-1} Q_0(F)} e^{\langle \mu, H_Q(\delta h) \rangle}.$$

The behaviour of the partial Eisenstein series on Siegel domains is of importance for us. To understand how a partial Eisenstein series decreases when one goes to infinity on such a Siegel domain, we will study its *constant term*.

The constant term Θ_{Q_0} of a function $\Theta : H(\mathbf{A}) \to \mathbf{C}$, which is left-invariant under $V_0(F)$ and lies in $L_{1,\mathrm{loc}}(H(\mathbf{A}))$, is defined by

$$\Theta_{Q_0}(h) = \int_{V_0(F) \backslash V_0(\mathbf{A})} \Theta(vh) dv,$$

where the measure dv on $V_0(F) \backslash V_0(\mathbf{A})$ is the quotient of the Haar measures and is normalized by $\int_{V_0(F) \backslash V_0(\mathbf{A})} dv = 1$.

The following statement is the key to determine the abcissa of convergence of the height zeta functions of the strata. We put $\rho_0 = \rho_{Q_0}$.

Proposition 2.5. — *Let $\mathscr{F} \subset H(\mathbf{A})$ and $\mathscr{C} \subset 2\rho_Q + X^*(Q)_{\mathbf{R}}^+$ be relatively compact subsets ($\mathscr{C}$ relatively compact in $X^*(Q)_{\mathbf{R}}$), and fix $t_0 \in A_{M_0}$.*

a) There is a constant $c > 1$, such that for all $\mu \in \mathscr{C}$, $w \in \mathscr{W}^M$, $t \in A_{M_0}(t_0)$ and $v \in \mathscr{F}$

$$c^{-1} E_{Q,Q_0}^{Qw^{-1}Q_0}(\mu, 1_H) \le e^{-\langle w(\mu - \rho_0) + \rho_0, H_{Q_0}(t) \rangle} E_Q^{Qw^{-1}Q_0}(\mu, tv)$$

$$\le c E_{Q,Q_0}^{Qw^{-1}Q_0}(\mu, 1_H).$$

b) Now let $w = w_H^M$. Then there is a constant $c > 1$, such that for all $\mu \in 2\rho_Q + X^*(Q)_{\mathbf{R}}^+$

$$E_Q^H(s\mu, 1_H) \le c E_Q^{Qw^{-1}Q_0}(s\mu, 1_H).$$

Proof. — a) Let V_0 be the unipotent radical of Q_0, and let $\mathscr{F}_0 \subset V_0(\mathbf{A})$ be a compact subset such that $V_0(F)\mathscr{F}_0 = V_0(\mathbf{A})$. Because
$$\{t^{-1}vt \mid t \in A_{M_0}(t_0),\, v \in \mathscr{F}_0\}$$
is relatively compact, Lemma 2.1 ensures the existence of a $c > 1$ such that for all $h \in H(\mathbf{A})$, $t \in A_{M_0}$, $v \in \mathscr{F}$, $v_0 \in \mathscr{F}_0$ one has
$$c^{-1} e^{\langle \mu, H_Q(ht(t^{-1}v_0 t))\rangle} \leq e^{\langle \mu, H_Q(htv)\rangle} \leq c e^{\langle \mu, H_Q(ht(t^{-1}v_0 t))\rangle}.$$
Therefore
$$c^{-1} \int_{V_0(F)\backslash V_0(\mathbf{A})} E_Q^{Qw^{-1}Q_0}(\mu, v_0 t) dv_0 \leq E_Q^{Qw^{-1}Q_0}(\mu, tv)$$
$$\leq c \int_{V_0(F)\backslash V_0(\mathbf{A})} E^{Qw^{-1}Q_0}(\mu, v_0 t) dv_0.$$
Moreover,
$$\int_{V_0(F)\backslash V_0(\mathbf{A})} E_Q^{Qw^{-1}Q_0}(\mu, v_0 t) dv_0 = \int_{(V_0 \cap wV_0w^{-1})(\mathbf{A})\backslash V_0(\mathbf{A})} e^{\langle \mu, H_{Q_0}(w^{-1}v_0 t))\rangle} dv_0$$
$$= e^{\langle w\mu + \rho_0 - w\rho_0, H_Q(t)\rangle} E_{Q,Q_0}^{Qw^{-1}Q_0}(\mu, 1_H).$$
And this proves assertion a).

b) Because $H(F)$ is dense in H and $Qw^{-1}Q_0$ is open (and dense) in H, there are elements $\delta_1, \ldots, \delta_n \in H(F)$ such that
$$H(F) = \cup_{i=1}^{n} Q(F)w^{-1}Q_0(F)\delta_i.$$
Therefore,

(2.5.1) $$E_Q^H(\mu, 1_H) \leq \sum_{i=1}^{n} E_Q^{Qw^{-1}Q_0}(\mu, \delta_i).$$

By Lemma 2.1, there exists $c_1 > 1$ such that for all $\mu \in \mathscr{C}$, $\delta \in H(F)$ and $i = 1, \ldots, n$
$$e^{\langle \mu, H_Q(\delta \delta_i)\rangle} \leq c_1 e^{\langle \mu, H_Q(\delta))\rangle}.$$
The sum on the right of (2.5.1) can be bounded by $nc_1 E_Q^{Qw^{-1}Q_0}(s\mu, 1_H)$. □

2.6. The following lemma explains the significance of assumption 1.4: the geometric assertion that $\check{\eta}$ maps the cone in $X^*(T_0)_\mathbf{R}$ generated by $-\Delta_0^H$ into the cone $X^*(P)_\mathbf{R}^+ \subset X^*(P)_\mathbf{R}$, has as analytic consequence that there exists a fixed Siegel domain such that η maps the "P-part" of each rational element $\gamma \in G(F)$ into this fixed Siegel domain. To be more precise, put
$$T_0(\mathbf{A})^1 = M_0(\mathbf{A})^1 \cap T_0(\mathbf{A}),$$

cf. §2.2. By the product formula we have $T_0(F) \subset T_0(\mathbf{A})^1$, and the quotient $T_0(F)\backslash T_0(\mathbf{A})^1$ is compact. Let $\Omega_0 \subset T_0(\mathbf{A})^1$ be a compact subset such that $T_0(\mathbf{A})^1 = T(F)\Omega_0$.

Lemma 2.7. — *There is $t_0 \in A_{M_0}$, such that for all $\gamma \in G(F)$ one has*
$$\eta(p_\gamma) \in T_0(\mathbf{A})^1 A_{M_0}(t_0),$$
where $\gamma = p_\gamma k_\gamma$, $p_\gamma \in P(\mathbf{A})$, $k_\gamma \in \mathbf{K}_G$.

Proof. — For $\beta \in \Delta_0^H$ we have
$$e^{\langle \beta, H_{Q_0}(\eta(p_\gamma))\rangle} = e^{\langle \beta \circ \eta, H_P(\gamma)\rangle} = H_{\mathbf{L}_{-\beta \circ \eta}}(P(F)\gamma).$$

By hypothesis, the line bundle $\mathbf{L}_{-\beta \circ \eta}$ lies in the cone of effective divisors, hence the height function associated to $\mathbf{L}_{-\beta \circ \eta}$ is bounded from below. □

2.8. Before proving our result about the abcissa of convergence we have to introduce the following constants. For $(\mu, \lambda) \in X^*(Q)_\mathbf{R} \oplus X^*(P)_\mathbf{R}$ and $w \in \mathscr{W}^M$ put

$\check{a}_{\mu,\lambda}^w = \inf\{a \mid \text{for all } a' \geq a : a'(\lambda + \check{\eta}(w\mu)) - \check{\eta}(w\rho_0 - \rho_0) \in 2\rho_P + X^*(P)_\mathbf{R}^+\}$,
$a_{\mu,\lambda}^w = \max\{a_\mu, \check{a}_{\mu,\lambda}^w\}$.

In general, $\check{a}_{\mu,\lambda}^w$ may not be finite (by convention, the infimum of the empty set is ∞). For $w = w_H^M$ put $a_{\mu,\lambda}^\circ = a_{\mu,\lambda}^w$. The stratification of X gives rise to a stratification of Y because all strata $X^w = Q\backslash Qw^{-1}Q_0$ are stable under the action of T_0 from the right:
$$Y = \coprod_{w \in \mathscr{W}^M} Y^w \text{ with } Y^w = (X^w \times G)/P.$$

The open stratum, corresponding to $w = w_H^M$ will be denoted by Y°. The corresponding height zeta functions take the following form:
$$Z_{Y^w}(\mathbf{L}_\mu^Y \otimes \pi^* \mathbf{L}_\lambda, s) = \sum_{\gamma \in P(F)\backslash G(F)} e^{\langle s\lambda, H_P(\gamma)\rangle} E_Q^{Qw^{-1}Q_0}(s\mu, \eta(p_\gamma)),$$
with the notations from Lemma 2.7.

Proposition 2.9. — *a) If $\lambda + \check{\eta}(w\mu)$ is not contained in $X^+(P)_\mathbf{R}^+$, then the series $Z_{Y^w}(\mathbf{L}_\mu^Y \otimes \pi^* \mathbf{L}_\lambda, s)$ does not converge for any $s > 0$.*
b) If $\lambda + \check{\eta}(w\mu)$ is in $X^*(P)_\mathbf{R}^+$, then the abcissa of convergence of $Z_{Y^w}(\mathbf{L}_\mu^Y \otimes \pi^* \mathbf{L}_\lambda, s)$ is in the interval $[\check{a}_{\mu,\lambda}^w, a_{\mu,\lambda}^w]$.
c) Suppose that $L_\mu^Y \otimes \pi^* L_\lambda$ lies in $\Lambda(Y)_{\text{eff}}^\circ$. Then the abcissa of convergence of $Z_{Y^\circ}(\mathbf{L}_\mu^Y \otimes \pi^* \mathbf{L}_\lambda, s)$ is $a_{\mu,\lambda}^\circ$.

Proof. — a) Let us fix t_0 such that the assertion of Lemma 2.7 holds. Then we can write $\eta(p_\gamma) = \vartheta_\gamma o_\gamma a_\gamma$ with $\vartheta_\gamma \in T(F)$, $o_\gamma \in \Omega_0$ and $a_\gamma \in A_{M_0}(t_0)$. Since all partial Eisenstein series are left-invariant under $T(F)$ one has

$$E_Q^{Qw^{-1}Q_0}(s\mu, \eta(p_\gamma)) = E_Q^{Qw^{-1}Q_0}(s\mu, a_\gamma o_\gamma).$$

For each real number $a > a_\mu$ there is by Proposition 2.5 a $c_1 > 0$ such that for all $w \in \mathscr{W}^M$ and all $s \in (a_\mu, a]$

$$c_1 E_{Q,Q_0}^{Qw^{-1}Q_0}(s\mu, 1_H) \leq e^{-\langle w(s\mu - \rho_0) + \rho_0, H_{Q_0}(a_\gamma)\rangle} E_Q^{Qw^{-1}Q_0}(s\mu, \eta(p_\gamma))$$

$$\leq c_1^{-1} E_{Q,Q_0}^{Qw^{-1}Q_0}(s\mu, 1_H),$$

where $E_{Q,Q_0}^{Qw^{-1}Q_0}$ is the constant term of the partial Eisenstein series:

$$E_{Q,Q_0}^{Qw^{-1}Q_0}(s\mu, 1_H) = \int_{V_0(F)\backslash V_0(\mathbf{A})} E_Q^{Qw^{-1}Q_0}(s\mu, v_0) dv_0.$$

For $s \in (a_\mu, a]$ this function is bounded from below by a positive number c_2. For such s we have then

$$Z_{Y^w}(\mathbf{L}Y_\mu \otimes \pi^*\mathbf{L}_\lambda, s) \geq c_1 E_{Q,Q_0}^{Qw^{-1}Q_0}(s\mu, 1_H)$$

$$\times \sum_{\gamma \in P(F)\backslash G(F)} e^{\langle s\lambda, H_P(a_\gamma)\rangle} e^{\langle sw\mu - w\rho_0 + \rho_0, H_{Q_0}(\eta(p_\gamma))\rangle}$$

$$\geq c_1 c_2 E_P^G(s(\lambda + \check{\eta}(w\mu)) - \check{\eta}(w\rho_0 - \rho_0), 1_G).$$

It is well-known that $\rho_0 - w\rho_0$ lies in the closed cone generated by the positive simple roots, and by assumption 1.4 we conclude that $\check{\eta}(w\rho_0 - \rho_0)$ lies in the closure of $X^*(P)_\mathbf{R}^+$. Therefore, if $\lambda + \check{\eta}(w\mu)$ is not contained in $X^*(P)_\mathbf{R}^+$, the same is true for $s(\lambda + \check{\eta}(w\mu)) - \check{\eta}(w\rho_0 - \rho_0)$, for all $s > 0$. In this case, $Z_{Y^w}(\mathbf{L}_\lambda \otimes \pi^*\mathbf{L}_\mu, s)$ does not converge for any $s > 0$.

b) Suppose now that $\lambda + \check{\eta}(w\mu)$ is an element of $X^*(P)_\mathbf{R}^+$. The above estimate shows that the abcissa of convergence of $Z_{Y^w}(\mathbf{L}_\lambda \otimes \pi^*\mathbf{L}_\mu, s)$ is greater or equal to $\check{a}_{\mu,\lambda}^w$. On the other hand, using the estimate above, we get that

$$Z_{Y^w}(\mathbf{L}_\mu^Y \otimes \pi^*\mathbf{L}_\lambda, s) \leq c_1^{-1} E_{Q,Q_0}^{Qw^{-1}Q_0}(s\mu, 1_H) E_P^G(s(\lambda + \check{\eta}(w\mu)) - \check{\eta}(w\rho_0 - \rho_0), 1_G).$$

Hence, this function converges for $s > \max\{a_\mu, \check{a}_{\mu,\lambda}^w\} = a_{\mu,\lambda}^w$, by Proposition 2.3 a).

c) Let $w = w_H^M$. By Proposition 2.5 b) and the estimate above, we have for $s \in (a_\mu, a]$

$$E_{Q,Q_0}^{Qw^{-1}Q_0}(s\mu, 1_H) \geq c_1 E_Q^{Qw^{-1}Q_0}(s\mu, 1_H) \geq c_3 E_Q^H(s\mu, 1_H),$$

for some $c_3 > 0$. This shows that $Z_{Y^\circ}(\mathbf{L}^Y_\mu \otimes \pi^* \mathbf{L}_\lambda, s)$ converges only for $s > a^\circ_{\mu,\lambda}$. By part b) we conclude that the abcissa of convergence is exactly $a^\circ_{\mu,\lambda}$. □

3. Arithmetic stratification

In this section we determine the arithmetic stratification of $Y = (Q \backslash H \times G)/P$ with respect to line bundles $L^Y_\mu \otimes \pi^* L_\lambda$ which lie in the interior of the cone of effective divisors $\Lambda(Y)^\circ_{\text{eff}}$. Otherwise, there is no open non-empty subset for which the corresponding height zeta function converges for all $s \gg 0$. First we need a lemma.

Lemma 3.1. — *If $X^{w'} \subset \overline{X^w}$, then $a^{w'}_{\mu,\lambda} \geq a^w_{\mu,\lambda}$.*

Proof. — The assertion follows if we can show that for all $a > a_\mu$ one has

$$a(\lambda + \check{\eta}(w\mu)) - \check{\eta}(w\rho_0 - \rho_0) - a(\lambda + \check{\eta}(w'\mu)) + \check{\eta}(w\rho_0 - \rho_0) \in \overline{X^*(P)^+_{\mathbf{R}}}.$$

Assumption 1.4 on η ensures us that this is true if we have

$$aw'\mu - w'\rho_0 - aw\mu + w\rho_0 \in \sum_{\beta \in \Delta_0^H} \mathbf{R}_{\geq 0} \beta$$

for all $a > a_\mu$. Note that

$$aw'\mu - w'\rho_0 - aw\mu + w\rho_0$$
$$= w'(a\mu - 2\rho_Q) - w(a\mu - 2\rho_Q) + w'(2\rho_Q - \rho_0) - w(2\rho_Q - \rho_0).$$

Now let w_M be the longest element in $\mathcal{W}_M$. Then we have $w_H = w_H^M w_M$ and $2\rho_Q - \rho_0 = w_M \rho_0$. Therefore,

$$aw'\mu - w'\rho_0 - aw\mu + w\rho_0 = w'(a\mu - 2\rho_Q) - w(a\mu - 2\rho_Q) + w'w_M\rho_0 - ww_M\rho_0.$$

By [J], II, 13.8 (4), the assumption $X^{w'} \subset \overline{X^w}$ implies that $w' \leq w$ with respect to the Bruhat ordering on $\mathcal{W}_H$. It follows from the description of the Bruhat ordering given in [J], II, Proposition 13.7, together with

$$l(w'w_M) = l(w') + l(w_M), \quad l(ww_M) = l(w) + l(w_M)$$

that moreover

$$w'w_M \leq ww_M.$$

Hence we are done, if we can show that whenever $w_1 \leq w_2$ and $x \in \overline{X^*(Q_0)^+_{\mathbf{R}}}$ one has

$$w_1 x - w_2 x \in \sum_{\beta \in \Delta_0^H} \mathbf{R}_{\geq 0} \beta.$$

For a root β of H with respect to T_0 we denote by s_β the reflection along β. By [**BGG**], Proposition 2.8, there exist positive roots $\beta_1,\ldots,\beta_k$ such that with $s_i = s_{\beta_i}$

$$w_2 = s_k \cdot s_{k-1} \cdot \ldots \cdot s_1 \cdot w_1,$$
for all $i = 1,\ldots,k$: $l(s_i \cdot s_{i-1} \cdot \ldots \cdot s_1 \cdot w_1) = l(s_{i-1} \cdot \ldots \cdot s_1 \cdot w_1) + 1$.

Because

$$w_1 x - w_2 x = \sum_{i=1}^k s_{i-1} \cdot \ldots \cdot s_1 \cdot w_1 x - s_i \cdot s_{i-1} \cdot \ldots \cdot s_1 \cdot w_1 x,$$

we can assume that $w_2 = s_\beta w_1$ with a positive root β and $l(w_2) = l(w_1) + 1$. By [**Bou**], ch. VI, §1, Proposition 17, this implies $w_2^{-1}\beta < 0$, i.e. $w_1^{-1}\beta > 0$. Now suppose x lies in the closure of $X^*(Q_0)_{\mathbf{R}}^+$. Then one has

$$w_1 x - w_2 x = w_1 x - \left(w_1 x - 2\frac{(w_1 x, \beta)}{(\beta, \beta)}\beta\right) = 2\frac{(x, w_1^{-1}\beta)}{(\beta, \beta)}\beta,$$

and this proves the assertion. □

To describe the arithmetic stratification, consider those $a_{\mu,\lambda}^w$ which are finite and order them:

$$\{a_{\mu,\lambda}^w \mid w \in \mathscr{W}^M, a_{\mu,\lambda}^w < \infty\} = \{a_0, a_1, \ldots, a_{r_{\mu,\lambda}}\},$$

with $a_0 > a_1 > \ldots > a_{r_{\mu,\lambda}}$ and put formally $a_{-1} = \infty$.

Theorem 3.2. — *Let $Y = Y_\eta$ be the fibre bundle over $W = P\backslash G$ defined in (1.1), and suppose η fulfills the assumption of (1.4).*

a) If $L_\mu^Y \otimes \pi^* L_\lambda$ is not an element of $\Lambda(Y)_{\mathrm{eff}}^\circ$, then there is no open nonempty subset such that the corresponding height zeta function converges for $s \gg 0$.

b) Suppose $L_\mu^Y \otimes \pi^* L_\lambda$ lies in $\Lambda(Y)_{\mathrm{eff}}^\circ$. For $i = 0, 1, \ldots, r = r_{\mu,\lambda}$ put

$$Y_i = Y - \bigcup_{w \in \mathscr{W}^M,\, a_{\mu,\lambda}^w = a_{i-1}} \overline{Y^w}.$$

Then the arithmetic stratification of Y with respect to $L_\mu^Y \otimes \pi^* L_\lambda$ is

$$Y \supset Y_0 \supset Y_1 \supset \ldots \supset Y_r.$$

Proof. — First we prove the second assertion. Lemma 3.1 implies that

$$Y_i = \bigcup_{w \in \mathscr{W}^M,\, a_{\mu,\lambda}^w \leq a_i} \overline{Y^w}.$$

Hence we conclude that Y_{i+1} is properly contained in Y_i ($i = 0, ..., r-1$). Let $\sigma_{Y_i}(L)$ be the abcissa of convergence of $Z_{Y_i}(\mathbf{L}, s)$ with $\mathbf{L} = \mathbf{L}_\mu^Y \otimes \pi^* \mathbf{L}_\lambda$. We shall show that $\sigma_{Y_i}(L) = a_i$. Obviously,

$$Z_{Y_i}(\mathbf{L}, s) = \sum_{w \in \mathscr{W}^M,\, a_{\mu,\lambda}^w \leq a_i} Z_{Y^w}(\mathbf{L}, s).$$

By Proposition 2.9 b), $Z_{Y^w}(\mathbf{L}, s)$ converges for $s > a_{\mu,\lambda}^w$. This shows that $Z_{Y_i}(\mathbf{L}, s)$ converges for $s > a_i$. On the other hand, there is a $w \in \mathscr{W}^M$ such that $a_{\mu,\lambda}^w = a_i$ and in order for $Z_{Y^w}(\mathbf{L}, s)$ to converge it is necessary that $s > \breve{a}_{\mu,\lambda}^w$. Because the closure of X° is X, we have $a_{\mu,\lambda}^\circ = a_r$, by Lemma 3.1, and $Z_{Y^\circ}(\mathbf{L}, s)$ is a summand of $Z_{Y_i}(\mathbf{L}, s)$. By Proposition 2.5 c), we know that for $Z_{Y^\circ}(\mathbf{L}, s)$ to converge it is necessary that $s > a_\mu$. Hence, for $Z_{Y_i}(\mathbf{L}, s)$ to converge it is necessary that $s > \max\{\breve{a}_{\mu,\lambda}^w, a_\mu\} = a_{\mu,\lambda}^w = a_i$ and we conclude that the abcissa of convergence of $Z_{Y_i}(\mathbf{L}, s)$ is a_i.

Fix $i \in \{-1, \ldots, r\}$ and consider an open subset $Y' \subset Y_i$ which contains Y_{i+1} properly ($Y_{-1} = Y$, $Y_{r+1} = \emptyset$). The case $i = -1$ can only occur if $Z_Y(\mathbf{L}, s)$ does not converge for any $s > 0$. We have to show that $Z_{Y'}(\mathbf{L}, s)$ converges only for $s > a_i$ (and then this sum converges for all $s > a_i$). By hypothesis, Y' properly contains the open subset Y_{i+1}, hence there is a $w \in \mathscr{W}^M$ such that $a_{\mu,\lambda}^w = a_i$ und $Y' \cap Y^w \neq \emptyset$. $Y' \cap Y^w$ is an open and dense subset of Y^w. Therefore, there are open and dense subsets $X' \subset X^w$, $G' \subset G$, such that the image of $X' \times G'$ under the canonical projection $X \times G \to Y$ is contained in $Y' \cap Y^w$. Moreover, we may and we will assume that $PG' \subset G'$. Let $H' \subset H$ be the preimage of X' under the projection $H \to X$. For $h \in H(\mathbf{A})$ put

$$E_Q^{H'}(s\mu, h) = \sum_{\delta \in Q(F) \backslash H'(F)} e^{\langle s\mu, H_Q(\delta t)\rangle}.$$

Then we have

(3.2.1) $\quad Z_{Y'}(\mathbf{L}, s) \geq \sum_{\gamma \in P(F) \backslash G'(F)} e^{\langle s\lambda, H_P(\gamma)\rangle} E_Q^{H'}(s\mu, \eta(p_\gamma)).$

For each $\vartheta \in T_0(F)$ the set $H'\vartheta$ is open in $Qw^{-1}Q_0$, hence $\tilde{H} = \cup_{\vartheta \in T_0(F)} H'\vartheta$ is an open subset of $Qw^{-1}Q_0$ and there exist $\vartheta_1, \ldots, \vartheta_m \in T_0(F)$ such that $\tilde{H} = \cup_{1 \leq j \leq m} H'\vartheta_j$. For $h \in H(\mathbf{A})$ we define as above

$$E_Q^{\tilde{H}}(s\mu, h) = \sum_{\delta \in Q(F) \backslash \tilde{H}(F)} e^{\langle s\mu, H_Q(\delta t)\rangle}.$$

Fix $a > a_\mu$. By Lemma 2.1 there is a $c_1 > 0$ such that for all $s \in (a_\mu, a]$ and $t \in T_0(\mathbf{A})$

(3.2.2)
$$E_Q^{H'}(s\mu, t) \geq c_1 E_Q^{\tilde{H}}(s\mu, t).$$

Note that for all $\vartheta \in T_0(F)$ and $h \in H(\mathbf{A})$ one has $E_Q^{\tilde{H}}(s\mu, \vartheta h) = E_Q^{\tilde{H}}(s\mu, h)$. For each $\eta \in Q_0(F)$ the set $\tilde{H}\eta$ is open in $Qw^{-1}Q_0$. Therefore, there are $\eta_1, \ldots, \eta_n \in Q_0(F)$ such that $Qw^{-1}Q_0 = \cup_{1 \leq k \leq n} \tilde{H}\eta_k$. Write $\eta(p_\gamma) = \vartheta_\gamma a_\gamma o_\gamma$ with $\vartheta_\gamma \in T_0(F)$, $o_\gamma \in \Omega_0$ and $a_\gamma \in A_{M_0}(t_0)$, where t_0 is as in Lemma 2.7. The set $\{a_\gamma^{-1} \eta_j a_\gamma o_\gamma \,|\, \gamma \in P(F)\backslash G'(F), j = 1, \ldots n\,\}$ is relatively compact in $H(\mathbf{A})$. So we can find $c_2 > 0$ such that for all $\gamma \in P(F)\backslash G'(F)$ and $j = 1, \ldots, n$ we have
$$E_Q^{\tilde{H}}(s\mu, \eta_j a_\gamma o_\gamma) \geq c_2 E_Q^{\tilde{H}}(s\mu, a_\gamma o_\gamma) = c_2 E_Q^{\tilde{H}}(s\mu, \eta(p_\gamma)).$$
Replacing nc_2 by c_2 we even get the following estimate

(3.2.3)
$$E_Q^{\tilde{H}}(s\mu, \eta(p_\gamma)) \geq c_2 E_Q^{Qw^{-1}Q_0}(s\mu, \eta(p_\gamma))$$

Putting (3.2.1), (3.2.2) and (3.2.3) together and using Proposition 2.5 we have
$$Z_{Y'}(\mathbf{L}, s) \geq c_1 c_2 \sum_{\gamma \in P(F)\backslash G'(F)} e^{\langle s\lambda, H_P(\gamma)\rangle} E_Q^{Qw^{-1}Q_0}(s\mu, \eta(p_\gamma))$$
$$\geq c_3 \sum_{\gamma \in P(F)\backslash G'(F)} e^{\langle s(\lambda + \check{\eta}(w\mu)) - \check{\eta}(w\rho_0 - \rho_0), H_P(\gamma)\rangle},$$

with a suitable $c_3 > 0$. Because G' is dense in G there is $c_4 > 0$ such that
$$\sum_{\gamma \in P(F)\backslash G'(F)} e^{\langle s(\lambda + \check{\eta}(w\mu)) - \check{\eta}(w\rho_0 - \rho_0), H_P(\gamma)\rangle} \geq c_4 E_P^G(s(\lambda + \check{\eta}(w\mu)) - \check{\eta}(w\rho_0 - \rho_0)).$$

(The argument is the same as in the proof of Prop. 2.5 b).) This shows that the abcissa of convergence of $Z_{Y'}(\mathbf{L}, s)$ is not less than $\check{a}_{\mu,\lambda}^w$. Because Y' is dense in Y, the intersection $Y' \cap Y^\circ$ is not empty. By the same reasoning as above we find $c_5 > 0$ such that
$$Z_{Y'}(\mathbf{L}, s) \geq c_5 \sum_{\gamma \in P(F)\backslash G'(F)} e^{\langle s\lambda, H_P(\gamma)\rangle} E_Q^{Q(w_H^M)^{-1} Q_0}(s\mu, \eta(p_\gamma)).$$

Again by Proposition 2.5 b), $E_Q^{Q(w_H^M)^{-1} Q_0}(s\mu, \cdot)$ converges only for $s > a_\mu$, so we can conclude that $Z_{Y'}(\mathbf{L}, s)$ converges only for $s > \max\{\check{a}_{\mu,\lambda}^w, a_\mu\} = a_{\mu,\lambda}^w = a_i$, and this proves the second statement.

To prove the first assertion, observe that we have just seen that the height zeta function of any open non-empty subset can be bounded from below by the height zeta function of Y°. But this series does not converge for any $s > 0$ if $L_\mu^Y \otimes \pi^* L_\lambda$ is not in the interior of the positive cone, by proposition 2.9. $\square$

References

[BGG] I. N. Bernstein, I. M. Gelfand and S. I. Gelfand, *Schubert cells and the cohomology of the spaces G/P*, Russian Math. Surveys **28** (1973), no. 3, 1–26.

[BM] V. V. Batyrev and Yu. I. Manin, *Sur le nombre des points rationnels de hauteur bornée des variétés algébriques*, Math. Ann. **268** (1990), 27–43.

[Bou] N. Bourbaki, *Groupes et algèbres de Lie, chapitres 4,5 et 6*, Hermann, Paris, 1968.

[FMT] J. Franke, Yu. I. Manin and Yu. Tschinkel *Rational points of bounded height on Fano varieties*, Invent. math. **95** (1989), 421–435.

[Go] R. Godement, *Introduction à la théorie de Langlands*, Séminaire Bourbaki, exposé 321 (1966/67).

[J] J.C. Jantzen, *Representations of algebraic groups*, Academic press, Orlando, Florida, 1987.

[MW] C. Mœglin and J.-L.Waldspurger *Décomposition spéctrale et séries d'Eisenstein*, Birkhäuser, Basel, 1994.

[P] E. Peyre, *Hauteurs et mesures de Tamagawa sur les variétés de Fano*, Duke Math. Journal **79** (1995), 101–218.

[Sa] J.-J. Sansuc, *Groupe de Brauer et arithmétique des groupes algébriques linéaires sur un corps de nombres*, Journal f. d. reine u. angewandte Math. **327** (1981), 12–80.

[St] M. Strauch, *Rational points on twisted products of flag varieties*, in preparation.

Rational points on algebraic varieties
(E. PEYRE, Y. TSCHINKEL, ed.), p. 357–404

WEAK APPROXIMATION AND R-EQUIVALENCE ON CUBIC SURFACES

Sir Peter Swinnerton-Dyer

Abstract. — Let V be a nonsingular cubic surface defined over an algebraic number field K, and assume that V has points in every completion K_v. There is a long-standing problem of finding the obstructions to the Hasse principle and to weak approximation on V, the conjecture in each case (due to Colliot-Thélène and Sansuc) being that the obstruction is just the Brauer-Manin obstruction. The latter is known to be computable, though the algorithm is somewhat ugly and a heuristic process is usually preferable. Another way of phrasing the same problem is to ask what is the adelic closure of the set $V(K)$.

A partial answer to this question is given by the following theorem: to each place v of bad reduction one can associate a finite disjoint union

$$V(K_v) = \cup W_j^{(v)}$$

which is easily computable in any particular case. The v-adic closure of any R-equivalence class in $V(K)$ is a set U_{iv} which is the union of some of the $W_j^{(v)}$; and the adelic closure of any R-equivalence class is of the form $\prod' U_{iv} \times \prod'' V(K_v)$, where i depends on v, the first product is over all places of bad reduction and the second product is over all places of good reduction for V. Thus the adelic closure of $V(K)$ is a union of sets $\prod' W_j^{(v)} \times \prod'' V(K_v)$. For specific V a search program will give those products which contain a point of $V(K)$ and in which points of $V(K)$ are therefore everywhere dense. For a product which appears not to contain a point of $V(K)$, it is reasonable to hope that there is a Brauer-Manin obstruction. For all V for which this process has been used, it turns out that one can indeed find the exact adelic closure of $V(K)$ in this way. This is illustrated in the final section.

1. Introduction

The origin of this paper is a remark of Heath-Brown [2], that for an explicitly given cubic surface defined over $\mathbf{Q}$ it should be possible to show that the Brauer-Manin obstruction is the only obstruction to weak approximation — provided that statement is in fact true. He suggested that this could be done by studying particular parametric solutions — or, which comes to almost the same thing, particular R-equivalence classes; and we shall see that this is true (and indeed true over an arbitrary algebraic number field), though in general more than one class is needed.

Let V be a nonsingular cubic surface defined over an algebraic number field K; then R-equivalence on V is defined as the finest equivalence relation such that two points given by the same parametric solution are equivalent. Let ℓ be any line not lying in V and meeting V in three points P_1, P_2, P_3, each defined over K but not necessarily distinct. Following Manin, if $\mathscr{C}_i$ is the R-equivalence class of P_i we shall write $\mathscr{C}_3 = \mathscr{C}_1 \circ \mathscr{C}_2$. Let P be a point of an equivalence class $\mathscr{C}$ which does not lie on any of the lines of V, and let Γ be the intersection of V with the tangent plane to V at P. Thus Γ is an irreducible cubic with a singularity at P. Every point P_1 of $\Gamma(K)$ other than P is in $\mathscr{C} \circ \mathscr{C}$, and by taking ℓ to be the tangent to Γ at P_1 we obtain

$$(1) \qquad (\mathscr{C} \circ \mathscr{C}) \circ (\mathscr{C} \circ \mathscr{C}) = \mathscr{C} \circ \mathscr{C}.$$

Still following Manin, we define *universal equivalence* to be the finest equivalence relation in $V(K)$ such that $P_1 \sim P_2$ whenever $P \circ P_1 \sim P \circ P_2$ for some P. This is not trivial, because $Q_1 \circ Q_2$ is many-valued if $Q_1 = Q_2$ or if Q_1 and Q_2 lie on the same line of V.

The main object of this paper is to describe the closure of any R-equivalence class $\mathscr{C}$ in the adelic topology on V.

Theorem 1. — *Let V be a nonsingular cubic surface defined over an algebraic number field K. There is a finite set $\mathscr{B}$ of places of K and for each v in $\mathscr{B}$ a partition*

$$(2) \qquad V(K_v) = U_{1v} \cup \ldots \cup U_{nv},$$

where the U_{iv} are disjoint open sets in the v-adic topology and n depends on v, with the following property. If $\mathscr{C}$ is any R-equivalence class in $V(K)$ then the closure of $\mathscr{C}$ in the adelic topology has the form $\prod U_{iv} \times \prod V(K_v)$, where the first product is over the places in $\mathscr{B}$, the second product is over the places not in $\mathscr{B}$, and each i depends on v and $\mathscr{C}$.

In more low-brow language, Theorem 1 asserts that for the problem of weak approximation within $\mathscr{C}$ only the places in $\mathscr{B}$ matter and there is no interaction between distinct places. It follows that the first of these statements holds also for weak approximation over K, though it is well known that the second does not. It is implicit in this (and the corresponding local result is a necessary step in the proof) that if $\mathscr{C}_1$ and $\mathscr{C}_2$ are any two R-equivalence classes their adelic closures are either disjoint or identical. In the former case $\mathscr{C}_1$ and $\mathscr{C}_2$ are easily separated by congruence conditions. Colliot-Thélène has pointed out to me that there are known to be Châtelet surfaces on which there are distinct R-equivalence classes with the same adelic closure, and from these one can derive cubic surfaces with the same property. Indeed except in very special cases it is not even known whether the number of R-equivalence classes is finite.

The set $\mathscr{B}$ in Theorem 1 is actually the set of places v of bad reduction of V in the following sense. Let $\mathfrak{p}$ be a finite prime in K. After multiplying the equation of V by a suitable constant (depending on $\mathfrak{p}$) we can assume that all its coefficients are integers at $\mathfrak{p}$ and at least one of them is a unit at $\mathfrak{p}$. We define $\tilde{V} = \tilde{V}_\mathfrak{p}$ to be the surface defined over the residue field mod $\mathfrak{p}$ whose equation is the reduction mod $\mathfrak{p}$ of the equation of V. We now say that v is a *place of bad reduction* for V if either

(i) : v is finite and $\tilde{V}$ has a singularity or a multiple component, or
(ii) : v is a real infinite place and $V(K_v)$ has more than one component.

The set of places of bad reduction for V may be changed if one makes a linear transformation in the ambient space; but clearly this would not affect the smallest set $\mathscr{B}$ for which Theorem 1 holds. There is scope for a more sophisticated approach, which would involve developing a theory of Néron models for cubic surfaces.

In this paper I only give the proof of Theorem 1 when $\mathscr{B}$ is the larger set which also contains the primes of Norm 2,3 or 4; as so often, much of the difficulty in the proof comes from the small primes. This raises the further question: if $\tilde{V}$ is not a cone and does not contain a line of singular points, and if P_1, P_2 are points of $V(K_\mathfrak{p})$ whose reductions mod $\mathfrak{p}$ are nonsingular on $\tilde{V}$, does it follow that they lie in the same component of the partition (2)? This is certainly true except perhaps when $\mathfrak{p}$ is small.

Each element of the partition (2) is either the closure in $V(K_v)$ of some R-equivalence class or the set of points of $V(K_v)$ not in the closure of $V(K)$. I do not know an algorithm for constructing the partition (2); but there is a wholly explicit procedure for constructing a finite refinement of it, and this is good enough for applications. It is probable that the algorithm used does not generally give the actual partition (2); when it does not do so, how to generate the correct coarsening is an interesting and probably difficult open question. The main complication comes from the possible existence of sporadic parametric solutions. For example, over the algebraic closure $\bar{K}$ there are in general ∞^5 quadrics which touch V at 4 points, and whose intersection with V is therefore a parametrisable curve; but nothing is known about which if any of them are defined over K.

In what follows, if L, Q or C is a polynomial in the coordinates $X_0, \ldots, X_3$ then L will always be linear, Q quadratic and C cubic, and their coefficients will be integral at any relevent prime; but each of these letters is also used for other purposes. All constants implicit in the O notation will be absolute. Except in the purely geometric §2, we shall also adopt the following conventions. If $\mathfrak{p}$ is a finite prime of K, we denote by $\mathfrak{O}_\mathfrak{p}$ the ring of elements of K integral at $\mathfrak{p}$, by k the residue field $\bmod \mathfrak{p}$ and by q the number of elements of k; and a tilde will denote the reduction $\bmod \mathfrak{p}$ of anything defined over K. When we write an equation for V, it will always be assumed that all the coefficients are in $\mathfrak{O}_\mathfrak{p}$. At least one of them will be a unit at $\mathfrak{p}$; so we obtain the equation of $\tilde{V}$ simply by reducing all the coefficients $\bmod \mathfrak{p}$. If we define something over k otherwise than by reduction $\bmod \mathfrak{p}$, we shall denote it by a lower case letter; and the corresponding capital letter will denote some lift of it to K. We use the same conventions occasionally for objects defined over extensions of K and k; but this will always be made clear.

The structure of this paper is as follows. In §2 we prove a miscellany of geometric results which are needed later in the paper; some of them are certainly known but not well known. The next four sections cover the local theory. In §3 we consider the archimedean places, for which the partition (2) is simply into the connected components of $V(K_v)$. For finite primes a crucial weapon in the argument is the process of moving a curve sideways on V; §4 describes this process and provides quantitative estimates of how nearly parallel the new curve is to the old one. The process enables one to show that if $\mathscr{C}$ is an R-equivalence class in $V(K)$ then its closure in the p-adic topology is open. This, and the fact proved in Lemma 11 that the closures of two R-equivalence classes are either coincident or disjoint, are needed to prove the existence and key properties of the partition (2). In §5 we prove that if $\tilde{V}$ is nonsingular and $V(K)$ is not empty then every point of $\tilde{V}(k)$ is liftable to $V(K)$. At the end of §5 we prove that under the same conditions every point of $\tilde{V}(k)$ is liftable to each class of $V(K)$.

We shall say that a point p of $\tilde{V}(k)$ is *densely liftable* to an R-equivalence class $\mathscr{C}$ if $\mathscr{C}$ is dense in the inverse image of p under the map $V(K_\mathfrak{p}) \to \tilde{V}(k)$; note that to say simply that p is densely liftable means that it is densely liftable to some class $\mathscr{C}$ and not simply to $V(K)$. In §6, which depends on the techniques of §4, we prove that if $\tilde{V}$ is nonsingular and $V(K)$ is not empty then every point of $\tilde{V}(k)$ is densely liftable. It now follows from the final result of §5 that every point of $\tilde{V}(k)$ is actually densely liftable to each class of $V(K)$; at the end of §6 we give an alternative proof, which contains ideas which may be valuable even when $\tilde{V}$ is singular. Thus the partition (2) is trivial when $\tilde{V}$ is nonsingular. The proof of Theorem 1 is now reduced to the global analogue of Lemma 11; this occupies §7. In the final section of the paper we specialize to the surfaces

(3) $$f = X_1^3 + X_2^3 + X_3^3 - dX_0^3 = 0$$

which occur for $d = 2, 3$ in Heath-Brown's paper; in particular we show that for Heath-Brown's surfaces the Brauer-Manin obstruction is indeed the only obstruction to weak approximation.

I am indebted to Jean-Louis Colliot-Thélène for a number of valuable comments.

2. Geometric background

This section contains various geometric statements which will be needed elsewhere in the paper. Throughout it, W will be an absolutely irreducible cubic surface, not necessarily nonsingular, defined over a field L. We use the language of classical geometry, so that the objects introduced may not all be defined over L.

Lemma 1. — *Let Λ be a line on W and Π any plane through Λ. If Π is the tangent to W at more than two points of Λ then Λ contains a singular point of W.*

Proof. Take coordinates so that Λ is $X_2 = X_3 = 0$ and Π is $X_3 = 0$. The equation of W will have the form

(4) $$X_0^2 L_0(X_2, X_3) + X_0 X_1 L_2(X_2, X_3) + X_1^2 L_1(X_2, X_3) + \ldots = 0$$

where the remaining terms are at most linear in X_0, X_1 together. If Π is the tangent to W at $(x_0, x_1, 0, 0)$ then

$$x_0^2 L_0(1, 0) + x_0 x_1 L_2(1, 0) + x_2^2 L_1(1, 0) = 0.$$

Thus Π can only be the tangent to W at more than two points of Λ if the equation (4) reduces to $X_3 Q(X_0, X_1) + \ldots = 0$; and in this case W has a singular point where $Q(X_0, X_1) = X_2 = X_3 = 0$. □

Let P be a point of W; what can we say about the set of points P_1 on W such that the line PP_1 touches W at P_1? Take coordinates so that the equation of W is $f(X_0, \ldots, X_3) = 0$ and P is $(1, 0, 0, 0)$. The points P_1 are those which satisfy $\partial f/\partial X_0 = 0$. If this does not hold identically, it defines a (possibly decomposable) quadric; and the intersection of this with W breaks up into at most six curves, each of which has absolutely bounded genus. In particular, if L is the finite field of q elements then there are only $O(q)$ such points P_1 in $W(L)$.

If $\partial f/\partial X_0$ vanishes identically then the choice of P implies that f cannot contain a term of the form aX_0^3, so we have just two possibilities:

(i) : f is independent of X_0, so that W is a cone with vertex P;
(ii) : char $L = 2$ and after a linear transformation on X_1, X_2, X_3 we can take

$$f = X_0^2 X_3 + C(X_1, X_2, X_3).$$

In both these cases P_1 can be any point of W. If (i) holds then P is a singular point of W. If (ii) holds then the singular points of W are given by

$$\partial C/\partial X_1 = \partial C/\partial X_2 = \partial C/\partial X_3 + X_0^2 = 0,$$

so that W has singular points though P is not one of them. In particular we have proved

Lemma 2. — *Suppose that W is nonsingular and let P be a point on W. Then the points P_1 on W such that the tangent to W at P_1 passes through P all lie on a certain (possibly decomposable) quadric. If L is the finite field of q elements then there are only $O(q)$ such points in $W(L)$.* □

Lemma 3. — *If W is nonsingular, P is on W and ℓ is a line through P on W then ℓ cannot be a multiple component of the intersection of W with the tangent to W at P.*

Proof. If not, take coordinates so that P is $(1, 0, 0, 0)$, ℓ is $X_2 = X_3 = 0$, the tangent to W at P is $X_3 = 0$ and its intersection with W contains the line ℓ with multiplicity at least 2. The equation of W has the form

$$X_3 Q(X_0, \ldots, X_3) + X_2^2 L(X_0, X_1, X_2) = 0;$$

and any point $(x_0, x_1, 0, 0)$ with $Q(x_0, x_1, 0, 0) = 0$ is singular on W. □

Recall that an Eckardt point of W is a nonsingular point P of W such that the tangent plane to W at P meets W in three lines through P. If W has a singular point on the tangent plane, these lines need not be distinct.

Lemma 4. — *Suppose that W is nonsingular and Λ is a line on W. Then Λ contains at most 5 Eckardt points if $\text{char}(L) = 2$ and at most 2 otherwise.*

Proof. Take coordinates so that Λ is $X_2 = X_3 = 0$ and contains the two Eckardt points $P_0 = (1, 0, 0, 0)$ and $P_1 = (0, 1, 0, 0)$. We need to distinguish two cases, according as the tangents to W at P_0 and P_1 are the same or different. If they are the same, we can take them both to be $X_3 = 0$; it follows that the equation of W must have the form

(5) $$X_3 Q(X_0, \ldots, X_3) + X_2^3 = 0$$

where the coefficients of X_0^2 and X_1^2 in Q are nonzero. But now the intersection of W with the tangent at P_0 (or P_1) is the line Λ counted three times and this contradicts Lemma 3. If instead the tangents at P_0 and P_1 are distinct then we can take them to be $X_3 = 0$ and $X_2 = 0$ respectively. Now the equation of W must have both the forms

$$X_3 Q_0'(X_0, \ldots, X_3) + C_0(X_1, X_2) = X_2 Q_1'(X_0, \ldots, X_3) + C_1(X_0, X_3) = 0$$

where the coefficients of X_0^2 in Q_0' and X_1^2 in Q_1' are nonzero. Hence the equation of W has the form

(6) $$f = X_3 Q_0(X_0, X_3) + X_2 Q_1(X_1, X_2) + X_2 X_3 L(X_0, \ldots, X_3) = 0$$

where c_0 and c_1, the coefficients of X_0^2 in Q_0 and of X_1^2 in Q_1, are nonzero. Conversely, P_0 and P_1 are Eckardt points on (6). The tangent to (6) at a point $(a_0, a_1, 0, 0)$ is $c_0 a_0^2 X_3 + c_1 a_1^2 X_2 = 0$; so the line $X_2 = X_3 = 0$ contains no singular points. If $(a_0, a_1, 0, 0)$ is another Eckardt point on Λ, so that a_0, a_1 are nonzero, then we could also write f in the form

(7) $$Y Q_0^*(X_0, Y) + X_2 Q_1^*(a_0 X_1 - a_1 X_0, X_2) + Y X_2 L^*(X_0, \ldots, X_3)$$

where $Y = c_0 a_0^2 X_3 + c_1 a_1^2 X_2$. This forces $\mathrm{char}(L) = 2$, because otherwise (7) would contain a term in $X_0 X_1 X_2$ whereas (6) does not. But if $\mathrm{char}(L) = 2$ then (7) is equivalent to

$$L^* = L^\sharp(a_0 X_1 - a_1 X_0, X_2, X_3), \quad c_0^2 a_0^3 d_1 + c_1^2 a_1^3 d_0 = 0$$

where d_0, d_1 are the middle coefficients of Q_0, Q_1. In general the first condition allows only one value of a_0/a_1, but if $L^* = L^\sharp(X_2, X_3)$ the first condition is identically satisfied. If also d_0, d_1 are both nonzero there are five distinct Eckardt points on $X_2 = X_3 = 0$. If d_0, d_1 both vanish then W has the form

$$c_0 X_0^2 X_3 + c_1 X_1^2 X_2 + C(X_2, X_3) = 0$$

and every point of $X_2 = X_3 = 0$ is an Eckardt point; but now

$$c_0 X_0^2 + \partial C/\partial X_3 = c_1 X_1^2 + \partial C/\partial X_2 = 0$$

determines a line of singular points of W. $\square$

Lemma 5. — *Suppose that W is not a cone and does not contain a line of singular points. Then W contains at most 27 lines and at most 4 singular points; it contains at most 45 Eckardt points unless it has the special form (5), and at most 18 Eckardt points if $\operatorname{char}(L) \neq 2$.*

Proof. Let $X_0 = X_1 = 0$ be a line on W; the general plane Π through it has the form $\mu X_0 = \lambda X_1$, so we can write $X_0 = \lambda X_4$, $X_1 = \mu X_4$ and use X_2, X_3, X_4 as coordinates in Π. The residual intersection of Π with W is obtained by substituting for X_0, X_1 in the equation of W and dividing by X_4, so its equation has the form

(8) $\quad a_{22} X_2^2 + a_{23} X_2 X_3 + a_{33} X_3^2 + a_{24} X_2 X_4 + a_{34} X_3 X_4 + a_{44} X_4^2 = 0$

where a_{22}, a_{23}, a_{33} are linear in λ and μ, a_{24} and a_{34} are quadratic and a_{44} is cubic. This equation defines a union of lines if and only if its discriminant vanishes, and the latter is of degree 5 in λ, μ. Thus either there are at most 5 planes through $X_0 = X_1 = 0$ whose intersection with W is a union of lines or every such plane has this property — the latter being the case when the discriminant vanishes identically.

Suppose first that the discriminant does vanish identically. We distinguish five cases, according as for general λ, μ the equation (8) represents

(i) : two distinct lines whose common point is not on $X_4 = 0$,
(ii) : two distinct lines meeting on $X_4 = 0$, neither being $X_4 = 0$,
(iii) : a double line other than $X_4 = 0$,
(iv) : the line $X_4 = 0$ and another line,
(v) : the line $X_4 = 0$ taken twice.

In case (i) let P be the point of intersection of the two lines; then in the original coordinates P has the form $(\lambda f_1, \mu f_1, f_2, f_3)$ where the f_i are in $L[\lambda, \mu]$ and f_1 does not vanish identically. Let C be the locus of P as λ/μ varies; Π is transversal to C when P is not on $X_4 = 0$, so C is nonsingular at P and the tangent line to C at P does not lie in Π. Thus P is a singular point of W and C is a curve of singular points of W; since the join of any two singular points of W lies in W and W is irreducible, C must be a line, contrary to hypothesis. In case (ii) the discriminant of $a_{22} X_2^2 + a_{23} X_2 X_3 + a_{33} X_3^2$ must vanish; but this form cannot vanish identically since we are not in case (iv) or (v). There are now two alternatives: either the form is the product of a linear form in $L[\lambda, \mu]$ and the square of a linear form in $L[X_2, X_3]$ or $\operatorname{char}(L) = 2$, $a_{23} = 0$ and a_{22}/a_{33} is not constant. If the first alternative holds, then by a linear transformation on X_2, X_3 with constant coefficients we can make $a_{23} = a_{33} = 0$, whence the fact that (8) is a pair of lines implies $a_{34} = 0$; now the equation of W does not involve X_3, so that W is a cone. If the second alternative holds, then (8) can only factorize if $(a_{24}X_2 + a_{34}X_3)^2$ is proportional to $a_{22}X_2^2 + a_{33}X_3^2$; this requires $a_{24} = a_{34} = 0$ and we would be in case (iii). In case (iii) we have the same alternatives as in case (ii), and the reason for rejecting the first alternative remains valid; hence as above we can assume that char$(L) = 2$ and that the equation of W has the form

$$X_0X_2^2 + aX_1X_3^2 = f(X_0, X_1)$$

for some $a \neq 0$. But now every point of the ambient space with

$$X_2^2 = \partial f/\partial X_0, \quad aX_3^2 = \partial f/\partial X_1$$

is a singular point of W; so we have a curve of singular points, and as before this is impossible. In cases (iv) and (v) $a_{22} = a_{23} = a_{33} = 0$, so every point of the line $X_0 = X_1 = 0$ is singular on W.

Henceforth we can assume that given any line ℓ on W there are at most 5 planes through ℓ whose intersection with W is a union of lines; and there is at least one such plane. Among the planes which meet W in a union of lines, let Π_0 be one which involves the smallest number $n \leqslant 3$ of distinct lines. Any line ℓ which lies in W but not in Π_0 meets Π_0 in a point which lies on some line ℓ_0 in $\Pi_0 \cap W$; and we have already seen that there are at most 8 lines on W which meet ℓ_0 but do not lie in Π_0. Hence there are altogether at most $8n + n \leqslant 27$ lines on W. Except in the special case (5), an Eckardt point on W lies on at least n distinct lines of W, and there are at most 5 Eckardt points on any line, and at most 2 if char$(L) \neq 2$. Hence there are at most $5(9n)/n = 45$ Eckardt points on W, and at most $2(9n)/n = 18$ if char$(L) \neq 2$. I have no reason to think that either of these estimates is best possible.

If W has more than one singular point, then we can take coordinates so that two of them are $P_0 = (1, 0, 0, 0)$ and $P_1 = (0, 1, 0, 0)$. Thus none of the monomials which appear in the equation of W is of degree greater than 1 in X_0 or in X_1, and the entire line P_0P_1 lies in W. If the equation of W contains a term $aX_0X_1X_2$ or $aX_0X_1X_3$, no other point of P_0P_1 is singular; otherwise every point of P_0P_1 is singular. Thus under the hypotheses of the Lemma W cannot contain three collinear singular points. If W contained four coplanar singular points $P_0, \ldots, P_3$ the plane of these points would meet W in at least the six distinct lines P_iP_j, which is impossible. Finally, suppose that W contains five singular points $P_0, \ldots, P_4$ no four of which are coplanar. The point Q at which the line P_3P_4 meets the plane $P_0P_1P_2$ cannot for example lie on P_1P_2 because that would imply that $P_1, \ldots, P_4$ are coplanar; so W meets the plane $P_0P_1P_2$ in three lines and at least one additional point Q, which again is impossible. □

If P is a nonsingular point of $W(L)$, we shall need various constructions for generating further such points. The importance of the first one is that it works even if P is an Eckardt point and none of the three lines through it is defined over L. Note that though we assume $\mathrm{char}(L) = 0$ for the conclusion of the Lemma, we do not use this in the construction. Indeed, we are also interested in using this construction when L is a finite field; see Lemma 12 and the end of the proof of Theorem 5. However we then need to assume or demonstrate the existence of a line ℓ defined over L and having the properties stated in the second sentence of the proof; and we need to show that ℓ_1 can be chosen so that the resulting point R is not simply P. The proof follows a remark of Manin ([3], p2). The idea goes back to Richmond and in principle to Ryley; see the notes to §13.6 in [1].

Lemma 6. — *Suppose that W is not a cone, that there is a point P of $W(L)$ nonsingular on W and that $\mathrm{char}(L) = 0$; then $W(L)$ is Zariski dense in W, and the same holds for each R-equivalence class in $W(L)$.*

Proof. We may assume that W does not contain a line of singular points; for if $X_0 = X_1 = 0$ were such a line then the equation of W would be inhomogeneous linear in X_0, X_1 together and it would be trivial to parametrize W. By hypothesis, the general line ℓ through P does not lie on or touch W, and it does not meet any line of W except perhaps at P. Since L is infinite we can find such a line ℓ which is defined over L. Denote by P', P'' the other points where ℓ meets W, so that P', P'' are distinct and distinct from P. Both P' and P'' are nonsingular on W, because they have multiplicity one in $\ell \cdot W$. They are either each defined over L or each defined over a quadratic extension of L and conjugate over L. Let Π' be the tangent plane to W at P' and Γ' the intersection of Π' with W, and similarly for Π'', Γ''; by hypothesis Γ', Γ'' are absolutely irreducible cubic curves with singularities at P', P'' respectively. Let ℓ_1 be any line in the ambient space not passing through P' or P'' nor lying on Π' or Π''; and let Q', Q'' be the intersections of ℓ_1 with Π', Π'' respectively. Let R' be the third intersection of $P'Q'$ with Γ' and similarly for R''; we assume ℓ_1 so chosen that neither R' nor R'' is one of the points where $\Pi' \cap \Pi''$ meets W, nor does the line $R'R''$ lie in W. Let R be the third intersection of $R'R''$ with W; if ℓ_1 is defined over L then so are $R'R''$ and R.

Conversely, let R be in general position on W and let Γ_1 be the projection of Γ' from R onto Π''. The intersections of Γ_1 and Γ'' at the points where Π' meets Γ'' contribute only $\deg(\Gamma' \cdot \Gamma'') = 3$ to $\deg(\Gamma_1 \cdot \Gamma'') = 9$; so there are points R' on Γ' and R'' on Γ'' with $R' \neq R''$ and $R'R''$ passing through R. Moreover R', R'' must be in general position on Γ', Γ'' respectively. By taking $R'R''$ to be ℓ_1, we see that the set of R constructed in this way contains a Zariski open subset of W. Hence the same holds for the set of all R obtained when ℓ_1 is in general position. But since L is infinite, the set of lines ℓ_1 defined over L is

Zariski dense in the set of all lines. Because the space of lines in $\mathbf{P}^3$ is rational over the prime field, all the points R constructed in the proof of Lemma 6 lie in the same R-equivalence class $\mathscr{C}_0$. Projecting from a point of $\mathscr{C}_0 \circ \mathscr{C}$ now shows that each class $\mathscr{C}$ is Zariski dense on W. □

Remark. Suppose that W and P are defined over an algebraic number field K, and let $\mathfrak{p}$ be a finite prime of K. In the first instance we choose $L = K_\mathfrak{p}$; then provided P' and P'' are each defined over $K_\mathfrak{p}$ we can simplify the description of this construction. For in this case R' and R'' are defined over $K_\mathfrak{p}$. Conversely if R' is any point of $\Gamma'(K_\mathfrak{p})$ other than P', and similarly for R'', then we can find a line ℓ_1 defined over $K_\mathfrak{p}$ which gives rise to this pair R', R''; indeed we can simply take ℓ_1 to be $R'R''$. In this way we obtain a point R in $W(K_\mathfrak{p})$. But we can approximate arbitrarily closely to ℓ and ℓ_1 over K; so we can by this construction obtain a point of $W(K)$ arbitrarily close to R. Moreover, for fixed P the points of $W(K)$ generated in this way as ℓ, ℓ_1 vary all lie in the same R-equivalence class. Now let U be an open subset of $W(K_\mathfrak{p})$. If for fixed P we can generate every point of U in this way, working over $K_\mathfrak{p}$ and requiring that P', P'' are each defined over $K_\mathfrak{p}$, it follows that there is an R-equivalence class of $W(K)$ which is dense in U. We shall make use of this argument in §8.

In the statement and proof of Lemma 7, all P_i will be points of W and all Q_i points of $\mathbf{P}^1$. Let P_0 be a point of W which does not lie on any of the lines on W and let Γ_0 be the intersection of W with the tangent plane to it at P_0; denote by $\varphi : \mathbf{P}^1 \to \Gamma_0$ the desingularization of Γ_0. Let $Z \subset W \times \mathbf{P}^1 \times \mathbf{P}^1 \times W$ be the locus of points $P_1 \times Q_2 \times Q_3 \times P_4$ such that $P_1, \varphi(Q_2), P_4$ are the three intersections of some line with W and P_4 lies on the tangent plane to W at $\varphi(Q_3)$. For P_1 on W, define $Z(P_1)$ by

$$Z \cdot (P_1 \times \mathbf{P}^1 \times \mathbf{P}^1 \times W) = P_1 \times Z(P_1).$$

In other words, $Z(P_1)$ is the fibre of $Z \to W$ at P_1.

Lemma 7. — *If W is nonsingular and P_1 is in general position on W, then $Z(P_1)$ is an absolutely irreducible curve in $\mathbf{P}^1 \times \mathbf{P}^1 \times W$.*

Proof. Let Y be the locus of points $P \times P'$ on $W \times W$ such that P lies on the tangent plane to W at P'. By Lemma 2, if $Y \cdot (P \times W) = P \times Y(P)$ then $Y(P)$ is the intersection of W with a quadric. If P is not in Γ_0 then $Y(P)$ does not contain Γ_0 because it does not contain P_0; and if P is in Γ_0 the intersection of $Y(P)$ with Γ_0 consists of P, P_0 and at most two other points, so again $Y(P)$ does not contain Γ_0. Hence $Y(P) \cdot \Gamma_0$ is well-defined and therefore $\deg(Y(P) \cdot \Gamma_0) = 6$. Now suppose that $P_1 \times Q_2 \times Q_3 \times P_4$ is in Z with Q_2 in general position on $\mathbf{P}^1$; then P_1 and Q_2 determine P_4 uniquely and hence $\varphi(Q_3)$

must be in $Y(P_4) \cdot \Gamma_0$. Thus $Z(P_1)$ has dimension 1. Moreover P_4 can only lie on Γ_0 if P_4 coincides with $\varphi(Q_2)$ and is in $Y(P_1) \cdot \Gamma_0$. We shall subsequently need the fact that Q_3 is not determined by P_1. For otherwise the cone with vertex P_1 and base Γ_0 would contain the intersection of W with the tangent plane at $\varphi(Q_3)$, which is impossible.

Let $C(P_1)$ be the projection of $Z(P_1)$ on $\mathbf{P}^1 \times \mathbf{P}^1$. We have already proved that if Q is on $\mathbf{P}^1$ then

$$\deg\{(Q \times \mathbf{P}^1) \cdot C(P_1)\} = \deg\{Y(P') \cdot \Gamma_0\} = 6 \tag{9}$$

where P' is the remaining intersection with W of the line joining P_1 and $\varphi(Q)$. The cone with vertex P_1 and base Γ_0 meets W again in the intersection of W with a quadric, on which P_4 lies, so we also have

$$\deg\{(\mathbf{P}^1 \times Q) \cdot C(P_1)\} = 6. \tag{10}$$

These two results determine the class of $C(P_1)$ as a divisor on $\mathbf{P}^1 \times \mathbf{P}^1$.

We next investigate the singularities of $C(P_1)$. Let $Q_2 \times Q_3$ be a point of $C(P_1)$ and let P_4 be the third intersection of W with the line joining P_1 and $\varphi(Q_2)$. A necessary and sufficient condition for $Q_2 \times Q_3$ to be a singular point of $C(P_1)$ is that it has multiplicity greater than 1 as a component of both $C(P_1) \cdot (Q_2 \times \mathbf{P}^1)$ and $C(P_1) \cdot (\mathbf{P}^1 \times Q_3)$. By (9), the first of these happens if and only if $Y(P_4)$ touches Γ_0 at $\varphi(Q_3)$; so it can only happen if P_4 lies on a certain not necessarily irreducible curve Γ_1 which depends only on P_0. To express the second condition, let C_1 be the intersection of W with the tangent plane to it at $\varphi(Q_3)$ and let C_2 be the remaining intersection of W with the cone of vertex P_1 and base Γ_0; then the second condition is equivalent to saying that C_1 and C_2 touch at P_4 — in other words, that the tangent lines to C_1 at P_4 and to Γ_0 at $\varphi(Q_2)$ meet. Suppose that both conditions hold for a particular P_4 not on Γ_0; then $\varphi(Q_3)$ lies in $Y(P_4) \cdot \Gamma_0$ which is finite, whence $\varphi(Q_2)$ belongs to a finite set and therefore so does P_1. Since P_4 lies on Γ_1, this implies that P_1 is not in general position. Hence P_4 must lie on Γ_0, and it now follows from the first condition that $P_4 = \varphi(Q_3)$, and from the first paragraph of the proof that $P_4 = \varphi(Q_2)$. Moreover P_4 is one of the points of $Y(P_1) \cdot \Gamma_0$. Conversely, any such point is indeed a singularity of $C(P_1)$.

We have therefore shown that, for P_1 in general position, the singularities of $C(P_1)$ are precisely at the points where $C(P_1)$ meets Δ, the diagonal of $\mathbf{P}^1 \times \mathbf{P}^1$. There are just six of these points, because they correspond to the points where $Y(P_1)$ meets Γ_0. On the other hand, (9) and (10) imply $\deg\{C(P_1) \cdot \Delta\} = 12$; so each point occurs with intersection multiplicity 2 and is therefore an ordinary double point or a cusp. But, using (9) and (10) again, the only way that $C(P_1)$ can be reducible and yet have only six singularities, each of which is a double point or a cusp, is if it has just 2 components and one of them has the form $Q \times \mathbf{P}^1$ or $\mathbf{P}^1 \times Q$; and this is impossible because the singularities lie on Δ and hence cannot all lie on a component of this form. □

When the context makes it necessary, we shall require that every point which we introduce has a fixed coordinate representation $(X_0, \ldots, X_3)$ in which the X_i are all integers at $\mathfrak{p}$ and at least one of them is a unit there. Using this representation, we write arithmetic operations on such quadruplets as the associated arithmetic operations on points. Let $f(X_0, \ldots, X_3) = 0$ be the equation of W, where the coefficients of f are in $\mathfrak{O}_\mathfrak{p}$ and not all divisible by $\mathfrak{p}$; and for any points T_1, T_2 in $\mathbf{P}^3$ write

$$(11) \qquad f_2(T_1, T_2) = \sum_{i=0}^{3} X_i(T_1) \frac{\partial f(T_2)}{\partial X_i}.$$

The value of $f_2(T_1, T_2)$ here depends on the coordinate representations of T_1, T_2. If T_1, T_2 lie in W and T_2 is nonsingular then $f_2(T_1, T_2) = 0$ if and only if T_1 is on the tangent to W at T_2. Now let ℓ be a line which does not touch W and let T_1, T_2, T_3 be the three points of intersection of ℓ and W; thus

$$(12) \qquad a_1 T_1 + a_2 T_2 + a_3 T_3 = 0$$

for some nonzero a_1, a_2, a_3. Because f_2 is linear in its first argument

$$a_1 f_2(T_1, T_3) + a_2 f_2(T_2, T_3) = -a_3 f_2(T_3, T_3) = 0;$$

and because f is homogeneous cubic and vanishes at T_1 and T_2

$$3 a_1 a_2^2 f_2(T_1, T_2) + 3 a_1^2 a_2 f_2(T_2, T_1) = f(-a_3 T_3) = 0.$$

Hence

$$(13) \qquad -\frac{a_1}{a_2} = \frac{f_2(T_2, T_3)}{f_2(T_1, T_3)} = \frac{f_2(T_1, T_2)}{f_2(T_2, T_1)}$$

and in particular

$$(14) \qquad f_2(T_1, T_3) f_2(T_1, T_2) = f_2(T_2, T_3) f_2(T_2, T_1).$$

We shall only need these results when $\operatorname{char}(L) = 0$. The proof above fails when $\operatorname{char}(L) = 3$, but because (13), (14) are identities which we have already shown hold when $\operatorname{char}(L) = 0$, they must hold always.

3. Approximation at an infinite prime

In this section v will denote an infinite place of K, so that K_v is either $\mathbf{C}$ or $\mathbf{R}$. Our object is to prove

Theorem 2. — *If v is an infinite place of K and $V(K)$ is not empty then each R-equivalence class of $V(K)$ is contained in a connected component of $V(K_v)$ and is dense in it.*

Proof. R-equivalence is the finest equivalence relation such that if $\varphi : \mathbf{P}^1 \to V$ is defined over K as are Q_1, Q_2 on $\mathbf{P}^1$, then $\varphi(Q_1)$ is R-equivalent to $\varphi(Q_2)$. Each R-equivalence class lies in a connected component of $V(K_v)$ because $\mathbf{P}^1(K_v)$ is connected and hence so is its image. To complete the proof it is enough to show that if $\mathscr{C}$ is any R-equivalence class then the closure of $\mathscr{C}$ is open in $V(K_v)$.

Let P be a point of $V(K_v)$ in the closure of $\mathscr{C}$ and choose a point Q_0 in $V(K)$ such that the tangent plane to V at Q_0 meets V in an irreducible curve Γ_0 which does not pass through P; this is possible because $V(K)$ is dense in the Zariski topology. For any point Q_1 of $\Gamma_0(K)$ other than Q_0, let Q'_1 be the other intersection of PQ_1 with V and let Γ_1 be the residual intersection of V with the cone of vertex Q'_1 and base Γ_0; thus Γ_1 is birationally equivalent to $\mathbf{P}^1$ over K_v. It is easy to see that as Q_1 varies on Γ_0 the curves Γ_1 are not all tangent to one another at P. Choose points Q_2, Q_3 of $\Gamma_0(K)$ so that the corresponding curves Γ_2, Γ_3 are not tangent at P; and choose local coordinates ξ, η on V near P so that the tangents to Γ_2, Γ_3 at P are $\eta = 0, \xi = 0$ respectively. By continuity we can choose $\varepsilon > 0$ with the following property. Let P' be any point of $V(K_v)$ with $|\xi| < 3\varepsilon, |\eta| < 3\varepsilon$ and, starting from Q_2, Q_3 and Γ_0, construct curves Γ'_2, Γ'_3 through P' in the same way as we constructed Γ_2, Γ_3 through P. Near P, Γ'_2 is nearly parallel to Γ_2 and similarly for Γ'_3; so for all points of Γ'_2, Γ'_3 in $|\xi| < 2\varepsilon, |\eta| < 2\varepsilon$ we have

$$(15) \qquad |d\eta/d\xi| < \tfrac{1}{10} \text{ for } \Gamma'_2, \quad |d\xi/d\eta| < \tfrac{1}{10} \text{ for } \Gamma'_3.$$

Let $P^* = (\xi^*, \eta^*)$ be a point of $V(K_v)$ with $|\xi^*| < \varepsilon, |\eta^*| < \varepsilon$; to complete the proof of the theorem it is enough to show that P^* is in the closure of $\mathscr{C}$. Choose $P_0 = (\xi_0, \eta_0)$ in $\mathscr{C}$ with $d(P, P_0) < \tfrac{1}{2}\varepsilon$ and therefore $d(P^*, P_0) < 2\varepsilon$, where d is the metric induced by ξ, η. Without loss of generality we can assume that $|\xi^* - \xi_0| \leqslant |\eta^* - \eta_0|$. Starting with Γ_0 and Q_3 construct $\Gamma_3^\sharp$ passing through P_0 in the same way as we constructed Γ_3 passing through P. There is a non-constant map $\mathbf{P}^1 \to \Gamma_3^\sharp$ defined over K, and $\Gamma_3^\sharp$ is almost parallel to Γ_3. Hence there is a point of $\Gamma_3^\sharp$ at which $\eta = \eta^*$; so we can find a point $P_1 = (\xi_1, \eta_1)$ in

$\Gamma_3^\sharp(K)$, and therefore R-equivalent to P_0, such that $|\eta^* - \eta_1| \leq \frac{1}{10}|\eta^* - \eta_0|$. It follows from (15) that
$$|\xi_0 - \xi_1| \leq \frac{1}{10}|\eta_0 - \eta_1| \leq \frac{1}{9}|\eta^* - \eta_0|;$$
hence $|\xi^* - \xi_1| \leq |\xi^* - \xi_0| + \frac{1}{9}|\eta^* - \eta_0|$ and so $d(P^*, P_1) \leq \frac{4}{5}d(P^*, P_0)$. Iterating this construction, we obtain a sequence of points $P_0, P_1, \ldots$ in $\mathscr{C}$ which tend to P^*. □

4. Approximation at a finite prime

If $\mathfrak{p}$ is a finite prime of K then $V(K_\mathfrak{p})$ is totally disconnected, so the analogue of Theorem 2 is worthless. The methods which were used to prove Theorem 2 will show that any R-equivalence class of $V(K)$ is dense in some open subset of $V(K_\mathfrak{p})$, but for applications we shall need the more specific statement in Theorem 3 below. In this and the next two sections we shall be concerned with one finite prime $\mathfrak{p}$. We shall denote by v the normalized additive valuation associated with $\mathfrak{p}$ and by π a uniformizing parameter for $\mathfrak{p}$ in K. Expressions such as $O(\mathfrak{p}^n)$ will denote elements of K or $K_\mathfrak{p}$ divisible by $\mathfrak{p}^n$. Note that if Q_1, Q_2 are points of V such that $\widetilde{Q_1} = \widetilde{Q_2}$ is nonsingular on V then $f_2(Q_1, Q_2) = O((Q_1 - Q_2)^2)$ in the notation of (11).

An important procedure in several parts of this paper is the translation of a smooth curvilinear arc in V. Let P_0 be a point of $V(K)$ such that $\widetilde{P_0}$ is nonsingular on $\tilde{V}$ and let $\Gamma_0 \subset V$ be an arc through P_0. Let P_1 be another point of $V(K)$ such that $\widetilde{P_1} = \widetilde{P_0}$. Let R be a point of $V(K)$ such that $\tilde{R}$ is nonsingular on $\tilde{V}$ and not on the tangent plane to $\tilde{V}$ at $\widetilde{P_0}$, and neither P_0 nor P_1 is on the tangent plane to V at R. For $i = 0, 1$ let S_i be the third intersection of $P_i R$ with V. Denote by Γ' the residual intersection of V with the cone of vertex S_0 and base Γ_0, and by Γ_1 the residual intersection of V with the cone of vertex S_1 and base Γ'; thus R lies on Γ' and P_1 on Γ_1. We call Γ_1 a *translation* of Γ_0 from P_0 to P_1. In applications, Γ_0 will be an arc of a rational curve defined over K; then this will also be true of Γ_1. We shall show that Γ_1 is approximately parallel to Γ_0, in a sense made explicit in (22) or (23) below.

Take coordinates such that P_0 is $(1, 0, 0, 0)$, R is $(0, 1, 0, 0)$ and the tangent to V at P_0 is $X_1 = 0$. Let $X = X_2/X_0$ and $Y = X_3/X_0$, so that X, Y are local coordinates at P_0 on V. If $S_0 = aP_0 + bR$ the fact that neither $\tilde{R}$ nor $\widetilde{S_0}$ is equal to $\widetilde{P_0}$ implies that a, b are in $\mathfrak{O}_\mathfrak{p}$ and b is a unit at $\mathfrak{p}$. Normalizing, we can write $S_0 = C_0 P_0 + R$, where $C_0 \neq 0$ because P_0 is not on the tangent to V at R. Write $N = v(C_0)$; thus N is independent of the representations of

P_0, R chosen, though C_0 is not. Denote by $\mathcal{N}$ the subset of the inverse image of $\widetilde{P_0}$ in $V(K)$ at which $v(X) \geq N+1, v(Y) \geq N+1$. There is a one-one correspondence between the points of the inverse image of $\widetilde{P_0}$ in $V(K_\mathfrak{p})$ and the pairs X, Y in $K_\mathfrak{p}$ with $\pi|X, \pi|Y$. Let A, B be such that $AX + BY = 0$ touches Γ_0 at P_0, where A, B are in $\mathfrak{O}_\mathfrak{p}$ and at least one of them is a unit; and write

(16) $$\xi = AX + BY, \quad \eta = A'X + B'Y$$

for some A', B' in $\mathfrak{O}_\mathfrak{p}$ such that $AB' - A'B$ is a unit at $\mathfrak{p}$. Thus η is a local variable at P_0 on Γ_0 and $v(\eta) \geq N+1$ in $\mathcal{N}$. It follows that on Γ_0 we have a power series expansion $\xi = \sum_2^\infty a_n \eta^n$ convergent in some neighbourhood of $\eta = 0$. We assume Γ_0 so chosen that this series converges in $v(\eta) \geq N+1$ and that $v(a_n) > -(N+1)(n-1)$ for all n; thus ξ is small compared to η on Γ_0 provided $v(\eta) \geq N+1$. This will certainly hold if the arc Γ_0 has good reduction mod $\mathfrak{p}$. In this section, for any variable ζ we shall denote by $\Omega(\zeta)$ a power series of the form $\sum_1^\infty c_n \zeta^n$ which converges in $v(\zeta) \geq N+1$ and satisfies $v(c_n) > -n(N+1)$ for all n; here Ω need not be the same from one occasion to the next. Thus $\xi = \eta \Omega(\eta)$ on Γ_0.

Lemma 8. — *If P_1 is in $\mathcal{N}$ then Γ_1 has the form $\xi - \xi_1 = (\eta - \eta_1)\Omega(\eta - \eta_1)$ where ξ_1, η_1 are the values of ξ, η at P_1.*

Proof. Because $\tilde{R}$ is not on the tangent to $\tilde{V}$ at $\widetilde{P_0}$, $f_2(R, P_0)$ is a unit at $\mathfrak{p}$. Since any point of $\mathcal{N}$ has a representation which differs from that of P_0 by $O(\mathfrak{p}^{N+1})$, $f_2(R, P)$ is a unit at $\mathfrak{p}$ for any P in $\mathcal{N}$, and similarly for $f_2(S_0, P)$ and $f_2(S_1, P)$. Now $S_0 = R + C_0 P_0$ can be written

(17) $$S_0 = R - \{f_2(P_0, R)/f_2(R, P_0)\}P_0 = R + \{f_2(P_0, S_0)/f_2(S_0, P_0)\}P_0$$

in view of (13), so that $v(f_2(P_0, R)) = v(f_2(P_0, S_0)) = N$; hence as above $v(f_2(P, R)) = v(f_2(P, S_0)) = N$ for all P in $\mathcal{N}$. But once we have fixed the representation of P_1 it follows from (13) that we can choose the representation of S_1 to satisfy

(18) $$S_1 = R - \{f_2(P_1, R)/f_2(R, P_1)\}P_1;$$

and as above $v(f_2(P, S_1)) = N$ for all P in $\mathcal{N}$. Now let P_0' be any point of $\mathcal{N} \cap \Gamma_0$ and let R' be the third intersection of $P_0' S_0$ with V and P_1' the third intersection of $R'S_1$ with V; thus P_1' is on Γ_1. We can take

(19) $$R' = S_0 - \{f_2(P_0', S_0)/f_2(S_0, P_0')\}P_0'.$$

It follows from (17) and (19) that

$$(20) \qquad R' - R = \frac{f_2(P_0, S_0)}{f_2(S_0, P_0)} P_0 - \frac{f_2(P_0', S_0)}{f_2(S_0, P_0')} P_0';$$

since the denominators are units, the coordinates of R' differ from those of R by $O(\mathfrak{p}^{N+1})$. A similar result holds for S_0 and S_1. Hence

$$C_0^{-1}(S_1 - R') = C_0^{-1}(S_0 - R) + O(\mathfrak{p}) = P_0 + O(\mathfrak{p})$$

is the coordinate representation of a point of $R'S_1$; since this is close to P_0 the reduction mod $\mathfrak{p}$ of $R'S_1$ is $\widetilde{P_0}\widetilde{R}$. Since this line is transversal to $\tilde{V}$ at $\widetilde{P_0}$, it follows from Hensel's Lemma that P_1' is within $O(\mathfrak{p})$ of P_0. Now let a, b be such that $P_1' = -aR' + bS_1$. Using (20) and the analogous estimate for $S_0 - S_1$,

$$(21) \qquad P_1' + O(a\mathfrak{p}^{N+1}) + O(b\mathfrak{p}^{N+1}) = -aR + bS_0 = (b-a)R + bC_0P_0.$$

If $v(b) > v(a)$ then $v(b-a) = v(a)$ and (21) is impossible since $(b-a)R$ is strictly larger than any of the other terms except perhaps P_1'; so we can assume that $v(b) \leq v(a)$. Because $\tilde{R} \neq \widetilde{P_0}$ each term on the right must be $O(1) + O(a\mathfrak{p}^{N+1}) + O(b\mathfrak{p}^{N+1})$. Applying this first to bC_0R_0 and remembering that $v(a) \geq v(b)$, we obtain $b = O(\mathfrak{p}^{-N})$; now applying it to $(b-a)R$ we obtain $b - a = O(\mathfrak{p})$ and so $v(b) = v(a) = -N$. Using (17), (18) and (19) in $P_1' = -aR' + bS_1$,

$$(22) \qquad P_1' = -aC_0P_0 + a\frac{f_2(P_0', S_0)}{f_2(S_0, P_0')} P_0' - b\frac{f_2(P_1, R)}{f_2(R, P_1)} P_1 + (b-a)R.$$

The coefficients of P_0', P_1 here are $aC_0 + O(\mathfrak{p})$ and $-bC_0 + O(\mathfrak{p})$ respectively, so

$$P_1' = (bC_0 + O(\mathfrak{p}))P_0 + O(\mathfrak{p}^{N+1}) + (b-a)R.$$

Since the term in R does not affect the values of X, Y at P_1', we deduce that P_1' is in $\mathcal{N}$.

We shall repeatedly use the fact that if any of the power series $\sum_1^\infty c_n\eta^n$ which appear in the following argument converges in $v(\eta) \geq N+1$ and has a sum which is $O(\mathfrak{p})$ for all such η, then it has the form $\Omega(\eta)$. To prove this, we note that because of the way in which these properties are proved they hold when η is merely in the maximum unramified extension of $K_\mathfrak{p}$. Since the residue field of this extension is infinite, the bound $\sum c_n\eta^n = O(\mathfrak{p})$ cannot depend on cancellation between terms individually larger; so $c_n\eta^n = O(\mathfrak{p})$ for each n. Now let η_0' be the value of η at P_0'. It follows from (20) that $R' - R$ has a power series expansion in η_0' convergent in $v(\eta_0') \geq N+1$ and with zero constant term. Since it is also $O(\mathfrak{p}^{N+1})$ we deduce $R' - R = C_0\Omega(\eta_0')$. By (18) and (19)

$$f_2(S_1, R') = f_2(R, R') - \frac{f_2(P_1, R)f_2(P_1, R')}{f_2(R, P_1)},$$

$$f_2(R', S_1) = f_2(S_0, S_1) - \frac{f_2(P_0', S_0)f_2(P_0', S_1)}{f_2(S_0, P_0')}.$$

In each of these equations, the first term on the right is $O(\mathfrak{p}^{2N+2})$ and the second is the sum of a constant and an expression of the form $C_0^2\Omega(\eta_0')$. It follows from the first equation that

$$f_2(S_1, R') = f_2(S_1, R) + C_0^2\Omega(\eta_0')$$

where the first term on the right has valuation $2N$ and is independent of η_0'. By the second equation $f_2(R', S_1) - f_2(R, S_1)$ is $O(\mathfrak{p}^{2N+1})$ and has a power series expansion convergent in $v(\eta_0') \geqslant N+1$; so

$$f_2(R', S_1) = f_2(R, S_1) + C_0^2\Omega(\eta_0')$$

where again the first term on the right has valuation $2N$ and is independent of η_0'. But by (13)

$$-b/a = f_2(S_1, R')/f_2(R', S_1) = c + \Omega(\eta_0')$$

where $c = f_2(S_1, R)/f_2(R, S_1)$ is a unit at $\mathfrak{p}$. Dividing (22) by aC_0, we find that the values of X, Y at P_1' can be expanded as power series in η_0' convergent in $v(\eta_0') \geqslant N+1$; and they satisfy

(23) $$X(P_1') = X(P_1) + X(P_0') + \Omega(\eta_0') + C_0^{-1}O({\eta_0'}^2)$$

together with a similar equation for Y. It follows that Γ_1 is smooth at P_1 and the tangent there is given by

$$A(X - X(P_1)) + B(Y - Y(P_1)) = 0,$$

so that $\eta - \eta_1$ is a local variable on Γ_1 at P_1. Moreover $\eta(P_1') - \eta_1 = \eta_0' + \Omega(\eta_0')$; inverting this we get $\eta_0' = \eta(P_1') - \eta_1 + \Omega(\eta(P_1') - \eta_1)$, and it follows that

$$\xi(P_1') - \xi_1 = \eta_0'\Omega(\eta_0') = (\eta(P_1') - \eta_1)\Omega(\eta(P_1') - \eta_1)$$

as claimed. $\square$

Theorem 3. — *Let P, R be points of $V(K)$ such that $\tilde{P}$ is nonsingular on $\tilde{V}$, $\tilde{R}$ is not on the tangent to $\tilde{V}$ at $\tilde{P}$ and P is not on the tangent to V at R. Let $S = R + C_0 P$ be the third intersection of PR with V and write $N = v(C_0)$. Let Γ_0, Γ_1 be arcs of rational curves on V defined over K and passing through P; assume that $\tilde{P}$ is nonsingular on both $\widetilde{\Gamma_0}$ and $\widetilde{\Gamma_1}$ and Γ_0, Γ_1 do not touch at P.*

Let X, Y be local variables on V at P whose reductions mod $\mathfrak{p}$ are local variables on $\tilde{V}$ at $\tilde{P}$. For $i = 0, 1$ let $A_i X + B_i Y = 0$ touch Γ_i at P, where A_i, B_i are in $\mathfrak{O}_\mathfrak{p}$ and at least one of them is a unit at $\mathfrak{p}$; and write $M = v(A_0 B_1 - A_1 B_0)$. If $\mathscr{C}$ is the R-equivalence class of P then the points of $\mathscr{C}$ are dense in the $\mathfrak{p}$-adic topology in the open neighbourhood

(24) $$v(X) \geq L, \quad v(Y) \geq L$$

where $L = 3M + N + 1$.

Proof. Denote the set (24) by $\mathcal{N}$ and let P^* be a point of $\mathcal{N}$ defined over $K_\mathfrak{p}$. To prove the Theorem, it is enough to construct a sequence of points $P_0 = P, P_1, \ldots$ in $\mathscr{C}$ such that $P^* - P_n = O(\mathfrak{p}^{L+n})$. The construction of P_{n+1} from P_n is as follows. Write $\xi_i = A_i X + B_i Y$ for $i = 0, 1$ and let A, B in $\mathfrak{O}_\mathfrak{p}$ be such that $AB_i - A_i B$ is a unit for $i = 1, 2$; thus $\eta = AX + BY$ is a local variable at P on each Γ_i and

(25) $$(A_0 B_1 - A_1 B_0)\eta = (AB_1 - A_1 B)\xi_0 - (AB_0 - A_0 B)\xi_1.$$

Let ξ_0^*, ξ_1^* be the values of ξ_0, ξ_1 at P^*. Let Γ_{0n} be the translation of Γ_0 from P to P_n, and choose a point Q_n in $\Gamma_{0n}(K)$ with

$$(\eta(Q_n) - \eta(P_n)) - \frac{A_0 B - A B_0}{A_0 B_1 - A_1 B_0}(\xi_1^* - \xi_1(P_n)) = O(\mathfrak{p}^{L+n+1}).$$

The second term on the left is $O(\mathfrak{p}^{L+n-M})$, so it follows from Lemma 8 that

$$\xi_0(Q_n) - \xi_0(P_n) = O(\mathfrak{p}^{2L+2n-2M-N})$$

and therefore $\xi_0^* - \xi_0(Q_n) = O(\mathfrak{p}^{L+n})$. Also (25) implies

$$\xi_1(Q_n) - \xi_1(P_n) = \xi_1^* - \xi_1(P_n) + O(\mathfrak{p}^{L+n+M+1}) + O(\mathfrak{p}^{2L+2n-2M-N}),$$

so that

$$\xi_1^* - \xi_1(Q_n) = O(\mathfrak{p}^{L+n+M+1}).$$

Now let Γ_{1n} be the translation of Γ_1 from P to Q_n and choose P_{n+1} in $\Gamma_{1n}(K)$ with

$$(\eta(P_{n+1}) - \eta(Q_n)) - \frac{AB_1 - A_1 B}{A_0 B_1 - A_1 B_0}(\xi_0^* - \xi_0(Q_n)) = O(\mathfrak{p}^{L+n+1}).$$

Much as before, we have

$$\xi_1^* - \xi_1(P_{n+1}) = O(\mathfrak{p}^{L+n+M+1}),$$
$$\xi_0^* - \xi_0(P_{n+1}) = O(\mathfrak{p}^{L+n+M+1});$$

and these give $P^* - P_{n+1} = O(\mathfrak{p}^{L+n+1})$ by the analogues of (25) for X, Y in terms of ξ_0, ξ_1. □

The statement of Theorem 3 naturally raises the question: if γ is a curve of genus 0 on $\tilde{V}$, under what conditions can it be lifted to an arc of a curve of genus 0 on V? If γ has simply been conjured up, this may well be impossible; but if γ has been constructed it will often be possible to lift the construction. A good example of this is implicit in the proof of Lemma 12. The simplest case is the following, which nevertheless involves certain subtleties.

Lemma 9. — *Let P be a point of $V(K)$ such that $\tilde{P}$ is nonsingular on $\tilde{V}$, and suppose that $\tilde{V}$ does not contain a plane through $\tilde{P}$. Let γ be the intersection of $\tilde{V}$ with the tangent to $\tilde{V}$ at $\tilde{P}$, and let p_1 be a point of $\gamma(k)$ which is nonsingular both on γ and on $\tilde{V}$. Then p_1 is liftable to a point P_1 and γ is liftable at p_1 to an arc of a curve Γ of genus 0 passing through P_1.*

Proof. The first step is to show that we can reduce to the case when P is not on any line of V. We proceed step-by-step, at each step replacing P by a less insalubrious point. Suppose first that P lies on a line of V not defined over K; each such line contains at most one point of $V(K)$, so we have only finitely many such P to consider. By Lemma 6 we can find a point R of $V(K)$ which does not lie on any line of V and is such that the tangent to V at R does not contain P; thus the tangent to V at R meets V in an irreducible cubic curve C. If S_0 is a general point of C, the tangent to V at S_0 does not pass through P because otherwise the tangent to V at R would pass through P. Choose such an S_0 in $C(K)$ and then choose S_1 in $C(K)$ close to S_0. Let PS_0 meet V again in T and let TS_1 meet V again in P_2. Then P_2 is close to P but not identical with it; in particular $\widetilde{P_2} = \tilde{P}$ but P_2 does not lie on a line of V not defined over K.

It can still happen that P_2 lies on a line Λ of V defined over K. By the proof of Lemma 5, there are at most 10 points of Λ at which the tangent to V meets V in a union of lines; so we can replace P_2 by a point P_3 which does not have this property but for which $\widetilde{P_3} = \widetilde{P_2}$. The residual part of the intersection of V with the tangent to V at P_3 is an irreducible conic passing through P_3; we can therefore choose a point P_4 on it and defined over K such that $\widetilde{P_4} = \widetilde{P_3}$ but P_4 does not lie on any line of V. We now start again with P_4 in place of P; the advantage of this is that we can now assume that the tangent to V at P meets V in an irreducible cubic curve Γ.

Now choose coordinates so that P is $(1,0,0,0)$ and the tangent to V at P is $X_3 = 0$. The equation of V has the form $X_3 X_0^2 + \ldots = 0$, and it contains at least one term not divisible by X_3 whose coefficient is a unit at $\mathfrak{p}$; for otherwise $\tilde{V}$ would contain the plane $X_3 = 0$. So we obtain the equation of γ by setting $X_3 = 0$ and reducing all coefficients mod $\mathfrak{p}$. By Hensel's Lemma we can lift

p_1 to a point P'_1 in $\Gamma(K_\mathfrak{p})$. Let Λ_1 be a line defined over K, lying in $X_3 = 0$, passing through P and close to PP'_1; then Λ_1 meets Γ again in a point P_1 defined over K and close to P'_1, and γ lifts to an arc of Γ. □

Corollary. — *Let ℓ be a line on $\tilde{V}$ defined over k and such that no point of ℓ is singular on $\tilde{V}$. If there is a point P of $V(K)$ such that $\tilde{P}$ lies on ℓ, then any point p_1 of $\ell(k)$ can be lifted to a point P_1 of $V(K)$, and at p_1 the line ℓ can be lifted to an arc of a curve Γ_1 of genus 0 passing through P_1.*

Proof. If $\tilde{V}$ were not irreducible then ℓ would contain a singular point, contrary to hypothesis. Let γ be the intersection of $\tilde{V}$ with the tangent to $\tilde{V}$ at $\tilde{P}$. It follows from the Lemma that the property holds for any p_1 in $\ell(k)$ except perhaps the singular points of γ on ℓ, which are $\tilde{P}$ and at most one other point. Since $\ell(k)$ consists of $q+1$ points, we can find at least one p_1 for which the property does hold. Now repeat the argument with P_1 instead of P, noting that if γ_1 is the intersection of $\tilde{V}$ with the tangent to $\tilde{V}$ at p_1 then no point of ℓ can be singular on both γ and γ_1. □

Because of the hypotheses involved, it might appear that Theorem 3 is not always applicable; but we can cope with this by rescaling. Let V be any nonsingular cubic surface and P any point of $V(K)$. By Lemma 6 we can find a point R in $V(K)$ such that the line PR does not lie in or touch V, and R does not lie on any line of V. We can take P to be $(1,0,0,0)$, the tangent to V at P to be $X_3 = 0$ and R to be $(0,0,0,1)$; thus the equation of V can be written

$$f(X_0, \ldots, X_3) = a_{003} X_0^2 X_3 + \ldots = 0$$

where the coefficients of f are in $\mathfrak{O}_\mathfrak{p}$ and $a_{003} \neq 0$. Multiply by a_{003} and replace X_0 by $a_{003}^{-1} X_0$; this forces $a_{003} = 1$, so that now $\tilde{P}$ is nonsingular on $\tilde{V}$ and $\tilde{R}$ does not lie on the tangent to $\tilde{V}$ at $\tilde{P}$. To construct the curves Γ_i let Γ' be the intersection of V with the tangent to V at R and Λ the line of intersection of the tangents at P and R; because R does not lie on a line of V, Γ' is an absolutely irreducible plane cubic. Let T be in general position on Γ'. If PT touched V then this would hold for every point of Γ', whereas it does not hold for R. Let T' be the remaining intersection of PT with V and let Γ be the residual intersection of V with the cone of vertex T' and base Γ'. If the tangent to Γ' at T meets Λ at U then the tangent to Γ at P is PU. Hence as T varies in $\Gamma(K)$, the tangent to Γ at P does vary; so we can take Γ_0, Γ_1 to be the Γ constructed from some T_0, T_1 in $\Gamma(K)$.

We do not yet know the behaviour of the $\tilde{\Gamma}_i$ at $\tilde{P}$. But let η be a common local variable for each Γ_i at P of the form $(AX_1 + BX_2)/X_0$, with A, B in $\mathfrak{O}_\mathfrak{p}$.

The coordinates of a point of Γ_i are given by power series in η convergent in some p-adic neighbourhood of $\eta = 0$. We can choose n so that for $i = 0, 1$ these series converge in $v(\eta) \geqslant n+1$ and their values are $O(\mathfrak{p})$ there. If we now write $\pi^n X_1, \pi^n X_2$ for X_1, X_2, the effect will be that the $\widetilde{\Gamma}_i$ are nonsingular at $\tilde{P}$.

We have therefore shown that to any P in any class $\mathscr{C}$ there exists an integer $L(P)$ such that if P^* in $V(K_\mathfrak{p})$ satisfies $P^* - P = O(\mathfrak{p}^{L(P)})$ in the original coordinate system then P^* is in the closure of $\mathscr{C}$ in the p-adic topology. We next show that we can choose an L which is valid for all P. Indeed we can apply the last construction with P^* instead of P, but with R, T_0 and T_1 still defined over K; and these will work for all P near enough to P^*. Thus there is a neighbourhood $\mathscr{N}$ of P^* such that if there is a point of some class $\mathscr{C}$ in $\mathscr{N}$ then points of $\mathscr{C}$ are dense in $\mathscr{N}$. Because $V(K_\mathfrak{p})$ is compact, it is covered by finitely many such neighbourhoods.

It is natural to ask for conditions under which we can assert that if P is in $V(K)$ then $\tilde{P}$ is densely liftable. The most useful result of this kind which I can prove are Theorem 4 and Lemma 10 and its Corollary.

Theorem 4. — *Let P, R be points of $V(K)$ such that $\tilde{P}, \tilde{R}$ are nonsingular on $\tilde{V}$ and $\tilde{P}\tilde{R}$ does not lie on or touch $\tilde{V}$.*

(i) : *Let Γ_0, Γ_1 be arcs of rational curves on V defined over K such that $\widetilde{\Gamma_0}, \widetilde{\Gamma_1}$ are nonsingular at $\tilde{P}$ and do not touch there. Then $\tilde{P}$ and $\tilde{R}$ are both densely liftable.*

(ii) : *Suppose that $\tilde{R}$ does not lie on any line on $\tilde{V}$, and denote by γ the intersection of $\tilde{V}$ with the tangent to $\tilde{V}$ at $\tilde{R}$. Let t_1, t_2 be points of $\gamma(k)$ distinct from each other and from $\tilde{R}$, and such that $\tilde{P}t_i$ does not lie in or touch $\tilde{V}$; and suppose that the tangents to γ at t_1, t_2 do not meet on the tangent to $\tilde{V}$ at $\tilde{P}$. Then $\tilde{P}$ and $\tilde{R}$ are both densely liftable.*

Proof. These are both special cases of Theorem 3 in which the hypotheses imply $M = N = 0$. If S is the third point of intersection of PR with V then $\tilde{S}$ is distinct from $\tilde{P}$ and $\tilde{R}$; hence $N = 0$. For (i) we can now replace Γ_0 and Γ_1 by their translations to P; Theorem 3 now shows that $\tilde{P}$ is densely liftable, and projection from S gives the same result for $\tilde{R}$. For (ii) we argue as follows. By hypothesis γ is an irreducible cubic; let Γ be the intersection of V with the tangent to V at R. Lift each $\tilde{R}t_i$ to a line through R in the plane of Γ; its third intersection T_i with Γ is a lift of t_i. Let S_i be the third intersection of PT_i with V; by hypothesis $\tilde{S}_i$ is not equal to $\tilde{P}$ or t_i. If Γ_i is the residual intersection of V with the cone with base Γ and vertex S_i then each $\widetilde{\Gamma}_i$ is nonsingular at $\tilde{P}$ and $\widetilde{\Gamma}_1, \widetilde{\Gamma}_2$ do not touch at $\tilde{P}$; thus $M = 0$. Theorem 3 now shows that $\tilde{P}$

is densely liftable; and again projection from S proves that $\tilde{R}$ is also densely liftable. □

The last part of this argument can be generalized. Let P, R be points of $V(K)$ such that $\tilde{P}, \tilde{R}$ are nonsingular on $\tilde{V}$, $\tilde{P}$ is densely liftable and $\tilde{R}$ is liftable. If $\tilde{P}\tilde{R}$ does not touch $\tilde{V}$ at $\tilde{R}$ then $\tilde{R}$ is also densely liftable.

Suppose that $\tilde{V}$ is not a cone and does not have a line of singularities, and let P be any point of $V(K)$ such that $\tilde{P}$ is nonsingular on $\tilde{V}$. Provided q is large enough, arguments like those of Lemma 12 below show that we can find R, t_1, t_2 satisfying all the conditions of Theorem 4. It follows that if q is large enough, every nonsingular point of $\tilde{V}(k)$ which is liftable is densely liftable. We shall see in Theorem 5 that, without any condition on q, if $\tilde{V}$ is nonsingular then every point of $\tilde{V}(k)$ is liftable; we are therefore motivated to ask what is then the obstruction to every point of $\tilde{V}(k)$ being densely liftable. We consider this question in §6. A similar investigation would be possible though laborious when $\tilde{V}$ is singular but is not a cone and does not have a line of singular points.

If $\tilde{V}$ is a cone or has a line of singular points, I have no reason to believe that every nonsingular point of $\tilde{V}(k)$ which is liftable is densely liftable. However, if $\mathfrak{p} \not| 3$ it is true (though not proved in this paper) that we can always rescale V so that reduction mod $\mathfrak{p}$ yields a more tractable surface over k than the original $\tilde{V}$. I believe this to hold even if $\mathfrak{p}|3$. Special cases of this assertion can be found in §8.

In the proof of Theorem 3 the key fact was that, given rational arcs Γ_i through P, through any point P' of $V(K)$ near enough to P we could find rational arcs Γ'_i whose reductions mod $\mathfrak{p}$ are the same as those of Γ_i. We achieved this by translation; but this is not the only method available — and indeed it may cease to be available if $\tilde{V}(k)$ is too small. We shall later need the following result. The construction underlying the proof is the same as that of Theorem 4(i), except that we no longer have to employ translations or mention R because the hypotheses of Lemma 10 provide us with the rational arcs through P which had to be constructed in the proof of Theorem 4(i).

Lemma 10. — *Suppose that p is nonsingular in $\tilde{V}(k)$ and liftable, and there are rational arcs γ_1, γ_2 on $\tilde{V}$, defined over k and transversal at p. Suppose that p is liftable and that for every point P in the inverse image of p in $V(K)$ there are rational arcs Γ_1, Γ_2 defined over K, passing through P and such that the reduction mod $\mathfrak{p}$ of each Γ_i is γ_i. Then p is densely liftable.* □

Corollary. — *Suppose that p is nonsingular on $\tilde{V}(k)$ and that the tangent to $\tilde{V}$ at p meets $\tilde{V}$ in a curve γ, not necessarily irreducible, which has an ordinary* double point at p. Suppose further that the two tangent directions to γ at p are each defined over k. Then p is densely liftable.

Proof. Denote by γ_1, γ_2 the two branches of γ at p. In the notation of the Lemma, we can only lift γ_1 and γ_2 over $K_{\mathfrak{p}}$, because if Γ is the intersection of V with the tangent to V at P then the two points above P on the desingularization of Γ are defined over $K_{\mathfrak{p}}$ by Hensel's Lemma but need not be defined over K. Denote the branches of Γ by Γ_1 and Γ_2. If P_1^* is a point of $\Gamma_1(K_{\mathfrak{p}})$ then we can construct P_1 in $\Gamma(K)$ and arbitrarily close to P_1^* by approximating to the line PP_1^* by a line Λ through P and in the plane of Γ, and taking P to be the third intersection of Λ with Γ. (The modification of this argument needed if Γ_1 is a line is obvious.) This is all we need for the successive approximation construction which underlies the proofs of Theorems 3 and 4 to work. □

Lemma 11. — *Let $\mathscr{C}_1, \mathscr{C}_2$ be R-equivalence classes in $V(K)$; then their closures in $V(K_{\mathfrak{p}})$ are either disjoint or identical.*

Proof. It is enough to show that if P^* in $V(K_{\mathfrak{p}})$ is in the closure of both $\mathscr{C}_1$ and $\mathscr{C}_2$, and Q^* in $V(K_{\mathfrak{p}})$ is in the closure of $\mathscr{C}_1$, then Q^* is in the closure of $\mathscr{C}_2$. By Lemma 6 we can choose R in $V(K)$ so that neither of RP^* and S^*Q^* lies in or touches V, where S^* in $V(K_{\mathfrak{p}})$ is the third intersection of RP^* with V. Let the third intersection of S^*Q^* with V be T^* in $V(K_{\mathfrak{p}})$.

Now let P_1 in $\mathscr{C}_1$ and P_2 in $\mathscr{C}_2$ be arbitrarily close to P^*, and for $i = 1, 2$ let P_iR meet V again in S_i; thus S_i is arbitrarily close to S^* and the class of S_i is $\mathscr{C} \circ \mathscr{C}_i$ where $\mathscr{C}$ is the class of R. Let Q_1 in $\mathscr{C}_1$ be arbitrarily close to Q^* and let Q_1S_1 meet V again in T; thus T is in $\mathscr{C}$ and is arbitrarily close to T^*. If S_2T meets V again in Q_2 then Q_2 is in $\mathscr{C}_2$ and arbitrarily close to Q^*. □

If $\mathscr{C}$ is an R-equivalence class, the closure of $\mathscr{C}$ in $V(K_{\mathfrak{p}})$ is a union of some of the neighbourhoods constructed in the proof of Theorem 3. This is the partition of $V(K_{\mathfrak{p}})$ which we claimed in the Introduction that we could construct explicitly. It will follow from Lemma 11 and the arguments of §7 that (2) is the coarser partition in which each set U_{iv} is either the closure of some R-equivalence class or the set of points of $V(K_{\mathfrak{p}})$ which are not in the closure of $V(K)$.

5. The lifting process

This and the next section are concerned with various aspects of the following question: if $\tilde{V}$ is nonsingular and $V(K)$ is not empty, in what circumstances can we lift a point of $\tilde{V}(k)$ to $V(K)$?

Lemma 12. — *There is an absolute constant q_0 with the following property. Suppose that $\tilde{V}$ is not a cone and does not have a line of singularities, and that P_0 is a point of $V(K)$ such that $\widetilde{P_0}$ is nonsingular on $\tilde{V}$. If $q > q_0$ there is a point of $V(K)$ in general position on V whose reduction mod $\mathfrak{p}$ does not lie on any of the lines on $\tilde{V}$.*

Proof. We mimic the construction in Lemma 6. Let P_0 be the given point of $V(K)$. Choose a line ℓ_1, defined over k and passing through $\widetilde{P_0}$, which does not touch $\tilde{V}$ either at $\widetilde{P_0}$ or elsewhere and which does not meet any of the lines on $\tilde{V}$ except perhaps at $\widetilde{P_0}$. This is possible if q is large enough, for there are q^2 lines through $\widetilde{P_0}$ defined over k which do not touch $\tilde{V}$ at $\widetilde{P_0}$; of these only $O(q)$ touch $\tilde{V}$ elsewhere and only $O(q)$ meet one of the lines on $\tilde{V}$ at a point other than $\widetilde{P_0}$. Let p_1' and p_1'' be the other points where ℓ_1 meets $\tilde{V}$; they are distinct and different from $\widetilde{P_0}$, and they are either each defined over k or conjugate over k and defined over k_1, the unique quadratic extension of k. Let π' be the tangent plane to $\tilde{V}$ at p_1' and γ' its intersection with $\tilde{V}$; and define π'', γ'' similarly. By the conditions above, γ' and γ'' are absolutely irreducible; they are also distinct, as are π' and π'', because π' is transversal to ℓ_1. Moreover, $\widetilde{P_0}$ does not lie on π' or π''.

Lift ℓ_1 to a line L_1 containing P_0 and defined over K. Let P_1', P_1'' be the other two intersections of L_1 with V; at worst they are defined over a quadratic extension K_1 of K and are conjugate over K, and by abuse of language we can say that their reductions are p_1', p_1'' respectively. (It can happen that p_1', p_1'' are defined over k but P_1', P_1'' are not defined over K; now $\mathfrak{p}$ splits in K_1 and we have to take the reductions modulo the appropriate one of its prime factors.) Let Π' be the tangent plane to V at P_1' and Γ' its intersection with V, and similarly for Π'', Γ''.

Let ℓ_2 be a line defined over k and not passing through p_1' or p_1'' nor lying on π' or π''. Let p_2', p_2'' be the intersections of ℓ_2 with π', π'' respectively and p_3', p_3'' the third intersections of $\tilde{V}$ with $p_1'p_2', p_1''p_2''$ respectively. We assume ℓ_2 so chosen that neither p_3' nor p_3'' lies on ℓ_3, the intersection of π' and π'', and that p_2' does not lie on one of the lines having triple contact with γ' at p_1' and similarly for p_2''. This is possible since q is large. We can generate in this way any pair of points p_3' on γ' and p_3'' on γ'' having the same rationality properties with respect to k as p_1' and p_1'' and distinct from p_1', p_1'' respectively and from the points where $\tilde{V}$ meets ℓ_3. We call these pairs p_3', p_3'' *allowable*. The line $p_3'p_3''$ is defined over k and hence so is p_4, the point where it meets $\tilde{V}$ again.

There are only $O(q)$ allowable pairs p_3', p_3'' for which $p_4 = \widetilde{P_0}$, because there are only $O(q)$ lines defined over k which meet γ' and contain $\widetilde{P_0}$. Also p_4 can never be p'_1, p''_1 or one of the points at which ℓ_3 meets $\tilde{V}$. I claim that any other point p_4 is generated by at most six allowable pairs. If p_4 is for example on π' then we need p'_3 to be p_4 and p''_3 to be one of the three points at which the tangent plane to $\tilde{V}$ at p_4 meets γ''; so assume that p_4 is on neither π' nor π''. Let γ^* be the projection of γ'' from p_4 onto π'; γ^* is different from γ' because the unique singularity of γ'' at p''_1 does not project onto the unique singularity of γ' at p'_1. Hence γ^* and γ' have at most six intersections outside ℓ_3; and these are the only points p'_3 which can give rise to p_4. But we have $q^2 - O(q)$ allowable pairs p'_3, p''_3; so we generate at least $\frac{1}{6}q^2 - O(q)$ distinct p_4. Hence if $q > q_0$ we can ensure that p_4 does not lie on any of the lines of $\tilde{V}$. Now lift ℓ_2 to any line L_2 defined over K. Let P'_2 be the intersection of L_2 with Π' and let P'_3 be the third intersection of $P'_1 P'_2$ with V; and similarly for P''_2, P''_3. Let Γ' be the intersection of Π' with V, and similarly for Γ''; then $P'_3 \times P''_3$ is in general position on $\Gamma' \times \Gamma''$. If P_4 is the third intersection of $P'_3 P''_3$ with V, then it is defined over K and is a lift of p_4. An argument like that earlier in this paragraph shows that any point P_4 other than P_0 comes from at most six pairs P'_3, P''_3; so P_4 is in general position on V. Here of course the Zariski open set in which P_4 is required to lie can depend on q. □

Theorem 5. — *With the notation of §1, if $\tilde{V}$ is nonsingular and $V(K)$ is not empty then every point of $\tilde{V}(k)$ can be lifted to a point of $V(K)$.*

Proof. Let K^* be a large enough extension of K in which $\mathfrak{p}$ remains prime, let k^* be the corresponding residue field mod $\mathfrak{p}$ and let q^* be the number of elements of k^*. The first step is to show that every point of $\tilde{V}(k^*)$ can be lifted to a point of $V(K^*)$ — in other words, that the Theorem holds if q is large enough. Let P be the given point of $V(K)$; after Lemma 12 we can assume that $\tilde{P}$ does not lie on any of the lines on $\tilde{V}$. Now take $W = \tilde{V}$ and $p_0 = \tilde{P}$ in Lemma 7; to say that p_1 is in general position on W is simply to say that it lies outside a certain proper subvariety W_1 of W. Since W_1 is independent of the choice of K^*, we can find a point p_1 in $W(k^*)$ but outside W_1. Let Z, φ, γ_0 be as in the proof of Lemma 7. Then $Z(p_1)$ is an absolutely irreducible curve of bounded genus, defined over k^*, and hence it contains a point $q_2 \times q_3 \times p_4$ defined over k^* with p_4 not on γ_0. Since q_3 is not determined by p_1, we can assume that it is not on one of the lines of $\tilde{V}$. By Lemma 9 $\varphi(q_2)$ and $\varphi(q_3)$ can be lifted to $V(K^*)$, and hence the same is true of p_4 because it lies on the tangent to $\tilde{V}$ at $\varphi(q_3)$; and since $p_1, \varphi(q_3), p_4$ are collinear points on $\tilde{V}$, this implies that we can lift p_1 also.

To come down from K^* to K, choose a sequence of quadratic extensions

$$K = K_0 \subset K_1 \subset K_2 \subset \ldots$$

in all of which $\mathfrak{p}$ remains prime, and let
$$k = k_0 \subset k_1 \subset k_2 \subset \ldots$$
be the corresponding sequence of residue fields mod $\mathfrak{p}$. By the previous paragraph, for large enough n every point of $\tilde{V}(k_n)$ can be lifted to a point of $V(K_n)$. Suppose that this holds for $n = N+1$; we proceed by downward induction, asking what are the obstacles for deducing it for $n = N$. This descent argument will involve a substantial number of cases, according to the rationality properties with respect to k_N of the lines on $\tilde{V}$; these cases fill all the paragraphs of the proof after this one. Write $M = 2^N$, so that k_N contains q^M elements. We are given a point P in $V(K) \subset V(K_N)$; let γ_0 be the intersection of $\tilde{V}$ with the tangent plane to $\tilde{V}$ at $\tilde{P}$. By Lemma 9 every point of $\gamma_0(k_N)$, including the singular point $\tilde{P}$, can be lifted to $V(K_N)$.

Suppose first that there is a line ℓ on $\tilde{V}$ which is defined over k_N; either it is contained in γ_0 or it meets γ_0 in a point defined over k_N and therefore liftable to $V(K_N)$. Using the Corollary to Lemma 9 we see that in either case every point of $\ell(k_N)$ can be lifted to $V(K_N)$.

Next suppose that $\tilde{V}$ contains a pair of skew lines defined, as a pair, over k_N and let p_1 be a point of $\tilde{V}(k_N)$ not lying on either line. Let p_2', p_2'' be the two points at which the unique transversal through p_1 to the two lines meets them. If each of p_2', p_2'' is defined over k_N we can lift them to points P_2', P_2'' of $V(K_N)$ because each of them lies on a line on $\tilde{V}$ defined over k_N; if not, they are defined over k_{N+1} and conjugate over k_N; so we can lift p_2' to a point P_2' of $V(K_{N+1})$ and p_2'' to its conjugate P_2'' over K_N. In either case the line $P_2'P_2''$ is defined over K_N and its third intersection with V is a lift of p_1 to $V(K_N)$. Hence in this case we can lift any point of $\tilde{V}(k_N)$.

Henceforth we can assume that $\tilde{V}$ does not contain such a pair of skew lines. Suppose that p_1 is a point of $\tilde{V}(k_N)$ and let ℓ_1 be a line through p_1, defined over k_N and not touching $\tilde{V}$ at p_1. There are three possibilities for the other intersections of ℓ_1 with $\tilde{V}$:

(i) : a point p_2 defined over k_N, counted twice;
(ii) : two points p_2 and p_3, each defined over k_N;
(iii) : two points p_2', p_2'' conjugate over k_N and defined over k_{N+1}.

Much as above, in case (iii) we can lift p_2', p_2'' to points of $V(K_{N+1})$ conjugate over K_N and thus lift p_1 to a point of $V(K_N)$. Similarly if we can lift p_2, p_3 in case (ii), or p_2 in case (i), to $V(K_N)$ then the same is true for p_1; in case (i) this uses Lemma 9. Hence if p_1 cannot be lifted then on each of the q^{2M} lines ℓ_1 there is another point of $\tilde{V}(k_N)$ which cannot be lifted; in particular, if there are any points of $\tilde{V}(k_N)$ which cannot be lifted, there are at least $q^{2M}+1$ of them.

Suppose first that $\tilde{V}$ contains three coplanar lines each defined over k_N; these lines contain at least $3q^M$ points of $\tilde{V}(k_N)$ all of which are liftable. If there are also unliftable points it follows from the tables in Swinnerton-Dyer [4], partly repeated in Manin [3], pp 176-7, that we are in case C_4 and there are no liftable points outside these three lines. But let p be a point of $\tilde{V}(k_N)$ lying on just one of these lines; p cannot be an Eckardt point because the tables show that there are no more lines of $\tilde{V}$ defined over k_{N+1}. So the tangent plane to $\tilde{V}$ at p meets $\tilde{V}$ again in an irreducible curve, and this contains at least one point of $\tilde{V}(k_N)$ which is liftable by Lemma 9 and does not lie on any of the three lines. This is a contradiction.

Suppose next that $\tilde{V}$ contains just one line ℓ defined over k_N and does contain unliftable points. Now $\tilde{V}(k_N)$ must contain at least $q^{2M}+q^M+2$ points, so the tables show that we must be in case C_{24}. As in the previous paragraph, no point of $\ell(k_N)$ can be Eckardt; so if p is any point of $\ell(k_N)$ the tangent plane to $\tilde{V}$ at p meets $\tilde{V}$ again in an irreducible conic, which contains at least q^M+1 liftable points. Since there are at least $\frac{1}{2}(q^M+1)$ distinct such conics, no two of which have a point in common, there are at least $\frac{1}{2}(q^M+1)^2$ liftable points on $\tilde{V}$. Counting the total number of points on $\tilde{V}(k_N)$ we obtain

$$q^{2M}+1+\tfrac{1}{2}(q^M+1)^2 \leqslant q^{2M}+2q^M+1$$

which is impossible.

Henceforth we can assume that $\tilde{V}$ contains no lines defined over k_N; it now follows from the table that the total number of points on $\tilde{V}(k_N)$ is

(26) $$q^{2M}+rq^M+1 \quad \text{with} \quad r \leqslant 2.$$

Suppose that $\tilde{V}(k_N)$ contains R unliftable points of which E are Eckardt. If p is an unliftable Eckardt point then there are no other points of $\tilde{V}(k_N)$ on the tangent plane to $\tilde{V}$ at p, and every line through p defined over k_N and not in the tangent plane contains one or two more points of $\tilde{V}(k_N)$; moreover, every point of $\tilde{V}(k_N)$ other than p lies on one such line. Hence there are exactly $q^{2M}-rq^M$ such lines which touch $\tilde{V}$ at a point $p_1 \neq p$, and each such p_1 is an unliftable point of $\tilde{V}(k_N)$. Similarly, if p is unliftable but not Eckardt the number of other points of $\tilde{V}(k_N)$ on the tangent plane to $\tilde{V}$ at p is $q^M+\varepsilon_p$ where $\varepsilon_p = 0$ or ± 1; so the number of points p_1 as above is

$$q^{2M}-(r-1)q^M+\varepsilon_p.$$

It follows that the number of pairs p, p_1 with p and p_1 distinct and unliftable and p in the tangent plane to $\tilde{V}$ at p_1 is
$$R(q^{2M} - rq^M) + (R - E)q^M + \sum \varepsilon_p.$$
On the other hand, the number of pairs p, p_1 with p and p_1 distinct, p in the tangent plane to $\tilde{V}$ at p_1 and p_1 unliftable is
$$(R - E)q^M + \sum \varepsilon_{p_1}.$$
Since the second of these numbers cannot be less than the first, we have $R(q^M - r) \leqslant 0$; so we can only have $R > 0$ if $q^M = r = 2$. But any $\tilde{V}$ with $q^M = r = 2$ which contains no lines defined over k must be in case C_{12} of [4] and therefore has a triple of conjugate coplanar lines. Since $\mathbf{P}^3(k)$ contains only 8 points outside the plane of these lines and $\tilde{V}(k)$ contains 9 points, these lines must meet in an Eckardt point. It follows that $\tilde{V}$ is
$$X_3(X_0^2 + X_1^2 + X_1 X_2 + X_2^2) + X_0 X_3^2 + X_1^3 + X_1 X_2^2 + X_2^3 = 0.$$
However, the methods which we have already used are enough to show that if one point defined over k_N on this surface is liftable then all of them are. (If the point known to be liftable is the Eckardt point $(1, 0, 0, 0)$ then we have to use the construction in the proof of Lemma 12.) Thus there is no obstruction to the downward induction. □

Corollary. — *Let $\mathscr{C}$ be a given R-equivalence class in $V(K)$; then under the conditions of the Theorem every point of $\tilde{V}(k)$ can be lifted to $\mathscr{C}$.*

Proof. It is enough to show that there is some class $\mathscr{C}_0$ to which every point of $\tilde{V}(k)$ can be lifted. For let P_1 be a point of $\mathscr{C} \circ \mathscr{C}_0$ and p any point of $\tilde{V}(k)$. If Q is a lift of $p \circ \overline{P_1}$ to $\mathscr{C}_0$ then $Q \circ P_1$ is a lift of p to $\mathscr{C}$. We use this idea at each stage of the argument.

In the first part of the proof of the Theorem, for which q^* is large, we exhibited a birational map $\psi : \mathbf{P}^4 \to V$ defined over K^* such that $\psi(V)$ contains lifts of every point of $\tilde{V}(k^*)$. Since all points of $\psi(V)$ are in the same R-equivalence class, this proves the Corollary in this case.

Now let P_2 be a point of $V(K_N)$ and let $\mathscr{C}_2$ be its class as a point of $V(K_{N+1})$. If P', P'' are points of $V(K_{N+1})$ conjugate over K_N and such that P' is in $\mathscr{C}_2$, then by taking conjugates in the linkage between P_2 and P' it follows that P'' is also in $\mathscr{C}_2$. If P in $V(K_N)$ is the third intersection of $P'P''$ with V, then the class $\mathscr{C}_0$ of P in $V(K_N)$ depends only on $\mathscr{C}_2$. (The special cases when $P' \circ P''$ is not singlevalued are not straightforward. But let Γ' be a rational curve defined over K_{N+1} and Γ'' its conjugate over K_N. We can take Q' generic on Γ' over K_{N+1} and therefore defined over a field $K_{N+1} \otimes K_N(s, t)$ where s, t

are independent transcendentals over K_N. Let Q'' be the conjugate of Q' over $K_N(s,t)$; then the locus of $Q' \circ Q''$ over K_N is an image of $\mathbf{P}^2$ and there is a value of $P' \circ P''$ in this image. Now we can appeal to Lemma 9.) Replacing P_2 by $P_2 \circ P_2$ and using (1) we can assume that $\mathscr{C}_2 \circ \mathscr{C}_2 = \mathscr{C}_2$; thus some point of $V(K_N)$ in $\mathscr{C}_0$ (and therefore every such point) is in $\mathscr{C}_2$ when considered as a point of $V(K_{N+1})$. Now the induction argument still works if instead of the set of liftable points of $\tilde{V}(k_N)$ we consider the set of points of $\tilde{V}(k_N)$ liftable to $\mathscr{C}_0$. □

6. The dense lifting process

Throughout this section, we assume that $\tilde{V}$ is nonsingular and $V(K)$ is not empty. Everything will be defined over k unless otherwise specified. We shall repeatedly use the fact that the order of $\tilde{V}(k)$ is congruent to 1 mod q. This holds because the characteristic roots of Frobenius in middle dimension are in this case q times roots of unity. Alternatively one may consult the tables in [4] or [3] already cited.

Theorem 5 shows that every point of $\tilde{V}(k)$ is liftable; the object of this section is to show that, with an explicit list of possible exceptions, every point of $\tilde{V}(k)$ is densely liftable. With the possible exception of (42), it follows from the Remark at the end of §5 that if one point of $\tilde{V}(k)$ is densely liftable, then for any prescribed R-equivalence class $\mathscr{C}$ every point of $\tilde{V}(k)$ is densely liftable to $\mathscr{C}$. There is a major bifurcation of the argument, according to whether $\tilde{V}$ contains a line defined over k or not. Until further notice, we suppose that $\tilde{V}$ contains at least one line defined over k.

Suppose first that there is a plane π_0 which meets $\tilde{V}$ in three lines ℓ_1, ℓ_2, ℓ_3 each defined over k (and distinct because $\tilde{V}$ is nonsingular) and that these lines do not meet in an Eckardt point. There are exactly $3q$ points of $\tilde{V}(k)$ on π_0; but the total number of points in $\tilde{V}(k)$ is congruent to 1 mod q, so there is at least one such point p not on π_0. The tangent to $\tilde{V}$ at p cannot pass through all three points like p_{ij} at which ℓ_i, ℓ_j meet; suppose it does not pass through p_{12}. By Theorem 4(i) and the Corollary to Lemma 9, both p_{12} and p are densely liftable. Henceforth we can assume that this configuration does not occur.

Suppose next that there is a plane π_0 which meets $\tilde{V}$ in three lines ℓ_1, ℓ_2, ℓ_3 each defined over k and all meeting in an Eckardt point p_0. It was shown in the proof of Lemma 5 that there are at most 5 planes through ℓ_1 which meet $\tilde{V}$ in a union of lines. If there is a plane π_1 through ℓ_1 and defined over k which meets $\tilde{V}$ residually in an irreducible conic γ_1 (which certainly happens if $q > 4$), then we can usually find a point p' on γ_1 but not on ℓ_1 such that the tangent to γ_1

at p' does not pass through p_0. Indeed this can only fail if $q = 3$; for there are 0 or 2 points of $\gamma_1(k)$ at which the tangent to γ_1 passes through p_0, and 0 or 2 other points of $\ell_1 \cap \gamma_1(k)$. By Theorem 4(i) and the Corollary to Lemma 9, p_0 and p' are densely liftable. By the remark after the proof of Theorem 4, any point of $\pi_0 \cap \tilde{V}(k)$ not on the tangent to $\tilde{V}$ at p' is densely liftable; and clearly there are such points on at least two of the lines ℓ_i.

We must still identify those $\tilde{V}$ for which the argument fails. Take coordinates so that p_0 is $(1,0,0,0)$, π_0 is $X_3 = 0$ and the ℓ_i are $X_2 = 0$, $X_1 = 0$ and $X_1 + X_2 = 0$ respectively. The equation of $\tilde{V}$ has the form

(27) $$X_1 X_2 (X_1 + X_2) + X_3 Q(X_0, \dots, X_3) = 0$$

where the coefficient of X_0^2 in Q is nonzero. There are two reasons why the argument of the previous paragraph might break down: because there is no suitable π_1 or because there is no suitable p'. Suppose first that there is no suitable π_1. Let p_1 be a point of $\ell_1(k)$ other than p_0. Because we have ruled out the previous case, p_1 must be Eckardt; so by Lemma 4 $q = 2$ or 4. If $q = 4$ the proof of Lemma 4 shows that $\tilde{V}$ has the form

$$X_3(c_0 X_0^2 + c_0^2 X_0 X_3 + e_0 X_3^2) + X_2(c_1 X_1^2 + c_1^2 X_1 X_2 + e_1 X_2^2)$$
$$+ X_2 X_3 L(X_2, X_3) = 0.$$

But it must also have the corresponding form with X_1 and X_2 interchanged, as well as having the form (27); so it must actually have the form

$$X_1 X_2 (X_1 + X_2) + X_3 (c_0 X_0^2 + c_0^2 X_0 X_3 + e_0 X_3^2) = 0.$$

If the last factor on the left splits, then we can use the argument of the previous paragraph, taking for p' any point with $X_2 = 0$, $X_3 \neq 0$. If it does not split, then after writing $c_0^{-1} X_3$ for X_3 we are reduced to the two surfaces

(28) $\quad X_1 X_2 (X_1 + X_2) + X_3 (X_0^2 + X_0 X_3 + \omega X_3^2) = 0 \quad (\omega^2 + \omega + 1 = 0)$

over $\mathbf{F}_4$. If $q = 2$ the proof of Lemma 4 shows that $\tilde{V}$ has the form

$$X_0 X_3 (X_0 + X_3) + X_1 X_2 (X_1 + X_2) + b X_2 X_3 (X_0 + X_1) + C(X_2, X_3) = 0.$$

But it must also have the corresponding form with X_1, X_2 interchanged; so $b = 0$ and $C(X_2, X_3)$ is simply a multiple of X_3^3. Using an argument like that just before (28), we are reduced to the surface

(29) $$X_1 X_2 (X_1 + X_2) + X_3 (X_0^2 + X_0 X_3 + X_3^2) = 0.$$

over $\mathbf{F}_2$. Suppose instead that γ_1 exists but contains no suitable p'; thus $q = 3$. We have excluded the previous case, and if γ_1 touches ℓ_1 then the argument above shows that a suitable p' exists. So the argument can only fail if one of the 3 points of ℓ_1 other than p_0 is Eckardt and γ_1 is a conic through the remaining two points of $\ell_1(k)$. Moreover, the remaining plane through ℓ_1 must meet $\tilde{V}$ residually in a pair of lines not individually defined over k; for if it met $\tilde{V}$ in an irreducible conic that conic would contain a point which could serve as p'. Thus in particular there must be just three points of $\tilde{V}(k)$ outside π_0. Much as before, $\tilde{V}$ must have the form (6) and also the corresponding form with the roles of X_1 and X_2 interchanged; and it must also have the form (27). Thus it can be written

$$X_1X_2(X_1 + X_2) + X_3Q_0(X_0, X_3) + bX_1X_2X_3 = 0.$$

But it also needs to have a second Eckardt point on $X_1 + X_2 = X_3 = 0$, so $b = 0$. Thus $\tilde{V}$ can only be

$$X_1X_2(X_1 + X_2) + X_3(X_0^2 + X_3^2) = 0;$$

and this is not a counterexample because we can take $p' = (1, 1, 1, 2)$.

Suppose next that $\tilde{V}$ contains two skew lines ℓ_1, ℓ_2 each defined over k. These account for $2q + 2$ points of $\tilde{V}(k)$; since the order of $\tilde{V}(k)$ is congruent to 1 mod q, there must be such points not on ℓ_1 or ℓ_2. Let p be one of them, and let the unique transversal through p to ℓ_1 and ℓ_2 meet these lines in p_1 and p_2 respectively. By Lemma 9 we can lift each p_i to a point P_i of $V(K)$ and each ℓ_i to an arc Γ_i of genus 0 through P_i. Let P be the third intersection of P_1P_2 with V, and let Γ_2' be the residual intersection of V with the cone of base γ_2 and vertex P. Apply Theorem 4(i) to the curves Γ_1, Γ_2' and the line PP_2 through P_1; thus P, P_1, P_2 are all densely liftable.

We now suppose that $\tilde{V}$ contains a line ℓ defined over k, but does not contain any of the configurations discussed above. Let p be a point of ℓ such that the tangent to $\tilde{V}$ at p meets $\tilde{V}$ in ℓ and an irreducible conic γ which meets ℓ in two distinct points p and p'. By the Corollary to Lemma 10, p and p' are densely liftable. This argument only fails if for every p_1 of $\tilde{V}(k)$ outside $\ell \cup \gamma$ the tangent at p_1 to $\tilde{V}$ contains ℓ. This implies that any point p_0 on ℓ other than p or p' must be Eckardt; so $q = 2, 3$ or 4. Moreover, we have now accounted for all $q + 1$ planes through ℓ, so $\tilde{V}(k)$ contains just one point outside $\ell \cup \gamma$. The tables already cited show that this cannot happen for $q = 4$. Now take ℓ to be $X_2 = X_3 = 0$, p and p' to be $(1, 0, 0, 0)$ and $(0, 1, 0, 0)$, the unique point p_1 of $\tilde{V}(k)$ outside $\ell \cup \gamma$ to be $(0, 0, 0, 1)$ and γ to be $X_3 = 0, X_0X_1 + X_2^2 = 0$; thus $\tilde{V}$ has the form

$$X_2(X_0X_1 + X_2^2) + cX_2X_3^2 + X_3Q(X_0, X_1, X_2) = 0,$$

where $Q(X_0, X_1, 0)$ is irreducible over k. Direct calculation now yields the following two possibilities:

(30)
$$X_2(X_0X_1 + X_2^2) + X_3(X_0^2 + X_0X_1 + X_1^2) + X_2X_3(X_0 + X_1 + X_2 + X_3) = 0$$

over $\mathbf{F}_2$, and

(31) $$X_2(X_0X_1 + X_2^2) + X_3(X_0^2 + X_1^2 + X_2^2 - X_2X_3) = 0$$

over $\mathbf{F}_3$.

Still supposing that $\tilde{V}$ contains a line ℓ defined over k, we can now assume that if p is any point of $\ell(k)$ then either p is Eckardt or the tangent to $\tilde{V}$ at p meets $\tilde{V}$ residually in an irreducible conic γ which touches ℓ. Thus the tangent planes at the $q + 1$ points of $\ell(k)$ are distinct and therefore exhaust the planes through ℓ. Take coordinates so that the line ℓ is $X_2 = X_3 = 0$; thus $\tilde{V}$ has the form

$$X_2 Q_2(X_0, X_1, X_2) + X_3 Q_3(X_0, X_1, X_3) + X_2 X_3 L(X_0, \ldots, X_3) = 0.$$

By considering the tangent planes at the points of ℓ, we see that each equation

$$\lambda Q_2(X_0, X_1, 0) + \mu Q_3(X_0, X_1, 0) = 0$$

(with λ, μ in k and not both zero) has exactly one root. Hence char$(k) = 2$ and Q_2, Q_3 have no terms in $X_0 X_1$. Suppose there are points p_1, p_2 on ℓ such that the residual intersection of $\tilde{V}$ with the tangent to $\tilde{V}$ at p_i is an irreducible conic γ_i; and suppose further that there are points r_i on γ_i but not on ℓ such that $r_1 r_2$ does not touch $\tilde{V}$ and the tangents to γ_1 at r_1 and to γ_2 at r_2 do not meet. (All this is certainly possible if $q \geq 8$. For we can find p_1, p_2 because there are at most 5 Eckardt points on ℓ; and there are at most $4q$ pairs r_1, r_2 which fail the first condition and none which fail the second.) Denote by r_0 the third intersection of $r_1 r_2$ with $\tilde{V}$ and by γ_0 the residual intersection of $\tilde{V}$ with the cone of base γ_2 and vertex r_0. By Lemma 9 each γ_i is liftable at r_i and hence γ_0 is also liftable at r_1. Hence by Theorem 4(i) r_1 and r_2 are densely liftable.

We must still find the $\tilde{V}$ for which this argument fails. We know already that $q = 2$ or 4. The first possibility is that all the points of $\ell(k)$ are Eckardt and that there are no points of $\tilde{V}(k)$ outside ℓ. The tables already cited show that this cannot happen for $q = 4$; for $q = 2$ calculations based on the proof of Lemma 4 show that $\tilde{V}$ must be

(32) $$X_0 X_3(X_0 + X_3) + X_1 X_2(X_1 + X_2) + X_2^3 + X_2^2 X_3 + X_3^3 = 0$$

over $\mathbf{F}_2$. Again, if $q = 4$ the proof of Lemma 4 shows that $\ell(k)$ cannot contain exactly 4 Eckardt points; but a calculation starting from (6) shows that $\ell(k)$ can have exactly 2 Eckardt points if $\tilde{V}$ is

(33) $\quad X_3(X_0^2 + X_0X_3 + X_3^2) + X_2(X_1^2 + X_1X_2 + X_2^2) + X_0X_2X_3 = 0$

over $\mathbf{F}_2$. Leaving these two cases aside, after a change of coordinates we can take p_1 to be $(1, 0, 0, 0)$ with tangent $X_3 = 0$ and p_2 to be $(0, 1, 0, 0)$ with tangent $X_2 = 0$; and by a further change of coordinates we can reduce $\tilde{V}$ to the form

$$c_2 X_2(X_1^2 + X_0X_2) + c_3(X_0^2 + X_1X_3) + X_2X_3L(X_0, \ldots, X_4) = 0$$

where c_2, c_3 are nonzero. Straightforward calculation shows that for $q = 2$ the argument fails only for the two surfaces

(34) $\quad X_2(X_1^2 + X_0X_2) + X_3(X_0^2 + X_1X_3)X_2X_3^2 = 0,$

(35) $\quad X_2(X_1^2 + X_0X_2) + X_3(X_0^2 + X_1X_3)X_2X_3(X_0 + X_1 + X_3) = 0,$

in both cases over $\mathbf{F}_2$. But if $q = 4$ there are no cases when the argument fails.

Now we can assume that $\tilde{V}$ contains no lines defined over k. Thus if p is in $\tilde{V}(k)$ and not Eckardt, the tangent to $\tilde{V}$ at p meets $\tilde{V}$ in an irreducible cubic γ. Suppose first that there is a point p such that γ has an ordinary double point with tangent directions defined over k. By the Corollary to Lemma 10, p is densely liftable. Thus henceforth we can assume there is no such point p.

Suppose however that there is a point p such that γ has a cusp or an ordinary double point at which the tangent directions are not defined over k. Let p_1 be any other point of $\gamma(k)$ and denote by γ_1 the intersection of $\tilde{V}$ with the tangent to $\tilde{V}$ at p_1; then the third intersection p_2 of γ and γ_1 is also the third intersection of γ with the tangent to γ at p_1. Take coordinates so that p is $(1, 0, 0, 0)$ and the plane of γ is $X_3 = 0$; then γ has the form

(36) $\quad X_0 Q(X_1, X_2) + C(X_1, X_2) = 0,$

and the general point on it is $(-C(\lambda, \mu), \lambda Q(\lambda, \mu), \mu Q(\lambda, \mu))$. In particular the three points given by (λ_i, μ_i) for $i = 1, 2, 3$ are collinear if and only if the determinant whose i^{th} row is

$$-C(\lambda_i, \mu_i) \quad \lambda_i Q(\lambda_i, \mu_i) \quad \mu_i Q(\lambda_i, \mu_i)$$

vanishes. Write $C = a_0 X_1^3 + \ldots + a_3 X_2^3$. If p is an ordinary double point, by working over $\mathbf{F}_{q^2}$ we can take $Q = X_1 X_2$ and now the collinearity condition becomes

$$a_0 \lambda_1 \lambda_2 \lambda_3 + a_3 \mu_1 \mu_2 \mu_3 = 0;$$

here a_0, a_3 are nonzero because (36) is nonsingular. In particular there are at most 3 points at which the tangent has triple contact with γ. But if γ has a cusp we can take $Q = X_2^2$ and now the collinearity condition becomes

$$a_0(\lambda_1\mu_2\mu_3 + \lambda_2\mu_1\mu_3 + \lambda_3\mu_1\mu_2) + a_1\mu_1\mu_2\mu_3 = 0,$$

where $a_0 \neq 0$ because (36) is nonsingular, and where $\mu = 0$ corresponds to p. There is just one point of triple contact when char$(k) \neq 3$; when char$(k) = 3$ there are no points of triple contact when $a_1 \neq 0$ and every point is a point of triple contact when $a_1 = 0$. Also if char$(k) = 2$ all tangents meet γ again in the same point, given by $a_0\lambda = a_1\mu$.

Suppose in the notation of the previous paragraph that we can choose p_1 so that $p_2 \neq p_1$; this is certainly possible unless $q = 2$ or char$(k) = 3$ and γ has a cusp. Then γ and γ_1 are transversal at p_2. If also there is a line through p_2 which meets $\tilde{V}$ at three distinct points, then p_2 is densely liftable. This is certainly possible, using a line in the plane of γ, unless $q < 4$ or char$(k) = 2$ and γ has a cusp; for there are at most 4 points $p_3 \neq p_2$ on γ such that p_2p_3 touches γ, and at most 3 such points if $q = 4$.

We have still to list the $\tilde{V}$ for which this argument fails. Take coordinates such that p is $(1,0,0,0)$ and the tangent there is $X_3 = 0$. If $q = 2$ direct calculation shows that the argument only fails for

(37) $$X_0^2 X_1 + X_1^2 X_2 + X_2^2 X_0 + X_0 X_1 X_2 + X_3^3 = 0,$$

and

(38) $$\begin{aligned}X_0^2 X_1 + X_1^2 X_2 + X_2^2 X_0 + X_0 X_1 X_2 + X_3(X_0 X_1 + X_1 X_2 + X_2 X_0) \\ + X_3^2(X_0 + X_1 + X_2 + X_3) = 0,\end{aligned}$$

with $k = \mathbf{F}_2$ in both cases. Next suppose that $q = 3$ and p is not a cusp. The two possibilities for γ are

$$X_0(X_1^2 + X_2^2) + X_1 X_2(X_1 + X_2) = 0 \quad \text{or} \quad X_0(X_1^2 + X_2^2) + X_1^2 X_2 = 0;$$

but in the first case we can take p_1 to be $(2,1,1,0)$ and in the second case we can take it to be $(0,0,1,0)$. There remains the possibility that char$(k) = 2$ or 3 and every point of $\tilde{V}(k)$ is either an Eckardt point or a cusp. Suppose first that char$(k) = 2$; we can take the equation of γ to be $X_0 X_2^2 + X_1^3 + aX_1^2 X_2 = 0$. Thus p_2 is $(0, a, 1, 0)$ whatever choice of p_1 we make. But now the points of $\tilde{V}$ at which the tangent passes through p_2 are the points of γ and of one other plane, and this second plane must pass through p. Because we can assume that γ_1 has a cusp at p_1, if this second plane contains a point of γ_1 then it contains the whole of γ_1. Since $q \geq 4$ it is now easy to see that there is a point p_3 of $\tilde{V}(k)$ not in either of these planes nor in the tangent at p_2; and p_2p_3

is the line through p_2 which we need to complete the proof. Finally suppose that char$(k) = 3$ and that the equation of γ is $X_0 X_2^2 + X_1^3 = 0$; thus $p_2 = p_1$ whatever p_1 we choose. Now $\tilde{V}$ has the form

$$X_0 X_2^2 + X_1^3 + X_3 Q(X_0, X_1, X_2) + X_3^2 L(X_0, \ldots, X_3) = 0,$$

where the coefficient of X_0^2 in Q is nonzero and can be made 1 by rescaling. The coefficients of $X_0 X_2$ and X_2^2 in Q can be reduced to 0 by adding suitable multiples of X_3 to X_0, X_2. The condition that $(\lambda^3, -\lambda, 1)$ is a cusp for every λ is equivalent to $Q = X_0^2 + a X_0 X_1 - a X_1^2$ for some a. Applying the same argument with $(0,0,1,0)$ in place of p gives $a = 0$ because the equation of $\tilde{V}$ cannot contain a term in $X_1^2 X_3$, and also shows that L is simply a multiple of X_2. Rescaling and using the uniqueness of cube roots, we find that $\tilde{V}$ must be

$$X_0 X_2^2 + X_2 X_3^2 + X_3 X_0^2 + X_1^3 = 0.$$

But this has a singular point at $(1,0,1,1)$.

It only remains to find those $\tilde{V}$ for which every point of $\tilde{V}(k)$ is Eckardt and there are no lines on $\tilde{V}$ defined over k. Under these conditions, a line on $\tilde{V}$ cannot contain two Eckardt points defined over k, so the order of $\tilde{V}(k)$ is at most 9. Suppose first that this order is 1; then $k = 2$ and $\tilde{V}$ must be

(39)
$$\begin{aligned} & X_2^3 + X_2^2 X_3 + X_3^3 + X_1^2 X_2 + X_1 X_3^2 + X_1^3 + X_1 X_2 X_3 \\ & + X_0(X_1^2 + X_2^2 + X_3^2) + X_0^2(X_1 + X_2 + X_3) = 0 \end{aligned}$$

over $\mathbf{F}_2$. Any line of $\mathbf{P}^3(k)$ through two Eckardt points meets $\tilde{V}$ in a third distinct point, and this limits the configurations of $\tilde{V}(k)$. If $\tilde{V}(k)$ has order 3 then $q = 2$ and $\tilde{V}$ must be

(40)
$$\begin{aligned} & X_2^3 + X_2^2 X_3 + X_3^3 + (X_0 + X_1)(X_0 X_1 + X_2^2 + X_3^2) \\ & + (X_2 + X_3)(X_0^2 + X_1^2) = 0 \end{aligned}$$

over $\mathbf{F}_2$. The next possible configuration consists of 7 coplanar points, and this can only occur when char$(k) = 2$; the corresponding $\tilde{V}$ is

(41) $\quad X_1^2 X_2 + X_1 X_2^2 + X_1^2 X_3 + X_1 X_3^2 + X_2^2 X_3 + X_2 X_3^2 + X_0^3 = 0$

over $\mathbf{F}_2$. Finally, if the order of $\tilde{V}(k)$ is 9 then $\tilde{V}$ must be (42).

We have now achieved the objective stated in the first paragraph of this section: to show that unless $\tilde{V}$ belongs to one of a small number of exceptional cases every point of $\tilde{V}$ is densely liftable. These are in fact only exceptions to the method of proof and not to the conclusion; and for each of them the conclusion can be proved by exploiting the construction in Lemma 6. We illustrate this by considering the surface $\tilde{V}$ given by (29). Let ω in $\mathbf{F}_4$ be such that $1 + \omega + \omega^2 = 0$. The line $X_0 = X_1 = X_3$ meets $\tilde{V}$ in the three points

$$p_0 = (0, 0, 1, 0), \quad p'_1 = (1, 1, \omega, 1), \quad p''_1 = (1, 1, \omega, 1).$$

The tangent to $\tilde{V}$ at p'_1 is $X_0 + \omega^2 X_1 + X_2 = 0$, which meets $\tilde{V}$ in the line $X_1 = X_3$ and the irreducible conic

$$C' : X_0^2 + X_0 X_1 + X_0 X_3 + X_1^2 + X_1 X_3 + X_3^2 = 0.$$

The tangent to C' at $p'_2 = (1, \omega, 0, 0)$ is $X_0 + \omega^2 X_1 + X_2 = X_2 + \omega X_3 = 0$; so the plane through $p_3 = (1, 0, 0, 0)$ and this tangent, which is $X_2 + \omega X_3 = 0$, is transversal to its conjugate. Now choose any lift P_0 of p_0 to K, and any lift L of $p'_1 p''_1$ to K passing through P_0. Let P'_1, P''_1 be the other intersections of L with V; they are defined over a quadratic extension K_1 of K, and the prime $\mathfrak{p}$ in K is inert in K_1/K because its residue field in K is $\mathbf{F}_2$ and its residue field in K_1 contains $\mathbf{F}_4$. Let Γ' be the intersection of V with the tangent at P'_1; its reduction mod $\mathfrak{p}$ consists of a line and a conic, but the point p'_2 is nonsingular on it. Now let R be any lift of p_3 to $K_\mathfrak{p}$; by the geometric form of Hensel's Lemma and the transversality proved above, there is a unique line L_1 through R which meets Γ' at a point S' whose reduction is p'_2 and which meets Γ'' at a point whose reduction is p''_2. Since this line is unique, it is defined over $K_\mathfrak{p}$; hence S' and S'' are conjugate over $K_\mathfrak{p}$. Let T' be a point on Γ' defined over K_1 and arbitrarily close to S', and let T'' be its conjugate over K. Since T'' is arbitrarily close to S'', the line $T'T''$ meets V again in a point defined over K and arbitrarily close to R, and its R-equivalence class does not depend on the choice of T'. Hence p_3 is densely liftable, and the result for the other points of $\tilde{V}(k)$ follows immediately.

It now follows from the Corollary to Theorem 5 that, subject to the unproved cases listed earlier in this section, if $\mathscr{C}$ is an R-equivalence class of $V(K)$ then every point of $\tilde{V}(k)$ is densely liftable to $\mathscr{C}$. We now give an alternative proof of this result. It follows from Lemma 11 that if p in $\tilde{V}(k)$ is densely liftable to $\mathscr{C}_1$ and liftable to $\mathscr{C}_2$ then it is densely liftable to $\mathscr{C}_2$. We shall say that two points of $\tilde{V}(k)$ are *lift-equivalent* if they can be lifted to the same R-equivalence class in $V(K)$ — which is the same as saying that they can be densely lifted to the same class; and we shall say that two R-equivalence classes are *closure-equivalent* (in the v-adic topology) if their closures are the same. Clearly two points of $\tilde{V}(k)$ are lift-equivalent if and only if they can be lifted to the same closure-equivalence class.

The definition of universal equivalence was given at the end of §2. The main theorem of [5] states that, with the exception of the surface

(42) $$X_0^3 + X_1^3 + X_2^3 + \omega X_3^3 = 0 \quad (\omega^2 + \omega + 1 = 0)$$

over $\mathbf{F}_4$, there is only one class for universal equivalence in $\tilde{V}(k)$ provided $\tilde{V}$ is nonsingular. Clearly two points of $\tilde{V}(k)$ which are universally equivalent are also lift-equivalent; and since there is only one universal equivalence class in $\tilde{V}(k)$, there is also only one lift-equivalence class. When $\tilde{V}$ is given by (42), it is not hard to give a direct proof that there is only one lift-equivalence class, using the construction in the proof of Lemma 6; on the other hand, the proof of R-equivalence in [5] is not obviously liftable.

7. Adelic results

In this section we prove Theorem 1. Let $\mathscr{B}$ consist of the places of bad reduction of V, together with the primes that divide 6. The proof of Theorem 5 and the calculations in §6, taken together, show that if $\mathfrak{p}$ is not in $\mathscr{B}$ any point of $\tilde{V}(k)$ is densely liftable to each R-equivalence class. (The reader who does not wish to rely on heavy calculations will notice that this can be proved much more simply when q is large; see the remark at the end of §4.) If v is in $\mathscr{B}$ we define the set $U_v(\mathscr{C})$ for any R-equivalence class $\mathscr{C}$ to be the closure of $\mathscr{C}$ in $V(K_v)$ and we define U_v^* to be the complement of the closure of $V(K)$ in $V(K_v)$. By Lemma 11 the $U_v(\mathscr{C})$ as $\mathscr{C}$ varies are either disjoint or identical, and by the remarks after the proof of the Corollary to Lemma 9 there are only finitely many distinct $U_v(\mathscr{C})$ for fixed v. Hence we have defined a suitable partition (2) of $V(K_v)$.

The closure of $\mathscr{C}$ is clearly contained in the appropriate $\prod U_{iv} \times \prod V(K_v)$, so we have only to prove that the two are equal. For this it is enough to show that if $P = \prod P_v$ is in the adelic closure of $\mathscr{C}$ and we obtain P^* by replacing one P_{v_0} by a $P_{v_0}^*$ in the same U_{iv_0} or $V(K_{v_0})$, then P^* is also in the closure of $\mathscr{C}$. Fix adelic neighbourhoods $\mathscr{N} = \prod \mathscr{N}_v$ of P and $\mathscr{N}^* = \prod \mathscr{N}_v^*$ of P^*, where we can assume that $\mathscr{N}_v = \mathscr{N}_v^*$ for $v \neq v_0$, and let $\mathscr{S}$ be the finite set of places such that $\mathscr{N}_v = V(K_v)$ for v outside $\mathscr{S}$. Let $Q = \prod Q_v$ be in $\mathscr{C} \cap \mathscr{N}$. Since P_v^* is in the v-adic closure of $\mathscr{C}$, there is a point $Q_{v_0}^*$ in $\mathscr{C} \cap \mathscr{N}_{v_0}^*$; so there are finite sequences of points R_i in $\mathscr{C}$ and curves γ_i of genus 0 on V and defined over K such that

(i) : $R_0 = Q_{v_0}$ and $R_n = Q_{v_0}^*$;
(ii) : if $\varphi_i : \mathbf{P}^1 \to \gamma_i$ is the desingularization of γ_i then R_{i-1} and R_i are both in $\varphi_i(\mathbf{P}^1(K))$.

By downward induction on i we choose v_0-adic neighbourhoods $\mathcal{N}_{iv_0}$ of each R_i such that $\mathcal{N}_{nv_0} = \mathcal{N}_{v_0}^*$ and if T_i is in $\mathcal{N}_{iv_0}$ then the image of R_{i+1} under the translation of γ_{i+1} from R_i to T_i lies in $\mathcal{N}_{(i+1)v_0}$; that we can do this follows from the fact that, for fixed auxiliary points, translation is continuous in T_i and R_{i+1}. By upward induction on i, we now define a sequence of points S_i in $\mathscr{C}$ and curves γ_i' of genus 0 on V defined over K such that

(i) : $S_0 = R_0$ and S_i is in $\mathcal{N}_{iv_0}$ for $i > 0$;
(ii) : S_i is in $\mathcal{N}_v$ for each $v \neq v_0$;
(iii) : if $\psi_i : \mathbf{P}^1 \to \gamma_i'$ is the desingularization of γ_i' then S_{i-1} and S_i are both in $\psi_i(\mathbf{P}^1(K))$.

Then S_n will be in $\mathcal{N}^*$ and Theorem 1 will be proved.

The case $i = 1$ is slightly special; we take γ_1' to be the same as γ_1 and, using weak approximation on $\mathbf{P}^1(K)$, choose S_2 in $\mathcal{N}_{2v_0}$ and in $\mathcal{N}_v$ for each $v \neq v_0$. (This last condition is trivial for v outside $\mathcal{S}$.) If $i > 1$ we can suppose that we have already chosen S_i and γ_i' to satisfy the conditions above. Take γ_{i+1}' to be the translation of γ_{i+1} from R_i to S_i, and using weak approximation on $\mathbf{P}^1(K)$ choose S_{i+1} to be in $\mathcal{N}_{(i+1)v_0}$ and in $\mathcal{N}_v$ for each $v \neq v_0$.

8. Surfaces $X_1^3 + X_2^3 + X_3^3 - dX_0^3 = 0$

In this section we suppose for simplicity that $K = \mathbf{Q}$; our aim is to prove results for the special case where V is given by (3), though some of our results are more general. In (3) we can assume that d is a cube-free integer. The primes of bad reduction are those which divide $3d$; and it follows from the results of §6 that if p is not a prime of bad reduction then every point of $\tilde{V}(k)$ is densely liftable to any R-equivalence class. For any d, the surface (3) contains three obvious points such as $(0, 1, -1, 0)$; but these are all Eckardt points. Non-Eckardt points can be found by the methods of Lemma 6; the first such example was given by Ryley in 1825. The simplest non-Eckardt solution known to me is

(43) $\qquad (3(1 - 3d + 9d^2), 1 + 27d^3, -1 + 9d - 27d^3, 9d - 27d^2)$,

due to Richmond; see the notes to §13.6 in [1]. But in order to prove results about lifting at bad primes we make use of parametric solutions over $\mathbf{Q}_p$ generated by the construction in Lemma 6, and apply the Remark which follows that Lemma; because of the very simple form of the equation for V, this turns out to be simpler than using the methods of Theorems 3 and 4. By analogy with the notation introduced just before Lemma 8, we shall in this section denote by Ω a double power series in the variables A, B which

(i) : converges whenever A, B are both integral at p,
(ii) : has zero constant term,
(iii) : has all its coefficients integral at p;

Ω need not be the same from one appearance to the next, even within the same equation.

If $p \neq 3$ is a prime of bad reduction for V, then $\tilde{V}$ is a cone whose base

$$\gamma: X_1^3 + X_2^3 + X_3^3 = 0$$

is a nonsingular cubic curve defined over $\mathbf{F}_p$; and the vertex of $\tilde{V}$ is not liftable even to $\mathbf{Q}_p$. It is now convenient to generalize the previous definition of *densely liftable*. There is a map $\Phi : V(\mathbf{Q}_p) \to \gamma(\mathbf{F}_p)$ obtained by forgetting the value of X_0 and then reducing mod p. We shall say that a point r of $\gamma(\mathbf{F}_p)$ is *densely liftable* to an R-equivalence class $\mathscr{C}$ of $V(\mathbf{Q})$ if $\mathscr{C}$ is dense in $\Phi^{-1}(r)$. This definition can be extended to the general case when $\tilde{V}$ is a cone whose base is a nonsingular cubic, though it is probably only valuable when the vertex of $\tilde{V}$ is not liftable. The ideas behind the following proofs can also probably be extended to the general case, but the calculations required become considerably more tiresome.

Lemma 13. — *Let V be given by (3) and let $p \neq 3$ be a prime of bad reduction for V. Let $\mathscr{C}$ be the R-equivalence class of $P = (0, 1, -1, 0)$. Then the point $(1, -1, 0)$ on γ is densely liftable to $\mathscr{C}$.*

Proof. We use the notation of Lemma 6 with $L = \mathbf{Q}_p$; the mysterious powers of 3 in the following formulae have been inserted so that the same formulae can be used in the proofs of Lemmas 15 and 16. Take ℓ to be

$$X_0 = 3A(X_1 + X_2), \quad X_3 = -(1 + pB)(X_1 + X_2)$$

where A, B are elements of $\mathbf{Z}_p$. The points P', P'' where ℓ meets V again satisfy

$$X_1 X_2 = -(X_1 + X_2)^2 (9dA^3 + pB + p^2 B^2 + \tfrac{1}{3} p^3 B^3).$$

By Hensel's Lemma this is soluble in $\mathbf{Q}_p$, so that the Remark which follows Lemma 6 is applicable. Denote the solutions by

$$X_1/X_2 = \theta \text{ or } \theta^{-1} \quad \text{where} \quad \theta = -9dA^3(1 + p\Omega) - pB(1 + p\Omega).$$

Thus we can take P' to be given by

$$P' = (-3A(1 + \theta), -1, -\theta, (1 + pB)(1 + \theta));$$

for P'' we simply interchange X_1 and X_2. The tangent to V at P' is

(44) $\quad 9dA^2(1 + \theta)^2 X_0 - X_1 - \theta^2 X_2 - (1 + pB)^2 (1 + \theta)^2 X_3 = 0.$

Choose the line $P'R'$ to be the intersection of (44) and $X_2 = \theta X_1$, so that the general point on $P'R'$ can be written

$$(-3A(1+\theta) + s(1+pB)^2, -1, -\theta, (1+pB)(1+\theta) + 9sdA^2)$$

for some s. Substituting into the equation for V and noting that the resulting cubic in s has $s = 0$ as a double root, we find that R' is given by

$$s = \frac{9A(1+\theta)(1+pB)^4 + 243dA^4(1+pB)(1+\theta)}{(1+pB)^6 - 729d^2A^6} = 9A(1+p\Omega).$$

In contrast, choose $P''R''$ to be the intersection of $X_0 = 3A(1+\theta)X_2$ with the tangent to V at P'', so that the general point on $P''R''$ can be written

$$(-3A(1+\theta), -\theta + t(1+pB)^2(1+\theta)^2, -1, (1+pB)(1+\theta) - t\theta^2)$$

for some t. Substituting into the equation for V, we find that R'' is given by

$$t = \frac{3\theta(1+pB)^4(1+\theta)^4 - 3(1+pB)(1+\theta)\theta^4}{(1+pB)^6(1+\theta)^6 - \theta^6} = 3\theta(1+p\Omega).$$

In the notation of (3) and (12), and removing a factor 3, we have

$$R = \tfrac{1}{3}\{f_2(R'', R')R'' - f_2(R', R'')R'\}.$$

But direct clculation gives

$$\tfrac{1}{3}f_2(R'', R') = 1 + 2\theta(1+p\Omega) + 270dA^3(1+p\Omega),$$
$$\tfrac{1}{3}f_2(R', R'') = 1 - \theta(1+p\Omega) + 27dA^3(1+p\Omega),$$

whence

(45) $\quad R = (-9A(1+p\Omega), 1+p\Omega, -1+p\Omega, -3\theta(1+p\Omega) - 324dA^3(1+p\Omega)).$

It now follows from Hensel's Lemma that we can obtain in this way any point in $U = \Phi^{-1}((1, -1, 0))$. Using the approximation process described in the Remark after Lemma 6, we obtain a set of points of $V(\mathbf{Q})$ dense in U and all belonging to the same R-equivalence class $\mathscr{C}_1$. But the closure of $\mathscr{C}_1$ contains P, which is in $\mathscr{C}$; so by Lemma 11 the closure of $\mathscr{C}$ contains U. $\square$

Corollary. — *Any point of $\gamma(\mathbf{F}_p)$ which is liftable to V is densely liftable to V.*

Proof. Let R be a point of $V(\mathbf{Q})$; we have to show that $\Phi(R)$ is densely liftable. We can assume that $\Phi(R)$ is not one of the three points like $(1, -1, 0)$. Because $d \neq 1$ or 3, the tangent to γ at $\Phi(R)$ cannot pass through all these three points; so we can assume that it does not pass through $(1, -1, 0)$. Let R_2 be the third intersection of V with the join of R and $(0, 1, -1, 0)$. We have ensured that $\Phi(R_2)$ is neither $\Phi(R)$ nor $(1, -1, 0)$. Now let S be any point of $\mathscr{C} \cap \Phi^{-1}((1,-1,0))$ where $\mathscr{C}$ is as in the Lemma, and let S' be the third intersection of V with SR_2; the points S' obtained in this way are dense in $\Phi^{-1}(R)$ and they all lie in $\mathscr{C} \circ \mathscr{C}_2$ where $\mathscr{C}_2$ is the class of R_2. □

If p is large, we still appear to need a large number of R-equivalence classes — indeed potentially one for each point of $\gamma(\mathbf{F}_p)$. But this is not so, as the following result shows.

Lemma 14. — *Let V be given by (3) and let $p \neq 3$ be a prime of bad reduction for V. Let G be the group of linear equivalence classes of divisors on γ of the form $3s - \mathfrak{b}$ for some s in $\gamma(\mathbf{F}_p)$, where $\mathfrak{b}$ is the divisor of a linear section of γ. Let P be any point of $V(\mathbf{Q})$ and p_1 any point of $\gamma(\mathbf{F}_p)$ such that $p_1 - \Phi(P)$ is in G. Then p_1 is liftable (and hence densely liftable) to the class of P in $V(\mathbf{Q})$.*

Proof. Let s be such that $p_1 - \Phi(P) \sim 3s - \mathfrak{b}$. We again use the notation and construction of Lemma 6. Let P' defined over $\mathbf{Q}_p$ be in $\Phi^{-1}(s)$ and let P'' be the third intersection of PP' with V; thus $\Phi(P'') \sim \mathfrak{b} - \Phi(P) - s$. Because we can if necessary translate s by $(1,-1,0) - (0,1,-1)$, which corresponds to a 3-division point on the Jacobian, we can assume that $s \neq \mathfrak{b} - \Phi(P) - s$; this implies that $\Phi(P') \neq \Phi(P'')$ and therefore the condition in the Remark is satisfied. To choose the lines $P'R'$ and $P''R''$, let Π be a plane containing PP' but whose reduction mod p does not contain the vertex of $\tilde{V}$. Choose the line $P'R'$ to be the intersection of Π with the tangent to V at P'; thus $\Phi(R') \sim \mathfrak{b} - 2s$. Now write $P'' = (x_0, \ldots, x_3)$; we can assume without loss of generality that x_1 is a unit at p. Let $P''R''$ be the intersection of the tangent to V at P'' with $x_2 X_3 = x_3 X_2$; this last plane is defined because x_2, x_3 cannot both vanish. Since the coefficient of X_0 in the equation of the tangent is divisible by p, $\Phi(R'') = \Phi(P'') \sim \mathfrak{b} - \Phi(P) - s$. If R is the third intersection of $R'R''$ with V then

$$\Phi(R) \sim \mathfrak{b} - \Phi(R') - \Phi(R'') \sim 3s + \Phi(P) - \mathfrak{b} \sim p_1.$$

It can be shown that the points of $\gamma(\mathbf{F}_p)$ form 9 cosets of G if $p \equiv 1 \mod 3$ and 3 cosets if $p \equiv 2 \mod 3$. But it is not implied that all these cosets can be lifted, nor that we need as many distinct R-equivalence classes as there are liftable cosets. □

We have still to consider the case $p = 3$. In principle, the methods which we used to prove Lemma 13 are still applicable, but the results become more complicated because of the appearance of 3 as a binomial coefficient.

Lemma 15. — *Let V be given by (3), let $p = 3$ and suppose that $d \equiv \pm 2$ or $\pm 4 \mod 9$. For a, b in $\mathbf{Z}/(3)$ let $U(a,b)$ be the subset of $V(\mathbf{Q}_3)$ satisfying*

(46) $\qquad X_0 \equiv 3aX_1 \mod 9, \quad X_3 \equiv 3bX_1 \mod 9.$

Then any R-equivalence class is either disjoint from or dense in each $U(a,b)$; and for each $U(a,b)$ there is at least one class which is dense in it.

Proof. The argument in the proof of Lemma 13 remains valid up to and including (45); and for $a = b = 0$ the Lemma follows just as before. Analogously to (46), define subsets $U'(a,b)$, $U''(a,b)$ of $V(\mathbf{Q}_3)$ by

$$X_0 \equiv 3aX_2 \mod 9,\ X_1 \equiv 3bX_2 \mod 9 \quad \text{for} \quad U'(a,b),$$
$$X_0 \equiv 3aX_3 \mod 9,\ X_2 \equiv 3bX_3 \mod 9 \quad \text{for} \quad U''(a,b).$$

If for example we normalize a point of $U(a,b)$ by setting $X_1 = 1$ then

$$X_2 \equiv -1 - 9b^3 + 9da^3 \mod 27.$$

Now let P' be in $U'(a',b')$ and P'' in $U''(a'',b'')$, and normalize P' by $X_2 = 1$ and P'' by $X_3 = 1$; then

$$\tfrac{1}{3}f_2(P', P'') \equiv -1 + 3b' \mod 9, \quad \tfrac{1}{3}f_2(P'', P') \equiv 1 + 3b'' \mod 9$$

so that (12) and (13) imply that the third intersection of $P'P''$ with V is in $U(-a' - a'', -b' - b'')$. Similarly if P is any point of $U(-a' - a'', -b' - b'')$ and P' any point of $U'(a',b')$ then PP' meets V again in a point P'' in $U''(a'',b'')$; in other words, given such P, P' we can find P'' in $U''(a'',b'')$ such that $P'P''$ meets V again in P. This is a situation akin to translation in the sense defined in §4; and since the Lemma holds for $U'(0,0)$, each class must be either disjoint from or dense in each $U(a,b)$.

It only remains to show that each $U(a,b)$ contains a point of $V(\mathbf{Q})$. The example (43) shows that this is true for $U(1,0)$. Now let $P = (3u, 1, v, 3w)$ in $V(\mathbf{Q})$ with u, w integral at 3. A typical line through P in the tangent to V at P is the locus of

$$(3u\lambda t,\ 1 + 9t(\lambda du^2 - \mu w^2),\ v,\ 3w + \mu t)$$

as t varies; here the ratio λ/μ determines the line. The third intersection of V with this line is given by

$$t = \frac{9(du\lambda^2 - w\mu^2) - 243(du^2\lambda - w^2\mu)^2}{d\lambda^3 - \mu^3 - 729(du^2\lambda - w^2\mu)^3}.$$

If we take P to be in $U(1,0)$ then u is a unit at 3 and $3|w$; and if we choose $\lambda = 1$, $\mu = d$ then $3\|(d\lambda^3 - \mu^3)$, though this is the best we can do. Thus $3\|t$, so that there is a point of $V(\mathbf{Q})$ in one of $U(2,d)$ and $U(0,-d)$; denote its class by $\mathscr{C}$. By taking $\lambda = 1$, $\mu = 0$ we obtain a point in $U(1,0)$ which is also in $\mathscr{C}$. Being the class of the points on a singular cubic, this class satisfies $\mathscr{C} = \mathscr{C} \circ \mathscr{C}$. Using symmetry and the calculations in the previous paragraph, every $U(a,b)$ meets $V(\mathbf{Q})$; and the R-equivalence classes which meet $U(a,b)$ depend only on $da - b$. □

This Lemma does not exhaust $V(\mathbf{Q}_3)$ when $d \equiv \pm 2 \mod 9$, because we still have to consider the set U of points congruent mod 3 to $(-d, 1, 1, 0)$ and those obtained from them by symmetry. But if we can exhibit one point of $V(\mathbf{Q})$ satisfying this condition, then the "third intersection" construction will show that there is a class dense in U. If for example $d = 2$ then $(1,1,1,0)$ is such a point. We can now complete the argument for $d = 2$.

Theorem 6. — *If V is $2X_0^3 = X_1^3 + X_2^3 + X_3^3$ then the adelic closure of $V(\mathbf{Q})$ consists of all points for which one of X_1, X_2, X_3 is divisible by 6.*

Proof. Heath-Brown [2] has shown that for any rational point on V one of X_1, X_2, X_3 is divisible by 6. On the other hand, in Theorem 1 we can take $\mathscr{B} = \{2, 3\}$. We have shown that for $p = 2$ we have $n \leqslant 3$; and because of Heath-Brown's result Theorem 1 can only hold if $n = 3$ and, after renumbering, the set U_{i2} consists of all points with $2|X_i$. For $p = 3$ we do not know the decomposition (2), but we have shown that $V(\mathbf{Q})$ is dense in $V(\mathbf{Q}_3)$, so that there is no summand in (2) which does not meet $V(\mathbf{Q})$. Putting together these facts, we obtain the Theorem. Note that in addition to $(0, 1, -1, 0)$ and (43), we needed to use the point $(1, 1, 1, 0)$ on V. □

The case when $d \equiv \pm 1 \mod 9$ and is therefore a cube in $\mathbf{Q}_3$ is totally different; for now V contains a pair of skew lines together defined over $\mathbf{Q}_3$ and is therefore birationally equivalent over $\mathbf{Q}_3$ to a plane. We cannot lift this map to $\mathbf{Q}$, but we can imitate it. The lines

$$\ell_0' : d^{-1/3} X_0 - \omega X_3 = X_1 + \omega X_2 = 0,$$
$$\ell_0'' : d^{-1/3} X_0 - \omega^2 X_3 = X_1 + \omega^2 X_2 = 0$$

are conjugate over $\mathbf{Q}_3$, lie in V and do not meet. By Lemma 6 we can choose P in $V(\mathbf{Q})$ to be in general position on V; let ℓ be the unique transversal to ℓ_0', ℓ_0'' through P and let it meet ℓ_0' in P_0'. The tangent to V at P_0' meets V in ℓ_0' and a conic C' through P_0'. Let ℓ_1' be a line defined over $\mathbf{Q}_3(\omega)$, passing through P_0' and lying in the tangent to V at P_0', and suppose it meets C' again in P_1'. Given any point P_2 in $V(\mathbf{Q}_3)$, let ℓ_2 be the transversal to ℓ_0', ℓ_0'' through P_2 and let P_3' be the point where it meets ℓ_0'. Let $P_1' P_3'$ meet V again in P_4', which lies on C', and denote by ℓ_4' the line $P_0' P_4'$ which touches V at P_0'. Now let ℓ^* be a line defined over $\mathbf{Q}$ close to ℓ and passing through P; let R_0', R_0''

be its remaining intersections with V and let L be the quadratic extension of $\mathbf{Q}$ over which R_0' and R_0'' are defined. Then $L \otimes_\mathbf{Q} \mathbf{Q}_3 = \mathbf{Q}_3(\omega)$ and we can suppose that R_0' is close to P_0'. Choose lines in the tangent to V at R_0' close respectively to $P_1'P_3'$ and $P_1'P_4'$ and defined over L, and suppose they meet V again in R_3', R_4'. Let R_2' be the remaining intersection of $R_3'R_4'$ with V; thus each R_i' is close to the corresponding P_i'. Now let R_2'' be the conjugate of R_2' over $\mathbf{Q}$ and let R be the third intersection of $R_2'R_2''$ with V. Thus R is in $V(\mathbf{Q})$ and close to P_2. But every point P_2 in $V(\mathbf{Q}_3)$ can be obtained in this way; so the points R are dense in $V(\mathbf{Q}_3)$ and they all belong to the same R-equivalence class. Thus in this case the decomposition (2) contains only one summand.

Lemma 16. — *Let V be given by (3), let $p = 3$ and suppose that $3 \| d$. For a, b, c in $\mathbf{Z}/(3)$ define $U(a, b)$ by (46) and let $U^*(c)$ be the subset of $V(\mathbf{Q}_3)$ satisfying*

$$X_1 \equiv (1 + 3c)X_3 \bmod 9, \quad X_2 \equiv (1 - 3c)X_3 \bmod 9.$$

Then any R-equivalence class is either disjoint from or dense in each $U(a, b)$ and $V(\mathbf{Q})$ is either dense in or disjoint from each $U^*(c)$; the $U(a, 0)$ and $U^*(0)$ necessarily contain points of $V(\mathbf{Q})$, and all or none of the $U(a, b)$ with $b \neq 0$ and the $U^*(c)$ with $c \neq 0$ contain points of $V(\mathbf{Q})$.

Proof. If 3 divides one of X_1, X_2, X_3 then also $3 | X_0$; so the union of the $U(a, b)$ gives all points of $V(\mathbf{Q}_3)$ for which $3 | X_3$. Again, if 3 divides none of X_1, X_2, X_3 then $X_1 \equiv X_2 \equiv X_3 \bmod 3$ and $X_1 + X_2 - 2X_3 \equiv 0 \bmod 9$; so the union of the $U^*(c)$ gives all points of $V(\mathbf{Q}_3)$ for which none of X_1, X_2, X_3 is divisible by 3. The first two paragraphs of the proof of Lemma 15 remain valid, and as before (43) shows that $U(1, 0)$ is not empty; moreover the third intersection of V with the line joining any point of $U'(0, 0)$ and any point of $U'''(1, 0)$ lies in $U(2, 0)$.

We next show how to obtain a point of $U^*(0)$. In the notation of Lemma 6 with $L = \mathbf{Q}_3$, let ℓ be

$$X_0 = \tfrac{1}{2}\varepsilon(X_1 + X_2), \quad X_3 = \tfrac{1}{2}(X_1 + X_2)(1 + 3\lambda)$$

where $\varepsilon = \pm 1$ is such that $\varepsilon d \equiv 3 \bmod 9$ and λ will be chosen shortly. Thus ℓ passes through $P = (0, 1, -1, 0)$. The points P', P'' where ℓ meets V again are given by

$$8(X_1^2 - X_1X_2 + X_2^2) = (X_1 + X_2)^2(\varepsilon d - (1 + 3\lambda)^3).$$

The discriminant of this equation is $-96(2 + (1 + 3\lambda)^3 - \varepsilon d)$; if we define μ by $\varepsilon d = 3 + 9\mu$ and set $\lambda = \mu + 3\nu$ then this discriminant is congruent to

$-2^5 \cdot 3^4(\lambda^2 + \lambda^3 + \nu)$ mod 3^5 and we can therefore choose ν so that this is a 3-adic square. Denote the solutions by
$$X_1/X_2 = \theta \text{ or } \theta^{-1} \quad \text{where} \quad \theta \equiv 1 \text{ mod } 3;$$
they are in $\mathbf{Q}_3$, so that the Remark which follows Lemma 6 is applicable. We can take P' to be given by
$$P' = (\varepsilon(1+\theta),\, 2,\, 2\theta,\, (1+\theta)(1+3\lambda));$$
for P'' we simply interchange X_1 and X_2. The tangent to V at P' is
$$d(1+\theta)^2 X_0 - 4X_1 - 4\theta^2 X_2 - (1+\theta)^2(1+3\lambda)^2 X_3 = 0.$$
Choose the line $P'R'$ to be the intersection of this with $X_2 = \theta X_1$, so that the general point of $P'R'$ can be written
$$(\varepsilon(1+\theta) + s(1+3\lambda)^2,\, 2,\, 2\theta,\, (1+\theta)(1+3\lambda) + sd)$$
for some s. Substituting into the equation for V, we find that R' is given by
$$s = \frac{3d\varepsilon(1+\theta)(1+3\lambda)^4 - 3d^2(1+\theta)(1+3\lambda)}{-d(1+3\lambda)^6 + d^3} \equiv 3\varepsilon \text{ mod } 9.$$
In contrast, choose $P''R''$ to be the intersection of $\theta X_0 = \tfrac{1}{2}\varepsilon(1+\theta)X_1$ with the tangent to V at P'', so that the general point on $P''R''$ can be written
$$(\varepsilon(1+\theta),\, 2\theta,\, 2 + t(1+\theta)^2(1+3\lambda)^2,\, (1+\theta)(1+3\lambda) - 4t)$$
for some t. Substituting into the equation for V we find that R'' is given by
$$t = \frac{6(1+\theta)^4(1+3\lambda)^4 + 48(1+\theta)(1+3\lambda)}{64 - (1+\theta)^6(1+3\lambda)^6},$$
where the denominator is certainly divisible by 9. Thus
$$R'' \equiv (0,\, 0,\, (1+\theta)^2(1+3\lambda)^2,\, -4) \text{ mod } 9.$$
It follows that
$$\tfrac{1}{3}f_2(R'', R') \equiv -4(1-\theta) \text{ mod } 9, \quad \tfrac{1}{3}f_2(R', R'') \equiv 1 \text{ mod } 3$$
whence R is in $U^*(0)$. Approximating in the usual way gives a point R_0 in $V(\mathbf{Q}) \cap U^*(0)$.

A calculation like that in the second paragraph of the proof of Lemma 15 shows that if Q_1 is in $U(a,b)$ and Q_2 is in $U^*(c)$ then the third intersection of Q_1Q_2 with V is in $U^*(b-c)$. But reduction mod 3 shows that if Q' and Q'' are both congruent to $(\varepsilon, 1, 1, 1)$ mod 3 then the third intersection of $Q'Q''$ with V is not of this form, and therefore lies in some U, U' or U''. It follows that if R_1 is in $U^*(0)$ then the third intersection of R_0R_1 with V must be in some $U(a,0), U'(a,0)$ or $U''(a,0)$. Since $V(\mathbf{Q})$ is dense in each of these, it must be dense in $U^*(0)$. The remaining parts of the Lemma follow by very similar arguments. □

Remark. Further straightforward calculations, which are left to the reader, show that any R-equivalence class which meets $U^*(c)$ is dense in it and that for fixed a, b the three sets $U(a, b), U'(a, b)$ and $U''(a, b)$ meet the same R-equivalence classes. This last result is in interesting contrast with the situation in Lemma 15.

We can now complete the argument for $d = 3$.

Theorem 7. — *If V is $3X_0^3 = X_1^3 + X_2^3 + X_3^3$ then the adelic closure of $V(\mathbf{Q})$ consists of all points for which either one of X_1, X_2, X_3 is divisible by 9 or $X_1 \equiv X_2 \equiv X_3 \bmod 9$.*

Proof. Heath-Brown [2] has shown that any rational point of V satisfies these conditions. Now the Theorem follows at once from Theorem 1, since by Lemma 16 $V(\mathbf{Q})$ is dense in each $U(a, 0)$ and in $U^*(0)$. □

For completeness we now deal with the remaining case for which $p = 3$.

Lemma 17. — *Let V be given by (3), let $p = 3$ and suppose that $3^2 \| d$. For a, b in $\mathbf{Z}/(3)$ let $U(a, b)$ be the subset of $V(\mathbf{Q}_3)$ satisfying*

$$X_0 \equiv aX_1 \bmod 3, \quad X_3 \equiv 3bX_1 \bmod 9.$$

Then any R-equivalence class is either disjoint from or dense in each $U(a, b)$, and each $U(a, b)$ meets $V(\mathbf{Q})$ except perhaps when $\tfrac{1}{9}d \equiv \pm 1 \bmod 9$ and $b \neq 0$.

The proof of this is very similar to the first two paragraphs in the proof of Lemma 13, and is left to the reader. The analogues of (43) are

$$(1 - e + e^2, \ 1 + e^3, \ -1 + 3e - e^3, \ 3e - 3e^2),$$
$$(1 + e + e^2, \ -1 + e^3, \ 1 + 3e - e^3, \ 3e + 3e^2)$$

where $e = \tfrac{1}{9}d$ and a factor 3 can be taken out in one of the two cases. If $d = 9$, the Brauer-Manin obstruction shows that $V(\mathbf{Q})$ can only meet $U(a, b)$ if $b = 0$; but I do not know of any other cube-free value of d with $3^2 \| d$ for which any of the $U(a, b) \cap V(\mathbf{Q})$ is empty.

References

[1] G.H. HARDY and E.M. WRIGHT, *An Introduction to the Theory of Numbers* (Oxford, first published 1938).

[2] R. HEATH-BROWN, The density of zeros of forms for which weak approximation fails, Math. Comp. **59** (1992), 613–622.

[3] Y. MANIN, *Cubic Forms, Algebra, Geometry, Arithmetic* (North-Holland, 1974).

[4] H.P.F. SWINNERTON-DYER, *The zeta-function of a cubic surface over a finite field*, Proc. Cambridge Phil.Soc. **63** (1967), 55–71.

[5] H.P.F. SWINNERTON-DYER, *Universal equivalence for cubic surfaces over finite and local fields*, Symp. Math. **24** (1981), 111–143.

Rational points on algebraic varieties
(E. PEYRE, Y. TSCHINKEL, ed.), p. 405–446

HUA'S LEMMA AND EXPONENTIAL SUMS OVER BINARY FORMS

Trevor D. Wooley
Department of Mathematics, University of Michigan, East Hall, 525 East University Avenue, Ann Arbor, MI 48109-1109, U.S.A.
E-mail : wooley@math.lsa.umich.edu

Abstract. — We establish mean value estimates for exponential sums over binary forms of strength comparable with the bounds attainable via classical, single variable estimates for diagonal forms. These new mean value estimates strengthen earlier bounds of the author when the degree d of the form satisfies $5 \leqslant d \leqslant 10$, the improvements stemming from a basic lemma which provides uniform estimates for the number of integral points on affine plane curves in mean square. Exploited by means of the Hardy-Littlewood method, these estimates permit one to establish asymptotic formulae for the number of integral zeros of equations defined as sums of binary forms of the same degree d, provided that the number of variables exceeds $\frac{17}{16}2^d$, improving significantly on what is attainable either by classical additive methods, or indeed the general methods of Birch and Schmidt.

1. Introduction

Rather general versions of the Hardy-Littlewood method due to Birch [2] and Schmidt [13] offer remarkably successful approaches to estimating the number

2000 *Mathematics Subject Classification*. — 11D72, 11L07, 11E76, 11P55.
Key words and phrases. — Exponential sums, binary forms, diophantine equations.

*Packard Fellow and supported in part by NSF grant DMS-9970440.

of integral zeros of prescribed height satisfying a given homogeneous polynomial with integral coefficients. Both approaches require the polynomial under investigation to possess many variables in terms of its degree, and there are further hypotheses to be negotiated involving, directly or indirectly, the singular locus of the associated hypersurface. These unfortunate deficiencies of the method are significantly less pronounced when the polynomial under investigation is diagonal, which is to say, of the shape $a_1 x_1^d + \cdots + a_s x_s^d$ (see Chapter 9 of Vaughan [19]), and such is also the case when the polynomial diagonalises over $\mathbb{C}$ (see Birch and Davenport [3]). The availability of superior analytic methods for the diagonal situation motivates investigation of polynomials intermediate in complexity between the diagonal ones, and the quite general homogeneous polynomials investigated by Birch and Schmidt, the hope being that insight will be obtained relevant to the general situation. One such intermediate situation is that in which the polynomial splits as a sum of binary homogeneous polynomials, and such has been investigated with some success for cubic forms by Chowla and Davenport [7], and more recently by Brüdern and Wooley [6]. The author [22] has rather recently obtained analogues of Weyl's inequality and Hua's lemma for exponential sums over binary forms of higher degree, and thereby has made progress on problems involving sums of binary forms of arbitrary degree. This work was hindered by our lack of good uniform estimates for the number of integral points on affine plane curves. The object of this paper is to sharpen our earlier conclusions, and this we achieve by developing useful mean square estimates for the number of integral points on certain families of affine plane curves. It is to be hoped that progress will be stimulated in problems involving higher degree forms in many variables.

Before proceeding to the main thrust of this paper, it seems worthwhile to recall the conclusions stemming from the classical additive theory, and the work of Birch and Schmidt, so far as the density of integer points on hypersurfaces is concerned. First, on combining estimates of Weyl and Hua, one obtains the following classical conclusion (see Chapter 9 of Vaughan [19]).

Theorem A (classical). — *Let $a_1, \ldots, a_s \in \mathbb{Z} \setminus \{0\}$ and write*
$$F(\mathbf{x}) = a_1 x_1^d + \cdots + a_s x_s^d.$$
Then whenever $s > 2^d$, one has
$$\text{card}(\{\mathbf{x} \in [-B, B]^s \cap \mathbb{Z}^s \ : \ F(\mathbf{x}) = 0\}) \sim CB^{s-d},$$
where C denotes the "product of local densities" within the box $[-B, B]^s$.

In order to save space at this point, we avoid explaining what is meant by the term "product of local densities", and instead note merely that this number is positive and uniformly bounded away from zero whenever the equation $F(\mathbf{x}) = 0$ possesses non-singular real and p-adic solutions for every prime p. We refer the reader to Vaughan [17], [18], Heath-Brown [10] and Boklan [4] for the theory underlying the latest developments concerning the asymptotic formula in the diagonal situation. In order to describe Birch's theorem (see [2]), we recall that the singular locus of the hypersurface defined by the homogeneous equation $F(x_1, \ldots, x_s) = 0$ is the set of points $\mathbf{y} \in \mathbb{C}^s$ satisfying the equations

$$\frac{\partial F}{\partial x_1}(\mathbf{y}) = \cdots = \frac{\partial F}{\partial x_s}(\mathbf{y}) = 0.$$

Theorem B. — *Let $F(\mathbf{x}) \in \mathbb{Z}[x_1, \ldots, x_s]$ be homogeneous of degree d, and suppose that the variety defined by the equation $F(\mathbf{x}) = 0$ has a singular locus of dimension at most D. Then whenever $s - D > (d-1)2^d$, one has*

$$\mathrm{card}(\{\mathbf{x} \in [-B, B]^s \cap \mathbb{Z}^s : F(\mathbf{x}) = 0\}) \sim CB^{s-d},$$

where C denotes the "product of local densities" within the box $[-B, B]^s$.

Mention of the singular locus is removed by Schmidt [13] at the cost of introducing an invariant h associated with the polynomial under consideration. When $F(\mathbf{x}) \in \mathbb{Q}[x_1, \ldots, x_s]$ is a form of degree $d > 1$, write $h(F)$ for the least number h such that F may be written in the form

$$F = A_1 B_1 + A_2 B_2 + \cdots + A_h B_h,$$

with A_i, B_i forms in $\mathbb{Q}[\mathbf{x}]$ of positive degree for $1 \leqslant i \leqslant h$.

Theorem C. — *Let d be an integer exceeding 1, and write $\chi(d) = d^2 2^{4d} d!$. Let $F(\mathbf{x}) \in \mathbb{Z}[x_1, \ldots, x_s]$ be homogeneous of degree d, and suppose that $h(F) \geqslant \chi(d)$. Then one has*

$$\mathrm{card}(\{\mathbf{x} \in [-B, B]^s \cap \mathbb{Z}^s : F(\mathbf{x}) = 0\}) \sim CB^{s-d},$$

where C denotes the "product of local densities" within the box $[-B, B]^s$.

We reiterate that the relative simplicity and strength of Theorem A over Theorems B and C seems to us to justify the investment of further effort in investigations which carry successful elements of the classical methods over to more general situations. We are now at liberty to focus on the topics central to this paper.

Over sixty years ago, Hua [11] greatly simplified the analysis of the asymptotic formula in Waring's problem and allied additive problems with the introduction of a new mean value estimate which, to this day, remains central to the theory of exponential sums of small degree in a single variable. Roughly speaking, Hua observed that by Weyl differencing half of the exponential sums

in a suitable mean value, and interpreting the result in terms of the underlying diophantine equation, one obtains a recursive estimate for successive mean values in terms of divisor sum estimates of particularly simple type. The author has recently obtained a version of Hua's lemma for exponential sums of the type
$$\sum_{0\leqslant x,y \leqslant P} e(\alpha \Phi(x,y)),$$
in which $\Phi(x,y)$ is a non-degenerate binary form with integral coefficients, and as usual, we write $e(z)$ to denote $e^{2\pi i z}$ (see [22]). By means of a carefully orchestrated differencing procedure, we are able to engineer a recursion similar to that of Hua in the situation of a single variable. Unfortunately, however, the divisor sum estimates are complicated by the presence of estimates for the number of integral points on affine plane curves, and our relative ignorance of such matters somewhat weakens the ensuing mean value estimates. In this paper we sharpen our analogue of Hua's lemma by means of an enhanced treatment of the affine curves that arise from the differencing process at the heart of our treatment.

In order to describe our version of Hua's lemma, we require some notation. Suppose that $\Phi(x,y) \in \mathbb{Z}[x,y]$ is a binary form of degree d exceeding 1. Then we say that Φ is *degenerate* if there exist complex numbers α and β such that $\Phi(x,y)$ is identically equal to $(\alpha x + \beta y)^d$. It is easily verified that when $\Phi(x,y)$ is degenerate, then there exist integers a, b and c with $\Phi(x,y) = a(bx+cy)^d$. Finally, define the exponential sum

(1.1) $$f_\Phi(\alpha; P) = \sum_{0\leqslant x,y \leqslant P} e(\alpha \Phi(x,y)).$$

Theorem 1.1. — *Suppose that $\Phi(x,y) \in \mathbb{Z}[x,y]$ is a non-degenerate form of degree $d \geqslant 3$. Then the following estimates hold.*

(i) *When $d = 3$ or 4 and j is an integer with $1 \leqslant j \leqslant d$, or when $d \geqslant 5$ and $j = 1$ or 2, one has for each positive number ε the bound*
$$\int_0^1 |f_\Phi(\alpha; P)|^{2^{j-1}} d\alpha \ll P^{2^j - j + \varepsilon}.$$

(ii) *When $d = 5$, one has for each positive number ε the bounds*
$$\int_0^1 |f_\Phi(\alpha; P)|^4 d\alpha \ll P^{21/4+\varepsilon}, \quad \int_0^1 |f_\Phi(\alpha; P)|^8 d\alpha \ll P^{49/4+\varepsilon},$$
$$\int_0^1 |f_\Phi(\alpha; P)|^{10} d\alpha \ll P^{127/8+\varepsilon}, \quad \int_0^1 |f_\Phi(\alpha; P)|^{17} d\alpha \ll P^{29+\varepsilon}.$$

(iii) When $6 \leqslant d \leqslant 10$ and j is an integer with $3 \leqslant j \leqslant d-2$, then for each positive number ε one has

$$\int_0^1 |f_\Phi(\alpha; P)|^{2^{j-1}} d\alpha \ll P^{2^j - j + 1/(d-j+2) + \varepsilon}.$$

Also, when $6 \leqslant d \leqslant 10$, one has for each $\varepsilon > 0$ the bounds

$$\int_0^1 |f_\Phi(\alpha; P)|^{\frac{9}{32} 2^d} d\alpha \ll P^{\frac{9}{16} 2^d - d + 1 + \varepsilon}$$

and

$$\int_0^1 |f_\Phi(\alpha; P)|^{\frac{17}{32} 2^d} d\alpha \ll P^{\frac{17}{16} 2^d - d + \varepsilon}.$$

Of course, bounds for moments of $f_\Phi(\alpha; P)$ intermediate between those recorded in the statement of Theorem 1.1 may be obtained by applying Hölder's inequality to interpolate between those above. For comparison, Theorem 2 of Wooley [22] shows that when $d \geqslant 5$ and j is an integer with $1 \leqslant j \leqslant d-1$, one has

$$\int_0^1 |f_\Phi(\alpha; P)|^{2^{j-1}} d\alpha \ll P^{2^j - j + \frac{1}{2} + \varepsilon},$$

and also provides the estimates

$$\int_0^1 |f_\Phi(\alpha; P)|^{\frac{5}{16} 2^d} d\alpha \ll P^{\frac{5}{8} 2^d - d + 1 + \varepsilon} \quad \text{and} \quad \int_0^1 |f_\Phi(\alpha; P)|^{\frac{9}{16} 2^d} d\alpha \ll P^{\frac{9}{8} 2^d - d + \varepsilon}.$$

Case (iii) of Theorem 1.1 above plainly provides estimates superior to the latter bounds. On the other hand, case (i) of Theorem 1.1 is simply a restatement of the first estimate of [22, Theorem 2]. We note also that when d is greater than or equal to 11, it is possible to apply a trivial variant of Vinogradov's methods in order to obtain conclusions superior to those stemming from Theorem 1.1 (see [22, §8] for details). Since we are interested primarily in ideas likely to generalise successfully to homogeneous forms in many variables, we discuss Vinogradov's methods no further herein.

There are immediate consequences of the estimates recorded in Theorem 1.1 for the solubility of homogeneous diophantine equations which split as sums of binary forms. We confine ourselves here to a routine conclusion discussed in detail in [22].

Theorem 1.2. — *Let d be an integer with $3 \leqslant d \leqslant 10$, and define $s_0(d)$ by*

$$s_0(d) = \begin{cases} 2^{d-1}, & \text{when } d = 3, 4, \\ \frac{17}{32} 2^d, & \text{when } 5 \leqslant d \leqslant 10. \end{cases}$$

Let $s > s_0(d)$, and let $\Phi_j \in \mathbb{Z}[x,y]$ $(1 \leqslant j \leqslant s)$ be homogeneous forms of degree d with non-zero discriminants. Let $\mathcal{N}(B) = \mathcal{N}_s(B; \boldsymbol{\Phi})$ denote the number of solutions of the diophantine equation

(1.2) $$\Phi_1(x_1, y_1) + \cdots + \Phi_s(x_s, y_s) = 0,$$

subject to $|x_j| \leqslant B$ and $|y_j| \leqslant B$ $(1 \leqslant j \leqslant s)$. Then provided that the form $\Phi_1(x_1, y_1) + \cdots + \Phi_s(x_s, y_s)$ is indefinite, one has

$$\mathcal{N}_s(B; \boldsymbol{\Phi}) = \mathscr{C}\mathfrak{S}B^{2s-d} + O_{\boldsymbol{\Phi}}(B^{2s-d-\delta}),$$

for some positive number δ. Here, $\mathscr{C}$ denotes the volume of the $(2s-1)$-dimensional hypersurface determined by the equation (1.2) contained in the box $[-1,1]^{2s}$. Also, $\mathfrak{S}$ denotes the singular series $\prod_p v_p$, where the product is over prime numbers,

$$v_p = \lim_{h \to \infty} p^{h(1-2s)} M_s(p^h; \boldsymbol{\Phi}),$$

and $M_s(p^h; \boldsymbol{\Phi})$ denotes the number of solutions of the congruence

$$\Phi_1(x_1, y_1) + \cdots + \Phi_s(x_s, y_s) \equiv 0 \pmod{p^h},$$

with $1 \leqslant x_j, y_j \leqslant p^h$ $(1 \leqslant j \leqslant s)$.

We note that the expression $\mathscr{C}\mathfrak{S}$ explicitly describes the "product of local densities", for the problem at hand, previously mentioned in Theorems A, B and C. Given the existence of non-singular real and p-adic solutions of the equation (1.2), the proof of Theorem 1.2 follows precisely the argument of the proof of [22, Theorem 3], and hence we omit details in the interest of saving space.

Following some preliminary reductions in §2, we grapple with basic estimates for the number of integral points on affine plane curves in §3. We discuss the main induction in §4, thereby establishing the majority of the estimates recorded in Theorem 1.1. The closing stages of the induction have a different flavour, and this we defer to §5, completing the proof of Theorem 1.1.

Throughout this paper, implicit constants occurring in Vinogradov's notation $\ll$ and $\gg$ will depend at most on the coefficients of the implicit binary forms, a small positive number ε, exponents d and k, and quantities occurring as subscripts to the latter notations, unless otherwise indicated. We write $f \asymp g$ when $f \ll g$ and $g \ll f$. When x is a real number, we write $[x]$ for the greatest integer not exceeding x. Also, we use vector notation for brevity. Thus, for example, the s-tuple $(\Phi_1, \ldots, \Phi_s)$ will be abbreviated simply to $\boldsymbol{\Phi}$. In an effort to simplify our exposition, we adopt the convention that whenever ε appears in a statement, we are implicitly asserting that the statement holds for each

$\varepsilon > 0$. Note that the "value" of ε may consequently change from statement to statement.

The author is grateful to the referee for useful comments.

2. Preliminary reductions

Let k be an integer with $k \geqslant 3$ and let $\Phi(x,y) \in \mathbb{Z}[x,y]$ be a non-degenerate homogeneous polynomial of degree k. Let P be a large real number, and define the exponential sum $f(\alpha) = f_\Phi(\alpha; P)$ as in (1.1). We aim initially to transform $f(\alpha)$ into an associated exponential sum amenable to our differencing procedure, and the latter goal we achieve by following closely the argument of [22, §2].

When $\Phi(x,y) \in \mathbb{Z}[x,y]$, we describe the polynomial Ψ as being a *condensation* of Φ when the following condition ($\mathscr{C}$) is satisfied.

($\mathscr{C}$) We have $\Psi(u,v) \in \mathbb{Z}[u,v]$, and the coefficients of Ψ depend at most on those of Φ. Further, the polynomial $\Psi(u,v)$ has the same degree as $\Phi(x,y)$, and takes the shape

$$(2.1) \qquad \Psi(u,v) = Au^k + Bu^{k-t}v^t + \sum_{j=t+1}^{k} C_j u^{k-j} v^j,$$

with $AB \neq 0$ and $2 \leqslant t \leqslant k$.

Lemma 2.1. — *There is a condensation Ψ of Φ, and a positive real number X with $X \asymp P$, with the property that for every natural number s one has*

$$\int_0^1 |f_\Phi(\alpha; P)|^{2s} d\alpha \ll \int_0^1 |H_\Psi(\alpha; X)|^{2s} d\alpha,$$

where we write

$$(2.2) \qquad H_\Psi(\theta; X) = \sum_{|u| \leqslant X} \sum_{|v| \leqslant X} e(\theta \Psi(u,v)).$$

Proof. — This is [22, Lemma 2.3]. □

The work of [22, §5] takes care of certain special cases that arise in our treatment. We summarise the relevant conclusions of this discussion in the following two lemmata.

Lemma 2.2. — *Suppose that $k = 3$ or 4 and j is an integer with $1 \leqslant j \leqslant k$, or else that $k \geqslant 5$ and $j = 1$ or 2. Then for each positive number ε, one has*

$$\int_0^1 |f_\Phi(\alpha; P)|^{2^{j-1}} d\alpha \ll P^{2^j - j + \varepsilon}.$$

Proof. — This estimate is recorded as the first conclusion of [22, Theorem 2]. □

Lemma 2.3. — *Suppose that $\Psi(u,v) \in \mathbb{Z}[u,v]$ has the shape (2.1). Suppose also that $k \geqslant 5$, that X is a large real number, and that $H_\Psi(\alpha; X)$ is defined as in (2.2). Then for $1 \leqslant j \leqslant k$, and for each positive number ε, one has the upper bound*

$$\int_0^1 |H_\Psi(\alpha; X)|^{2^{j-1}} d\alpha \ll X^{2^j - j + \varepsilon},$$

provided either that $t = k$, or else that $t = k-1$ and $C_k = 0$. When $t = k-1$ and $C_k \neq 0$, meanwhile, then there is a condensation Υ of Ψ with the property that Υ has the shape

$$\Upsilon(x, y) = A' x^k + B' x^{k-2} y^2 + \sum_{j=3}^{k} C'_j x^{k-j} y^j,$$

with $A'B' \neq 0$, and there is a positive real number Y with $Y \asymp X$, and Υ and Y satisfy the property that for each natural number s, one has

$$\int_0^1 |H_\Psi(\alpha; X)|^{2s} d\alpha \ll \int_0^1 |H_\Upsilon(\alpha; Y)|^{2s} d\alpha.$$

Proof. — The situations in which $t = k$, or else $t = k - 1$ and $C_k = 0$, are dealt with, respectively, in Lemmata 5.2 and 5.3 of [22]. The alternative situation in which $t = k - 1$ and $C_k \neq 0$, on the other hand, is discussed in the preamble to Lemma 5.3 of [22]. □

Our deliberations are also greatly simplified through a manoeuvre that transforms a polynomial of the shape (2.1) with $t = k - 2$ into a corresponding polynomial in which $t = 2$ or 3. We begin with an analogue of Lemma 5.3 of [22]. Suppose, temporarily, that $\Psi(u,v)$ has the shape (2.1) with $t = k - 2$, so that for some integers a, b, c, d with $ab \neq 0$, one has

(2.3) $$\Psi(x, y) = ax^k + bx^2 y^{k-2} + cxy^{k-1} + dy^k.$$

Lemma 2.4. — *Suppose that $k \geqslant 4$, and that $\Psi(u,v) \in \mathbb{Z}[u,v]$ has the shape (2.3) with $ab \neq 0$ and $d = 0$. Define the exponential sum $H_\Psi(\alpha; X)$ as in (2.2). Then for $1 \leqslant j \leqslant k$, and for each positive number ε, one has the upper bound*

(2.4) $$\int_0^1 |H_\Psi(\alpha; X)|^{2^{j-1}} d\alpha \ll X^{2^j - j + \varepsilon}.$$

Proof. — Our argument is a variant of the proof of Lemma 5.3 of [22]. We abbreviate $H_\Psi(\alpha; X)$ simply to $H(\alpha)$. Also, when $1 \leqslant j \leqslant k$, we write

$$(2.5) \qquad I_j(X) = \int_0^1 |H(\alpha)|^{2^{j-1}} d\alpha.$$

The bound (2.4) is immediate from Lemma 2.2 when $j = 1, 2$. Suppose then that j is an integer with $2 \leqslant j \leqslant k-1$, and that the inequality (2.4) holds. We seek to show that (2.4) holds with j replaced by $j+1$, whence the desired conclusion follows for $1 \leqslant j \leqslant k$ by induction.

Observe first that

$$|H(\alpha)| \ll X + \sum_{1 \leqslant |x| \leqslant X} \left| \sum_{|y| \leqslant X} e(\alpha(ax^k + bx^2 y^{k-2} + cxy^{k-1})) \right|.$$

Define the exponential sum $h_l(\alpha) = h_l(\alpha; X)$ by

$$h_l(\alpha; X) = \sum_{|y| \leqslant X} e(\alpha(bly^{k-2} + cy^{k-1})).$$

Then it follows from (2.5) via Hölder's inequality that

$$(2.6) \qquad \begin{aligned} I_{j+1}(X) &\ll X^{2^{j-1}} I_j(X) + \int_0^1 \left(|H(\alpha)| \sum_{1 \leqslant |x| \leqslant X} |h_x(x\alpha)| \right)^{2^{j-1}} d\alpha \\ &\ll X^{2^{j-1}} I_j(X) + X^{2^{j-1}-1} N(X), \end{aligned}$$

where

$$(2.7) \qquad N(X) = \int_0^1 |H(\alpha)|^{2^{j-1}} \sum_{1 \leqslant |x| \leqslant X} |h_x(x\alpha)|^{2^{j-1}} d\alpha.$$

In the special situation in which $j = k-1$ and $c = 0$, we instead note that by Cauchy's inequality, one has

$$\left| \sum_{|y| \leqslant X} \sum_{1 \leqslant |x| \leqslant X} e(\alpha(ax^k + bx^2 y^{k-2})) \right|^2$$

$$\ll X \sum_{|y| \leqslant X} \left| \sum_{1 \leqslant |x| \leqslant X} e(\alpha(ax^k + bx^2 y^{k-2})) \right|^2$$

$$\ll X^3 + X \sum_{\substack{1 \leqslant |x_1|, |x_2| \leqslant X \\ x_1 \neq \pm x_2}} \left| \sum_{|y| \leqslant X} e(\alpha b(x_1^2 - x_2^2) y^{k-2}) \right|.$$

Thus, on applying Hölder's inequality within (2.5), we now obtain

$$I_k(X) \ll X^{3 \cdot 2^{k-3}} I_{k-1}(X)$$

(2.8)
$$+ X^{2^{k-3}} \int_0^1 |H(\alpha)|^{2^{k-2}} \left(\sum_{\substack{1 \leqslant |x_1|, |x_2| \leqslant X \\ x_1 \neq \pm x_2}} |h_{x_1^2 - x_2^2}(\alpha)| \right)^{2^{k-3}} d\alpha$$

$$\ll X^{3 \cdot 2^{k-3}} I_{k-1}(X) + X^{3 \cdot 2^{k-3} - 2} M(X),$$

where

(2.9) $$M(X) = \int_0^1 |H(\alpha)|^{2^{k-2}} \sum_{\substack{1 \leqslant |x_1|, |x_2| \leqslant X \\ x_1 \neq \pm x_2}} |h_{x_1^2 - x_2^2}(\alpha)|^{2^{k-3}} d\alpha.$$

By orthogonality, it follows from (2.7) that $N(X)$ is equal to the number of integral solutions of the equation

(2.10)
$$x \sum_{i=1}^{2^{j-2}} \left(b x(y_i^{k-2} - z_i^{k-2}) + c(y_i^{k-1} - z_i^{k-1}) \right) = \sum_{i=1}^{2^{j-2}} \left(\Psi(u_i, v_i) - \Psi(t_i, w_i) \right),$$

with $1 \leqslant |x| \leqslant X$, and with each of $y_i, z_i, u_i, v_i, t_i, w_i$ ($1 \leqslant i \leqslant 2^{j-2}$) bounded in absolute value by X. Let $N_0(X)$ denote the number of such solutions of (2.10) in which the right hand side of the equation is equal to zero, and let $N_1(X)$ denote the corresponding number of solutions with the latter expression non-zero. Then one has

(2.11) $$N(X) = N_0(X) + N_1(X).$$

We first estimate $N_0(X)$. On considering the underlying diophantine equations and recalling (2.5), we have

$$N_0(X) \ll I_j(X) \sum_{1 \leqslant |x| \leqslant X} \int_0^1 |h_x(\alpha)|^{2^{j-1}} d\alpha.$$

But a classical version of Hua's lemma (see Lemma 2.5 of Vaughan [19]) shows that for $2 \leqslant j \leqslant k-1$, one has

$$\int_0^1 |h_x(\alpha)|^{2^{j-1}} d\alpha \ll X^{2^{j-1} - j + 1 + \varepsilon},$$

uniformly in $x \neq 0$. Thus we deduce that for $2 \leqslant j \leqslant k-1$, one has

(2.12) $$N_0(X) \ll X^{2^{j-1} - j + 2 + \varepsilon} I_j(X).$$

In order to dispose of $N_1(X)$, we introduce some additional notation. For each integer l, we denote by $r_j(n;l)$ the number of representations of the integer n in the form

$$n = l \sum_{i=1}^{2^{j-2}} \left(bl(y_i^{k-2} - z_i^{k-2}) + c(y_i^{k-1} - z_i^{k-1}) \right),$$

with $|y_i| \leqslant X$ and $|z_i| \leqslant X$ ($1 \leqslant i \leqslant 2^{j-2}$). Similarly, for each integer n we write $R_j(n)$ for the number of representations of n in the form

$$n = \sum_{i=1}^{2^{j-2}} \left(\Psi(u_i, v_i) - \Psi(t_i, w_i) \right),$$

with each of u_i, v_i, t_i, w_i ($1 \leqslant i \leqslant 2^{j-2}$) bounded in absolute value by X. Then on writing $\gamma = (|b| + |c|)2^j$, we find that

$$N_1(X) \leqslant \sum_{1 \leqslant |n| \leqslant \gamma X^k} R_j(n) \sum_{\substack{l|n \\ |l| \leqslant X}} r_j(n;l).$$

On applying an elementary estimate for the divisor function, we therefore deduce from Cauchy's inequality that

(2.13)
$$N_1(X) \leqslant \left(\sum_{n \in \mathbb{Z}} R_j(n)^2 \right)^{1/2} \left(\sum_{1 \leqslant |n| \leqslant \gamma X^k} \left(\sum_{\substack{l|n \\ |l| \leqslant X}} r_j(n;l) \right)^2 \right)^{1/2}$$

$$\ll X^\varepsilon \left(\sum_{n \in \mathbb{Z}} R_j(n)^2 \right)^{1/2} \left(\sum_{n \in \mathbb{Z}} \sum_{1 \leqslant |l| \leqslant X} r_j(n;l)^2 \right)^{1/2}.$$

However, on considering the underlying diophantine equations, it is apparent from (2.13) that

$$N_1(X) \ll X^\varepsilon \left(I_{j+1}(X) \right)^{1/2} \left(\sum_{1 \leqslant |l| \leqslant X} \int_0^1 |h_l(\alpha)|^{2^j} d\alpha \right)^{1/2}.$$

But the classical version of Hua's lemma (see Lemma 2.5 of [19]) shows that for $1 \leqslant j \leqslant k-2$, one has

$$\int_0^1 |h_l(\alpha)|^{2^j} d\alpha \ll X^{2^j - j + \varepsilon},$$

uniformly in $l \neq 0$. Moreover, the latter conclusion remains valid for $j = k - 1$ whenever c is non-zero. In either circumstance, we deduce that

$$(2.14) \qquad N_1(X) \ll X^\varepsilon (I_{j+1}(X))^{1/2} \left(\sum_{1 \leqslant |l| \leqslant X} X^{2^j - j + \varepsilon} \right)^{1/2}.$$

On combining (2.6), (2.11), (2.12) and (2.14), we find that for $2 \leqslant j \leqslant k - 2$, and also when $j = k - 1$ and $c \neq 0$, one has

$$I_{j+1}(X) \ll \left(X^{2^{j-1}} + X^{2^j - j + 1 + \varepsilon} \right) I_j(X) + X^{2^j - (j+1)/2 + \varepsilon} (I_{j+1}(X))^{1/2},$$

whence our inductive hypothesis (2.4) leads to the upper bound

$$I_{j+1}(X) \ll X^{2^{j+1} - j - 1 + \varepsilon} + X^{2^j - (j+1)/2 + \varepsilon} (I_{j+1}(X))^{1/2}.$$

Thus the estimate (2.4) follows with $j + 1$ in place of j in the current circumstances, and so the conclusion of the lemma has been established in all cases but that in which $c = 0$ and $j = k - 1$.

We now turn to the final elusive case wherein $c = 0$ and $j = k - 1$. By orthogonality, it follows from (2.9) that $M(X)$ is equal to the number of integral solutions of the equation

$$(2.15) \qquad b(x_1^2 - x_2^2) \sum_{i=1}^{2^{k-4}} (y_i^{k-2} - z_i^{k-2}) = \sum_{i=1}^{2^{k-3}} (\Psi(u_i, v_i) - \Psi(t_i, w_i)),$$

with $1 \leqslant |x_1|, |x_2| \leqslant X$ and $x_1 \neq \pm x_2$, and with each of y_i, z_i $(1 \leqslant i \leqslant 2^{k-4})$, and u_i, v_i, t_i, w_i $(1 \leqslant i \leqslant 2^{k-3})$ bounded in absolute value by X. Let $M_0(X)$ denote the number of such solutions of (2.15) in which the right hand side of the equation is equal to zero, and let $M_1(X)$ denote the corresponding number of solutions with the latter expression non-zero. Then plainly one has

$$(2.16) \qquad M(X) = M_0(X) + M_1(X).$$

We first estimate $M_0(X)$. On considering the underlying diophantine equations and recalling (2.5), we have

$$M_0(X) \ll I_{k-1}(X) \sum_{1 \leqslant l,m \leqslant 2X} \int_0^1 |h_{lm}(\alpha)|^{2^{k-3}} d\alpha.$$

But a classical version of Hua's lemma shows that

$$\int_0^1 |h_{lm}(\alpha)|^{2^{k-3}} d\alpha \ll X^{2^{k-3} - k + 3 + \varepsilon},$$

uniformly in $lm \neq 0$, whence we obtain

$$(2.17) \qquad M_0(X) \ll X^{2^{k-3} - k + 5 + \varepsilon} I_{k-1}(X).$$

Meanwhile, recycling the notation introduced to treat $N_1(X)$, we see that

$$M_1(X) \leqslant \sum_{1 \leqslant |n| \leqslant \gamma X^k} R_{k-1}(n) \sum_{\substack{l|n \\ |l| \leqslant 2X}} \sum_{\substack{m|n \\ |m| \leqslant 2X}} T(n; lm),$$

where we write $T(n; \lambda)$ for the number of representations of the integer n in the form

$$n = b\lambda \sum_{i=1}^{2^{k-4}} (y_i^{k-2} - z_i^{k-2}),$$

with $|y_i| \leqslant X$ and $|z_i| \leqslant X$ ($1 \leqslant i \leqslant 2^{k-4}$). Again applying an elementary estimate for the divisor function, we deduce from Cauchy's inequality that

(2.18)
$$M_1(X) \leqslant \left(\sum_{n \in \mathbb{Z}} R_{k-1}(n)^2\right)^{1/2} \left(\sum_{1 \leqslant |n| \leqslant \gamma X^k} \left(\sum_{\substack{l|n \\ |l| \leqslant 2X}} \sum_{\substack{m|n \\ |m| \leqslant 2X}} T(n; lm)\right)^2\right)^{1/2}$$

$$\ll X^\varepsilon \left(\sum_{n \in \mathbb{Z}} R_{k-1}(n)^2\right)^{1/2} \left(\sum_{n \in \mathbb{Z}} \sum_{1 \leqslant |l| \leqslant 2X} \sum_{1 \leqslant |m| \leqslant 2X} T(n; lm)^2\right)^{1/2}.$$

On considering the underlying diophantine equations, we find from (2.18) that

$$M_1(X) \ll X^\varepsilon (I_k(X))^{1/2} \left(\sum_{1 \leqslant |l| \leqslant 2X} \sum_{1 \leqslant |m| \leqslant 2X} \int_0^1 |h_{lm}(\alpha)|^{2^{k-2}} d\alpha\right)^{1/2}.$$

Again applying the classical version of Hua's lemma, one has

$$\int_0^1 |h_{lm}(\alpha)|^{2^{k-2}} d\alpha \ll X^{2^{k-2}-k+2+\varepsilon},$$

uniformly in $lm \neq 0$, whence

(2.19) $$M_1(X) \ll X^\varepsilon (I_k(X))^{1/2} \left(\sum_{1 \leqslant |l| \leqslant 2X} \sum_{1 \leqslant |m| \leqslant 2X} X^{2^{k-2}-k+2+\varepsilon}\right)^{1/2}.$$

On combining (2.8), (2.16), (2.17) and (2.19), we find that when $c = 0$, one has

$$I_k(X) \ll \left(X^{3 \cdot 2^{k-3}} + X^{2^{k-1}-k+3+\varepsilon}\right) I_{k-1}(X) + X^{2^{k-1}-k/2+\varepsilon}(I_k(X))^{1/2},$$

whence our inductive hypothesis (2.4) with $j = k - 1$ leads to the upper bound

$$I_k(X) \ll X^{2^k - k + \varepsilon} + X^{2^{k-1}-k/2+\varepsilon}(I_k(X))^{1/2}.$$

We therefore conclude that (2.4) holds with $j = k$ even when $c = 0$, and this completes the proof of the lemma. □

Lemma 2.5. — *Suppose that $k \geqslant 4$, and that $\Psi(u,v) \in \mathbb{Z}[u,v]$ has the shape (2.3) with $abd \neq 0$. Define the exponential sum $H_\Psi(\alpha; X)$ as in (2.2). Then there is a condensation Υ of Ψ with the property that Υ has the shape*

$$(2.20) \qquad \Upsilon(x,y) = A'x^k + B'x^{k-t}y^t + \sum_{j=t+1}^{k} C'_j x^{k-j} y^j,$$

with $A'B' \neq 0$ and $2 \leqslant t \leqslant 3$, and there is a positive real number Y with $Y \asymp X$, and Υ and Y satisfy the property that for each natural number s, one has

$$(2.21) \qquad \int_0^1 |H_\Psi(\alpha; X)|^{2s}\, d\alpha \ll \int_0^1 |H_\Upsilon(\alpha; Y)|^{2s}\, d\alpha.$$

Proof. — By hypothesis, the coefficient d is non-zero, and thus we may make the non-singular change of variable $u = kdy + cx$, $v = x$. Write

$$(2.22) \qquad \Upsilon(u,v) = \Psi(kdv, u - cv),$$

so that one has $\Upsilon(u,v) = (kd)^k \Psi(x,y)$. Then it follows from the argument of the proof of Lemma 2.3 of [**22**] that for some positive real number Y with $Y \asymp X$, and for every natural number s, one has the upper bound (2.21). The proof of the lemma will therefore be completed on establishing that the polynomial $\Upsilon(x,y)$, defined in (2.22), has the shape (2.20). In order to establish the latter conclusion, we apply Taylor's theorem to determine whether or not various coefficients of $\Upsilon(u,v)$ vanish.

Write ∂_z for the differential operator $\partial/\partial z$. Then the coefficient of u^k in $\Upsilon(u,v)$ is equal to

$$\frac{1}{k!} \partial_u^k \Psi(kdv, u - cv) \Big|_{(u,v) = (0,0)}.$$

On writing $\Psi_{i,j}$ for

$$\partial_x^i \partial_y^j \Psi(x,y) \Big|_{(x,y) = (0,0)},$$

one finds by the chain rule, therefore, that the coefficient of u^k in $\Upsilon(u,v)$ is equal to

$$(2.23) \qquad \frac{1}{k!} \Psi_{0,k} = d,$$

and this is non-zero by hypothesis. Similarly, the coefficient of $u^{k-1}v$ in $\Upsilon(u,v)$ is equal to

(2.24)
$$\frac{1}{(k-1)!}\partial_u^{k-1}\partial_v\Psi(kdv, u-cv)\Big|_{(u,v)=(0,0)} = \frac{1}{(k-1)!}(kd\Psi_{1,k-1} - c\Psi_{0,k})$$
$$= kdc - ckd = 0.$$

Next, the coefficient of $u^{k-2}v^2$ in $\Upsilon(u,v)$ is equal to

$$\frac{1}{2!(k-2)!}\partial_u^{k-2}\partial_v^2\Psi(kdv, u-cv)\Big|_{(u,v)=(0,0)}$$

(2.25)
$$= \frac{1}{2!(k-2)!}\left((kd)^2\Psi_{2,k-2} - 2kdc\Psi_{1,k-1} + c^2\Psi_{0,k}\right)$$
$$= bk^2d^2 - \tfrac{1}{2}k(k-1)dc^2.$$

Finally, the coefficient of $u^{k-3}v^3$ in $\Upsilon(u,v)$ is equal to

(2.26)
$$\frac{1}{3!(k-3)!}\partial_u^{k-3}\partial_v^3\Psi(kdv, u-cv)\Big|_{(u,v)=(0,0)}$$
$$= \frac{1}{3!(k-3)!}\left((kd)^3\Psi_{3,k-3} - 3(kd)^2c\Psi_{2,k-2} + 3kdc^2\Psi_{1,k-1} - c^3\Psi_{0,k}\right)$$
$$= -bk^2(k-2)d^2c + \tfrac{1}{3}k(k-1)(k-2)dc^3.$$

When $c = 0$, one finds from (2.25) that the coefficient of $u^{k-2}v^2$ is bk^2d^2, and this is non-zero by hypothesis. When $c \neq 0$, on the other hand, it follows from (2.25) and (2.26) that when the coefficients of both $u^{k-2}v^2$ and $u^{k-3}v^3$ are zero, then necessarily

$$2kbd = (k-1)c^2 \quad \text{and} \quad 3kbd = (k-1)c^2,$$

whence $bd = c = 0$, contrary to hypothesis. We therefore conclude from equations (2.23)–(2.26) that $\Upsilon(x,y)$ does indeed take the shape (2.20), wherein $A'B' \neq 0$ and $t = 2$ or 3. This completes the proof of the lemma. □

We next recall the Weyl differencing lemma. Let Δ_j denote the jth iterate of the forward differencing operator, so that for any function Ω of a real variable α, one has

$$\Delta_1(\Omega(\alpha); \beta) = \Omega(\alpha + \beta) - \Omega(\alpha),$$

and when j is a natural number,

$$\Delta_{j+1}(\Omega(\alpha); \beta_1, \ldots, \beta_{j+1}) = \Delta_1(\Delta_j(\Omega(\alpha); \beta_1, \ldots, \beta_j); \beta_{j+1}).$$

We adopt the convention that $\Delta_0(\Omega(\alpha); \beta) = \Omega(\alpha)$.

Lemma 2.6. — *Let X be a positive real number, and let $\Omega(x)$ be an arbitrary arithmetical function. Write*
$$T(\Omega) = \sum_{|x| \leqslant X} e(\Omega(x)).$$
Then for each natural number j there exist intervals $I_i = I_i(\mathbf{h})$ $(1 \leqslant i \leqslant j)$, possibly empty, satisfying
$$I_1(h_1) \subseteq [-X, X] \quad \text{and} \quad I_i(h_1, \ldots, h_i) \subseteq I_{i-1}(h_1, \ldots, h_{i-1}) \quad (2 \leqslant i \leqslant j),$$
with the property that
$$|T(\Omega)|^{2^j} \leqslant (4X+1)^{2^j - j - 1} \sum_{|h_1| \leqslant 2X} \cdots \sum_{|h_j| \leqslant 2X} T_j,$$
and here we write
$$T_j = \sum_{x \in I_j \cap \mathbb{Z}} e(\Delta_j(\Omega(x); h_1, \ldots, h_j)).$$

Proof. — This trivial variant of Lemma 2.3 of Vaughan [19] is recorded as Lemma 3.2 of [22]. □

We must also make use of a two dimensional forward differencing operator $\Delta_{i,j}$ defined as follows. When $\Omega(x, y)$ is a function of the real variables x and y, one defines
$$\Delta_{1,0}(\Omega(x,y); \beta) = \Omega(x + \beta, y) - \Omega(x, y)$$
and
$$\Delta_{0,1}(\Omega(x,y); \gamma) = \Omega(x, y + \gamma) - \Omega(x, y).$$
When i and j are non-negative integers, one then defines
$$\Delta_{i,j}(\Omega(x,y); \beta_1, \ldots, \beta_i; \gamma_1, \ldots, \gamma_j)$$
by taking $\Delta_{0,0}(\Omega(x,y); \beta; \gamma) = \Omega(x, y)$, and in general by means of the relations
$$\Delta_{i+1,j}(\Omega(x,y); \beta_1, \ldots, \beta_{i+1}; \gamma_1, \ldots, \gamma_j)$$
$$= \Delta_{1,0}(\Delta_{i,j}(\Omega(x,y); \beta_1, \ldots, \beta_i; \gamma_1, \ldots, \gamma_j); \beta_{i+1})$$
and
$$\Delta_{i,j+1}(\Omega(x,y); \beta_1, \ldots, \beta_i; \gamma_1, \ldots, \gamma_{j+1})$$
$$= \Delta_{0,1}(\Delta_{i,j}(\Omega(x,y); \beta_1, \ldots, \beta_i; \gamma_1, \ldots, \gamma_j); \gamma_{j+1}).$$

3. Integral points on affine plane curves

Essential to the main body of our argument are estimates for the number of integral points on affine plane curves, and in this section we record the estimates required for later use. Our basic tool is the following result of Bombieri and Pila [5].

Lemma 3.1. — *Let $\mathscr{C}$ be the curve defined by the equation $F(x,y) = 0$, where $F(x,y) \in \mathbb{R}[x,y]$ is an absolutely irreducible polynomial of degree $d \geqslant 2$. Also, let $N \geqslant \exp(d^6)$. Then the number of integral points on $\mathscr{C}$, and inside a square $[0,N] \times [0,N]$, does not exceed*
$$N^{1/d} \exp(12(d \log N \log \log N)^{1/2}).$$

Proof. — This is Theorem 5 of Bombieri and Pila [5]. We note that slightly sharper estimates are now available through work of Pila [12], though these new estimates have no impact on the present work. □

At the request of the referee, we point out that applications of this result of Bombieri and Pila (of a rather different flavour, involving slicing arguments) may be found in [1], [14], [15] and [16]. An application more akin to that at hand may be examined in §3 of [21] (the argument therein was in fact inspired by the proof of Lemma 3.2 below). We avoid detailed discussion of the absolute irreducibility condition occurring in the above lemma by careful averaging. Here the initial stages of our argument are modelled closely on the method of the proof of [22, Lemma 4.2].

Lemma 3.2. — *Let X denote a large real number. Suppose that $F(x,y) \in \mathbb{Z}[x,y]$ is a non-degenerate polynomial of degree $d \geqslant 2$, and that X is sufficiently large in terms of d. Suppose also that for some fixed positive number A, one has that the coefficients of F are each bounded in absolute value by X^A. Given a polynomial $T(x,y) \in \mathbb{R}[x,y]$, denote by $r_T(n;X)$ the number of solutions of the diophantine equation $T(x,y) = n$, with $(x,y) \in [-X, X]^2 \cap \mathbb{Z}^2$. Then one of the following two situations must occur, and in each of the bounds which follows, implicit constants depend at most on d, ε and A, and otherwise are independent of the coefficients of F.*

(i) There exist polynomials $G(x,y) \in \mathbb{Z}[x,y]$ and $g(t) \in \mathbb{Q}[t]$ satisfying the following conditions.

(a) G is non-degenerate of degree exceeding 1;
(b) g has degree exceeding 1;
(c) the equation $F(x,y) = g(G(x,y))$ is satisfied identically;

(d) one has

$$\sum_{n\in\mathbb{Z}} r_F(n;X)^2 \ll X^{2+1/d+\varepsilon} + X^\varepsilon \sum_{n\in\mathbb{Z}} r_G(n;X)^2.$$

(ii) No polynomials G, g exist satisfying the conditions (a), (b), (c), (d) above. Then one has

$$\sum_{n\in\mathbb{Z}} r_F(n;X)^2 \ll X^{2+1/d+\varepsilon}.$$

Proof. — Consider an integer $n \in \mathbb{N}$ with $r_F(n;X) \neq 0$. In view of the hypotheses of the statement of the lemma, we may suppose that for some fixed positive number B, one has $|n| \leqslant X^B$. When i is a non-negative integer, write

$$\mathscr{Z}_i = \{n \in \mathbb{Z} : |n| \leqslant 2^i X^B\}.$$

Also, let $\mathscr{N}_1$ denote the set of integers $n \in \mathscr{Z}_0$ for which the polynomial $F(x,y) - n$ is absolutely irreducible. Then an application of Lemma 3.1 reveals that for each $n \in \mathscr{N}_1$, one has $r_F(n;X) = O(X^{1/d+\varepsilon})$, whence

$$(3.1) \qquad \sum_{n\in\mathscr{N}_1} r_F(n;X)^2 \ll X^{1/d+\varepsilon} \sum_{n\in\mathscr{Z}_0} r_F(n;X) \ll X^{2+1/d+\varepsilon}.$$

Suppose next that $n \notin \mathscr{N}_1$, so that the polynomial $F(x,y) - n$ factors as a product of absolutely irreducible factors, say

$$F(x,y) - n = \prod_{j=1}^{l} g_j(x,y) \prod_{k=1}^{m} h_k(x,y),$$

where $l + m \geqslant 2$, and where $g_j(x,y) \in \mathbb{R}[x,y]$ $(1 \leqslant j \leqslant l)$, and

$$h_k(x,y) = u_k(x,y) + v_k(x,y)\sqrt{-1} \quad (1 \leqslant k \leqslant m),$$

with $u_k, v_k \in \mathbb{R}[x,y]$ for each k. Since $h_k(x,y)$ is presumed to be absolutely irreducible, we may suppose that $u_k(x,y)$ and $v_k(x,y)$ have no nontrivial polynomial common divisor over $\mathbb{C}[x,y]$. It therefore follows from Bezout's theorem that the number of solutions of the simultaneous equations $u_k(x,y) = v_k(x,y) = 0$ is bounded above by d^2. By considering real and imaginary components, therefore, the number of integral solutions of the equation $h_k(x,y) = 0$ is also bounded above by d^2. Next, if $g_j(x,y)$ is not some constant multiple of a $\mathbb{Q}$-rational polynomial, then since $g_j(x,y)$ is necessarily a constant multiple of a polynomial with algebraic coefficients, we deduce that the number of integral solutions of the equation $g_j(x,y) = 0$ is at most d^2. For we may remove the aforementioned constant factor and consider components with respect to some basis for the field extension containing the coefficients of

$g_j(x, y)$. Then since $g_j(x, y)$ is not a constant multiple of a $\mathbb{Q}$-rational polynomial, we find that the integral zeros of $g_j(x, y) = 0$ necessarily satisfy at least two linearly independent $\mathbb{Q}$-rational polynomial equations of degree at most d, whence the desired conclusion follows as in the complex case. Let $\mathcal{N}_2$ denote the set of integers $n \in \mathcal{Z}_0 \setminus \mathcal{N}_1$ for which the polynomial $F(x, y) - n$ possesses no non-trivial, absolutely irreducible $\mathbb{Q}$-rational polynomial factor. Then the above argument shows that for each $n \in \mathcal{N}_2$, one has $r_F(n; X) = O(1)$, whence

$$(3.2) \qquad \sum_{n \in \mathcal{N}_2} r_F(n; X)^2 \ll \sum_{n \in \mathcal{Z}_0} r_F(n; X) \ll X^2.$$

Suppose next that the set $\mathcal{Z}_0 \setminus (\mathcal{N}_1 \cup \mathcal{N}_2)$ is non-empty, so that there exists some integer $n_0 \in \mathcal{Z}_0$ with the property that $F(x, y) - n_0$ possesses a non-trivial, absolutely irreducible $\mathbb{Q}$-rational polynomial factor. Since $F(x, y)$ has integer coefficients, it follows that $F(x, y) - n_0$ may be written as a product

$$(3.3) \qquad F(x, y) - n_0 = \psi_1(x, y) \ldots \psi_m(x, y),$$

with each $\psi_i(x, y) \in \mathbb{Z}[x, y]$ irreducible of degree d_i, say. Moreover, we may suppose without loss of generality that $m \geqslant 2$ and that $d_1 + \cdots + d_m = d$. Furthermore, on writing

$$R(\mathbf{u}; \boldsymbol{\varphi}; X) = R(u_1, \ldots, u_m; \varphi_1, \ldots, \varphi_m; X)$$

for the number of integer solutions of the system of equations

$$(3.4) \qquad \varphi_i(x, y) = u_i \quad (1 \leqslant i \leqslant m),$$

with $|x|, |y| \leqslant X$, it follows from (3.3) that when $\mathcal{Z}_0 \setminus (\mathcal{N}_1 \cup \mathcal{N}_2)$ is non-empty, one has

$$\sum_{n \in \mathcal{Z}_0 \setminus \{n_0\}} r_F(n; X)^2 \leqslant \sum_{n \in \mathcal{Z}_1 \setminus \{0\}} \left(\sum_{u_1 \ldots u_m = n} R(\mathbf{u}; \boldsymbol{\psi}; X) \right)^2.$$

Notice here that on the right hand side of the last inequality, we are implicitly applying a shift by $-n_0$ to $\mathcal{Z}_0$, and then we note that this shifted set is contained in $\mathcal{Z}_1$. Thus, on combining an application of Cauchy's inequality with an elementary estimate for the divisor function, we obtain

$$(3.5) \qquad \sum_{n \in \mathcal{Z}_0} r_F(n; X)^2 \ll r_F(n_0; X)^2 + X^\varepsilon \sum_{n \in \mathcal{Z}_1 \setminus \{0\}} \sum_{u_1 \ldots u_m = n} R(\mathbf{u}; \boldsymbol{\psi}; X)^2$$

$$\ll X^2 + X^\varepsilon \sum_{\mathbf{u} \in \mathcal{Z}_1^m} R(\mathbf{u}; \boldsymbol{\psi}; X)^2.$$

Suppose now that $m \geqslant 2$, and that for $1 \leqslant i \leqslant m$ the polynomials $\varphi_i(x, y) \in \mathbb{Z}[x, y]$ have degree $d_i \geqslant 1$. Suppose also that these polynomials satisfy the condition that $d_1 + \cdots + d_m \leqslant d$, that $F(x,y)$ is a polynomial in $\varphi_1,\ldots,\varphi_m$, and that for some j with $1 \leqslant j < d$, one has the upper bound

(3.6) $$\sum_{n \in \mathcal{X}_0} r_F(n;X)^2 \ll X^{2+\varepsilon} + X^\varepsilon \sum_{\mathbf{u} \in \mathcal{X}_j^m} R(\mathbf{u};\varphi;X)^2.$$

Note that by (3.3) and (3.5), this condition is already met when $\varphi = \psi$, wherein we take $j = 1$. It is possible that the intersection (3.4) is proper for every available choice of $\mathbf{u}$, by which we mean that the intersection over $\mathbb{C}$ consists of isolated points only, and in such circumstances an application of Bezout's theorem leads to the bound $R(\mathbf{u};\varphi;X) = O(1)$ uniformly in $\mathbf{u}$, whence

$$\sum_{\mathbf{u} \in \mathcal{X}_j^m} R(\mathbf{u};\varphi;X)^2 \ll \sum_{\mathbf{u} \in \mathcal{X}_j^m} R(\mathbf{u};\varphi;X) \ll X^2.$$

If, on the other hand, there exists a choice of $\mathbf{u}$ in the summation for which the intersection defined by (3.4) is improper, say $\mathbf{u} = \mathbf{u}^*$, then the polynomials $\varphi_i - u_i^*$ ($1 \leqslant i \leqslant m$) must possess a non-trivial common factor $\chi_{m+1} \in \mathbb{Z}[x,y]$. Denote by $\chi_1,\ldots,\chi_m \in \mathbb{Z}[x,y]$ the quotient polynomials satisfying the equations

(3.7) $$\varphi_i(x,y) - u_i^* = \chi_{m+1}(x,y)\chi_i(x,y) \quad (1 \leqslant i \leqslant m).$$

Then it is apparent that

$$\sum_{\mathbf{u} \in \mathcal{X}_j^m} R(\mathbf{u};\varphi;X)^2 \ll \sum_{i=1}^m R(u_i^*;\varphi_i;X)^2 + \sum_{\mathbf{u} \in (\mathcal{X}_{j+1}\setminus\{0\})^m} \left(\sum_{\substack{\mathbf{v} \in \mathcal{X}_{j+1}^{m+1} \\ v_i v_{m+1} = u_i \ (1 \leqslant i \leqslant m)}} R(\mathbf{v};\chi;X) \right)^2,$$

whence by combining Cauchy's inequality with an elementary divisor function estimate, one obtains

(3.8) $$\sum_{\mathbf{u} \in \mathcal{X}_j^m} R(\mathbf{u};\varphi;X)^2 \ll X^2 + X^\varepsilon \sum_{\mathbf{u} \in (\mathcal{X}_{j+1}\setminus\{0\})^m} \sum_{\substack{\mathbf{v} \in (\mathcal{X}_{j+1}\setminus\{0\})^{m+1} \\ v_i v_{m+1} = u_i \ (1 \leqslant i \leqslant m)}} R(\mathbf{v};\chi;X)^2$$

$$\leqslant X^2 + X^\varepsilon \sum_{\mathbf{v} \in \mathcal{X}_{j+1}^{m+1}} R(\mathbf{v};\chi;X)^2.$$

Let $\{\chi_{i_1},\ldots,\chi_{i_l}\}$ denote the subset of $\{\chi_1,\ldots,\chi_{m+1}\}$ in which constant polynomials are omitted. Then it is apparent from (3.7) and our initial hypothesis that $F(x,y)$ is a polynomial in $\chi_{i_1},\ldots,\chi_{i_l}$. If the degrees of the latter polynomials are respectively $e_1,\ldots,e_l$, then it is clear from (3.7) also that

$$e_1 + \cdots + e_l < d_1 + \cdots + d_m \leqslant d.$$

Also, on combining the hypothesis (3.6) with (3.8), one deduces that

(3.9) $$\sum_{n \in \mathscr{L}_0} r_F(n;X)^2 \ll X^{2+\varepsilon} + X^\varepsilon \sum_{\mathbf{w} \in \mathscr{X}_{j+1}^l} R(\mathbf{w};\chi_{i_1},\ldots,\chi_{i_l};X)^2.$$

In view of the above discussion, therefore, we infer from the hypotheses concluding with (3.6) either that

(3.10) $$\sum_{n \in \mathscr{L}_0} r_F(n;X)^2 \ll X^{2+\varepsilon},$$

or that (3.9) holds with $l = 1$, or else that these initial hypotheses again hold, but with j replaced by $j+1$, and with the m-tuple φ replaced by an m'-tuple of polynomials with strictly smaller degree in the sense that their sum of degrees is strictly smaller. Since the sum of the degrees of the φ_i must always be at least 1, we conclude that repeated application of this reduction must terminate after at most d steps either with the conclusion (3.10), or else with the conclusion that (3.9) holds with $l = 1$ and $j = d$. In the former case we deduce that

(3.11) $$\sum_{n \in \mathbb{Z}} r_F(n;X)^2 \ll X^{2+\varepsilon}.$$

In the latter situation, meanwhile, we may conclude that polynomials $G(x,y) \in \mathbb{Z}[x,y]$ and $g(t) \in \mathbb{Q}[t]$ exist satisfying the conditions (b), (c) of the statement of Lemma 3.2. If $G(x,y)$ has degree 1, or else is degenerate of degree exceeding 1, moreover, then it follows from conditions (b) and (c) that $F(x,y)$ is itself degenerate, contrary to our earlier hypotheses. Thus condition (a) is also satisfied. Furthermore, our above discussion also yields the bound

(3.12) $$\sum_{n \in \mathbb{Z}} r_F(n;X)^2 \ll X^{2+\varepsilon} + X^\varepsilon \sum_{n \in \mathbb{Z}} r_G(n;X)^2.$$

On combining the estimates (3.1), (3.2), (3.11) and (3.12), we find that the conclusion of the lemma follows in all cases. □

In the later stages of our argument we are reduced to equations quadratic with respect to a subset of the variables. These we handle with the aid of the following elementary estimate.

Lemma 3.3. — *Let a, b, c be integers with $abc \ne 0$, and let $S(a, b, c; P)$ denote the number of integral solutions of the equation $ax^2 + by^2 = c$, with $|x| \le P$ and $|y| \le P$. Then for each positive number ε, one has $S(a, b, c; P) \ll 1 + (|abc|P)^\varepsilon$.*

Proof. — This well-known estimate can be found in Estermann [8] or Vaughan and Wooley [20, Lemma 3.5]. □

We now provide the refinement of Lemma 3.2 of such utility in quadratic cases, basing our argument on that occurring in the proof of Lemma 7.1 of [22].

Lemma 3.4. — *Let X denote a large real number. Suppose that $F(x, y) \in \mathbb{Z}[x, y]$ is a non-degenerate polynomial of degree $d \ge 2$, and suppose also that $F(x, y)$ has degree precisely 2 in terms of x. Suppose in addition that for no rational numbers λ and μ is it true that there exists a polynomial $f(x, y) \in \mathbb{Z}[x, y]$ for which the equation*

$$F(x, y) = \lambda f(x, y)^2 + \mu$$

is satisfied identically. Further, suppose that for some fixed positive number A, the coefficients of F are each bounded in absolute value by X^A. Then in the notation defined in the statement of Lemma 3.2, one has

$$\sum_{n \in \mathbb{Z}} r_F(n; X)^2 \ll X^{2+\varepsilon}.$$

Proof. — We may rewrite the polynomial $F(x, y)$ in the form

(3.13) $$F(x, y) = \alpha(y) x^2 + \beta(y) x + \gamma(y),$$

where $\alpha(y)$ is a polynomial in y with integral coefficients which is not identically zero, though possibly constant, and $\beta(y), \gamma(y) \in \mathbb{Z}[y]$. Let $R_1(X)$ denote the number of solutions of the equation

(3.14) $$F(x_1, y_1) = F(x_2, y_2),$$

with $|x_i| \le X$, $|y_i| \le X$ ($i = 1, 2$), in which $\alpha(y_i) = 0$ for $i = 1$ or 2. Define the polynomial $\Delta(y)$ by

(3.15) $$\Delta(y) = \beta(y)^2 - 4\alpha(y)\gamma(y),$$

and let $R_2(X)$ denote the corresponding number of solutions of (3.14) in which $\alpha(y_i) \ne 0$ ($i = 1, 2$), and one has that $\Delta(y)$ is identically zero as a polynomial in y. Let $R_3(X)$ denote the corresponding number of solutions in which $\alpha(y_i) \ne 0$ ($i = 1, 2$), and $\Delta(y)$ is not identically zero as a polynomial in y, and moreover one has

(3.16) $$\alpha(y_2)\Delta(y_1) = \alpha(y_1)\Delta(y_2).$$

Finally, let $R_4(X)$ denote the corresponding number of solutions with $\alpha(y_i) \neq 0$ ($i = 1, 2$), and for which the equation (3.16) does not hold. Then plainly,

$$(3.17) \qquad \sum_{n \in \mathbb{Z}} r_F(n; X)^2 \leqslant \sum_{i=1}^{4} R_i(X).$$

We first bound $R_1(X)$. Suppose that $\alpha(y_i) = 0$ for $i = 1, 2$. Since $\alpha(y)$ is not identically zero, it follows that there are at most d^2 permissible choices for **y**. Since there are trivially $O(X^2)$ possible choices for **x**, we find that the contribution to $R_1(X)$ from this first class of solutions is $O(X^2)$. Consider next the remaining solutions for which $\alpha(y_i) = 0$ for at most one value of i. By relabelling variables, we may suppose that $\alpha(y_1) = 0$. There are consequently at most d choices permissible for y_1. Fix any one such choice, and also fix any one of the $O(X)$ available choices for x_1. Since $\alpha(y_2)$ is non-zero, it follows from (3.13) and (3.14) that the latter equation is explicit in both x_2 and y_2, whence a simple counting argument reveals that the number of possible choices for x_2 and y_2 satisfying (3.14) is at most $O(X)$. There are thus $O(X^2)$ solutions of this second type, whence

$$(3.18) \qquad R_1(X) \ll X^2.$$

Consider next the solutions counted by $R_2(X)$. There exist non-trivial polynomials $\alpha_1(y), \alpha_2(y) \in \mathbb{Z}[y]$ with the property that $\alpha(y) = \alpha_1(y)\alpha_2(y)^2$, and $\alpha_1(y)$ has no repeated factors over $\mathbb{C}[y]$. Since $\alpha(y)$ is a non-trivial polynomial in y, it follows from (3.15) that if $\Delta(y)$ is identically zero as a polynomial in y, then $\beta(y)$ is divisible by the polynomial $\alpha_1(y)\alpha_2(y)$. Such is immediate when $\gamma(y)$ is non-zero, and when $\gamma(y)$ is equal to zero one has $\beta(y) = 0$, and the desired conclusion again follows. But if $\beta(y)$ is divisible by $\alpha_1(y)\alpha_2(y)$, then the vanishing of $\Delta(y)$ ensures, by (3.15), that $\gamma(y)$ is divisible by $\alpha_1(y)$. We therefore deduce that for some non-zero integers κ_1, κ_2, and some polynomial in y with integral coefficients, say $\delta(y)$, one has

$$(3.19) \qquad \kappa_1 F(x, y) = \alpha_1(y)(\kappa_2 \alpha_2(y)x + \delta(y))^2$$

identically as a polynomial in x and y. We observe here that since $\alpha_1(y)$ and $\alpha_2(y)$ are divisors of $\alpha(y)$, it follows that their coefficients have absolute values at most $O(X^A)$ (see, for example, Granville [9]). One finds in like manner that the coefficients of $\delta(y)$, and also κ_1 and κ_2, may be chosen with absolute values at most $O(X^{2A})$. Notice also that our hypothesis that F is not a rational multiple of the square of a polynomial ensures that $\alpha_1(y)$ is not a constant polynomial. Let x_2 and y_2 be any one of the $O(X^2)$ permissible choices counted by $R_2(X)$. Since, by an elementary counting argument, the number of solutions of the equation $F(x, y) = 0$ with $|x| \leqslant X$ and $|y| \leqslant X$ is

$O(X)$, the total number of solutions $\mathbf{x},\mathbf{y}$ counted by $R_2(X)$ with $F(x_2,y_2)=0$ is $O(X^2)$. We may therefore suppose that our aforementioned choice of x_2, y_2 satisfies the condition that $F(x_2, y_2) \neq 0$, whence $\kappa_1 F(x_2, y_2) \neq 0$. But it follows from (3.14) and (3.19) that $\alpha_1(y_1)$ and $\kappa_2\alpha_2(y_1)x_1 + \delta(y_1)$ are both divisors of the fixed non-zero integer $\kappa_1 F(x_2, y_2)$. By elementary estimates for the divisor function, therefore, there are at most $O(X^\varepsilon)$ possible choices for integers d_1 and d_2 with $\alpha_1(y_1) = d_1$ and $\kappa_2\alpha_2(y_1)x_1 + \delta(y_1) = d_2$. Since $\alpha_1(y)$ is not a constant polynomial, the first of the latter equations shows that there are at most d possible choices for y_1. Given any one fixed such choice of y_1, on noting that the non-vanishing of $\alpha(y_1)$ ensures also that $\alpha_2(y_1) \neq 0$, one finds that x_1 is uniquely determined from the second of these equations. Thus we deduce that

$$(3.20) \qquad R_2(X) \ll X^{2+\varepsilon}.$$

Consider next the solutions $\mathbf{x},\mathbf{y}$ counted by $R_3(X)$. If, on the one hand, the polynomial equation (3.16) is non-trivial in y_1 and y_2, then a simple counting argument shows that there are $O(X)$ permissible choices for y_1 and y_2 satisfying (3.16). Given any one such choice of $\mathbf{y}$, in view of the presumed non-vanishing of $\alpha(y_i)$ $(i=1,2)$, it follows from (3.13) that the equation (3.14) is non-trivial in x_1 and x_2, whence there are $O(X)$ permissible choices of x_1 and x_2 satisfying (3.14). Thus the total number of solutions of this type is $O(X^2)$. If, on the other hand, the polynomial equation (3.16) is trivial in y_1 and y_2, then it follows that $\Delta(y)$ is a non-zero constant multiple of $\alpha(y)$, say $\Delta(y) = \lambda\alpha(y)$. We may again write $\alpha(y) = \alpha_1(y)\alpha_2(y)^2$, with α_1 and α_2 defined as in the treatment of $R_2(X)$. An inspection of (3.15) now reveals that

$$\lambda\alpha_1(y)\alpha_2(y)^2 = \beta(y)^2 - 4\alpha_1(y)\alpha_2(y)^2\gamma(y),$$

whence $\beta(y)$ is a multiple of $\alpha_1(y)\alpha_2(y)$. Write $\beta(y) = \mu^{-1}\beta_1(y)\alpha_1(y)\alpha_2(y)$, where μ is a non-zero integer and $\beta_1(y) \in \mathbb{Z}[y]$. Note here that, as in the above discussion, one may suppose that the coefficients of $\beta_1(y)$, $\alpha_1(y)$ and $\alpha_2(y)$, together with the integer μ, have absolute values at most $O(X^{2A})$. We thus infer that

$$4\mu^2\gamma(y) = \alpha_1(y)\beta_1(y)^2 - \lambda\mu^2.$$

On substituting into (3.13), we find that

$$4\mu^2 F(x,y) = \alpha_1(y)(2\mu\alpha_2(y)x + \beta_1(y))^2 - \lambda\mu^2.$$

In particular, our hypothesis that $F(x,y)$ is not a translation of a rational multiple of a square of a polynomial ensures that $\alpha_1(y)$ is not a constant polynomial. In this way, it follows that the equation (3.14) takes the shape

$$\alpha_1(y_1)(2\mu\alpha_2(y_1)x_1 + \beta_1(y_1))^2 = \alpha_1(y_2)(2\mu\alpha_2(y_2)x_2 + \beta_1(y_2))^2.$$

A comparison between the polynomial $\alpha_1(y)(2\mu\alpha_2(y)x + \beta_1(y))^2$ and that on the right hand side of (3.19) reveals that we may now apply the argument concluding the treatment of $R_2(X)$ above in order to conclude that the number of solutions of this type is $O(X^{2+\varepsilon})$. Thus we have

(3.21) $$R_3(X) \ll X^{2+\varepsilon}.$$

Finally, we discuss the solutions counted by $R_4(X)$. Let $\mathbf{x}, \mathbf{y}$ be any solution of (3.14) of the latter type. Then on recalling (3.13), (3.14) and (3.15), we deduce that

(3.22) $$\alpha(y_2)(2\alpha(y_1)x_1 + \beta(y_1))^2 - \alpha(y_1)(2\alpha(y_2)x_2 + \beta(y_2))^2 \\ = \alpha(y_2)\Delta(y_1) - \alpha(y_1)\Delta(y_2).$$

But in view of our hypotheses relevant to $R_4(X)$, for each of the $O(X^2)$ permissible values of $\mathbf{y}$, one has that the right hand side of (3.22) is a non-zero integer, say N. Fix any one such choice of $\mathbf{y}$, and note that our hypotheses ensure also that $\alpha(y_i) \neq 0$ ($i = 1, 2$). But by Lemma 3.3, the number of solutions of the equation

$$\alpha(y_2)\xi^2 - \alpha(y_1)\eta^2 = N,$$

with ξ and η each bounded in absolute value by a fixed power of X, is $O(X^\varepsilon)$. Consequently, the number of possible x_i ($i = 1, 2$) is also $O(X^\varepsilon)$, and thus we conclude that

(3.23) $$R_4(X) \ll X^{2+\varepsilon}.$$

The conclusion of the lemma now follows immediately on collecting together the estimates (3.18), (3.20), (3.21), (3.23) with (3.17). □

We note that the treatment of $\mathcal{K}_3(X; \mathbf{h})$ in the proof of Lemma 7.1 of [22] contains an oversight in that $\beta(y; \mathbf{h})$ was presumed to be necessarily zero, as a consequence of the argument presented therein. The treatment of $R_3(X)$ above takes care of this oversight, and the diligent reader will find that there are no substantive difficulties encountered here. Indeed, one may assume in the above treatment that $\alpha_2(y)$ is identically equal to 1 when applying this argument in the context of the treatment of $\mathcal{K}_3(X; \mathbf{h})$ in the aforementioned work.

Before proceeding to the main inductive part of our argument, we require still another estimate of simpler type than those embodied in Lemmata 3.2 and 3.4.

Lemma 3.5. — *Let X denote a large real number. Suppose that $F(x, y) \in \mathbb{Z}[x, y]$ is a non-degenerate polynomial of total degree 2, and suppose also that $F(x, y)$ has degree precisely 1 in terms of x. Further, suppose that for some*

fixed positive number A, the coefficients of F are each bounded in absolute value by X^A. Then in the notation defined in the statement of Lemma 3.2, one has
$$\sum_{n \in \mathbb{Z}} r_F(n; X)^2 \ll X^{2+\varepsilon}.$$

Proof. — We may rewrite the polynomial $F(x, y)$ in the shape
$$F(x, y) = \alpha(y)x + \beta(y),$$
where $\alpha(y)$ is a linear polynomial in y with integral coefficients that is not identically zero, and $\beta(y)$ is a quadratic polynomial in y with integral coefficients. By considering the putative coefficient of x^2, it is apparent that $F(x, y)$ cannot be the translate of a rational multiple of the square of a polynomial. Consequently, when $\beta(y)$ has non-vanishing leading coefficient we may reverse the roles of x and y, and appeal to Lemma 3.4 in order to establish the conclusion of the lemma. When the leading coefficient of $\beta(y)$ is zero, on the other hand, it follows that for some integers a, b, c and d, with $a \neq 0$, one has
$$F(x, y) = axy + bx + cy + d,$$
whence
$$aF(x, y) = (ax + c)(ay + b) + ad - bc.$$
But then one has
(3.24) $$\sum_{n \in \mathbb{Z}} r_F(n; X)^2 = \sum_{n \in \mathbb{Z}} r_G(n; X)^2,$$
where $G(x, y) = (ax + c)(ay + b)$.

When n is non-zero and $G(x, y) = n$, an elementary divisor function estimate shows that there are $O(X^\varepsilon)$ possible choices for $ax + c$ and $ay + b$, whence also for x and y. When n is zero, on the other hand, one has that $ax + c = 0$ or else that $ay + b = 0$, so that the corresponding number of solutions is $O(X)$. Consequently,
$$\sum_{n \in \mathbb{Z}} r_G(n; X)^2 \ll r_G(0; X)^2 + X^\varepsilon \sum_{n \in \mathbb{Z} \setminus \{0\}} r_G(n; X) \ll X^{2+\varepsilon},$$
and the desired conclusion is again immediate, in view of (3.24). □

4. The inductive step

We are now equipped to discuss the main inductive step in the proof of Theorem 1.1. Consider a non-degenerate binary form $\Psi(x, y)$ of the shape

(2.1), and define the exponential sum $H_\Psi(\theta; X)$ as in (2.2). When X is a large real number and s is a positive number, define

$$\mathscr{I}_s(X) = \int_0^1 |H_\Psi(\alpha; X)|^s d\alpha.$$

Lemma 4.1. — *Let $\Psi(x, y) \in \mathbb{Z}[x, y]$ be a non-degenerate form of degree k, with $3 \leqslant k \leqslant 10$, of the shape discussed above. Then one of the following statements is true.*
(i) For $1 \leqslant j \leqslant k$, and for each positive number ε, one has

$$\mathscr{I}_{2^j-1}(X) \ll X^{2^j-j+\varepsilon}.$$

(ii) For each positive number s, and for each integer j with $1 \leqslant j \leqslant k-3$, one has for each $\varepsilon > 0$ the upper bound

$$\mathscr{I}_{s+2^j}(X) \ll X^{2^{j+1}-1}\mathscr{I}_s(X) + X^{2^{j+1}-\frac{1}{2}(j+2-\delta)+\varepsilon}(\mathscr{I}_{2s}(X))^{1/2},$$

where $\delta = \delta(j)$ is defined by

$$\delta(j) = \begin{cases} 1/(k-j), & \text{when } 1 \leqslant j < k-3, \\ 0, & \text{when } j = k-3. \end{cases}$$

Proof. — We begin by noting that the conclusion (i) of the lemma is immediate from Lemma 2.2 when $k = 3$ or 4, and also when $k \geqslant 5$ and $j = 1$ or 2. When $k \geqslant 5$ and $t \geqslant k-1$, moreover, the conclusion of Lemma 2.3 demonstrates either that conclusion (i) holds, or else that conclusion (ii) will follow provided we establish the validity of the latter when Ψ is replaced by a condensation Υ of Ψ of the shape (2.1) wherein $t = 2$. There is therefore no loss of generality in supposing that $2 \leqslant t \leqslant k-2$. Similarly, when $k \geqslant 5$ and $t = k-2$, the conclusions of Lemmata 2.4 and 2.5 ensure either that conclusion (i) holds, or else that conclusion (ii) will follow provided we establish the validity of the latter when Ψ is replaced by a condensation Υ of Ψ of the shape (2.1) wherein $t = 2$ or $t = 3$. We therefore deduce that the conclusion of the lemma follows by establishing the inequality recorded in (ii) for those polynomials Ψ for which either $k = 5$ and $t = 2$ or 3, or else $6 \leqslant k \leqslant 10$ and $2 \leqslant t \leqslant k-3$. We suppose henceforth that the latter conditions do indeed hold.

We now modify the argument applied in §§6 and 7 of [22], applying a more elaborate differencing procedure, and considering also moments other than the even ones. Let w be a parameter to be chosen later satisfying the inequalities

(4.1) $$\max\{1, j-k+t+2\} \leqslant w \leqslant \min\{j, t-1\}.$$

The significance of these inequalities will become clear in due course. For the moment we remark only that our hypotheses concerning j, t and k ensure that an integral value of w can always be found satisfying (4.1).

We first view the exponential sum $H(\theta) = H_\Psi(\theta; X)$ as an exponential sum over v, so that on applying Hölder's inequality to (2.2), and then making use of Lemma 2.6, we deduce that

(4.2)
$$|H(\theta)|^{2^w} \ll X^{2^w-1} \sum_{|u|\leqslant X} \left| \sum_{|v|\leqslant X} e(\theta\Psi(u,v)) \right|^{2^w}$$
$$\ll X^{2^{w+1}-w-2} \sum_{\mathbf{h}\in[-2X,2X]^w} \sum_{v\in I(\mathbf{h})} K(\theta;\mathbf{h};v),$$

where $I = I(h_1,\ldots,h_w)$ is an interval of integers contained in $[-X, X]$, and

$$K(\theta;\mathbf{h};v) = \sum_{|u|\leqslant X} e(\Delta_{0,w}(\theta\Psi(u,v);\mathbf{h})).$$

Next applying Lemma 2.6 to the latter exponential sum, we obtain

(4.3) $\quad |K(\theta;\mathbf{h};v)|^{2^{j-w}} \ll X^{2^{j-w}-j+w-1} \sum_{\mathbf{g}\in[-2X,2X]^{j-w}} \sum_{u\in J(\mathbf{g})} e(\theta p(v;u;\mathbf{g};\mathbf{h})),$

where $J = J(g_1,\ldots,g_{j-w})$ is an interval of integers contained in $[-X, X]$, and the polynomial $p(v;u;\mathbf{g};\mathbf{h})$ is defined by

(4.4) $\qquad p(v;u;\mathbf{g};\mathbf{h}) = \Delta_{j-w,w}(\Psi(u,v);\mathbf{g};\mathbf{h}).$

We note for future reference that, on recalling (4.1), and considering the term $Bu^{k-t}v^t$ in (2.1), it is apparent that the polynomial $p(v;u;\mathbf{g};\mathbf{h})$ is not identically zero.

On combining (4.2) and (4.3) via Hölder's inequality, we conclude that

(4.5) $\qquad |H(\theta)|^{2^j} \ll X^{2^{j+1}-j-2} \mathscr{G}(\theta),$

where

(4.6) $\qquad \mathscr{G}(\theta) = \sum_{\mathbf{m}\in[-2X,2X]^j} \sum_{u\in J(\mathbf{g})} \sum_{v\in I(\mathbf{h})} e(\theta p(v;u;\mathbf{g};\mathbf{h})),$

and here, and throughout, we adopt the convention that

(4.7) $\quad \mathbf{m} = (m_1,\ldots,m_j), \quad \mathbf{h} = (m_1,\ldots,m_w) \quad \text{and} \quad \mathbf{g} = (m_{w+1},\ldots,m_j).$

Define the exponential sum $\mathscr{G}_1(\alpha)$ by

$$\mathscr{G}_1(\alpha) = \sum_{\mathbf{m}} \sum_{u,v} e(\alpha p(v;u;\mathbf{g};\mathbf{h})),$$

where the summation is restricted to the values of $\mathbf{m}, u, v$ satisfying

(4.8) $\qquad \mathbf{m} \in [-2X, 2X]^j, \quad u \in J(\mathbf{g}), \quad v \in I(\mathbf{h}),$

with

(4.9) $\qquad p(v; u; \mathbf{g}; \mathbf{h}) \neq 0.$

Since $p(v; u; \mathbf{g}; \mathbf{h})$ is not identically zero, it follows from an elementary argument that the number of choices of $\mathbf{m}, u, v$ satisfying (4.8) and $p(v; u; \mathbf{g}, \mathbf{h}) = 0$ is at most $O(X^{j+1})$. Consequently, one has

$$|\mathscr{G}(\alpha) - \mathscr{G}_1(\alpha)| \ll X^{j+1}.$$

Then in view of (4.5), we obtain

$$\mathscr{I}_{s+2^j}(X) = \int_0^1 |H(\alpha)|^{s+2^j} d\alpha \ll X^{2^{j+1}-j-2} \int_0^1 \mathscr{G}(\alpha)|H(\alpha)|^s d\alpha$$

$$\ll X^{2^{j+1}-1} \int_0^1 |H(\alpha)|^s d\alpha + X^{2^{j+1}-j-2} \int_0^1 |\mathscr{G}_1(\alpha) H(\alpha)^s| d\alpha.$$

Let $\mathscr{T}$ denote the mean value

(4.10) $\qquad \mathscr{T}(X) = \int_0^1 |\mathscr{G}_1(\alpha)|^2 d\alpha.$

Then an application of Schwarz's inequality leads us to the estimate

(4.11) $\qquad \mathscr{I}_{s+2^j}(X) \ll X^{2^{j+1}-1} \mathscr{I}_s(X) + X^{2^{j+1}-j-2}(\mathscr{T}(X))^{1/2}(\mathscr{I}_{2s}(X))^{1/2}.$

Next observe that, in view of (4.8) and (4.9), and on considering the diophantine equation underlying (4.10), the mean value $\mathscr{T}(X)$ is bounded above by $\mathscr{K}(X)$, where $\mathscr{K}(X)$ denotes the number of integral solutions of the equation

(4.12) $\qquad p(x_1; y_1; \mathbf{g}_1; \mathbf{h}_1) = p(x_2; y_2; \mathbf{g}_2; \mathbf{h}_2),$

with $\mathbf{m}_i \in [-2X, 2X]^j$, in the sense of (4.7), also with $|x_i|, |y_i| \leq X$ ($i = 1, 2$), and subject to the conditions $p(x_i, y_i; \mathbf{g}_i; \mathbf{h}_i) \neq 0$ ($i = 1, 2$). A comparison between (4.11) and the estimate claimed in the statement of the lemma therefore reveals that the desired conclusion follows immediately from the upper bound

(4.13) $\qquad \mathscr{K}(X) \ll X^{j+2+\delta+\varepsilon}.$

We henceforth concentrate our efforts on establishing (4.13).

On recalling (4.4), a modicum of computation reveals that

(4.14) $\qquad p(x; y; \mathbf{g}; \mathbf{h}) = m_1 \ldots m_j F(x, y; \mathbf{m}),$

where

(4.15) $$F(x, y; \mathbf{m}) = \sum_{i=t}^{k} D_i \varphi_{k-i}(y; \mathbf{m}) \psi_i(x; \mathbf{m}),$$

in which D_i is an integer for $t \leqslant i \leqslant k$, and $D_t \neq 0$, and in which each $\psi_i(x; \mathbf{m})$ is a polynomial with integral coefficients of degree $i - w$ with respect to x, and each $\varphi_{k-i}(y; \mathbf{m})$ is a polynomial with integral coefficients of degree $k - i - j + w$ with respect to y. In view of (4.1), one has $2 \leqslant k - t - j + w \leqslant k - j - 1$ and $t - w \geqslant 1$. Thus $F(x, y; \mathbf{m})$ has degree at least 1 with respect to x, and degree at least 2 and at most $k - j - 1$ with respect to y. We note also for future reference that when $w = 1$ and $j = 1$, one may take

(4.16) $$\varphi_{k-t}(y; m) = y^{k-t} \quad \text{and} \quad \psi_t(x; m) = m^{-1}((x+m)^t - x^t).$$

Finally, we observe that the argument surrounding equations (6.17) and (6.18) of [22] easily establishes that when $m_l \neq 0$ $(1 \leqslant l \leqslant j)$, then one has that the polynomial $F(x, y; \mathbf{m})$ is non-degenerate with respect to x and y.

When $\mathbf{m} \in [-2X, 2X]^j$, let $\rho(n; \mathbf{m})$ denote the number of integral solutions of the equation $p(x; y; \mathbf{g}; \mathbf{h}) = n$, with $|x|, |y| \leqslant X$. Then it follows from (4.14) and (4.12) that

$$\mathscr{K}(X) = \sum_{n \in \mathbb{Z}\setminus\{0\}} \left(\sum_{\substack{|m_1| \leqslant 2X \\ m_1 | n}} \cdots \sum_{\substack{|m_j| \leqslant 2X \\ m_j | n}} \rho(n; \mathbf{m}) \right)^2.$$

Consequently, on applying Cauchy's inequality in combination with an elementary estimate for the divisor function, one obtains

(4.17) $$\begin{aligned} \mathscr{K}(X) &\ll X^\varepsilon \sum_{n \in \mathbb{Z}\setminus\{0\}} \sum_{\substack{\mathbf{m} \in [-2X, 2X]^j \\ m_i \neq 0 \ (1 \leqslant i \leqslant j)}} \rho(n; \mathbf{m})^2 \\ &\ll X^{j+\varepsilon} \max_{\substack{\mathbf{m} \in [-2X, 2X]^j \\ m_i \neq 0 \ (1 \leqslant i \leqslant j)}} \sum_{n \in \mathbb{Z}\setminus\{0\}} \rho(n; \mathbf{m})^2 \\ &= X^{j+\varepsilon} \max_{\substack{\mathbf{m} \in [-2X, 2X]^j \\ m_i \neq 0 \ (1 \leqslant i \leqslant j)}} \mathscr{M}(X; \mathbf{m}), \end{aligned}$$

where $\mathscr{M}(X; \mathbf{m})$ denotes the number of solutions of the equation

(4.18) $$F(x_1, y_1; \mathbf{m}) = F(x_2, y_2; \mathbf{m}),$$

with $|x_i|, |y_i| \leqslant X$ $(i = 1, 2)$.

We recall at this point that, by hypothesis, one has either

$$k = 5, \quad t = 2 \text{ or } 3 \quad \text{and} \quad j = 1 \text{ or } 2,$$

or else

$$6 \leqslant k \leqslant 10, \quad 2 \leqslant t \leqslant k-3 \quad \text{and} \quad 1 \leqslant j \leqslant k-3.$$

We now divide our argument into a number of cases, our aim being to make a choice of w, satisfying the condition (4.1), for which the estimates of §3 prove effective.

(a) (k,t,j) satisfies $j = k-3$. We take $w = t-1$. In this situation it is apparent that

(4.19) $$k - t - j + w = 2,$$

and also that

(4.20) $$(k - j, t - w) = 1.$$

Consider the shape of the polynomial $F(x,y; \mathbf{m})$ when the conditions (4.19) and (4.20) hold. We isolate the monomial of highest degree with respect to y that has highest degree with respect to x. In view of (4.15) and the associated discussion, this monomial has the shape

(4.21) $$D_t x^{t-w} y^{k-t-j+w}.$$

Suppose, if possible, that there exist polynomials $G(x,y) \in \mathbb{Z}[x,y]$ and $g(t) \in \mathbb{Q}[t]$ satisfying the conditions (a) and (b) of the statement of Lemma 3.2, and also satisfying the condition that $F(x,y) = g(G(x,y))$. Then the monomial of highest degree with respect to y that has highest degree with respect to x in the polynomial $g(G(x,y))$, must necessarily have the shape $Cx^{hl}y^{hm}$, where h is the degree of $g(t)$. But the condition (4.20) implies that

(4.22) $$(k - t - j + w, t - w) = 1,$$

and so the latter conclusion contradicts (4.21). It follows, in particular, that for no rational numbers λ and μ is it true that there exists a polynomial $f(x,y) \in \mathbb{Z}[x,y]$ for which the equation

$$F(x,y; \mathbf{m}) = \lambda f(x,y)^2 + \mu$$

is satisfied identically in x and y. But in view of (4.21) and (4.19), the polynomial $F(x,y; \mathbf{m})$ has degree precisely 2 with respect to y. Since, moreover, the coefficients of F are each bounded in absolute value by a fixed power of X, it follows from Lemma 3.4 that in this case one has

(4.23) $$\mathscr{M}(X; \mathbf{m}) \ll X^{2+\varepsilon}.$$

On recalling (4.17), we find that (4.13) holds with $\delta = 0$, and thus the proof of the lemma in the case $j = k - 3$ is complete.

(b) (k, t, j) satisfies $t - 1 \leqslant j < k - 3$. We take $w = t - 1$, and find that (4.20), and hence also (4.22), remain true. In view of the discussion in case (a) above, it follows that there can exist no polynomials g and G that satisfy the hypotheses (a), (b), (c) of Lemma 3.2(i). We therefore deduce from Lemma 3.2(ii) that in this case one has

(4.24) $$\mathscr{M}(X; \mathbf{m}) \ll X^{2+1/(k-j)+\varepsilon}.$$

Recalling (4.17) again, we now find that (4.13) holds with $\delta = 1/(k - j)$, and hence the proof of the lemma follows in the case currently under consideration.

(c) $(k, t, j) = (5, 3, 1)$. We take $w = 1$, and find that (4.19) holds. Then it follows from (4.21) that in the polynomial $F(x, y; \mathbf{m})$, the monomial of highest degree with respect to y, that has highest degree with respect to x, has the shape Cx^2y^2. It is possible that Lemma 3.4 succeeds in supplying the bound (4.23). If such is not the case, then there exists a polynomial $f(x, y) \in \mathbb{Z}[x, y]$, and rational numbers λ and μ, for which $F(x, y; \mathbf{m}) = \lambda f(x, y)^2 + \mu$. Moreover, it is apparent that λ must be non-zero, and our previous discussion ensures that $f(x, y)$ must be non-degenerate of total degree 2, with degree precisely one in terms of y. Thus we deduce that

$$\mathscr{M}(X; \mathbf{m}) \leqslant \sum_{n \in \mathbb{Z}} r_f(n; X)^2,$$

whence by Lemma 3.5 one again obtains the conclusion (4.23). Recalling (4.17), we now find that (4.13) holds with $\delta = 0$, and thus the proof of the lemma again follows.

By combining the conclusions of cases (a) and (b) above, one finds that when $k = 5$ and $j = 1$ or 2, our hypothesis that $t = 2$ or 3 leaves only the situation in which $(k, t, j) = (5, 3, 1)$ to consider. But the latter case is resolved in case (c) above, and so henceforth we may suppose that $6 \leqslant k \leqslant 10$. In the latter circumstances, cases (a) and (b) also dispose of all cases in which $t = 2$, and also all cases wherein $j \geqslant t - 1$. Thus we may suppose henceforth that

$$6 \leqslant k \leqslant 10, \quad 3 \leqslant t \leqslant k - 3 \quad \text{and} \quad 1 \leqslant j \leqslant t - 2.$$

We treat the remaining allowable cases by hand.

(d) $(k, t, j) = (6, 3, 1), (7, 4, 2)$. We take $w = j$, and find that (4.20) holds, since $(5, 2) = 1$. The argument of part (b) therefore yields the bound (4.24), and hence also (4.13).

(e) $(k,t,j) = (8,3,1), (8,4,1), (8,5,1)$. We take $w = 1$, and find that (4.20) holds, since $(7, t-1) = 1$ for $t = 3, 4, 5$. The argument of part (b) therefore yields the bound (4.24), and hence also (4.13).

(f) $(k,t,j) = (8,5,3)$. We take $w = 3$, and find that (4.20) holds, since $(5,2) = 1$. The argument of part (b) therefore yields the bound (4.24), and hence also (4.13).

(g) $(k,t,j) = (9,4,1), (9,6,1)$. We take $w = 1$, and find that (4.20) holds, since $(8, t-1) = 1$ for $t = 4, 6$. The argument of part (b) therefore yields the bound (4.24), and hence also (4.13).

(h) $(k,t,j) = (9,4,2), (9,5,2), (9,6,2)$. We take $w = 2$, and find that (4.20) holds, since $(7, t-2) = 1$ for $t = 4, 5, 6$. The argument of part (b) therefore yields the bound (4.24), and hence also (4.13).

(i) $(k,t,j) = (10,3,1), (10,5,1), (10,6,1)$. We take $w = 1$, and find that (4.20) holds, since $(9, t-1) = 1$ for $t = 3, 5, 6$. The argument of part (b) therefore yields the bound (4.24), and hence also (4.13).

(j) $(k,t,j) = (10,4,2), (10,6,2)$. We take $w = 1$, and find that (4.20) holds, since $(8, t-1) = 1$ for $t = 4, 6$. The argument of part (b) therefore yields the bound (4.24), and hence also (4.13).

(k) $(k,t,j) = (10,5,2), (10,7,2)$. We take $w = 2$, and find that (4.20) holds, since $(8, t-2) = 1$ for $t = 5, 7$. The argument of part (b) therefore yields the bound (4.24), and hence also (4.13).

(l) $(k,t,j) = (10,5,3), (10,6,3), (10,7,3)$. We take $w = 3$, and find that (4.20) holds, since $(7, t-3) = 1$ for $t = 5, 6, 7$. The argument of part (b) therefore yields the bound (4.24), and hence also (4.13).

(m) $(k,t,j) = (9,6,4)$. We take $w = 3$, and find that (4.19) and (4.20) both hold, since $(5,3) = 1$. The argument of part (a) now establishes the bound (4.23), and hence also (4.13).

(n) $(k,t,j) = (10,7,5)$. We take $w = 4$, and find that (4.19) and (4.20) both hold, since $(5,3) = 1$. The argument of part (a) now establishes the bound (4.23), and hence also (4.13).

(o) $(k,t,j) = (7,4,1)$. We take $w = 1$, and find from (4.21) that in the polynomial $F(x, y; \mathbf{m})$, the monomial of highest degree with respect to y, that has highest degree with respect to x, has the shape Cx^3y^3. It is possible that Lemma 3.2 succeeds in supplying the bound (4.24). If such is not the case, then there exist polynomials $G(x,y) \in \mathbb{Z}[x,y]$ and $g(t) \in \mathbb{Q}[t]$ satisfying the conditions (a), (b), (c) of Lemma 3.2. It is evident, moreover, that in such circumstances the degree of g must be 3, and the total degree of $G(x,y)$ must be 2, and also the degree of $G(x,y)$ with respect to y must be 1. In the latter

circumstances, it follows from Lemma 3.5 that one has the estimate
$$\sum_{n \in \mathbb{Z}} r_G(n; X)^2 \ll X^{2+\varepsilon}, \tag{4.25}$$
whence by Lemma 3.2(i),
$$\mathscr{M}(X; \mathbf{m}) \ll X^{2+1/(k-j)+\varepsilon} + X^\varepsilon \sum_{n \in \mathbb{Z}} r_G(n; X)^2 \tag{4.26}$$
$$\ll X^{2+1/(k-j)+\varepsilon}.$$
Then in any case, one has the upper bound (4.24), and hence (4.13).

(p) $(k, t, j) = (8, 5, 2)$. We take $w = 2$, and find from (4.21) that in the polynomial $F(x, y; \mathbf{m})$, the monomial of highest degree with respect to y, that has highest degree with respect to x, has the shape Cx^3y^3. The desired bound (4.24), and hence (4.13), now follows by the argument of case (o).

(q) $(k, t, j) = (9, 5, 3)$. We take $w = 2$, and find from (4.21) that we may again apply the argument of case (p).

(r) $(k, t, j) = (9, 6, 3)$. We take $w = 3$, and find from (4.21) that we may again apply the argument of case (p).

(s) $(k, t, j) = (10, 6, 4)$. We take $w = 3$, and find from (4.21) that we may again apply the argument of case (p).

(t) $(k, t, j) = (10, 7, 4)$. We take $w = 4$, and find from (4.21) that we may again apply the argument of case (p).

(u) $(k, t, j) = (7, 3, 1)$. We take $w = 1$, and find from (4.21) that in the polynomial $F(x, y; \mathbf{m})$, the monomial of highest degree with respect to y, that has highest degree with respect to x, has the shape Cx^2y^4. It follows that if there exist polynomials $G(x, y) \in \mathbb{Z}[x, y]$ and $g(t) \in \mathbb{Q}[t]$ satisfying the conditions (a), (b), (c) of Lemma 3.2, then g must have degree 2. Moreover, in the polynomial $G(x, y)$, the monomial of highest degree with respect to y, that has highest degree with respect to x, must have the shape $C'xy^2$. In such circumstances, one may apply the argument of case (a) above to obtain the bound (4.25). Then the estimate (4.24) follows in all circumstances from Lemma 3.2, and this suffices to establish (4.13).

(v) $(k, t, j) = (8, 4, 2)$. We take $w = 2$, and find from (4.21) that in the polynomial $F(x, y; \mathbf{m})$, the monomial of highest degree with respect to y, that has highest degree with respect to x, has the shape Cx^2y^4. The desired bound (4.24), and hence (4.13), now follows by the argument of case (u).

(w) $(k, t, j) = (9, 5, 1)$. We take $w = 1$, and find from (4.21) that in the polynomial $F(x, y; \mathbf{m})$, the monomial of highest degree with respect to y, that has highest degree with respect to x, has the shape Cx^4y^4. It follows that if there exist polynomials $G(x, y) \in \mathbb{Z}[x, y]$ and $g(t) \in \mathbb{Q}[t]$ satisfying the conditions (a), (b), (c) of Lemma 3.2, then g must have degree either 2 or 4. When the degree of g is 4, the polynomial $G(x, y)$ must have total degree 2, and the degree of $G(x, y)$ with respect to y must be 1. In these circumstances, Lemma 3.5 establishes the estimate (4.25). When the degree of g is 2, meanwhile, we may suppose without loss of generality that there are no rational numbers λ and μ for which a polynomial $f(x, y) \in \mathbb{Z}[x, y]$ exists satisfying $G(x, y) = \lambda f(x, y)^2 + \mu$ (if such a polynomial f were to exist, then we would be in the situation already considered wherein the degree of g was presumed to be 4). But then, in the polynomial $G(x, y)$, the monomial of highest degree with respect to y, that has highest degree with respect to x, has the shape $C'x^2y^2$. The hypotheses of Lemma 3.4 are therefore satisfied with F replaced by G, and the upper bound (4.25) again follows. We may therefore conclude from Lemma 3.2(i) that in either case, one has the estimate (4.26). If no such polynomials G and g exist, on the other hand, then the estimate (4.26) is immediate from Lemma 3.2(ii).

(x) $(k, t, j) = (10, 4, 1)$. We take $w = 1$, and find from (4.21) that in the polynomial $F(x, y; \mathbf{m})$, the monomial of highest degree with respect to y, that has highest degree with respect to x, has the shape Cx^3y^6. It follows that if there exist polynomials $G(x, y) \in \mathbb{Z}[x, y]$ and $g(t) \in \mathbb{Q}[t]$ satisfying the conditions (a), (b), (c) of Lemma 3.2, then g must have degree 3. Moreover, in the polynomial $G(x, y)$, the monomial of highest degree with respect to y, that has highest degree with respect to x, has the shape $C'xy^2$. Thus we may proceed as in case (u) to obtain the desired estimate (4.13).

(y) $(k, t, j) = (9, 3, 1)$. We take $w = 1$, and find from (4.21) that in the polynomial $F(x, y; \mathbf{m})$, the monomial of highest degree with respect to y, that has highest degree with respect to x, has the shape Cx^2y^6. It follows that if there exist polynomials $G(x, y) \in \mathbb{Z}[x, y]$ and $g(t) \in \mathbb{Q}[t]$ satisfying the conditions (a), (b), (c) of Lemma 3.2, then g must have degree 2, and $G(x, y)$ must have the shape

(4.27) $$G(x, y) = \alpha xy^3 + \beta y^3 + H(x, y),$$

with $H(x, y)$ of degree at most 2 with respect to y. In view of (4.16), one finds that with a suitable non-zero constant K, one has that

(4.28) $$F(x, y; \mathbf{m}) = Ky^6(3x^2 + 3xm + m^2) + I(x, y; \mathbf{m}),$$

where $I(x, y; \mathbf{m})$ is a polynomial of degree at most 5 with respect to y. Since we may suppose that g has degree 2, it follows from (4.27) and (4.28) that there is a non-zero number a with

$$a(\alpha xy^3 + \beta y^3)^2 = Ky^6(3x^2 + 3xm + m^2).$$

On equating coefficients of powers of x, we find that
$$a\alpha^2 = 3K, \quad 2a\alpha\beta = 3Km, \quad a\beta^2 = Km^2,$$
whence
$$9K^2m^2 = 4(a\alpha^2)(a\beta^2) = 12K^2m^2.$$
This yields a contradiction whenever $m \neq 0$, as we may suppose. In this way we find that no such polynomials G, g exist, and hence Lemma 3.2(ii) establishes that the estimate (4.26) holds.

(z) $(k, t, j) = (10, 7, 1)$. We take $w = 1$, and find from (4.21) that in the polynomial $F(x, y; \mathbf{m})$, the monomial of highest degree with respect to y, that has highest degree with respect to x, has the shape Cx^6y^3. It follows that if there exist polynomials $G(x, y) \in \mathbb{Z}[x, y]$ and $g(t) \in \mathbb{Q}[t]$ satisfying the conditions (a), (b), (c) of Lemma 3.2, then g must have degree 3, and $G(x, y)$ must have the shape

(4.29) $$G(x, y) = \alpha yx^2 + \beta yx + \gamma y + H(x),$$

with $H(x)$ a polynomial independent of y. In view of (4.16), one finds that with a suitable non-zero constant K, one has that

(4.30)
$$F(x, y; m) = Ky^3(7x^6 + 21x^5m + 35x^4m^2 + 35x^3m^3 + 21x^2m^4 + 7xm^5 + m^6) + I(x, y; m),$$

where $I(x, y; m)$ is a polynomial of degree at most 2 with respect to y. Since we may suppose that g has degree 3, it follows from (4.29) and (4.30) that there is a non-zero number a with

$$a(\alpha x^2 y + \beta xy + \gamma y)^3$$
$$= Ky^3(7x^6 + 21x^5m + 35x^4m^2 + 35x^3m^3 + 21x^2m^4 + 7xm^5 + m^6).$$

On equating coefficients of powers of x, we find that
$$a\alpha^3 = 7K, \quad 3a\alpha^2\beta = 21Km, \quad a(3\alpha^2\gamma + 3\alpha\beta^2) = 35Km^2, \quad a\gamma^3 = Km^6.$$
Thus we deduce that
$$27\alpha^6 a^2 Km^6 = a^3(3\alpha^2\gamma)^3 = (35Km^2 - 3a\alpha\beta^2)^3,$$
whence
$$3^3 7^5 (Km)^6 = (35Km^2(a\alpha^3) - 3(a\alpha^2\beta)^2)^3$$
$$= (245K^2m^2 - 147K^2m^2)^3 = (98K^2m^2)^3.$$

Since $3^3 7^5 \neq 98^3$, we obtain a contradiction whenever $m \neq 0$, as we may suppose. In this way, we find that no such polynomials G, g exist, and hence Lemma 3.2(ii) establishes that the estimate (4.26) holds.

On collecting together the conclusions of cases (a)–(z), we find that the estimate (4.13) holds in all circumstances under consideration. The conclusion of the lemma now follows immediately from (4.11). □

5. The completion of the proof of Theorem 1.1

We are now prepared to complete the proof of Theorem 1.1. We begin with an induction based on the use of Lemma 4.1 for the small moments.

Lemma 5.1. — *Let $\Psi(x,y) \in \mathbb{Z}[x,y]$ be a non-degenerate form of degree k, with $5 \leqslant k \leqslant 10$, of the shape discussed in the opening paragraph of §4. Then for each j with $3 \leqslant j \leqslant k-2$, and for each positive number ε, one has*

$$\int_0^1 |H_\Psi(\alpha; X)|^{2^{j-1}} d\alpha \ll X^{2^j - j + 1/(k-j+2) + \varepsilon}.$$

Proof. — The conclusion of the lemma is either immediate from part (i) of Lemma 4.1, or else we may apply part (ii) of that lemma. By part (i) of Theorem 1.1, which we have already established in Lemma 2.2, one has

$$\int_0^1 |H_\Psi(\alpha; X)|^2 d\alpha \ll X^{2+\varepsilon}.$$

Thus we have
$$\mathscr{I}_2(X) \ll X^{2+\varepsilon}.$$

Suppose that, in fact, one has the estimate

(5.1) $\qquad \mathscr{I}_{2^{j-1}}(X) \ll X^{2^j - j + 1/(k-j+2) + \varepsilon},$

for $2 \leqslant j \leqslant J$, where J is an integer with $2 \leqslant J \leqslant k-3$. We apply part (ii) of Lemma 4.1 with $j = J-1$ and $s = 2^{J-1}$ in order to obtain

$$\mathscr{I}_{s+2^{J-1}}(X) \ll X^{2^J - 1} \mathscr{I}_s(X) + X^{2^J - \frac{1}{2}(J+1-\delta) + \varepsilon}(\mathscr{I}_{2s}(X))^{1/2},$$

with $\delta = 1/(k-J+1)$. On employing the inductive hypothesis (5.1), we obtain

$$\mathscr{I}_{2^J}(X) \ll X^{2^{J+1} - J - 1 + 1/(k-J+2) + \varepsilon} + X^{2^J - \frac{1}{2}(J+1-\delta) + \varepsilon}(\mathscr{I}_{2^J}(X))^{1/2},$$

whence
$$\mathscr{I}_{2^J}(X) \ll X^{2^{J+1} - J - 1 + 1/(k-J+1) + \varepsilon}.$$

This establishes the inductive hypothesis for $j = J+1$, and thus the conclusion of the lemma follows by induction. □

Lemma 5.2. — *With the hypotheses of the statement of Lemma 5.1, one has*
$$\int_0^1 |H_\Psi(\alpha; X)|^{\frac{9}{32}2^k} d\alpha \ll X^{\frac{9}{16}2^k - k + 1 + \varepsilon}.$$

Proof. — As in the proof of the previous lemma, the desired conclusion is either immediate from part (i) of Lemma 4.1, or else we may apply part (ii) of that lemma. By the conclusion of Lemma 5.1 with $j = k - 2$, one has

(5.2) $$\mathscr{I}_{2^{k-3}}(X) = \int_0^1 |H_\Psi(\alpha; X)|^{2^{k-3}} d\alpha \ll X^{2^{k-2} - k + 9/4 + \varepsilon}.$$

On applying part (ii) of Lemma 4.1 with $s = 2^{k-3}$ and $j = k - 3$, one obtains
$$\mathscr{I}_{2^{k-2}}(X) \ll X^{2^{k-2} - 1} \mathscr{I}_{2^{k-3}}(X) + X^{2^{k-2} - (k-1)/2 + \varepsilon} (\mathscr{I}_{2^{k-2}}(X))^{1/2},$$
whence by (5.2),

(5.3) $$\mathscr{I}_{2^{k-2}}(X) \ll X^{2^{k-1} - k + 5/4 + \varepsilon}.$$

An application of Hölder's inequality establishes the upper bound
$$\mathscr{I}_{\frac{5}{32}2^k}(X) = \int_0^1 |H_\Psi(\alpha; X)|^{\frac{5}{32}2^k} d\alpha$$
$$\leq \left(\int_0^1 |H_\Psi(\alpha; X)|^{2^{k-2}} d\alpha\right)^{1/4} \left(\int_0^1 |H_\Psi(\alpha; X)|^{2^{k-3}} d\alpha\right)^{3/4}.$$

Thus, by (5.2) and (5.3) we deduce that

(5.4) $$\mathscr{I}_{\frac{5}{32}2^k}(X) \ll X^{\frac{5}{16}2^k - k + 2 + \varepsilon}.$$

A second application of part (ii) of Lemma 4.1, now with $s = \frac{5}{32}2^k$ and $j = k - 3$, gives the estimate
$$\mathscr{I}_{\frac{9}{32}2^k}(X) \ll X^{2^{k-2} - 1} \mathscr{I}_{\frac{5}{32}2^k}(X) + X^{2^{k-2} - (k-1)/2 + \varepsilon} \left(\mathscr{I}_{\frac{5}{16}2^k}(X)\right)^{1/2}.$$

But a trivial estimate for $H_\Psi(\alpha; X)$ demonstrates that
$$\mathscr{I}_{\frac{5}{16}2^k}(X) \ll X^{2^{k-4}} \int_0^1 |H_\Psi(\alpha; X)|^{\frac{9}{32}2^k} d\alpha = X^{2^{k-4}} \mathscr{I}_{\frac{9}{32}2^k}(X).$$

In view of (5.4), therefore, we obtain
$$\mathscr{I}_{\frac{9}{32}2^k}(X) \ll X^{\frac{9}{16}2^k - k + 1 + \varepsilon} + X^{\frac{9}{32}2^k - (k-1)/2 + \varepsilon} \left(\mathscr{I}_{\frac{9}{32}2^k}(X)\right)^{1/2},$$

and the conclusion of the lemma follows immediately. □

The large moments are estimated via the Hardy-Littlewood method by means of a treatment contained, in all essentials, within the proof of Lemma 7.4 of [22]. We include an account of the proof for the sake of completeness. We first require a major arc estimate stemming from our version of Weyl's inequality.

Lemma 5.3. — *Suppose that $\Phi(x,y) \in \mathbb{Z}[x,y]$ is a non-degenerate form of degree $d \geq 3$, and let $\alpha \in \mathbb{R}$.*
(i) Suppose that there exist $r \in \mathbb{Z}$ and $q \in \mathbb{N}$ with $(r,q) = 1$ and $|\alpha - r/q| \leq q^{-2}$. Then for each $\varepsilon > 0$, one has

$$\sum_{1 \leq x \leq X} \sum_{1 \leq y \leq X} e(\alpha \Phi(x,y)) \ll X^{2+\varepsilon} \left(q^{-1} + X^{-1} + qX^{-d}\right)^{2^{2-d}}.$$

(ii) Whenever $r \in \mathbb{Z}$ and $q \in \mathbb{N}$ satisfy $1 \leq q \leq X$ and $|q\alpha - r| \leq X^{1-d}$, one has

$$\sum_{1 \leq x \leq X} \sum_{1 \leq y \leq X} e(\alpha \Phi(x,y)) \ll X^{2+\varepsilon}(q + X^d|q\alpha - r|)^{-2^{2-d}}.$$

Proof. — The first conclusion is immediate from Theorem 1 of [22], and the second conclusion is Lemma 7.3 of [22]. □

Lemma 5.4. — *With the hypotheses of the statement of Lemma 5.1, one has*

$$\int_0^1 |H_\Psi(\alpha; X)|^{\frac{17}{32} 2^k} d\alpha \ll X^{\frac{17}{16} 2^k - k + \varepsilon}.$$

Proof. — For the sake of concision, we abbreviate $H_\Psi(\alpha; X)$ to $H(\alpha)$. When $r \in \mathbb{Z}$ and $q \in \mathbb{N}$, write

$$\mathfrak{M}(q,r) = \{\alpha \in [0,1) : |q\alpha - r| \leq X^{1-k}\}.$$

Take $\mathfrak{M}$ to be the union of the intervals $\mathfrak{M}(q,r)$ with $0 \leq r \leq q \leq X$ and $(r,q) = 1$. Note that the intervals occurring in the latter union are disjoint. Also, write $\mathfrak{m} = [0,1) \setminus \mathfrak{M}$. Since Lemma 5.3(i) yields the estimate

$$\sup_{\alpha \in \mathfrak{m}} |H(\alpha)| \ll X^{2 - 2^{-k} + \varepsilon},$$

and Lemma 5.2 establishes that

$$\int_0^1 |H(\alpha)|^{\frac{9}{32} 2^k} d\alpha \ll X^{\frac{9}{16} 2^k - k + 1 + \varepsilon},$$

we deduce that

(5.5)
$$\int_{\mathfrak{m}} |H(\alpha)|^{\frac{17}{32}2^k} d\alpha \ll \left(\sup_{\alpha \in \mathfrak{m}} |H(\alpha)|\right)^{2^{k-2}} \int_0^1 |H(\alpha)|^{\frac{9}{32}2^k} d\alpha$$
$$\ll X^{\frac{17}{16}2^k - k + \varepsilon}.$$

On making use of Lemma 5.3(ii) and the definition of $\mathfrak{M}$, on the other hand, we obtain

(5.6)
$$\int_{\mathfrak{M}} |H(\alpha)|^{\frac{17}{32}2^k} d\alpha \ll X^{\frac{17}{16}2^k + \varepsilon} \sum_{\substack{1 \leq q \leq X \\ (a,q)=1}} \sum_{a=1}^{q} \int_{|\beta| \leq (qX^{k-1})^{-1}} (q + X^k q|\beta|)^{-2} d\beta$$
$$\ll X^{\frac{17}{16}2^k - k + \varepsilon} \sum_{\substack{1 \leq q \leq X \\ (a,q)=1}} \sum_{a=1}^{q} q^{-2}$$
$$\ll X^{\frac{17}{16}2^k - k + 2\varepsilon}.$$

Consequently, on combining the estimates (5.5) and (5.6), we arrive at the upper bound

$$\int_0^1 |H(\alpha)|^{\frac{17}{32}2^k} d\alpha = \int_{\mathfrak{M}} |H(\alpha)|^{\frac{17}{32}2^k} d\alpha + \int_{\mathfrak{m}} |H(\alpha)|^{\frac{17}{32}2^k} d\alpha$$
$$\ll X^{\frac{17}{16}2^k - k + 2\varepsilon},$$

and so the conclusion of the lemma follows immediately. □

On recalling the conclusion of Lemma 2.1, and noting that all of the moments occurring in the statement of Theorem 1.1(iii) are even, one finds that the upper bounds provided in Theorem 1.1(iii) are immediate from Lemmata 5.1, 5.2 and 5.4. The same is true also for the first three bounds recorded in Theorem 1.1(ii), but in this case, for the second two estimates, one combines Lemmata 5.1, 5.2 and 5.4 via Hölder's inequality in the respective shapes

$$\int_0^1 |f_\Phi(\alpha; P)|^8 d\alpha \ll \int_0^1 |H_\Psi(\alpha; X)|^8 d\alpha$$
$$\ll \left(\int_0^1 |H_\Psi(\alpha; X)|^4 d\alpha\right)^{1/5} \left(\int_0^1 |H_\Psi(\alpha; X)|^9 d\alpha\right)^{4/5},$$

and

$$\int_0^1 |f_\Phi(\alpha;P)|^{10}d\alpha \ll \int_0^1 |H_\Psi(\alpha;X)|^{10}d\alpha$$
$$\ll \left(\int_0^1 |H_\Psi(\alpha;X)|^9 d\alpha\right)^{7/8} \left(\int_0^1 |H_\Psi(\alpha;X)|^{17}d\alpha\right)^{1/8}.$$

The final estimate of Theorem 1.1(ii), on the other hand, may be established along the lines of the proof of Lemma 5.4, now working from the 10th moment

$$\int_0^1 |f_\Phi(\alpha;P)|^{10} d\alpha \ll P^{127/8+\varepsilon},$$

together with the minor arc bound

$$\sup_{\alpha \in \mathfrak{m}} |f_\Phi(\alpha;P)| \ll P^{15/8+\varepsilon},$$

which is immediate from Lemma 5.3.

References

[1] M. A. Bennett, N. P. Dummigan and T. D. Wooley, *The representation of integers by binary additive forms*, Compositio Math. **111** (1998), 15–33.

[2] B. J. Birch, *Forms in many variables*, Proc. Roy. Soc. Ser. A **265** (1961), 245–263.

[3] B. J. Birch and H. Davenport, *Note on Weyl's inequality*, Acta Arith. **7** (1961/62), 273–277.

[4] K. D. Boklan, *The asymptotic formula in Waring's problem*, Mathematika **41** (1994), 329–347.

[5] E. Bombieri and J. Pila, *The number of integral points on arcs and ovals*, Duke Math. J. **59** (1989), 337–357.

[6] J. Brüdern and T. D. Wooley, *The addition of binary cubic forms*, R. Soc. Lond. Philos. Trans. Ser. A. Math. Phys. Eng. Sci. **356** (1998), 701–737.

[7] S. Chowla and H. Davenport, *On Weyl's inequality and Waring's problem for cubes*, Acta Arith. **6** (1961/62), 505–521.

[8] T. Estermann, *Einige Sätze über quadratfrei Zahlen*, Math. Ann. **105** (1931), 653–662.

[9] A. Granville, *Bounding the coefficients of a divisor of a given polynomial*, Monatsh. Math. **109** (1990), 271–277.

[10] D. R. Heath-Brown, *Weyl's inequality, Hua's inequality, and Waring's problem*, J. London Math. Soc. (2) **38** (1988), 216–230.

[11] L.-K. Hua, *On Waring's problem*, Quart. J. Math. Oxford **9** (1938), 199–202.

[12] J. Pila, *Density of integer points on plane algebraic curves*, Intern. Math. Res. Notices, **18** (1996), 903–912.

[13] W. M. Schmidt, *The density of integer points on homogeneous varieties*, Acta Math. **154** (1985), 243–296.

[14] C. M. Skinner and T. D. Wooley, *Sums of two kth powers*, J. Reine und Angew. Math. **462** (1995), 57–68.

[15] C. M. Skinner and T. D. Wooley, *On the paucity of non-diagonal solutions in certain diagonal diophantine systems*, Quart. J. Math. Oxford (2) **48** (1997), 255–277.

[16] W. Y. Tsui and T. D. Wooley, *The paucity problem for simultaneous quadratic and biquadratic equations*, Math. Proc. Cambridge Philos. Soc. **126** (1999), 209–221.

[17] R. C. Vaughan, *On Waring's problem for cubes*, J. Reine Angew. Math. **365** (1986), 122–178.

[18] R. C. Vaughan, *On Waring's problem for smaller exponents*, Proc. London Math. Soc. (3) **52** (1986), 445–463.

[19] R. C. Vaughan, *The Hardy-Littlewood Method, second edition*, Cambridge University Press, 1997.

[20] R. C. Vaughan and T. D. Wooley, *Further improvements in Waring's problem*, Acta Math. **174** (1995), 147–240.

[21] R. C. Vaughan and T. D. Wooley, *Further improvements in Waring's problem, IV: higher powers*, Acta Arith. **94** (2000), 203–285.

[22] T. D. Wooley, *On Weyl's inequality, Hua's lemma, and exponential sums over binary forms*, Duke Math. J. **100** (1999), 373–423.

Printed in the USA
CPSIA information can be obtained
at www.ICGtesting.com
CBHW050247071024
15472CB00004B/149

9 783764 366124